ISLE OF WIGHT CENTRE FOR THE COASTAL ENVIRONMENT

Instability
Planning and Management

Seeking sustainable solutions to ground movement problems

Proceedings of the international conference organised by the Centre for the Coastal Environment, Isle of Wight Council, and held in Ventnor, Isle of Wight, UK on 20–23rd May 2002

Edited by Robin G. McInnes and Jenny Jakeways

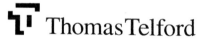

Thomas Telford

Conference organised by the Centre for the Coastal Environment, Isle of Wight Council, UK.

Organising Committee:

Mr Robin G. McInnes (Chairman), Centre for the Coastal Environment, Isle of Wight Council, UK;
Dr Maceo-Giovanni Angeli, National Research Council, IRPI Perugia, Research Institute for
Hydrogeological Protection in Central Italy;
Dr Christophe Bonnard, Swiss Federal Institute of Technology (EPFL), Lausanne, Switzerland;
Dr David Brook, Department for Transport, Local Government and the Regions, Planning Directorate,
UK;
Dr Alan Clark, High-Point Rendel, London, UK;
Mr Stuart Fraser, Senior Accounting Manager, Isle of Wight Council, UK;
Ms Jenny Jakeways, Centre for the Coastal Environment, Isle of Wight Council, UK;
Mr Mark Lee, University of Newcastle-upon-Tyne, Department of Marine Sciences and Coastal
Management, UK;
Dr Eric Leroi, Bureau des Recherches Géologiques et Minières (BRGM), Land Use Planning and Natural
Risks Division, France;
Mr Andrew Price, Head of Planning, Dorset County Council, UK; Vice Chairman Minerals and Waste
Topic Group, Planning Officers Society (POS), UK.

Published by Thomas Telford Publishing, Thomas Telford Ltd, 1 Heron Quay, London E14 4JD.
www.thomastelford.com

First published 2002

Distributors for Thomas Telford books are
USA: ASCE Press, 1801 Alexander Bell Drive, Reston, VA 20191-4400, USA
Japan: Maruzen Co. Ltd, Book Department, 3–10 Nihonbashi 2-chome, Chuo-ku, Tokyo 103
Australia: DA Books and Journals, 648 Whitehorse Road, Mitcham 3132, Victoria

A catalogue record for this book is available from the British Library

ISBN: 0 7277 3132 7

Printed and bound in Great Britain by MPG Books, Bodmin

Preface

The International Conference on *'Instability – Planning and Management'* held in May 2002 in the town of Ventnor on the Isle of Wight, UK, follows on from the very successful conference on *'Slope Stability Engineering – Developments and Applications'* held at Shanklin on the Isle of Wight in 1991. The proposal to hold this second international geotechnical conference arose from a major study entitled *'Coastal Change, Climate and Instability'* which received financial support from the European Union LIFE Environment Programme (L'Instrument Financier de L'Environnement). This wide-ranging three-year demonstration project involved academics, field scientists and practitioners engaged in the fields of coastal, palaeo-environmental and geotechnical studies. It highlighted the importance of implementing coastal and landslide management strategies and integrating the research findings into strategic planning and development control policies.

The purpose of this conference, therefore, is to disseminate as widely as possible the findings from the European project and to bring together a range of interest groups including scientists and engineers, as well as practitioners involved in planning and administration within local authorities, in order to assist the implementation of sound, sustainable management strategies.

The exceptionally high winter rainfall of 2000/2001 resulted in major ground instability problems on the Isle of Wight and coincided with the completion of important research on the predicted impacts of climate change on unstable coastal and mountainous areas. These events highlighted to those engaged in related professions the seriousness of the challenges that we face now and in the future.

The organisers very much hope that this conference and the field visits will provide an opportunity to allow exchange of experiences and ideas within the unique environment of the Isle of Wight. It is hoped that the town of Ventnor, located within the Isle of Wight Undercliff, the largest urban landslide complex in north-western Europe, will provide a memorable setting for the conference. I am sure that the Isle of Wight will gain enormous value from the papers, presentations and exchange of ideas at this conference. I would like to thank all the authors whose contributions form this *'Instability – Planning and Management'* conference proceedings volume.

Eur Geol. Robin G McInnes, FICE, FGS, FRSA
Chairman of the Organising Committee; Coastal Manager, Isle of Wight Council
Ventnor, Isle of Wight, UK. February 2002

Acknowledgements

The organisers of the *'Instability – Planning and Management'* International Conference (held in Ventnor, Isle of Wight, UK, 20–23rd May 2002) would like to thank all those whose hard work and support have made the production of this proceedings volume and the holding of this conference possible, including the keynote speaker, the authors and presenters, the session chairs, Thomas Telford Ltd. and the continued support of colleagues within the Isle of Wight Centre for the Coastal Environment and in particular the Isle of Wight Council.

The conference keynote address is being made by Professor J.N. Hutchinson of Imperial College, University of London, UK. This short introduction provides an appropriate opportunity to highlight the enormous contribution that John Hutchinson has made to the understanding of the Isle of Wight Undercliff Landslide Complex. The keynote paper by Professor Hutchinson and Professor E.N. Bromhead of Kingston University, UK, forms a very significant contribution to this volume.

These proceedings and the conference itself could not have been arranged without the invaluable assistance of my colleague Jenny Jakeways, Coastal Geomorphologist for the Isle of Wight Council. The support of the International Conference Organising Committee and of Emma Davies and other colleagues within the Isle of Wight Council's Centre for the Coastal Environment has enabled the preparation of what we hope will prove to be an innovative and stimulating conference programme and proceedings volume. In particular I would like to thank the members of the International Organising Committee for their invaluable advice and assistance:

- Dr Maceo-Giovanni Angeli, National Research Council, IRPI Perugia, Research Institute for Hydrogeological Protection in Central Italy;
- Dr Christophe Bonnard, Swiss Federal Institute of Technology (EPFL), Lausanne, Switzerland;
- Dr David Brook, Department for Transport, Local Government and the Regions, Planning Directorate, UK;
- Dr Alan Clark, High-Point Rendel, London, UK;
- Stuart Fraser, Senior Accounting Manager, Isle of Wight Council, UK;
- Jenny Jakeways, Centre for the Coastal Environment, Isle of Wight Council, UK;
- Mark Lee, University of Newcastle-upon-Tyne, Department of Marine Sciences and Coastal Management, UK;
- Dr Eric Leroi, Bureau des Recherches Gèologiques et Minières (BRGM), Land Use Planning and Natural Risks Division, France;
- Andrew Price, Head of Planning, Dorset County Council, UK; Vice Chairman Minerals and Waste Topic Group, Planning Officers Society (POS), UK.

Eur Geol. Robin G McInnes, FICE, FGS, FRSA
*Chairman of the Organising Committee; Coastal Manager, Isle of Wight Council
Ventnor, Isle of Wight, UK. February 2002*

Contents

SESSION 3: Hazard identification and risk assessment

SESSION 4: Handling information relating to unstable land

SESSION 5: Instability, planning and the natural environment

SESSION 6: Coastal and climate change and instability

Keynote paper

Isle of Wight landslides

Keynote Paper: Isle of Wight landslides

PROFESSOR J.N.HUTCHINSON, Imperial College, University of London, UK, and
PROFESSOR E.N.BROMHEAD, School of Engineering, Kingston University, UK

INTRODUCTION

The Isle of Wight lies just off the southern coast of England, from which it is separated by a sound called The Solent, several kilometres across (Fig. 1). In 1920, the then Director of the Geological Survey, John Flett, remarked that "no district of England of equal size is more interesting to the geologist than the Isle of Wight, alike from the variety of its formations, the excellence of the exposures and the abundance of fossils" (White, 1921). Its landslides, particularly around the coast, are similarly varied and fascinating.

Our brief from the Organising Committee was to give an overview of the situation with regard to ground instability, chiefly coastal, in the Isle of Wight, highlighting in particular the investigations and measures that have been pursued over many years in the Undercliff and the more recent work at Cowes, drawing general conclusions where possible. Accordingly, the emphasis of the paper is directed towards landslides which most strongly affect coastal communities.

The main emphasis of the present Conference is on Planning and Management in relation to ground instability. However, we have no doubt that few would disagree that the first essential in such activities is the possession and use of the best available scientific, technical and historical knowledge within an overall holistic approach, and it is on some of these matters that the present paper is concentrated.

OUTLINE OF PHYSICAL SETTING, ON- AND OFF-SHORE

With coastal landslides it is clearly desirable and often important to integrate the off-shore geological and other conditions with those on-shore. This area, often neglected in the past, is still little explored.

Solid geology: stratigraphy, lithology and structure

The solid rocks which form the Isle of Wight and its immediate offshore zone extend in age from the Wealden Beds of the Lower Cretaceous up to the Hamstead Beds of the Oligocene (White, 1921). A stratigraphical summary, for both the pre-1995 and the post-1995 terminology, is provided in Table 1. Lithologically these

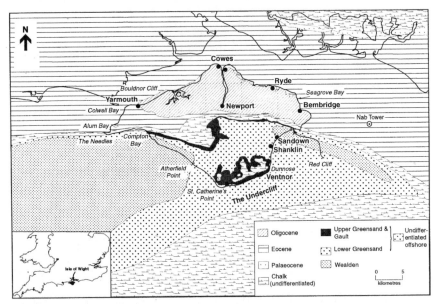

**Figure 1. Geology of Isle of Wight, off-shore and on-shore
(after British Geological Survey, 1995)**

Table 1. Summary of stratigraphy for the Isle of Wight

Chronostratigraphy and Biostratigraphy			Lithostratigraphy	Approx. thickness (m)	BGS 1995 terms
PALAEOCENE	Oligocene	Sannoisian	Hamstead Beds	78	SOLENT GROUP
		Priabonian (= Ludian)	Bembridge Marls / Bembridge Oyster Beds	25	
			Bembridge Limestone		
			Osborne Beds	23	
			Headon Beds	43	
	Eocene		Barton Sand	76	BARTON GRP
		Marinesian	Barton Clay		
		Auversian / Lutetian / Cuisian	Bracklesham Group	224	BRACKE-SHAM GRP
		Sparnacian	Bagshot Sands	10	THAMES GRP
			London Clay	81	
	Palaeo-cene	Thanetian	Reading Clay	26	LAMBETH GRP
		Montian + Danian	unconformity		
CRETACEOUS	Upper Cretaceous	Maastrichtian / Campanian / Santonian / Coniacian	Upper Chalk	380	CHALK GRP
		Turonian	(Bicavea Bed) / Spurious Chalk Rock		
			Middle Chalk	64	
		Cenomanian	Plenus Marls		
			Lower Chalk	56	
	Lower Cretaceous	Albian	Upper Greensand	44	LOWER GREEN-SAND GRP
			Gault	44	
		Aptian	Carstone	5	
			Sandrock	55	
			Ferruginous Sands	90	
			Atherfield Clay	22	
		Barremian	Wealden Shales	55	WEALDEN GRP
			Wealden Marls	170	

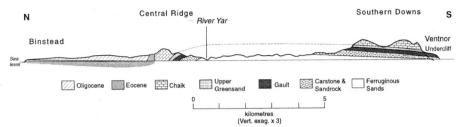

Figure 2. Geological section (N-S) of the Isle of Wight (after Institute of Geological Sciences, 1976)

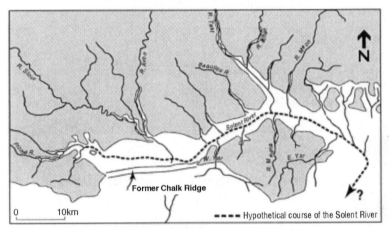

Figure 3. The Solent river (after Reid, 1905)

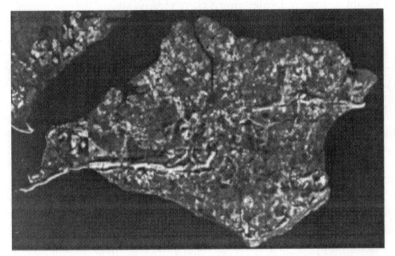

Figure 4. Satellite image of the Isle of Wight (with acknowledgements to ASTER greyscale satellite image). The scarps of the two half-grabens are marked by the broadly W-E Chalk outcrop.

rocks range from overconsolidated clays, marls and shales through weakly cemented sandstones to strong sandstones, limestones and chert beds. They also include the Chalk.

The structure of the Isle of Wight is at first sight simple. It is dominated by the sharp, east-west monoclinal feature in which dips in the Chalk approach the vertical (Fig.2). Away from this central Chalk ridge, to north and south, the dips of the beds are usually not more than a few degrees. While the monocline is chiefly associated with the Alpine orogeny, it has a more complicated history, outlined below, extending back to the close of the Carboniferous. The main effects of the tectonics on slope stability are discussed later.

Quaternary geology

The Isle of Wight has never been glaciated, but has been exposed to several phases of severe periglacial conditions, and widespread periglacially-triggered landslides and periglacial solifluction deposits can thus be expected. The important features of this nature around the landward parts of the Southern Downs are discussed later. Preece and Scourse (1987) have found elevated, Middle Pleistocene estuarine deposits at Bembridge (Fig.1), which equate in age (second half of the "Cromerian Complex" Stage) and elevation (c. +40 m) with those at Boxgrove, West Sussex (Preece, et al., 1990). Thus, earlier periglacial deposits on the island up to about +40 m above sea level are likely to have been eroded away. Above this level inland, where least affected by stream erosion, a relatively full sequence of Pleistocene periglacial deposits may exist.

During the Pleistocene, prior to the breaching of the former Chalk ridge between The Needles and Ballard Point, near Swanage, it is inferred that the Solent River, a major eastward-flowing stream, flowed out of Poole Harbour, across Christchurch Bay and through the East and West Solent to the sea (Darwin-Fox, 1862; Velegrakis, et al., 1999; Tomalin, 2000a; Allen and Gibbard (1993) (Fig. 3). The talweg of this system in the Solent has been identified and surveyed by Dyer (1975) and discussed by Nicholls (1987); the locations of the associated palaeochannels running out into the English Channel between the eastern Isle of Wight and the mainland are summarised by Velegrakis (2000). During periods of low sea level, these were confluent with the Channel River system, which originated with the breaching of the Weald-Artois anticline following the Anglian glaciation (Gibbard and Allen, 1994). Breaching of the Needles – Ballard Point Chalk ridge brought about a beheading of the Upper Solent River and other comprehensive changes in river flow patterns which, combined with major sea level fluctuations, led to the formation of further, broadly north-south palaeochannels between The Needles and Swanage, Dorset. Buried valleys in and around the mouths of the present rivers, discussed subsequently, were also formed during times of reduced sea level.

Marine environment

The northward-facing Solent coasts of the Isle of Wight are not exposed to severe swell waves and are relatively sheltered. The south-east facing coast of the island is quite exposed, with fetches of around 100 to 200 km into the English Channel. The most exposed facet of the island, however, is that facing between about south-west and just west of west-south-west, with much greater fetches into the Atlantic. The Undercliff of the southern Isle of Wight picks up waves from both the above directions (Fig. 1). Recent studies of wave conditions at Bonchurch, at the east end of the Undercliff, concluded that the majority of large waves, with heights up to 3-4m, occurred in a sector between bearings of 210 and 270°. For a 50-year return period, extreme conditions there were estimated to exhibit waves between 5.8 and 6.6 m high, with periods of 9.2 and 8.6 seconds, respectively (HR Wallingford, 1991).

The tides on the outer (SW and SE-facing) coasts of the Isle of Wight are those to be expected for the English Channel. Because of an amphidrome inland from Weymouth, there is a significant gradient in tidal range from a mean of 1.2 m in Christchurch Bay, to the west, to 3.0 m at Chichester Harbour, to the east. Phase differences and interactions between the tides in the English Channel and in the Solent result in a double high water at Calshot and an extended high water near Spithead (Webber, 1980; Price and Townend, 2000).

In coastal engineering projects nowadays, it is routine to make an assessment of the effects of wave refraction on the inshore wave climate and wave power as the incoming waves react with the shape of the seabed. As these factors are very sensitive to wave direction and period, a smoothed statistical analysis is generally made on the basis of numerous scenarios.

Dredging is carried out mainly for the maintenance of shipping channels and for the winning of marine aggregates. The effects of these operations on coastal stability are now closely monitored and controlled by a system of licensing, often aided by advice from HR Wallingford and others, in which each dredging proposal is judged on its own particular merits. In broad terms, the winning of aggregate may be allowed in "fossil" deposits which are not taking a significant part in the present sediment transport system as they lie below the depth of wave influence, provided that fishing, and the natural history and archaeology of submerged land surfaces is not put at risk (Tomalin, 2000b). In the range of water depths usually of interest to commercial dredging companies in the UK, from about 15 to 30 m below low water level, sand is often moved across the sea floor, but shingle and gravel will generally be immobile (Anon, 1995). Consideration is also always given as to whether the inshore wave climate would be adversely affected as a result of the dredging. If this proves to be the case, the application is likely to be rejected by the Department for Environment, Food and Rural Affairs (DEFRA), which plays a key role as a consultee in view of the Department's coastal defence function.

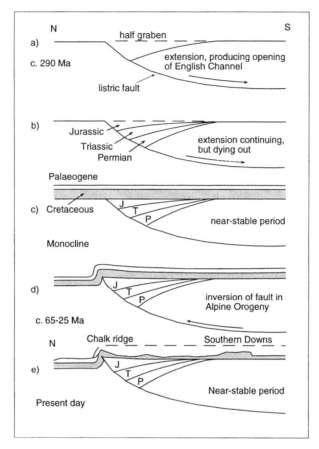

Figure 5. Tectonic background: a) Formation of half-grabens; b) Filling of half-graben with sediments; c) Blanketing by Cretaceous and Palaeogene; d) Inversion of half-grabens; e) Erosion

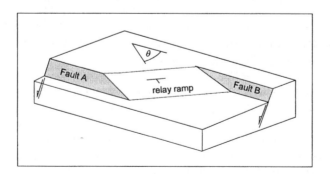

Figure 6. Morphology of relay ramps (after Ferrill and Morris, 2001)

SOME FACTORS AFFECTING LANDSLIDING

Tectonic background

From discussions with Dr J. W. Cosgrove, the following picture emerges. Plate movements during the Carboniferous, led to the Variscan orogeny which formed Pangea and produced, in the area of southern Britain, a N-S compression. This switched during the succeeding Permian to Cretaceous Periods to an extension, which caused the break-up of the NW European Plate and the opening up of the Atlantic. The associated opening of the English Channel was effected on its north side by the formation of a series of southward-facing half-grabens, generated by listric faults. (These may be thought of as related to huge, partially developed compound landslides with their toes not day-lighting). Two of these, with their traces running east-west, run through the central part of the Isle of Wight (Fig. 4). A typical section of such a half-graben is shown in Fig. 5a. During the following Permian, Triassic and Jurassic Periods, the basin was still extending and deepening, but at the same time being filled by the sediments of these three Periods. By the Cretaceous, the N-S extension ceased and the basin was filled (Fig. 5b). At this stage the whole region became blanketed by the thick chalk sediments of the Cretaceous sea and the succeeding Palaeogene deposits (Fig.5c).

Then, in the Late Cretaceous to Miocene, the Alpine orogeny took place, again turning the south British area into a N-S compression zone. One effect of this was to cause an inversion (a reversal of movement) on the listric faults. These movements thrust upwards the Chalk and Palaeogene mantle to the south of the emergent fault scarps, to produce the sharp, broadly E-W trending monocline which forms the Central Ridge of the Isle of Wight (Fig.5d). Subsequent sub-aerial erosion has modified this to give the situation which we have today (Fig.5e).

The actual three-dimensional nature of the listric faults leads to some interesting variations on the above two-dimensional model. The outcrop or scarp of each such fault in the Isle of Wight-Purbeck area is typically between 10 and 20 km long and slightly arcuate. The two scarps which occupy the central Isle of Wight, approaching each other south of Newport, are *en echelon*, that to the W being offset some 4 to 6 km southward of that to the E (Fig. 4). The secondary structures resulting from such an offset are termed relay ramps (Fig.6) (Ferril and Morris, 2001). These are characterised predominantly by a local flattening of dips and corresponding widening of outcrops, both well exhibited in the west-central Isle of Wight. The orientation, θ, of the axis of the relay ramp in plan (Fig. 6) depends on the degree of overlap or underlap of the two fault scarps concerned: it is around 30° in the above Isle of Wight case (Fig. 7).

The relay ramp of the central Isle of Wight developed during the Permian to Cretaceous extensional tectonics. The slight flexure of this ramp to produce the St Lawrence syncline is thought to be the result of the succeeding Alpine compressional phase. The St Lawrence syncline was first explicitly identified by

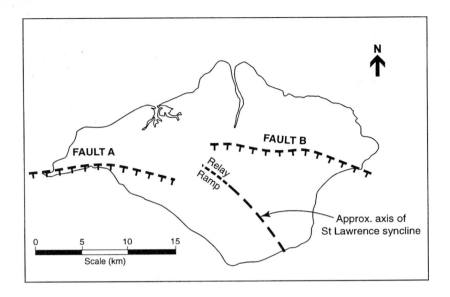

Figure 7. Approximate axis of St Lawrence Syncline in relation to central Isle of Wight relay ramp.

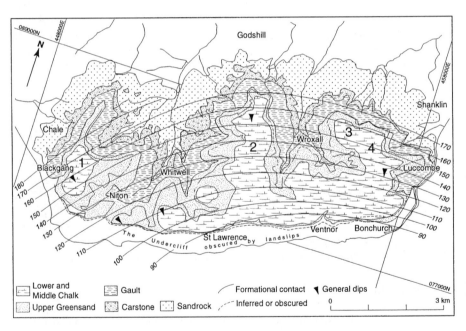

Figure 8. Surveyed plan of St Lawrence syncline in the Southern Downs (after Chandler, 1984).
1=St Catherine's Hill; 2=Stenbury Down;
3=St Martin's Down; 4=Shanklin Down.

Hutchinson (1965) in the Undercliff, particularly for its important influence on the landslides there. It was accurately surveyed by Chandler (1984) in the Southern Downs area from the available outcrops (Fig. 8). Its inland course may be reflected in an apparent sag in the Chalk skyline near Five Barrows, Chillerton (Dr B. Denness, pers. comm.), but more geological investigation is desirable. The course of the nearby Upper Medina may also provide some clues. The approximate position of the axis of the St Lawrence syncline is shown tentatively, in relation to the central Wight relay ramp, in Fig. 7.

Flexural slip
As noted, the geological structure of the island is dominated by the east-west trending monocline which bisects it. To the north and south of this, dips rarely exceed a few degrees; in its vicinity they can become nearly vertical. In such a situation, important bedding plane displacements linked to flexural slip can be expected, especially at and near the monocline. Even in the gently dipping beds such displacements may be significant. They are rarely reported, probably because a good degree of exposure is required to prove their presence. Some indication of their probable scale and variation across the island is provided by Fig. 9. Using the model of Hutchinson (1995), for Brittleness Index $I_B = 0.5$, this shows the estimated amounts of flexural slip displacement for a notional couplet of a competent and an incompetent bed, totalling 20 m in thickness, for the N-S Section 2 of the 1976 geological map. The estimates are also made for a couplet 5 m in total thickness. Taking 100 mm as the minimum shear displacement needed in the field to bring the strength down to residual, it is evident that, even with the thinner couplet, pre-existing shears at residual strength can be expected to be widespread and may have significant influences on slope stability. They were anticipated, but not proven, for the St Catherine's Point landslide (Hutchinson, 1995).

As described above, a further flexure, the St Lawrence syncline, is superimposed on the north-south pattern of the main folding in the area between west-central Wight and the Undercliff. Estimation of the degrees of flexural slip associated with this syncline are made in Fig.10 for two thicknesses of rock couplet, 10 and 50 m. The higher thickness of 50 m is justified for this more local study because of the considerable combined thickness of the Upper Greensand and the Gault couplet, around 70 m even if only the more plastic Upper Gault (noted later) is considered. The results indicate that appreciable ENE-WSW flexural slip is to be expected, which should be combined with that estimated earlier for the N-S direction.

Dipping strata
The amount and direction of the dip of bedding surfaces in relation to slope angle and attitude is always important in slope stability (Hutchinson, 2001). On the Isle of Wight, this is demonstrated most clearly at its western end on the two faces of Main Bench (Fig. 11a). On the south face the chalk is dipping northwards at angles of 45° or more into the cliff and modes of failure appear to involve block subsidence,

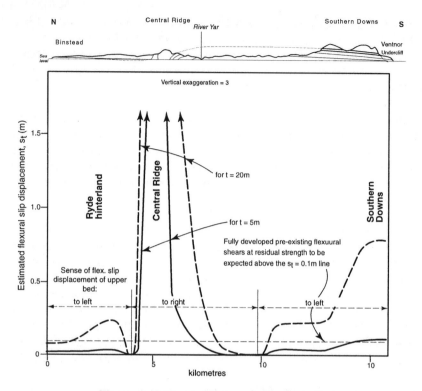

Figure 9. Estimated flexural slip - N to S

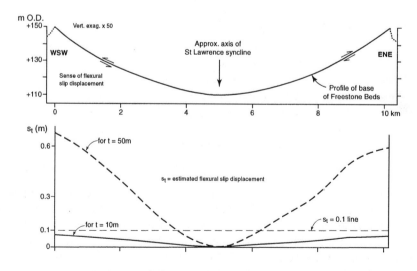

Figure 10. Estimated flexural slip - E to W

wedge failures and toppling. On the north face, however, the dip is out of the cliff at around 70 to 75° and rock-sliding on bedding planes is the dominant form of failure (Figs. 11b and 12). Related problems to those on the south face of Main Bench occur in the Afton Down cliffs to the east, where they affect the Military Road (Barton, 1991; McIntyre and McInnes, 1991).

Very subtle changes in strata with low dips can also be important. Thus, in the Undercliff, changes of a degree or two in the component of seaward dip are thought to be a significant factor in the abrupt change in the style of landsliding between the St Catherine's Point slides and those just to the NW, on Gore Cliff. In these the seaward components of dip controlling the basal slip surfaces are, respectively, about +1.5° and −0.1°.

Other tectonic effects

Hardening of the Chalk, associated with the tectonics, has an important influence on its failure modes. Tectonic hardening of the Isle of Wight Chalk, particularly in the vicinity of the monocline has been studied by several authors. Clayton (1983) states that in this situation "the sustained pressures which resulted in the folding of the beds were sufficient to cause compaction and shear distortion, and by pressure solution to provide additional cement for densification and induration". Dry density (or alternatively porosity) is a convenient measure of tectonic or other hardening. Some values for the Isle of Wight Chalk, from Clayton (1978), are given in Table 2:

Table 2. Dry densities for some Chalk zones

Locality (W to E)	Palaeontological zone	Dry density Mg/cm^3
Alum Bay	*Belemnitella mucronata*	1.77
Freshwater Bay	*Marsupites testudinarius*	2.00
.. ..	*Micraster coranguinum*	1.81
.. ..	*Micraster cortestudinarium*	2.11
Culver Cliff	*Belemnitella mucronata*	2.05
.. ..	*Actinocamax quadratus*	1.86
.. ..	*Marsupites testudinarius*	2.03
.. ..	*Micraster coranguinum*	1.83

Mimran (1975), working on the westward continuation of the monoclinal fold into the Isle of Purbeck, Dorset, made the approximate assumption that the degree of tectonic hardening in chalk is related positively to its dip. This is borne out to some degree by his evidence, although the original lithology is clearly also a factor. Subsequently, Clayton and Matthews (1987) enlarged this work to include the Isle of Wight and W Surrey. Their summary plot is reproduced in Fig.13.

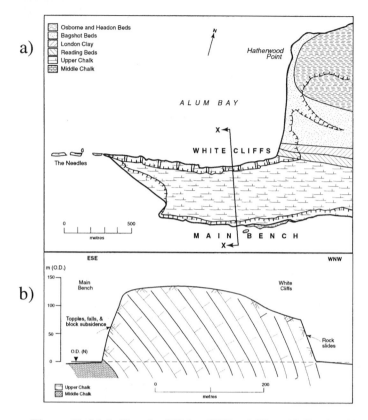

Figure 11. Main Bench - White Cliffs: a) Plan, b) Section X-X

Figure 12. Photo; rockslides in Chalk, Alum Bay White Cliffs, 1996, (looking SW)

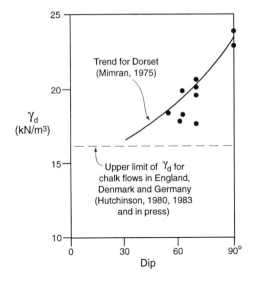

**Figure 13. Relationship of Chalk dry density to dip
(after Mimram 1975, Clayton and Matthews, 1987)**

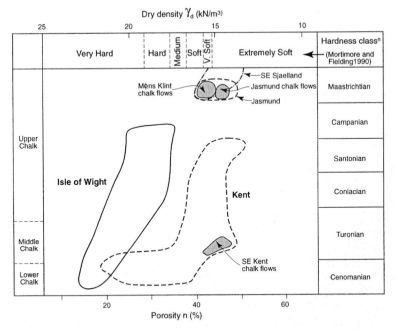

**Figure 14. Relationship of chalk flow occurrence to dry density and
stratigraphy for the Isle of Wight, Kent, Møns Klint (Denmark), and Jasmund
(Germany) (after Hutchinson, in press)**

Chalk flows are a particularly dangerous type of Chalk fall in which the debris flows rapidly outwards to reach run-outs of up to five times the cliff height. They occur in cliffs of soft chalk, generally more than 40 m or so in height, by a mechanism termed "impact collapse" which generates high undrained pore-water pressures. The lowest porosity, or the highest dry density, at which such failures take place is estimated to be about 40%, or 1.61 Mg/m^3, respectively (Hutchinson, 1980, 1983; in press). This conclusion is drawn from field records and laboratory measurements throughout NW Europe.

From Table 2 and Fig.14, it can be concluded that chalk flows should not occur on the Isle of Wight where, although the highest chalk cliffs (Main Bench) rise to about 125 m above sea level, chalk densities are too high. This prediction is borne out by the available records. While several dozen chalk flows have been identified on the sea cliffs of NW Europe, none has been recorded in the Isle of Wight. Despite their very destructive potential, no deaths are known to have been caused by these NW European chalk flows (as distinct from falls of associated Pleistocene materials, as at Rügen (Jasmund) and Møns Klint), probably because they tend to happen in the winter (Hutchinson, in press).

Fissuring and jointing of strata are generally linked partly to tectonics, but are not pursued here.

Lithology

As outlined broadly by Table 1, the rocks of the Isle of Wight consist of unevenly spaced alternations of competent and incompetent strata. Those which are most resistant to shear failure are the Freestones and Chert Beds of the Upper Greensand, the harder parts of the Chalk (particularly the indurated zones of the Upper Chalk) and the Bembridge Limestone. It should be added that all these rocks are jointed and thus weak in tension and accordingly commonly take part in landslides as caprocks. The strata most prone to shear failure are, in the Cretaceous; the Wealden Marls, the Wealden Shales, the Atherfield Clay, thin clay layers in the Sandrock and Ferruginous Sands, and the Gault, and in the Palaeogene; the Headon Beds, the Osborne Beds, the Bembridge Marls and the Hamstead Beds. Other beds in this part of the stratigraphic column that are landslide-prone elsewhere in the UK are insufficiently developed in the Isle of Wight to cause significant problems.

The potentially great value of biostratigraphy linked with geotechnics in a engineering geological context was well shown, for example, during the Folkestone Warren stabilisation works (Varley, *et al.*, 1996a) and in the investigations for and the construction of the Channel Tunnel (Varley, *et al.*, 1996b), particularly in the Gault and the Chalk. These tools have not yet been properly applied in the Undercliff or the Isle of Wight generally.

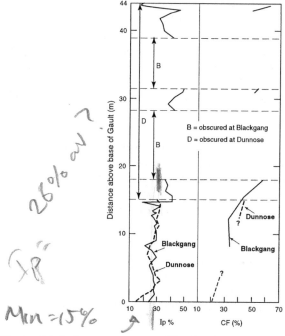

Figure 15. Index property profiles for the Gault, Isle of Wight Undercliff

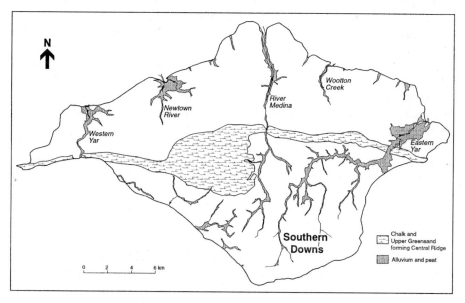

**Figure 16. Isle of Wight rivers and alluvial spreads
(after Institute of Geological Sciences, 1976)**

The classic ammonite zonal scheme of Beds I to XIII for the Gault of Folkestone (Price, 1879, slightly modified by later workers) only applies as far as a short distance W of that locality and thus not in the present area. In Isle of Wight, the stratigraphy of the Gault has been worked out in detail, using an ammonite zonal scheme (Beds 1 to 14) on the basis of exposures at: Rookley Brickworks, Arreton; Compton Bay, NW of Compton Chine; Blackgang, NW of Cliff Cottage (now destroyed) and Gore Cliff (Owen, 1971). A further, very useful account of the lithostratigraphy of the Isle of Wight Gault, based essentially on the Redcliff section (Fig. 1) and identifying 20 Beds, is given by Gale, *et al.* (1996). In neither of these cases were index properties or clay fractions measured. At Redcliff, however, sediment grades were recorded approximately and detailed clay mineralogical determinations were carried out, as noted below.

Some index properties in the Gore Cliff Gault were measured and related to microfossil zones by Denness (1969), but not published. Index properties in the available (incomplete) exposures of the Gault at Gore Cliff and at Dunnose, *i.e.* at the W and E ends of the Undercliff, were measured by Chandler (1984) and by Bromhead, *et al.*,(1991) and are summarised in Fig. 15. By this means, Bromhead, *et al.* (1991) divided the Gault in the Undercliff into an upper, more plastic stratum and a lower, less plastic, with a slip-prone horizon at the base of the upper zone, between 15 and 18 m above the base of the Gault, there about 44 m thick (Chandler, 1984). No associated fossil zoning was carried out. Some support for the identification of a more plastic upper Gault bed is provided by the work of Gale, *et al.*, (1996) in that, at Redcliff, they find a clay-rich layer extending from a metre or so below the top of the Gault to 21 m above its base. In addition, their clay mineralogical determinations show that from the top of the Gault to about 19 m above its base, 66 to 88% of the $< 2\mu$ fraction consists of smectite. From this horizon down to 8m above the base of the Gault, the smectite percentage lies around 60%. From the 8 m level down to the Gault base, this percentage decreases markedly and is only 7-8% at the base of the Gault. It should be borne in mind that at Redcliff the total thickness of the Gault is only just over 35 m, presumably largely as a result of tectonic thinning.

There is clearly scope in the future to pursue linked geotechnical/palaeontological studies of the Gault and other important strata to try and identify slide-prone horizons.

The Quaternary
Subaerial effects
The main subaerial effects of the Quaternary on landsliding in the Isle of Wight were: freeze-thaw weathering, erosion and deposition by mass movements, and stimulation of fluvial and marine erosion, chiefly as a result of sea level changes. In this regard, the most striking effects are seen around the flanks of the Southern Downs. While the valleys cutting into the northern side of these at Whitwell and

Wroxall (Fig. 8) are chiefly due to fluvial headwater erosion, the great embayments NNE of St Catherine's Hill, N and NW of Stenbury Down, N of St Martin's Down and E of Shanklin Down are situated largely between the stream tracks (Fig. 16) and are very probably the result of periglacially triggered deep-seated landslides with associated periglacial solifluction. There are also well-developed dry valleys in the Chalk of the Southern Downs. Some of those on their southern face, truncated by the Undercliff, influence the associated landslides, as described below.

An outline of the likely late Quaternary history of the coastal landslides of the Undercliff is given by Hutchinson (1987). In brief, it is suggested that the coastal slopes would have been mantled by thick deposits of soliflucted debris during the periglacial periods and that these would be largely or completely eroded away by the rising sea levels of the interglacials. At that time the presence of such periglacial deposits had not been proven. However, a recent section studied at Watcombe Bottom (Preece, *et al.*, 1995), about 400 m to the north of the rear scarp of the Undercliff landslides there, shows a sequence of much broken chalk debris deposits sloping seaward at between 5 and 10°. These exhibit, about 0.7 m below ground level, a palaeosol of Allerød (Late-glacial Interstadial) age. This indicates that the lower deposits and the bulk of the upper ones are periglacial. It is probable that these deposits, now-truncated by the Undercliff rear scarp, previously ran out seaward, during the Devensian and probably also earlier glacial periods, to form an apron blanketing the landslides. From the grading of the chalky debris (broadly of medium gravel size) below the Watcombe Bottom palaeosol, it is clear that, in the absence of recementation, it would indeed have readily been removed by marine erosion. Furthermore,the load of this periglacial mantle on the upper parts of the earlier landslides will have had a significant destabilising effect once the lower part of the apron had been removed.

A further section in the rear scarp of the Undercliff landslides, exposing many metres of Upper Greensand and Chalk debris at its head, has appeared this year as a result of the failure of February/March 2001 just SW of "The Landslip" at Dunnose (Grid Ref. 581786). Dating of an organic deposit about a metre below the present topsoil shows this debris to have been emplaced in Pre-Roman times (Dr R. Preece, pers. comm.).

Marine and fluvial effects: stimulation of erosion through sea level changes
It is helpful to subdivide the results of this erosion into palaeovalleys offshore, palaeovalleys in estuaries and the lower reaches of rivers, termed here buried valleys, and current bathymetric lows.

Palaeovalleys offshore
Over the past 20 years or so, much has been learnt about the sea bed around the Isle of Wight. This is scored by numerous palaeovalleys, infilled with drift, which fall broadly into three groups (Allen and Gibbard, 1993; Velegrakis, 2000):

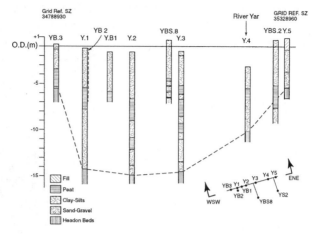

Figure 17. Buried valley of the Western Yar (after Devoy, 1987)

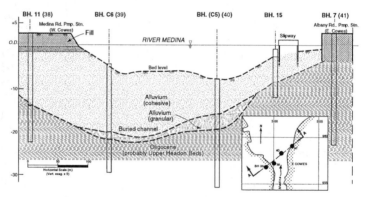

Figure 18. Buried valley of the Medina River
(after W. Hodges, pers. comm.)

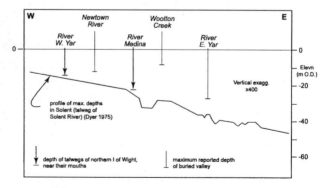

Figure 19. Relation of the talweg of the Solent River (after Dyer, 1975) to those
of the buried valleys of the northern Isle of Wight

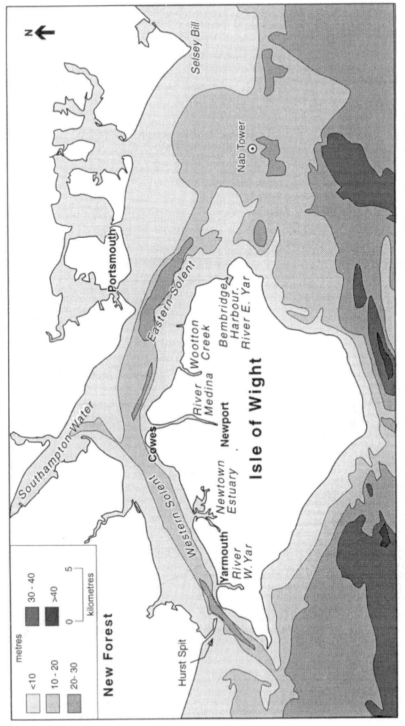

Figure 20. Present bathymetry around the Isle of Wight (after Collins and Ansell, 2000)

1. A system well to the E of the island, associated with the "Channel River", originating from the breakthrough of the Dover-Artois ridge,
2. A system N and E of the island, associated with the Solent River, both before and after its beheading in Christchurch Bay, and
3. A system W and SW of the island, associated with the rivers which effected the breaching of the former Needles to Swanage Chalk ridge and the beheading of the Solent River.

These infilled palaeovalleys appear to be too far offshore to have had much effect upon past or present coastal instability in the Isle of Wight. Groups 1 and/or 2 may have been instrumental in contributing to the erosion of St Catherines's Deep.

Palaeovalleys inshore - buried valleys
The more important of the Isle of Wight rivers flow northwards into the Solent (Fig. 16). In their lower reaches and estuaries they have paleovalleys, or buried valleys in the older nomenclature, which connect with the paleovalleys of Group 2 offshore. Such buried valleys are a normal feature of virtually all river mouths, reflecting the enhanced fluvial downcutting at times of general depression of world sea levels.

From W to E along the N coast of the island, the main rivers are: the Western Yar, the Newtown River, the River Medina, Wootton Creek and the Eastern Yar (Fig.16). Of these, Newtown River and Wootton Creek, do not cut the central Chalk ridge and thus currently have catchments of modest extent, limited to the north side of that ridge. Available data on the paleochannels of these five rivers are summarised in Table 3.

Table 3. Some data on the main northward flowing rivers of the Isle of Wight

| River | Cuts Chalk ridge | Catchment (km^2) | | Depth buried valley (m O. D.) |
		Present	Former	
River W. Yar	Yes	25.0	Much larger	- 13.9†
Newtown River	No	53.5	Similar*	below - 12.2††
Medina River	Yes	82.5	Similar*	- 22.3‡
Wootton Creek	No	22.5	Similar*	below - 9.0**
River E. Yar	Yes	95.0	Much larger	below c. - 28.0 ‡‡

* Wooldridge and Linton (1955, Fig. 20), from the evidence of wind-gaps, infer that these three streams (and the 9 other northward flowing streams of the island) all had more extensive catchments south of the Chalk ridge during the Pliocene.
† (Fig. 17) after Devoy (1987).
†† in East Spit (Tomalin, 2000c).
‡ (Fig.18) after W Hodges (pers. comm.).

** for a "submerged land surface" at Wootton-Quarr (Tomalin, 2000b). The base of
the Quaternary deposits is likely to be deeper.
‡‡ at St Helen's Fort (Whitaker, 1910).

Of the three rivers cutting the Chalk ridge, the Medina is currently has the largest
flow and a catchment, extending to only 2 km from the SE coast, which has
apparently not been modified appreciably by late Quaternary coast erosion. In
contrast, both the Western and Eastern Yar clearly had much more extensive
catchments in the past, on the S side of the chalk ridge (White, 1921). The evidence
for this (Fig. 16) consists, for the Western Yar, of W and SW flowing streams and
associated alluvium from Shippard's Chine to Whale Chine (or even Blackgang
Chine), now truncated by erosion of the SW coast of the island. The corresponding
evidence for the Eastern Yar comprises the spread of alluvium near Yaverland,
Sandown Bay, truncated by the SE coast.

The Eastern Yar currently has the largest catchment on the island (Table 3), reaching
to within 1.5 km of the S coast in Niton (Fig. 16). We are not yet aware of the full
depth of its buried valley or valleys, though alluvium extends to about - 14m O.D. in
a borehole just NE of the Sandown Treatment Works, less than 1.5 km from the SE
coast of the island (W. Hodges, pers. comm.), and to about −28m OD in the well at
St Helen's Fort.

A longitudinal profile of the maximum depths in the Solent, corresponding
approximately to the talweg of the Solent River, is provided by Dyer (1975). This
suggested that this river flowed to the east and incised to a base level falling from
about - 15m O.D. in the Hurst Castle area to at least - 46m O.D. between the eastern
Isle of Wight and Selsey Bill. The validity of the western extrapolation of the Solent
River talweg from about Hurst Point to Poole Harbour is questioned by Nicholls
(1987), but he accepts Dyer's estimate of this talweg in our present area of interest,
from just W of Yarmouth to the east end of the Isle of Wight. The relationship of the
known buried channels of the northern Isle of Wight rivers to the above talweg (Fig.
19) suggests that, locally to each, it formed their controlling base levels. On this
basis, the buried valley of the E. Yar looks likely to be the deepest of the five.

The question here is whether the erosion of these buried valleys has caused any
landsliding? Evidence for this is hard to come by, as such slides tend now to be
buried or partially obscured by later alluvial infill. In the Medina, detached blocks
of Bembridge Limestone, typically 2.0 x 1.0 x 0.75 m in size, encountered during
dredging for the heavy unloading facility for the East Cowes Generating Station (*i.e.*
well below its outcrop level), were considered to be landslip debris (C. Gaches, pers.
comm.). On the other side of the Medina, in West Cowes, the Bembridge Limestone
along the shore is also believed to be landslipped (W. Hodges, pers. comm.). In the
Eastern Yar, past landslipping on the south side of Bembridge Harbour is reported
(W. Hodges, pers. comm.).

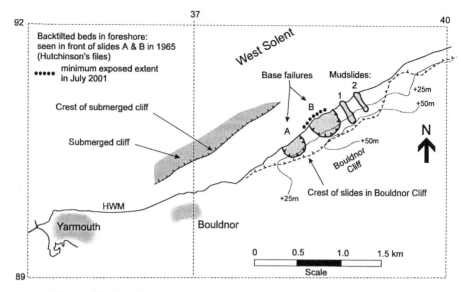

Figure 21. Bouldnor Cliff; plan of terrestrial and offshore features

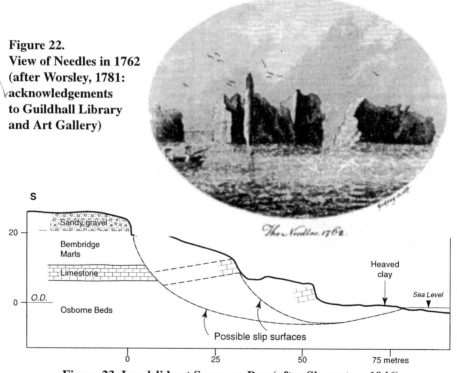

**Figure 22.
View of Needles in 1762
(after Worsley, 1781:
acknowledgements
to Guildhall Library
and Art Gallery)**

Figure 23. Landslide at Seagrove Bay (after Skempton, 1946)

Current bathymetric lows

The locations of the current bathymetric lows are summarised on Fig.20. These generally appear to be sufficiently far from the present shoreline to have no influence on the present coastal landslides. The two places where deep channels approach the coast most closely are between NW Osborne Bay and Old Castle Point, where there are landslides in the Osborne Beds, and from just E of Sconce Point to Cliff End, where there are sizeable landslides in the Bembridge Marls and Osborne Beds. The 20 m depth contour (relative to Chart Datum, therefore about - 22m O.D.) approaches to about 500 m from the coast at the former site and about 250m at the latter. A scour hole 56.7 m deep between Cliff End and Hurst Castle spit is reported by White (1921). It remains to be explored, bearing in mind the possible presence of pre-existing flexural slip shears, whether at either site, and particularly at that by Sconce Point, the current coastal landslides are influenced by the current offshore erosion, or are possibly inherited from slides generated when the shoreline was further seaward.

Recent investigations have revealed the presence of a submarine cliff, offshore of Bouldnor. It consists of a steep slope (*c.* 50°) about 8-9 m high, extending between about - 4 and - 12 m O.D., and consisting of Holocene silty clays with three main interbedded peats. Radiocarbon dates on these range from 7640 ± 70 years B.P. (Beta-140102) in the Lower Peat to 5580 ± 60 years B.P. (Beta-140102) in the Upper Peat (Long and Tooley, 1995; Dix, 2000; Scaife, 2000; Velegrakis, 2000). The beds appear to be in place and these dates are concordant with the known sea level rise curve for the Solent. As the adjacent foreshore, backed by Bouldnor Cliff, is formed of the Hamstead Beds (underlain by Bembridge Marls at the ENE end of the cliff), the above Holocene deposits can be expected to lie against a buried cliff or slope of these Oligocene Beds (not yet revealed by the investigations).

The present terrestrial Bouldnor Cliff is about 2 km long and up to 61 m high. It is occupied by extensive landslides. These are mainly rotational slope failures in the lower WSW and ENE extremities of the cliff, with some important mudslides to the ENE (Bhandari and Hutchinson, 1982), and deeper-seated rotational base failures in its higher, middle part. The approximate extent of the base failures in the cliff and the associated, strongly back-tilted Hamstead Beds in the foreshore (dips are sub-vertical in places), partly obscured by beach deposits, is shown in relation to the submerged cliff in Fig. 21. It is noted that the deep-seated failures, not surprisingly, coincide approximately with the higher parts of the cliff, with respect to which the main mapped part of the submerged cliff is offset some 1.5 to 2 km to the WSW. The crest of the latter lies some 200-500 m seawards of the foot of the terrestrial cliffs. This may suggest that these deep-seated landslides have occurred without any influence from the offshore erosion associated with the submerged cliff, but this question needs to be explored further. Some linkage between the erosion of the now submerged Oligocene clay cliff and the deep-seated failures of the terrestrial cliff may have occurred in the past when coastal erosion had progressed less far and the latter cliff was closer to the submerged one. Sections through the Bouldnor Cliff base failures and out into the Solent would be a useful first step in exploring this.

The most striking low is St Catherine's Deep, situated about 3 km offshore of, and parallel to, the Undercliff. The possible role of this in the origin of the Undercliff landslides is discussed subsequently.

Current cliff erosion and recession
Cliff erosion
Recession measurements differ, of course, from measurements of lateral erosion, being often strongly influenced by the erosional and depositional effects of landsliding.

The Zenkovich (1967) model postulates that marine erosion commonly consists of two components; downward erosion of foreshores, V, and lateral erosion of the associated cliffs, H, related by the expression $V = H \tan \alpha$, where α = average slope of foreshore. Hutchinson (1986, 2001) argues for the primacy of the downward erosion of the foreshore, because the waves strike there first, the associated lateral erosion occurring at the rate which the former permits.

On the Isle of Wight, no observations are known of the downward erosion of foreshores. A study of the lowering of the chalk shore platform of East Sussex, around 50 to 60 km to the east, using 44 micro-erosion meter sites over three years, indicates an average rate of lowering over this period of 2.62 mm/year, with a range from about 1 to 10 mm/year (Ellis, 1986). Assuming an average shore platform gradient, α, of 2°, this would theoretically be associated, following Zenkovich (1967), with an average lateral erosion rate of 2.62 x cot 2° = 75 mm/year.

The relative effects on coastal instability of downward foreshore erosion and lateral cliff erosion have been explored by appropriate stability analyses for the St Catherine's point landslide. Both for the landslide as a whole and for the seaward parts of its apron, the effects of typical annual marine erosion are calculated to be small, but cumulative. Their principal manifestation is the continuing slow adjustments of the landslide complex.

Cliff recession
From the general form of the coastline of the island in relation to its geology and degree of exposure to marine attack, some crude inferences can be made. Thus, much of the Undercliff and the monoclinal Chalk ridge appear to be quite highly resistant to recession compared to the older Lower Greensand and Wealden Beds forming Brixton and Sandown Bays. In the Undercliff, this is probably due to the good natural sea defences provided over much of its length by fallen blocks of the Freestones and, particularly, the Chert Beds of the Upper Greensand. The Chalk

ridge has, as discussed below, been hardened tectonically: it would seem to be more resistant at its western end, in The Needles, than at Culver Cliff in the east. The name "Needles" is believed to spring from an earlier pinnacle of Chalk called "Lot's Wife" (Fig. 22), which fell in about 1764 (Worsley, 1781). The relatively soft clays and sands of the Tertiaries forming the northern coasts of the island may generally have receded less than the Lower Greensand and Wealden Beds in Brixton and Sandown Bays, but this is probably because of their more sheltered situation. From the advances made in offshore geology and archaeology in the past decade or so (*e.g.* papers in Collins and Ansell, 2000; McInnes *et al.*, 2000), it may well be possible to improve considerably our knowledge of past coastal recession rates if submerged sites and sediments are effectively identified and investigated.

The straightness of the main, SSE-facing facet of the Undercliff, truncating the Southern Downs, is notable. It is probably joint-controlled. It contrasts markedly with the fretted northern face of the Southern Downs, produced mainly by periglacial and fluvial erosion.

Much general information regarding average historical rates of lateral cliff recession is obtainable from comparisons of surveys, of good quality and scale, of positions of cliff crest and cliff foot at various times. A comprehensive critical survey of this is understood to have recently been commenced by the Council's Centre for the Coastal Environment. The Centre is integrating a wide range of data to create a 'coastal evolution and risks map' to provide informal advice to the planning department (R. McInnes, pers. comm.). In the sandy Ferruginous Sands cliffs of Chale Bay, average rates of cliff top recession ranged from 0.22 to 0.89 m/year over the period from 1861 to 1980 (Hutchinson, *et al.*, 1981). Within the Undercliff at Blackgang, Moore, *et al.* (1998) report an average annual rate of retreat of the cliff base, in the lowermost unit of the Sandrock Beds and the underlying upper Ferruginous Sands, as 0.73 m/year, accelerating from 0.34 m/year between 1861 and 1907 to 1.54 m/year between 1980 and 1994. Elsewhere within the Undercliff, the following general average rates of recession are given for undefended cliffs (R. McInnes, pers. comm.): 0.3 to 0.4 m/year in cliffs east of Monks Bay, predominantly in Gault over Carstone and Sandrock; 0.2 to 0.3 m/year in cliffs of Chalk and Upper Greensand debris west of Steephill Cove; and 0.5 to 6 m/year in weak Ferruginous Sands cliffs to the west of St Catherine's Point. Across the 118 m frontage of the St Catherine's Lighthouse compound, cut in the debris apron of the St Catherine's Point landslide containing many blocks of resistant Chert Beds and Freestones, the average recession of the crest of the sea cliff was 14.9 m between 1868 and 2001, giving an average rate of 0.11 m/year with a range of 0.07 to 0.18 m/year (Hutchinson, *et al.*, 2002).

Reefs of resistant strata produced by various flexures are found just offshore and reduce the intensity of wave attack locally. On the south-western coast, Hanover Point, composed of relatively soft Wealden Marls, owes its existence partly to a reef of hard sandstone about 1.8 m thick and the associated "Pine Raft" (White, 1921). Similarly Atherfield Point, composed of Wealden Shales and Atherfield Clay, is significantly protected by a Perna Bed reef, a tenacious grit around half a metre thick at the base of the latter (White, 1921). Conversely, gaps in offshore reefs allow virtually unimpeded wave attack in their vicinity. An example of this in Compton Bay just west of Hanover Point is given by Lord Mottistone (1937, p.18). In several places on the north-east and north-west coasts of the island, the varying degrees of protection offered by the resistant Bembridge Limestone in the cliff foot and shore platform are reflected in the nature and degree of activity of the coastal landslides. Where the cliffs are higher, however, as at Seagrove Bay (Fig. 23), this bed is disrupted by the landsliding (Skempton, 1946). It is likely that the middle parts of the Seagrove Bay failure surfaces will be bedding-controlled.

Future climatic trends and sea level rise

There seems to be general agreement that, due in part to anthropogenic factors, southern England in the 21st century will exhibit:

- a rise in sea level. (That for the past 7,000 years is shown in Fig. 24).
- an increase in temperature and in rainfall variability.
- less summer rainfall but an increase in winter rainfall, giving an overall increase in mean annual rainfall.
- an increasing incidence of storms and storm surges.

More particularly, the UK Climate Impacts Programme (UKCIP) (Wade, *et al.* eds, 1999), with additional data from Hosking and Moore, 2001, predicts for South-east England (including the Isle of Wight), for its "medium high scenario", the following by the year 2080:

- an increase in mean sea level in the English Channel of + 540 mm.
- an increase in mean temperature of up to + 4.7°C.
- a decrease in mean summer rainfall by up to 20%.
- an increase in mean winter rainfall by up to 23%.
- little change in wind and wave magnitude and direction.

Corresponding extreme events are predicted as follows:

1. extreme sea levels to rise about 840mm for the 1 in 50 year event.
2. the probability of a dry summer (50% of normal rainfall) will increase from 1% with present climate to 10%.
3. the probability of a wet winter (160% of normal rainfall) will increase from 1.7% with present climate to 11%.

Initially, an increase in shallow landsliding, both on the coast and inland, is to be expected from particularly items 1 and 3 above. A continuation of these may also re-

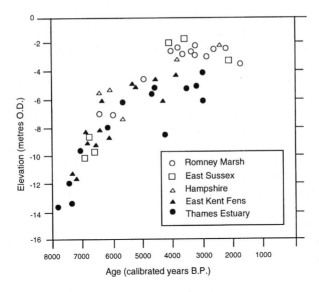

Figure 24. Changes in sea level in S. England since the late Quaternary (after Long and Tooley, 1995)

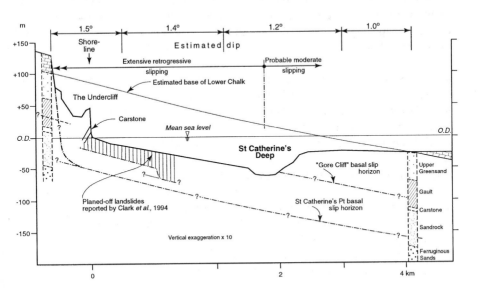

Figure 25. Relation of Undercliff to St Catherine's Deep (based upon Chandler, 1984, and British Geological Survey, 1995)

activate more deep-seated landslides. Some aspects of the rainfall and its correlation with landslide activity are explored by Ibsen and Brunsden (1997). This work was updated by Ibsen as part of the LIFE project (McInnes, *et al.*, 2000).

Stability calculations show that while the anticipated increases in rainfall will have a dramatic influence on the stability of shallow landslides and mudslides, their effect on the deep seated major Undercliff landslides is slight, but sufficient to cause fluctuations in the rate and magnitude of the slow deformations of the landslide system as it adjusts to coastal erosion.

SOME INLAND LANDSLIDES

Not unnaturally, the literature on the landslides of the Isle of Wight focuses predominantly on the coastal ones. For the inland landslides the best source of information on is the 1:50,000 Special Sheet (Drift Edition), Isle of Wight, geological map (Institute of Geological Sciences, 1976). The oldest beds affected are the Wealden Shales, in which White (1921) reports slips in the railway cutting near Sandown Station.

The most important inland landslides, probably triggered periglacially are found in the southern, Cretaceous part of the island, where very extensive slides in the Gault occupy much of the inland flanks of the Southern Downs. The principal localities are the western and northern slopes of St Catherine's Down, the northern slopes of Stenbury Down, the northern slopes of St Martin's Down and the eastern ones of Shanklin Down (Fig.16). In these cases, the slides are shown to be seated in the Gault, leaving steep high rear scarps of Upper Greensand. Whether any of these failures exploit clay layers within the Sandrock, as in the geologically related St Catherine's Point landslides on the coast (Hutchinson, *et al.*, 1991a), remains to be determined. In the toe area of the last locality, the brick-pits by the side of the railway between Cliff Farm, Shanklin, and Wroxall are dug in Gault that "has flowed down the hill-side" (White, 1921). On the geological map, the Gault is shown as undisturbed by sliding in the areas which form the headwaters of the River East Yar, specifically the Whitwell branch and the Wroxall branch, possibly because of gentler gradients there.

As a result of mudslide activity, inferred to be predominantly periglacial, the downslope extremities of the slides frequently extend well below the base of the Gault outcrop. The main cases mapped by the Institute of Geological Sciences (1976) are summarised in Table 4.

Table 4. Some mudslides on the Southern Downs

Locality	Approx. horizontal distance from outcrop of basal Gault to toe of "mudslide"
Gotten Manor Farm	650 m (oblique)
The Hermitage	450 m
Wydcombe	Two, 130 & 250 m
Gatcliff Farm	260 m
Godshill Park House	620 m
Yard Farm	140 m
Winstone *	430 m
Upper Hyde	550 m
Below Shanklin Down (to NE)	400 m

* Noted by Reid and Strahan (1889) as a "mud-river".

Two minor landslides just above the Lower Greensand/Gault contact are indicated. Both are elongate. One lies on the north-eastern part of Tenbury Down, above Appledurcombe House; the other in the valley just upstream of Whitwell. A further minor slide is mapped on the eastern facing slopes of Shanklin Down, between Nansen Hill and Shanklin Down and above Luccombe Chine. It appears to have involved the Lower Chalk and the overlying Angular Flint Gravel.

Also within the Cretaceous area of the island, although not mapped, is a slide in the Ferruginous Sands at Mottistone, which occurred on 27th November, 1703 (Phipps, 1992) and is the oldest historically recorded landslide in the Isle of Wight. Over a thousand tons of sandy earth moved down from the high bank above the Tudor/Elizabethan Mottistone Manor and buried the western part of its rear, northern wall. This was probably a long delayed result of oversteepening of the bank foot during construction of the manor, probably triggered by the great storm of 26/27th November, 1703. This "greatest recorded storm in British history" is described by Majdalany (1959). There is no evidence that the failure involved the underlying Atherfield Clay. The house was bought in 1926 by Lord Mottistone and dug out in 1927. 1300 tons of earth were removed (Mottistone, 1937, p. 308-310).

The only other inland slide mapped in the Cretaceous area of the island is at Newchurch. Subsequent re-examination, however, has shown that this is not a slide (Dr B. Denness, pers. comm.).

No inland landslides are mapped in the northern, Tertiary part of the island, apart from a continuation of the coastal landslides of Headon Point around onto the southern face of Headon Warren. These appear to involve chiefly the Oligocene

Osborne and Headon Beds, running down onto the Eocene Bagshot Beds. An unmapped landslide in the Hamstead Beds at Durton Farm, Newport, is reported by Hutchinson (1965). First seen in January, 1962, it occupied an approximately 10° slope, extending roughly 360 m across the slope and 45 m down it. The rear scarp was then fairly fresh and between 150 and 300 mm high.

THE WEST COWES LANDSLIDES

Because of increasing property damage, the IoW Council commissioned Halcrow (Moore *et al.*, 2000) to prepare a report on the Landslide Potential of a seaward area of Cowes from the Medina River towards Gurnard. The same methodology as used in the Undercliff was employed by essentially the same team of three geomorphologists, again producing a series of three maps at 1:2500 scale, namely; Geomorphology, Ground Behaviour and Planning Guidance. Although the area is much smaller and probably less complex than the Undercliff, it does not have the advantage of a well defined lithology and structure in the rear scarp, as provided by Chandler (1984) for the Undercliff.

The mapping has fulfilled its objective of identifying areas of past and present slope instability and, at least for the area of suspected abandoned cliff, supports the indications of the LIFE project (McInnes *et al.* 2000) that the present day coastal slopes are partly the result of erosion by the former Solent River. Clearly, in any future extension of this study, the possible destabilising role of the late-Quaternary buried channel of the Medina, discussed above, should be investigated.

As in all studies of a primarily geomorphological character, interpretation of the sub-surface conditions can provide little more than a framework on which to base further, subsurface, investigations; the four sections put forward for W. Cowes by Moore *et al.* (2000) should be regarded in this light. Hodges and Woodruff (2002) on the basis of a larger database of boreholes propose an alternative ground model. In order to further the understanding of these slopes, specifically targeted subsurface investigations should be given a high priority.

THE UNDERCLIFF LANDSLIDES
Possible origin

The period for which coastal landslides survive is a function partly of their scale and of the erosion resistance of the rocks involved. On both these counts the Undercliff landslides stand out as the oldest in the Isle of Wight. Elsewhere, with the partial exception of the Chalk ridge, the coasts are relatively soft and the current landslides are predominantly of recent date.

It is difficult to define the age of the "original" Undercliff landslides. They probably initiated some distance south of the present shoreline, on the dip slope in the Upper and Lower Cretaceous which declines from the present south coast to St Catherine's Deep, around 2.8 km off-shore and with a bottom level of about - 60 m O.D.

(Fig.25). The work of Clark, *et al.* (1994) shows this submarine slope is mantled by the planed off remnants of old, deep-seated landslides to a depth of at least 25 m and possibly 40 m. This would suggest that at least the northern slope and probably the bottom of the present St Catherine's Deep is formed of slide debris. Similar features are found in the shore platform/sea bed seaward of other landslides involving base failure, for example, at Warden Point and Folkestone Warren, both in Kent.

The trigger for the initiation of the forerunners to the Undercliff landslides may have been the erosion of St Catherine's Deep by fluvial erosion, possibly associated with the "Solent River" or the "Channel River", in combination with marine erosion associated with a past glacio-eustatically controlled sea level. Once formed, this landslide complex would tend to be reactivated during interglacials, when the rising sea levels were at their most destabilising level, and to retrogress progressively to the northward. Rates of such retrogression were no doubt variable and intermittent. It may be worth noting that assuming an average retrogression rate of 0.1 m/year during episodes of active erosion, *i.e.* about the minimum currently operating at the Undercliff, at least 28,000 years of such erosion would be required to create the present separation between the Undercliff and St Catherine's Deep. Taking into account periods of reduced or zero erosion during depressions of sea level to below the - 60 m O.D. level, the time of origin of the Undercliff landslides is likely to be considerably earlier than this.

Review of some past work
Engineering geology
The importance of the St Lawrence syncline has already been referred to. The original 1965 elevation of this from the sea, with the accompanying landslides, was up-dated by Hutchinson (1991). A simplified, further update of this, taking into account some of the recent boreholes, is given in Fig.26. This will probably need further revision as new borehole data are taken into account.

Inferred cyclicity of landsliding
From the geology and geomorphology of the Undercliff generally, more than 12 further compound slides, in various degrees of degradation, have been traced throughout its length (Hutchinson, 1991). All but the most recent large slides, at St Catherine's Point and Dunnose, have a zone of multiple compound-rotational slips behind them. The latter are inferred to be seated in the Gault. Putting together all this information enabled a landslide cycle to be discerned in the Undercliff, with a length estimated very roughly to be up to 6,000 years (Hutchinson, 1991). A cartoon illustrating this cycle for the St Catherine's Point slide is provided by Hutchinson, *et al.* (1991). However approximate, this is valuable in counteracting overly short-term views of the nature of the Undercliff landslides.

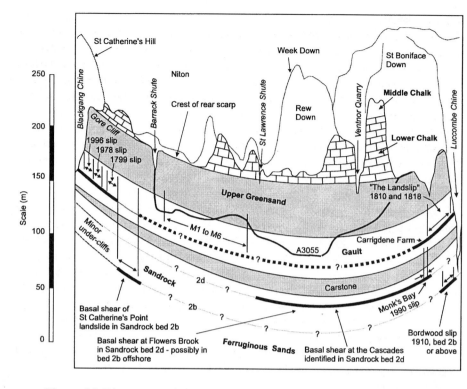

Figure 26. Diagrammatic elevation of Undercliff, in plane of rear scarp, showing stratigraphy and relative positions of the main slide-prone horizons (some tentative)

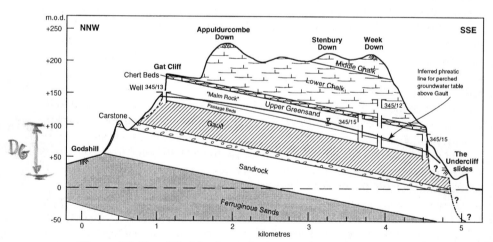

Figure 27. Simplified hydrogeological section of the Southern Downs (after Chandler, 1984)

Hydrogeology

Hydrogeologically, the Undercliff landslides are closely connected with the Southern Downs, to which the entire input of water is by precipitation. However, as the watershed of the Southern Downs lies very close to their southern edge, most of the surface and near-surface water runs northwards via the Whitwell and Wroxall valleys, into the Eastern Yar and, to a lesser extent to the NW into the Medina. The slight southerly dip carries the more deeply infiltrating remainder of this water towards the landslides of the Undercliff, where it is inferred to feed into the rear of the landslides chiefly via the spring line within the Passage Beds, generally obscured by debris (Chandler, 1984). Direct evidence of this hydraulic continuity has not yet been obtained.

A highly simplified section, modified after Chandler (1984), showing the relationship and hydrogeology of the Southern Downs and the Undercliff, is given in Fig. 27. The full lines indicate the ground profile and phreatic surface for a section between the Whitwell and Wroxall valleys. A section along the Whitwell valley may possibly be more representative. As a first approximation, the mass of the Southern Downs can be divided into the aquifers of the Chalk and Upper Greensand and of the Lower Greensand, separated by the Gault aquitard. The former aquifer is unconfined and perched on the Gault and Passage Beds: the latter is confined. In detail, there are minor aquitards in the Chalk, for example the *plenus* Marl (Whitaker, 1910) and in the Lower Greensand, where thin clay layers throw out lines of seepage in the lower cliffs around Blackgang (Chandler, 1984).

The degree of influence exerted by the confined Lower Greensand aquifer on the landslides has not been explored. It is particularly important to do this in view of the fact that the large, compound slides in the central Undercliff have their basal slip surfaces deep within the Sandrock, commonly well below sea level. Towards each end of the Undercliff, where the Lower Greensand is well exposed, outflowing groundwater may cause seepage erosion in the free faces, thus contributing to recession and undermining of slides in the overlying Gault (Chandler, 1984).

Landslide geometry and mechanics

The features forming the Undercliff of the Isle of Wight were first recognised to be landslides by Sir Richard Worsley (1781). Various mechanisms for the landsliding proposed subsequently, notably that by Conybeare and Phillips (1822), are reviewed by Hutchinson (1991). In the mid-1950s, Edmunds and Bisson (1954) and Toms (1955), suggested models much influenced by those at Folkestone Warren, predominantly multiple rotational landslides with their basal slip surface at the base of the Gault (Figs.8a and b of Hutchinson, 1991). That of Edmunds and Bisson is truer to nature in that, for the first time, it recognises the presence of translational (compound) slides in the seaward part of the Undercliff, with multiple rotational slides (more truly, multiple compound-rotational slides) in the landward part.

When we began our detailed research into the Undercliff slides in about 1980, while we were aware that the facies of the Gault and Lower Greensand in southern England are laterally variable, we were nevertheless unduly influenced initially by the Folkestone Warren model

The two subsurface investigations for which we had resources were those at the Gore Cliff and St Catherine's Point landslides. The results of these led to the abandonment of the Folkestone Warren model and the creation of two new models for the main landslides of the Undercliff, which have informed all subsequent work. These models are:

Gore Cliff landslide model

The investigation of the Gore Cliff slide, which occurred in the Gault in 1978 near the western extremity of the Undercliff, is described by Bromhead, *et al.*, 1991 (Fig. 28a). It approximated to a compound slide with its basal slip surface a bedding horizon about 18 to 15 m above the base of the Gault. It was shown, correspondingly, that the Upper Gault down to this level is significantly weaker and more plastic than the more silty Lower Gault These relationships were also demonstrated at the eastern end of the Undercliff, at Carrigdene Farm, Bonchurch, and are considered likely to hold in between.

St Catherine's Point landslide model

The investigation of this major old slide (Hutchinson, *et al.*, 1991a), the most complete and unspoilt of the Undercliff slides, situated about a kilometre SE of the Gore Cliff slide, showed it to be a compound slide with a collapsed graben scarp, cutting down through the Gault, the Carstone and most of the Sandrock to a basal slip surface nearly 50 m below the base of the Gault, which followed a thin clay layer forming a bedding plane within the Sandrock (Fig. 28b). As noted by Hutchinson (1991), Conybeare and Phillips (1822) correctly anticipated some aspects of this model in their first published section of the Undercliff.

Radiocarbon dating of a Yew branch buried in the seaward apron of landslide debris shows the latter to have been emplaced in two phases, one prior to about 4000 radiocarbon years B.P., possibly as early as around 8,000 years B.P., and one subsequent to 4,000 years B.P., and it is concluded that the St Catherine's Point slide occurred after the final emplacement of this apron (Hutchinson, *et al.*, 1991a). Recent archaeological work on the surface of the debris apron indicates human occupation from the early Bronze Age to medieval times. The linear features of Iron Age, Roman and medieval date are such that little appreciable ground disturbance or compression seems to have occurred during these periods, although this needs to be verified by thorough survey (Dr D. Tomalin, pers. comm.). From the morphology of the debris apron fronting the slide it was inferred that movements are still continuing and this has been confirmed by measurements by Bromhead *et al.* (1988).

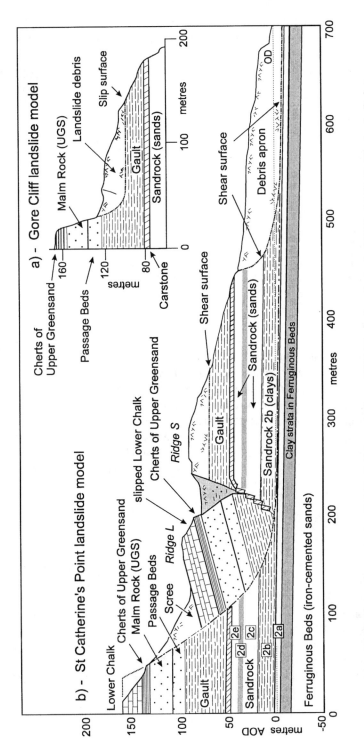

Figure 28. Models of Undercliff landslides: a) Gore Cliff (Bromhead, *et al.*, 1991); b) St Catherine's Point (Hutchinson, *et al.* 1991a)

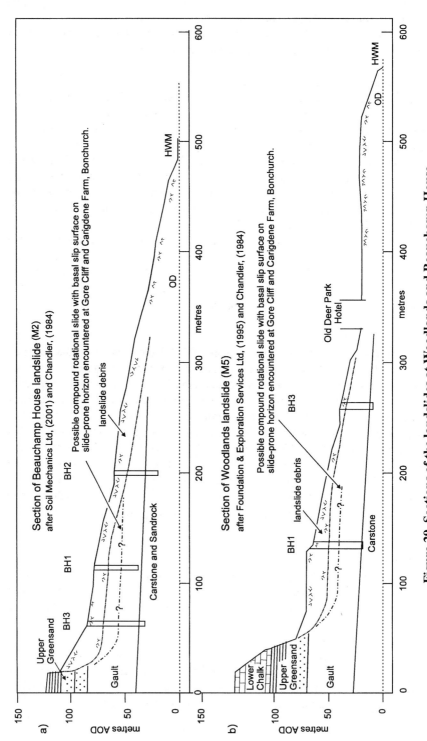

Figure 29. Sections of the landslides at Woodlands and Beauchamp House (after Foundation & Exploration Services, Ltd (1995), Soil Mechanics Ltd, (2001), and Chandler, (1984))

Development of additional landslide models

Two further landslide models have been developed from the above two. They are the Beauchamp House, Niton, and Ventnor Park landslide models.

Beauchamp House landslide model

The original Gore Cliff model has a basal slip surface in mid-Gault which daylights above a steep strongly eroding coastal slope. Any potential downslope development has been removed by erosion. Where Gore Cliff type slides take place above coastal slopes with restricted erosion, however, a system of shallow translational slides and mudslides can develop downslope from the main slide, often extending as far as the coastline. An example is provided by the recent section at "Beauchamp House" (Fig. 29a). A variant, in which the slide toe is located remote from the coast, at the rear of the frontal compound landslide system, is provided by the landslide at "Woodlands" (Fig. 29b). In the sections shown in Fig. 29, the base of the Upper Greensand derived debris is significantly higher than the slip-prone horizon identified at Gore Cliff and at Carrigdene Farm, Bonchurch, but there are indications that deeper slip surface are present in the Gault at around the position of this horizon. This matter needs to be further explored. For the purposes of the model, we assume that the lower slip surface position is the dominant one.

Ventnor Park landslide model

Several workers since Edmunds and Bisson (1954), have recognised that through most of the main, central Undercliff, a seaward zone of flat-lying ridges, sometimes multiple, deriving from deep-seated compound landslides, is backed by a system of multiple compound-rotational slides. Since the Gore Cliff and St Catherine's Point landslide models became available, this combined model has been confirmed and refined, particularly with regard to the depths of the basal slip surfaces (Fig. 32) (Hutchinson, *et al.*, 1991a; Hutchinson, *et al.*, 1991b; Rendel Geotechnics Ltd, 1993). This general combination of compound slides to seaward, seated in the Sandrock, and backed by multiple compound-rotational slides seated in the mid-Gault, is termed the "Ventnor Park" model. It can not be more precisely defined without further geological mapping and subsurface investigation.

Hazard zonation

   General

Hazard zonation is the delimitation of the nature, extent and frequency of potentially harmful physical processes affecting an area. Risk assessment, dealt with subsequently, examines the likely consequences of such hazards. In an urban area, such as Ventnor, the two tend to merge.

By 1961, Mr E. J. Alderton, the then Engineer of the former South Wight Borough Council, had already begun a simple, but useful hazard zonation for Ventnor. This involved making a map of the more serious damage to houses in the town, which in effect delineated the main intermediate movement surfaces of the landslip complex.

This information was then used to prevent future development in these hazardous zones. This map forms the basis of Fig. 7-8 in Chandler (1984).

A very simple preliminary zonation of the whole Undercliff, in terms of past landslide activity, was made by Hutchinson (1965). This identified the E and W ends of the Undercliff as by far the most active. The intervening straight part of the Undercliff was divided into two outer parts with medium past landslide activity and a central part, from about Binnel Point to just W of Steephill, with low activity.

Chandler (1984) made a morphological map of the whole Undercliff at 1:2,500 scale, summarised at 1: 6,666 scale. From this, supplemented by historical records of landslides and surveys of O.S Benchmarks to determine their displacement, he produced a landslide hazard zonation map at a scale of 1: 22,900 (Chandler, 1984). This map was supplemented by frequent cross sections giving the geology and estimated hydrogeology. The part of this map dealing with the Ventnor area was published by Chandler and Hutchinson (1984). An integrated summary of these various contributions is given by Hutchinson and Chandler (1991).

More recent geomorphological and related mapping
In 1988, the Department of the Environment commissioned a study of the land instability in Ventnor. The procedure adopted, building in a much more comprehensive way on the principles applied earlier by Alderton, and making use of the earlier work mentioned above, was firstly to produce a *Geomorphological* map of the Undercliff, showing surface water manifestations, at a scale of 1:2,500. A second map of similar scale was then made recording *Ground Behaviour, i.e.* distress to roads and structures, known landslides, etc. Finally, a third map, also at 1: 2,500 scale was made, entitled *Planning Guidance,* on the basis of the other two. This study was later extended by the former South Wight Borough Council (now the Isle of Wight Council) to cover the whole of the Undercliff from Luccombe to Niton. A final extension, commissioned by the Isle of Wight Council included the remaining area of the Undercliff to the west, as far as Blackgang. This valuable series of 7 x 3 maps was complemented by booklets giving guidance on "living with landslides" and by the opening of an information centre in Ventnor town centre. Maps 1 and 2 were prepared and published by Geomorphological Services Ltd (1991), Maps 3 to 5 by Rendel Geotechnics (1994) and Maps 6 and 7 by Rendel Geotechnics (1995). These maps and associated material have been invaluable in the planning process and have been much used by prospective house buyers. They also make a valuable contribution to the scientific study of the Undercliff. Ways of improving this type of mapping are put forward in the Concluding Remarks and Suggestions.

Mapping archaeological and palaeoenvironmental sites relating to ground behaviour

The European LIFE Project 1997-2000 included an assessment of the use of archaeological and palaeoenvironmental evidence in the investigation of both ground behaviour and coastal erosion (Tomalin and Scaife, 2000). In the Undercliff 121 sites were evaluated and 37 were found to offer potential evidence of ground behaviour. This study also recognised that at least one site outside the study area bore further relevant information and that the ongoing process of coastal erosion also meant that currently unrecognised sites would eventually be exposed and destroyed. The collective value of individual sites was also recognised and particular attention was given to the chronological significance of Palaeolithic implements found on the surface of the Undercliff landslide complex.

Monitoring and remedial works

The above work has been accompanied by a monitoring programme managed by the Isle of Wight Council. This comprises currently 45 instruments (piezometers, tiltmeters, settlement cells, crack meters and inclinometers at 10 sites, and a weather station. A further 10 instruments are planned). This programme is supported by visual inspections of slide activity, aerial photography and interpretation. Useful empirical relationships have been obtained between winter rainfall, its return period and the susceptibility of various parts of the Undercliff complex to ground movement (Lee, *et al.*, 1998). Relations between groundwater pressures and slide movement remain to be established. Efforts are being made to improve the handling and digestion of data, aided by the recent appointment of a fourth geologist to the Council team.

The LIFE Study recognised that certain sites offered the potential to resolve questions concerning climatic change, periodicity and the nature and rate of changes in local ground behaviour. For this reason planned monitoring was required, with particular attention being given to the archaeological and palaeoenvironmental inspection of cliffs, service trenches and ground works. The latter inspections were particularly recommended in Ventnor town.

The ruling philosophy so far in the Undercliff has been that the landslide complex as a whole is too big to stabilise and must be "lived with", and landslide stabilisation works, such as deep drainage or toe weighting, have been limited to coastal locations where, with the assistance of government coast protection grant aid, major defence works have been undertaken.

In addition, a number of measures have been taken which will tend to improve the situation with regard to slope stability, such as some coast protection works in the more built-up areas of Ventnor and Bonchurch, a major reconstruction of the main sewerage system for Ventnor and generally improved maintenance of water mains, sewers and cess pits (McInnes, 1996). The sewage from the most heavily built-up area between about Nat. Grid Ref. SZ 541767 (about the Ventnor-St Lawrence boundary) and Bonchurch, including Lowtherville to the rear of the Undercliff, is now collected behind East Ventnor Bay and pumped inland for treatment at Sandown. However, the more outlying areas of the Undercliff, NE of Bonchurch and W of the Ventnor-St Lawrence boundary, are still served by septic tanks or cesspits (R. McInnes, pers. comm.).

A component of existing background piezometric levels is likely to be the result of:

a)   leakages from water supplies.
b)   leakages from sewers.
c)   discharges into the ground from septic tanks, soakaways and highway drainage, etc.

While for the Undercliff as a whole, the anthropogenic water inputs are probably not the major contributor to the groundwater system, in the vicinity of concentrated discharges the effects are particularly telling, especially for shallow forms of failure. Clearly any reduction in these water inputs will act counter to the effects of natural climatic worsening.

The coast protection works, constructed over the area from Steephill Cove to Monks Bay between 1987 and 1999 at a total cost of nearly £14,000,000, are significant. In detail these are: Steephill Cove (1992), Castle Cove (1996), Western Cliffs (1992), Western Esplanade (1995), Eastern Esplanade I (1995), Eastern Esplanade II (1995), Wheeler's Bay (1999), Wheeler's Bay to Bonchurch (1987), Horseshoe Bay (1993) and Monks Bay (1991/2) (R. McInnes, pers. comm.). From Nat. Grid Ref. SZ 551768 at the W end of Steephill Cove, the coastline of the Undercliff to westward (and the rest of the Isle of Wight SW coast) is designated as a Heritage Coast, where sea defence works are unlikely to be permitted (except where economic and environmental criteria are particularly persuasive).

While the "living with landslides" approach has been, and continues to be, useful in much of the Undercliff, there are situations there, highlighted below, where we believe that it is inappropriate.

Recent deep boreholes
Since the 1991 Conference in Shanklin, four to five dozen deep cored instrumented boreholes have been put down in the Undercliff, especially in the Ventnor area, mostly in connection with the above major sewerage scheme. These provide very valuable data. Their primary purpose was in connection with the associated engineering works, but have only to a rather limited extent been digested and employed to further research on the Undercliff landslides. This is largely because these data have only just been put into the public domain (in October, 2001).

A brief but useful summary of some of the recent deep borehole information, combined with geophysics, is presented by Clark, *et al.* (1994). This indicates that, in west Ventnor, the Gore Cliff type slip surface about 18 m above the base of the Gault is present and that the slip surfaces in the Sandrock bed 2d are at roughly 20 to 25 m below the base of the Gault as shown above on Fig. 26. The slide-prone bed 2b of the Sandrock, exploited by the basal slip surface at St Catherine's Point, appears to be too deep below present sea level to form the main basal shear for the landslide complex, except perhaps offshore (Clark, *et al.*, 1994) as shown on Fig. 25.

It was suggested in 1991 (Hutchinson, 1991), that because of increasing passive resistance at the toe towards the axis of the St Lawrence syncline, there will be a tendency for the basal slip surface of slides in the Sandrock to shift upwards from the horizon (2b) exploited by the St Catherine's Point slide to higher clay layers within that stratum. This is supported by the recent borehole evidence (Rendel Geotechnics, 1993), showing the main seaward Ventnor landslides to be seated principally in horizon 2d. For sedimentological reasons (Dike, 1972), these layers are likely to be discontinuous.

Summary of broad styles of landsliding in the Undercliff

It will be evident from the foregoing that the Undercliff is an intriguing four-dimensional feature, with its various styles and activity of landsliding depending on an interplay between structure (especially the St Lawrence syncline), lithology, hydrogeology, wave attack, climate, geological history, particularly during the Quaternary, the debris of past landslides and anthropogenic effects.

The 1:2,500 scale Landslide Potential mapping of the whole Undercliff represents a major step forward in delineating the Undercliff landslides. However, there is naturally a danger of becoming immersed in the detail of this and perhaps missing some general patterns and relationships. Accordingly, a meso-scale review of the Undercliff, in which it is subdivided in 16 units, U1 to U16 (Fig. 30), is attempted. This is summarised briefly below.

Of the several relevant factors in determining the style and activity of the Undercliff landslides, the degree of toe loading by debris aprons is judged the most important. Such aprons are taken to include both loose debris and large compound landslide blocks in varying degrees of degradation. The following key is used to describe the nature of the debris aprons:

AP   =   apron present
AR   =   apron of restricted extent
AD   =   apron deficient or absent

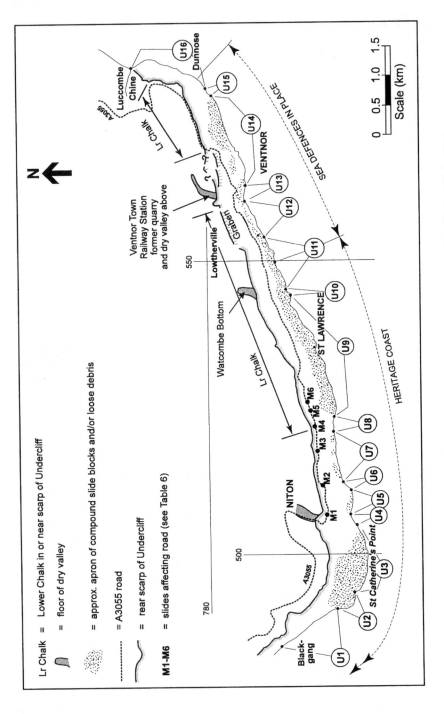

Figure 30. Map of Undercliff showing A3055 road and slide Units U1 to U16

The level of break-out of the deepest slip surfaces at the sea cliff is also noted as:

TF = toe or slope failure
BF = base failure.

The broad nature of the various units is:

U1, Blackgang Chine to Rocken End (NW) (AD, TF).

This unit exhibits very active marine erosion and landsliding particularly in the Gault which forms a bench beneath the precipitous rear scarp. The Gault slides are typified by the compound Gore Cliff slide of 1978 (Bromhead, *et al.*, 1991) which is the basis of the Gore Cliff landslide model (Fig. 28a). The rear scarp here is an old feature, with no known historical regression apart from a few rockfalls and topples. However, archaeological evidence (Preece, 1980) indicates that a considerable retrogression of this scarp, involving the loss of high ground ("Gore Hill") to seaward, has taken place since the second century A.D. Tomalin (2000d and in press) argues that this marked retrogression may have taken place in the early 14th century.

U2, Rocken End (AP, approx TF).

Although in the AP category, there is such strong erosion in this unit, and active rockfall and mudslide debris supply from above, that the area is very unstable. The feature is of interest in exemplifying how the debris aprons further East may have been emplaced, and in showing the power of such mudslides to deepen their channels (Hutchinson, 1970).

U3, Rocken End (SE) to Castlehaven (AP, BF).

The presence of the substantial debris apron seaward of the remains of the St Catherine's Point compound landslide (Hutchinson, *et al.*, 1991a), in which the sea cliffs are cut, renders this unit relatively stable, although small movements are occurring (Bromhead, *et al.*, 1988) and the western parts of the complex are beginning to reactivate. The above slide is an important example of a compound slide on a deep-seated clay layer within the Sandrock, and is the basis of the St Catherine's Point slide model (Fig. 28b).

Comment
After Unit U3, there is a 2km gap in the protective debris aprons from Castlehaven (west end of U4) to 150m W of Binnel Point (east end of U8) (with the partial exception of the eastern part of U5) with associated active shallow instability. In places this is affecting the A3055 road (Fig. 30).

U4, Castlehaven to mid-Reeth Bay (AD, TF).

This unit has steep cliffs of Sandrock, with the lower Gault scarp just above. From there, an extensive area of degraded shallow slides, deriving principally from the upper Gault and the Upper Greensand reach back towards compound-rotational landslides in these beds below the subdued, Upper Greensand rear scarp. In this length, the rear scarp is interrupted by the major Niton dry valley. The resultant diminution in the earlier supply of debris to the Undercliff is reflected in the general lack of a debris apron, the embayment of Reeth Bay and the associated instability. Coastal defences are planned for Castlehaven in 2002.

U5, mid-Reeth Bay to Puckaster Cove (AR, ?BF).

This unit also has steep sea-cliffs eroded in the Sandrock backed by apron remnants which, in places, improve the local stability. Above these are partly degraded, compound-rotational, slides (as in U4) below a more precipitous Upper Greensand rear scarp.

Comment
Much of Units U4 and U6-8 can be characterised approximately by the Beauchamp House landslide model (Fig. 29a). The apparent paradox of having an essentially in-situ Gault Clay scarp in the lower cliff slopes and Gore Cliff type landslides in the upper slopes, is resolved when the subdivision of the Gault into lower and upper members with an intervening slip horizon, is introduced. (This horizon is not shown on the Schematic Landslide Sections of, for example, Sheets 4 and 5 of the 1:2500 geomorphological maps).

U6, Puckaster Cove to 200m to the NE (AD, TF).

This unit is similar to U4, but has a more precipitous Upper Greensand rear scarp. The failure in 2001 of one of the rearward compound-rotational slides at Beauchamp House is taken as the basis of the similarly-named landslide model (Fig. 29a).

U7, 200m NE of Puckaster Cove to 450m W of Binnel Point (AR, ?BF).

Here, the sea cliffs are cut into slide debris. Above these, the lower slopes are occupied by the ridges of degraded compound landslides. The upper slopes and rear scarp are as described for U5.

U8, 450m to 150m W of Binnel Point (AD, approx TF).

The sea-cliffs are cut into mudslide debris except at the eastern end of the unit where they are composed of insitu Lower Greensand. The remains of old masonry sea defence walls (built by William Spindler after 1862) litter the foreshore. Shallow translational slides and mudslides extend to, or above, the middle of the Undercliff. The upper slopes and rear scarp are as described for U5.

Comment
The U8/U9 boundary represents an important change in the nature of the main SSE-facing Undercliff. To the west of this is the 2km gap in the debris aprons and the associated shallow instability; to the east debris aprons are present, occasionally restricted but generally substantial enough to provide some relative stability, which

**Figure 31. Aerial photograph of the Isle of Wight U8/U9 boundary in the
Undercliff, looking E (with acknowledgements to the Cambridge University
Committee for Aerial Photography)**

run through to the U14/U15 boundary. This abrupt transition from deficient to substantial aprons is well shown by the oblique aerial photograph of Fig. 31.

A significant factor affecting the size of the aprons is the amount of debris which has been supplied to them through failures of the Undercliff rear scarp, depending chiefly on its height and lithology. There is some correlation between the presence of a sizeable debris apron and the occurrence of the Lower Chalk as a capping to the rear scarp. The latter begins at the U8/U9 boundary, and runs eastwards, with a few small gaps, possibly to around Wheeler's Bay. The restriction of the aprons with associated increased instability, noted below former dry valleys and beneath lengths of rear scarp with little or no capping of Lower Chalk, is the reverse aspect of the same phenomenon. Better integration of the geology of the rear scarp area with that in the landslides of the Undercliff is needed before these correlations can usefully be explored further.

As a result of the St Lawrence syncline, all the units from U9 to U14 are characterised by base failure at the toe. The associated passive forces provide an additional stabilising influence throughout this length of the Undercliff. The effects of this syncline on the hydrogeology, especially in the Lower Greensand, have not yet been explored.

The above combination of passive resistance and toe loading/erosion protection provided by the debris aprons in Units U9 to U14 gives them generally a fairly good degree of stability, with the possible exception of unit U12 with its active graben. A further important feature of particularly the middle to upper slopes of units U9 to U14 is that exposures of Gault with associated energetic shallow sliding (as in most of units U4 to U8) are absent or very rare. This is probably because the aprons support the toes of the system of multiple compound-rotational slides which occupy the mid- to upper Undercliff and these, with their strong cappings of slipped Upper Greensand and Chalk, effectively armour and safely mask, the Gault. Not surprisingly, most of the built-up area of the Undercliff is situated on these units.

Coastal defence works constructed between Steephill Cove (U11) and Monk's Bay (U15) between 1987 and 1999 (R. McInnes, pers. comm.) have improved the erosion resistance of this length.

U9, 150 m W of Binnell Point to Orchard Bay W (AP, chiefly BF).

Here, the sea cliffs are cut in landslide debris. The lower half of the Undercliff is occupied by coastwise elongate ridges of Upper Greensand or Chalk and Upper Greensand from old, deep-seated compound landslides failing in the Sandrock, in various stages of degradation. The upper half of the Undercliff consists of multiple compound-rotational slides seated in the upper Gault, generally capped by Upper Greensand and frequently, the Lower Chalk. The precipitous rear scarp of Upper Greensand is generally capped by the Lower Chalk.

U10, Orchard Bay (AR, BF).
Broadly similar to U9, but exhibits some local diminution of the debris apron, doubtless as a result of the presence of the Watcombe Bottom dry valley. This gives rise to the erosion of Orchard Bay and environs and to some associated instability.

U11, Orchard Bay (E) to Castle Cove / Flowers Brook (AP to AR, BF).
Broadly similar to U9. The debris apron is somewhat restricted at the east end of this unit. Sea defences were provided there, at Steephill Cove and Castle Cove, in 1992 and 1996, respectively (R. McInness, pers. comm.).

U12, Castle Cove / Flowers Brook to Ventnor Bay W (AP, BF).
Generally like Unit 9, except that the rear scarp steps back (northwards) by about 150 m over this length of around 450 m to form the well-marked Lowtherville graben. The 550 m long, coastwise elongate ridge forming Ventnor Park lies almost directly downslope from the graben, with intervening multiple compound-rotational slides, and may represent a related former compound landslide. It should be noted that, in this unit, steady seawards movements are occurring (at the graben, and in some places downslope) despite the presence of substantial masses of debris at the slope toe.

U13, Ventnor Bay (AR, BF).
Here the set-back of the rear scarp continues eastwards, though the graben seems to peter out. In this unit, the mid- to upper coastal slopes are broadly similar to those in U12, but the coastal compound slide ridges in the lower slopes are less well developed, particularly in the western part of the Unit, allowing a shallow embayment to form in the coastline. This weakening of the apron is the probable reason for the persistent slow heave movements observed at and near the slope foot which continue despite the coastal defences. These features are directly below the mouth of the important dry valley in which Ventnor Station was built. The depression in the rear scarp produced by erosion of this valley, with the Lower Chalk and possibly some of the Upper Greensand missing, is believed to have led to this more restricted apron development. (The northward flowing tributary of the Eastern Yar through Wroxall has little influence in this regard).

Comment
It is particularly interesting that the deficiency of the aprons, and associated instability, in units U4 (western Reeth Bay), U13 (western Ventnor Bay) and, to a lesser degree, in U10 (Orchard Bay), can be directly related to the three main dry valleys running into the Undercliff from the north, that is, at Niton, that occupied by the former Ventnor railway station, and at Watcombe Bottom, respectively.

U14, Ventnor Bay E to Monks Bay SW (AP, BF).
The Undercliff here is also broadly similar to that in Unit U13, except that a broad bench, formed by the Chert Beds of the Upper Greensand, lies currently between the

foot of the Chalk Downs and the rear scarp of the Undercliff. However, some Chalky debris is involved in the landslides, and forms a few coastwise elongate ridges in the lower slopes, especially in the central and western parts of the Unit. Being well to the east of the axis of the St Lawrence syncline, the strata are rising to the ENE and the slides probably switch from base to toe failures just west of the east end of this Unit. Notably in this unit, the morphology of the Undercliff is much modified by past quarrying.

Comment
Units U9 to U14 may be broadly characterised by Ventnor Park landslide model; described earlier (Fig. 32).

U15, Monks Bay (AD, TF).

The sea cliffs are composed of in situ Sandrock. Above these, the lower slopes of the Undercliff are occupied by fairly shallow translational, possibly part rotational landslides in the Gault. The absence of a debris apron here led to shallow instability. Coastal defences and slope drainage have now been installed. The upper slopes consist of multiple compound-rotational landslides of Upper Greensand and Chalk, extending back to a rear scarp in the Chert beds. The bench of Chert Beds behind the rear scarp, mentioned under Unit U14, continues through this unit. In some ways this unit is the eastern equivalent of Units U4 and U6 to U8, but has a much more restricted coastwise extent. This probably arises more from the change in angle of the Undercliff in plan than from asymmetry in the St Lawrence syncline.

U16, Dunnose to Luccombe Chine (AD, TF).

The precipitous sea cliffs consist of the Sandrock. In contrast to the western end of the Undercliff, the Ferruginous Sands do not outcrop in the cliff foot here. In the northern third of this unit, a bench between about 10 and 20 m above sea level has developed in the lower cliff. Houses built there were destroyed by a landslide in 1910, doubtless seated on a clay layer within the Sandrock. In general, above the Sandrock cliffs the Carstone and the lower Gault can be seen in place beneath gentler slopes composed of shallow landslides. These involve chiefly the upper Gault, with Upper Greensand and Chalk debris. The mid-upper Undercliff consists of large slid blocks of Upper Greensand and Chalk, seated in the upper Gault and dating from the major slides of 1810 and 1818. These were unusual in causing some recession of the steep rear scarp, of predominantly the Upper Greensand, cutting out the bench of Chert Beds to landward (noted in U14 and U15), which now remains only in the western part of the unit. At Luccombe Chine, the eastern scarp of the Southern Downs is set back by stream erosion and the Undercliff landslides come to an end. This unit is most nearly equivalent, especially in its south-western part, to Unit U1, though the latter has not caused recession of the rear scarp within historical time. A modified first-time version of the Gore Cliff model might be appropriate here.

Using the size of the debris aprons as a diagnostic criterion for stability, we see that, in the Undercliff landslides, the AD units are generally the least stable, the AP units the most stable and the AR units in an intermediate condition. (Unit U12, possessing AP characteristics but with the slowly moving Lowtherville graben at its head, is discussed later).

Risk assessment
Within the area of the Undercliff landslides, both physical assets and the human population are exposed to various degrees of risk.

Fortunately there have been very few deaths and injuries from landsliding in the Undercliff, or the Isle of Wight generally. Discounting the half dozen or so skeletons of Iron Age to Romano-British date inferred to have been overwhelmed by landslips or falls in the Undercliff (Hutchinson, 1965), only three fatalities (one partly man-induced) in the Undercliff and one more elsewhere in the island (resulting from boys tunnelling in sand) are known during the past two centuries. Available details are given in Table 5 below.

Table 5. Fatalities and injuries caused by landsliding in the Isle of Wight

Locality	Date	Event	Deaths	Injuries	References
Gore Cliff	*c.* 19.2.1799	Fall, probably from rear scarp of Pitlands slide	2	2 to 3	Anon., 1799a, b; Mudie, 1838; Mew, 1934
Ventnor, S Grove Road	3.6.1985	Man buried by collapse of 15 m deep dry well in slide-disturbed Malm Rock	1	0	Hutchinson & Chandler, 1985
Alum Bay	31.8.1959	4 boys buried by fall of sand, three rescued	1	3	Anon., 1959
Shanklin	21.3.2001	Slide of talus beneath Ferruginous Sands cliffs pushed into rear of hotel*	0	4	BBC News Homepage, 22.3.2001

* Other talus slides of this nature, and rockfalls, at Shanklin are reported by Barton (1991).

Recent estimates put the cost of damage to buildings and infrastructure in the Undercliff at about £2 million per year (R. McInnes, pers. comm., 2001). Thus, the current emphasis on the assessment of risk to physical installations rather than human life is generally justified, subject to certain provisos given below.

The Undercliff is a complex of deep-seated and shallow landslides and, naturally, these have tended to dominate any risk assessments (hitherto predominantly qualitative).

In this connection, the speed and amount of displacements in individual slide events determines the degree of risk which they pose. Applying the debris apron criterion used above, the main, built-up area of the Undercliff, from just E of St Catherine's Point to Monks Bay (W end of Unit U4 to E end of Unit U15), discounting slides in the sea cliffs, can be divided broadly on the basis of present movements into:

- apron-deficient Units U4 to U8, with individual slide displacements of 10 m or more. (Unit U15 was probably in this category, but has been quietened down by coastal defences, drainage and other slope stabilisation measures).
- apron-protected Units U9 to U14, with individual slide displacements of a few hundred millimetres.

Two of the most serious of the above problems, the Lowtherville graben and downslope area, and the area of shallow sliding affecting the A3055 road between Castlehaven and Binnel Point, are discussed further below. This choice is confirmed by the plot of α, the average angle between the top of the slipped masses (base of rear scarp), against position along the Undercliff (Fig. 8 of Hutchinson and Chandler, 1991). Apart from the W and E extremities of the Undercliff, these areas have the greatest α values of around 12.8° and 15.0°, respectively.

In addition to the main landslides, it is important to pay proper attention to events of much smaller scale which do, however, present potential dangers to life. In particular, there are risks from rockfalls from the precipitous rear scarps, which can affect houses, roads and footpaths, and those arising from the collapse of old retaining walls, facing walls and fills, particularly where strained by slide movements. An example of the latter is provided by the collapse of a retaining wall and earth bank into Ventnor car park in March 1991. Attention was drawn to the dangers of the latter by Professor R. Fell (pers. comm.) in a recent visit, fresh from his experiences with the tragedy resulting from a collapsed fill at Thredbo, Australia and in Hong Kong (Fell and Hartford, 1997).

The rockfall danger is acknowledged in the existing Planning Guidance maps, but the advice could be checked and refined as indicated below. The risks attending collapsing fills and retaining walls are not dealt with by the existing Landslide Potential Mapping.

To cover these eventualities, all rock-falls should be documented as to location, time, size and run-out and all retaining walls, facing walls and fills above a minimum size should be examined, assessed and recorded as part of a quantitative risk assessment exercise. This will enable the above dangers to be prioritised and dealt with in due order. A good model for this is the Landslip Preventative Measures (LPM) programme of the Hong Kong Geotechnical Engineering Office (Wong, 2001).

Management
General
It is evident that proper management of landslides requires a full knowledge of the relevant facts. A major decision concerns whether the slides are benign enough to "live with" or in conjunction with future climatic changes, could generate a future unacceptable hazard and risk, to guard against which appropriate intervention would be needed. The impacts of predicted climate change scenarios on the landslide system are of particular concern to the Council which is commissioning its own research, and has coordinated a major study completed in September, 2001 (Hosking and Moore, 2001).

The view that the landslides are generally moving slowly enough to "live with" has hitherto been predominant in the Undercliff. With this philosophy, as noted, some well known measures have usefully been applied. These include diminishing man-induced inputs to the groundwater system by improving sewers and water mains and by replacing septic tanks by sealed cesspools, avoidance of destabilising excavations and fills, reduction of (chiefly lateral) coast erosion through the construction of flexible coastal defences (the downward erosion of foreshores is hard to prevent, except perhaps by offshore breakwaters), and discouraging the construction of buildings, retaining walls and fills of vulnerable design or construction. A major, pioneering effort to educate and involve the local population in the landslide situation has also been carried out.

The recent mapping of the "Coastal Landslide Potential" for the whole Undercliff, at a scale of 1:2500, outlined earlier, was a major undertaking (Geomorphological Services Ltd, 1991 and Rendel Geotechnics, 1994, 1995). It developed an initiative begun in the late 1950s by Alderton, a previous Borough Engineer of Ventnor Urban District Council, and is an important step forward. Particularly through the Planning Guidance maps, it has been of great service in the sensible ordering of developments in the Undercliff, most obviously in the avoidance of intermediate slip scarps. Some cautionary comments are made below.

Our ability, in most cases, to "live with" the landslides of the Undercliff depends predominantly on the fact that these are very largely slides on pre-existing slip surfaces. As a result, a majority of the movements are essentially non-brittle, responding to groundwater fluctuations and toe erosion, for instance, with modest displacements. Where first-time slides occur, for example where sliding retrogresses at the rear scarp, brittle behaviour can be expected, at least in cross-bedding parts of the failure surface and in accumulated debris. Where flexural slip has occurred it will tend to reduce the brittleness of retrogressive movements on bedding planes. However, even with slides totally or largely on pre-existing slip surfaces, unexpected brittleness can occur (Hutchinson, 1987), chiefly through toe brittleness and/or brittle internal shear. It is evidently vital to identify sites of this potential within the Undercliff.

In broad terms, discounting slides in the coastal cliffs, the current individual slide movements per single slide event in the Undercliff, as noted above, can be divided into those of 10 m or more and those of a few decimetres or less. The former occur most obviously in its western and eastern extremities, Units U1 to 2, and U16. Such energetic movements also take place at the heads of the relatively shallow translational slide and mudslide systems within the length of deficient debris aprons in the western part of the main Undercliff (Units U4 to U8), where, as discussed below, they disrupt the important A3055 road and threaten houses. Comparable movements do not appear currently to be affecting the equivalent unit at the eastern end of the Undercliff, U15, possibly because of improved standards of sewerage and surface-water drainage and coastal defence works there.

Again discounting landslides in the coastal cliffs, the generally deeper slide movements in the main, debris apron-protected or part-protected length of the Undercliff (Units U9 to U14), for the probable reasons given above, fortunately fall into the second category above. The largest slide displacement in a single event known there appears to be the 19 mm/day (for 7 days) in Bath Road during the wet period of 1960-1, eventually giving a total displacement in the slide event of 305 mm (Hutchinson and Chandler, 1991). As this is the length with the highest concentration of population, the crucial question is whether any slide movement there could depart from the prevailing pattern and be large enough and of sufficient displacement to constitute a serious risk? This is discussed more fully below.

Risk assessment for the Lowtherville graben and area downslope
A section through the west end of the Lowtherville Graben and the western end of the Ventnor Park ridge, adjacent to Flower's Brook (the western part of Unit U12), based on the recent deep boreholes (Rendel Geotechnics, 1993), is given in Fig. 32. This must be regarded as provisional, however, as there are, as yet, no boreholes in the upper slopes of the Undercliff and the situation in the lower slopes is complex.

The Undercliff area was mapped by the Geological Survey in 1856 and was revised on the Six-inch scale in 1886-87 (White, 1921). The early surveys, including the Six-inch of 1887, did not record the Lowtherville graben, though its landward scarp probably existed. The feature was perhaps first noted by Alderton on his "fissure map" of Ventnor and then incorporated in Chandler's (1984) mapping. Both there and on the Landslide Potential Map, it is shown as approaching 450 m in length. Chandler (in Hutchinson and Chandler, 1991) measured the movements on the counterscarp of the graben between January 1982 and February 1984. They were up to 45 mm/year vertically and 40 mm/year horizontally seaward. Readings recorded by the Council's monitoring instrumentation indicate that similar movements have continued to the present. At least twenty-four houses in the immediate vicinity of the graben have so far been demolished.

Although there are longer scarps than the Lowtherville graben within the Undercliff, none is so long and so active. There is, perhaps, a danger in that the mapping of a feature may make it seem, subconsciously, familiar and even acceptable. In fact, this graben probably constitutes the clearest evidence in the Undercliff of a potential slide of considerable size. Downslope from the graben is central and west central Ventnor, with well over 100 buildings and a network of roads and services and there is no guarantee that the present movement will not accelerate. Thus, we consider it inadvisable to continue to "live" with this particular landslide. A modern risk analysis (Fell and Hartford, 1997), on as quantitative a basis as possible, should be carried out on the area that could be involved in a slide headed by the graben to assess the merit of intervening with deep stabilisation measures. If this decision is taken, a year or more is likely to be needed for the considerable investigations that will be necessary to establish the engineering geology and mechanics of the potential slide. At present, its subsurface is unexplored above about mid-height of the Undercliff and it is not even known which stratum its basal slip surface is in, though the mid-Gault horizon would seem most probable. The available data are summarised on the section of Fig. 33. Even on this latter assumption, the width of the graben is relatively modest according to the guidelines of Cruden, et al., (1991). This suggests the possibility that a larger graben may tend to develop in the future, releasing a larger, more brittle movement.

Following Hutchinson (1987b), the most likely mechanisms by which the movements associated with the graben could accelerate are indicated diagrammatically in the approximate section of the Undercliff at the graben given in Fig. 33. They comprise:
  a) brittle failure of an emerging toe at the inherently weak transition between the rearward, compound-rotational slide system seated in the Gault, and the deep-seated compound slides to seaward.
  b) brittle failure in the rearward compound-rotational slide system on a second (or third) graben, not yet developed. Fig.34 shows a complex, three-graben structure in a slide on the French coast (Maquaire and Gigot, 1988).
  c) the development of a slide of St Catherine's Point type, seated on a clay horizon in the Sandrock (most probably 2d). This would exhibit a degree of brittleness in the "first-time", or rearward part of the bounding slip surface and in the shear forming the seaward scarp of the graben.

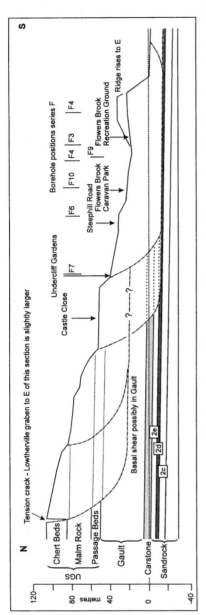

Figure 32. Provisional section of the Undercliff landslides near Ventnor Park (based on Rendels Geotechnics, 1993)

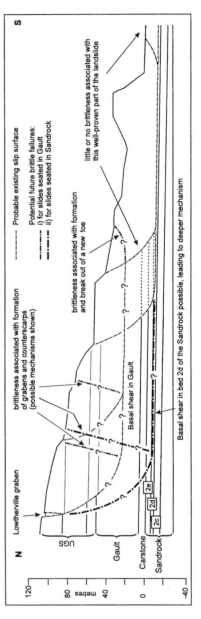

Figure 33. Diagrammatic section of the area of the graben and downslope

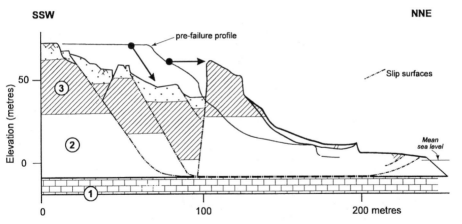

Figure 34. Section of the Bouffay landslide, Calvados coast, France
(after Maquaire and Gigot, 1988)

d)   since the Lowtherville graben to Ventnor Park landslide system is partly restrained by elements of the adjacent Undercliff, small scale ground movement which appears unconnected with this system, e.g. at Castle Court, might be an agent which releases the main slide.

Within a complex slowly sliding slope, areas of restricted movement may indicate the sites of "hang-ups", possibly building up to future sudden brittle failures. Monitored movement rates between the graben and the sea, provided by Moore, et al., (1991) and by Jones and Lee (1994), hint at such a development in relation to a possible type a) mechanism, but are inconclusive as the movement data are insufficiently precise with regard to directions and timings.

More information on the nature and probability of failure in these various mechanisms should be obtained as soon as feasible by closer monitoring of surface movements and by further sub-surface investigation.

Maintenance of the A3055 road

The geological reasons for the particular vulnerability of the Principal Road A3055 from just east of the mouth of the Niton dry valley to western St Lawrence to damage (Fig. 30) from the activity of shallow slides and mudslides has been described above. This has also long been evident from records of damage and disruption to the road. The available data on these events are given in Table 6.

The Landslide Potential mapping identified the three main points of attack on the A3055 that were known previously, *i.e.* M3 to M5. However, as the mapping did not clearly point to future problems at M1 and M2, the question of how many sites in this length may become unstable in the future remains open.

The recent boreholes at M2, M4 and M5, permit the preparation of the sections in Figs 29a, b which show that generally a Beauchamp House landslide model applies. Whilst, overall, toe unloading by marine erosion has played a large part in bringing about these failures, the fact that one slide, M5, toes out far from the sea, above the Old Park compound slide, indicates that rainfall can often be the trigger.

These failures pose a difficult problem. Apart from that just mentioned, they all involve a 300 to 500 m long mudslide or translational slide complex leading down to the coast, where coastal defence works are unacceptable on environmental grounds. If normal practice of arresting coast erosion before undertaking stabilisation works on the coastal slopes is abandoned, extensive drainage of the latter could win time, but at considerable expense. Eventually the road would again be threatened by the deterioration of these measures or by similar failures developing at new places. There is little scope left for realigning the road further back on the Undercliff and there will come a point, before long, where realignment of the road inland between Niton and St Lawrence becomes unavoidable.

Table 6 . Some details of slide damage to the A3055, Niton to St Lawrence

Code	Nat. Grid Ref. of head (SZ)	Date	Damage to road D=distressed; B= broken	Source
M1	5065 7595 (Puckaster Corner)	2001	Slight D	R. McInnes, pers. comm.
M2	5115 7600 (Beauchamp)	2001	B	R. McInnes, pers. comm.
M3*	5175 7610 (W of Mirables)	1960	rear scarp near road	Hutchinson, (Author's files)
M4*	5215 7615 (Caravan Park)	2001	B	R. McInnes, pers. comm.
		1960-61	rear scarp 6 m from road	Hutchinson, 1965
		1924 & 26	B	Hutchinson, (Author's files)
M5*	5245 7625 (Woodlands)	2001	D	R. McInnes, pers. comm.
		1961	D	Hutchinson, (Author's files)
M6	5260 7630	1951	D	Hutchinson, 1965

* indicates that these areas of instability were picked up by the Landslide Potential mapping in 1991 to 1995, most clearly on the Planning Guidance sheets.

CONCLUDING REMARKS AND SUGGESTIONS

 Following our investigations of Gore Cliff and St Catherine's Point landslides and Chandler's (1984) stratigraphical and mapping work in the 1980s, impressive progress has been made in the management of the Undercliff. This has largely been due to the efforts of Mr Robin McInnes and the Isle of Wight Council, latterly linked with the European Union, in conjunction with their consulting engineers. Our following remarks and suggestions are intended in no way to diminish this major achievement. Their aim is to identify possible improved ways forward.

1. The largely geomorphological mapping techniques applied so far to the urban landslide problems, as noted, have much merit but they tend to unduly reinforce, perhaps subconsciously, acceptance of a "living with landslides" philosophy. This has led to some imbalance in the overall approach, perhaps because intended, more expensive engineering geological/geotechnical work has sometimes been delayed. In future, more multidisciplinary teams could, with advantage, be used. We recommend that a risk assessment approach – made as quantitative as possible – be

applied to these problems. Initial targets for this are suggested below. In this way, a better-defined picture of the various risks can be obtained and priorities for further action soundly established.

2.   The following targets are proposed for the more quantitative risk assessments:

a)   the Lowtherville graben and its associated potential landslide. We regard it as inadvisable to continue to "live with" this, and suggest that a quantitative risk analysis be made for optimistic and pessimistic assumptions regarding the potential size of a landslide there, and for various assumptions of the magnitude of a possible future lurch forward. On this basis, decide whether and how to intervene with stabilisation measures. If a decision to intervene is made, comprehensive ground investigations of the area will be necessary to establish the relevant engineering geology and hydrogeology, with slip depths, mechanisms, displacements and groundwater pressures, as there are at present no boreholes in the upper half of the Ventnor Undercliff.

b)   the whole of the A3055 road, and particularly the section running between St Lawrence and Niton. Half a dozen actual or potential breaks in the road caused by shallow landslides are already known (some of long standing, evident since at least 1924) and the opportunities for further retreat of the road within the Undercliff are severely limited. On this basis, assess whether to repair and locally stabilise all these points (and possibly others), bearing in mind that in this length coastal defences are probably not an option, or to re-route the traffic inland. If the latter option is chosen, the need to maintain access to the considerable number of houses served by the present road will need to be costed in. It is understood that ground investigations, which will contribute to such an assessment, are under way (also at Bonchurch, where recent slips threaten the A3055 there).

c)   rockfalls and topples from the rear scarps. All information on these, including their locations, timing, nature and run-out should be entered into a database and used to check the danger zones shown on the existing Landslide Potential maps and, more generally, to define the problem and enable counter-actions to be prioritised.

d)   retaining walls, fills and facing walls. The risk of collapse of these is not covered by the existing mapping. Again, all relevant information should be collected and recorded for each future case – and retrospectively if the archives permit this – in order for the risk to be defined and actions to meet this prioritised.

3.   In the Coastal Visitor's Centre, establish comprehensive and accessible data bases for the Isle of Wight as a whole, for:

a)   all boreholes sunk in the Island (with the cooperation of the British Geological Survey).

b) all existing groundwater records (in cooperation with Southern Water Ltd), including those from wells, springs and piezometers.
c) all published references on landslides and related subjects.
d) relevant newspaper cuttings, videos, etc.

In all cases, past records should be incorporated to the fullest extent possible.

4.   Improve the current monitoring within the Undercliff (and elsewhere, as appropriate) of:

a) the hydrogeology of the Undercliff and its hinterland, in cooperation with Southern Water Services Ltd, both for water tables perched on the Gault and for those confined within the Lower Greensand.
b) the landslide movement and piezometric monitoring generally, trying to relate movements to changes in ground-water pressures. Aim to establish soundly-based warning thresholds for these factors.
c) the landslide movement and piezometric monitoring in the Lowtherville graben and the area downslope in order to define more closely the relevant mechanics and the extent and nature of a potential future breakout.
d) the archaeological and palaeoenvironmental deposits which attest to the timing and the nature of movement and environmental change on the surface of the landslide complex.
e) ground works and service trenches which present the opportunity of exposing or destroying sediment archives relevant to the understanding of past and present ground behaviour.

5.   In all monitoring, consolidate the steps already being taken to improve the accessibility, control and speed of digestion of the accumulated data.

6.   Regarding the mapping operations in the Undercliff and at W Cowes, we have the following comments and suggestions:

a) it is important to recognise that these maps have a limited life, as they are based on the assumption that future behaviour will be similar to that in the past, while it is known that, in general, conditions for stability are likely to worsen. This is of particular concern in areas where medium- to longer-term movements are continuing, such as at the Lowtherville graben, where a future brittle failure could locally invalidate the present mapping and its conclusions.
b) update the Landslide Potential maps as feasible, checking them against actual ground behaviour, making use of a GIS approach to facilitate this.
c) the present definitions of the various types of ground movement and their mapping tend to be over-complicated. This, combined with colour keys that are often hard to distinguish from one another and inadequate labelling of the National Grid lines, etc., renders the maps, particularly of the geomorphology,

sometimes hard to read, with a danger of not seeing the wood for the trees. These matters need to be attended to, again with the help of a GIS approach.

d) no geological map is provided and this omission is not adequately made good by sections. A map of the solid geology where exposed, the colluvial geology and more emphasis on the relevant Quaternary features (for example, the main dry valleys crossing the Undercliff rear scarp, as noted above, and the presence of a buried valley beneath the Medina at Cowes), would be valuable. Integrate the mapping more closely with the geology.

7. Continue the present policies of:

a) reducing the infiltration into areas of suspect stability of water leaking from water supply pipes and sewers, from road drains and from septic tanks.

b) reduce the marine erosion of landslide toes by building "soft sea" defences where geotechnically and environmentally acceptable. In addition, further means of diminishing erosion of the shore platform should be considered.

8. Continue to inform and educate the public on all relevant aspects of the Isle of Wight landslides and environmental issues related to the coastline.

9. Critically examine all safety and engineering procedures and ensure that the links between the geotechnical decision to raise an alarm, the passing of this to the local authorities and its further transmission to the public are effective and robust.

10. Form an Advisory Panel of Experts, meeting regularly to provide a strategic overview of ongoing and future coastal management. Such a measure is normal on most large projects, especially if long-running. In an urban landslide context, the panel set up recently by the GEO (Geotechnical Engineering Office) in Hong Kong is an excellent example.

ACKNOWLEDGEMENTS
The Authors wish especially to thank Robin McInnes for his major input into this Conference, and the Isle of Wight coastal problems generally, and for his unfailing patience and helpfulness in response to our many questions.

We are also grateful for the help of the following:
A. Brampton, M. Brown, K. Burgess, M. Chandler, A. Clark, C. Clayton, M. Collins, J. Cosgrove, E. Davies, B. Denness, R. Fell, C. Gaches, P. Gibbard, J. Hancock, P. Hopson, W. Hodges, J. Jakeways, D. Jones, M. Jones, M. Lee, P. Marsden, R. Moore, G. Momber, H. Owen, R. Preece, R. Scaife, D. Tomalin, T. Tutton and M. Woodruff. The Isle of Wight County Archaeological Unit is thanked for data from the Sites and Monuments Record, and the Remote Sensing Unit, Department of Earth Sciences and Engineering, Imperial College together with the University of London Intercollegiate Research Services, for the satellite image. The

work of R. Beaumont and P. Howard, who did the bulk of the graphics and photography is also much appreciated.

REFERENCES
Allen, L.G. and Gibbard, P.L.,1993. Pleistocene evolution of the Solent River of Southern England. *Quaternary Science Reviews,* **12**, 503-528.

Anon,1799a. Agricultural report for the Isle of Wight for February 1799. *Isle of Wight Magazine for 1799,* 88.

Anon.,1799b. Monthly occurrences: letter dated 9th February 1799, giving account of landslide near Niton. *Isle of Wight Magazine for 1799,* 96.

Anon.,1959. Boy dies after being buried in sand fall. *The Times,*1st September, 1959.

Anon.,1995. Effects of offshore dredging. MAFF Coastal Defence Forum, 4th April 1995. Wallingford: H R Wallingford Ltd.

Barton, M. E., 1991. The natural evolution of the soft rock cliff at Shanklin, Isle of Wight, and its planning and engineering implications. *International Conference on Slope Stability Engineering-Developments and Applications* (ed. R. J. Chandler), 181-188. London: Thomas Telford.

Barton, M. E. and McInnes, R. G., 1988. Experience with a tiltmeter-based early warning system on the Isle of Wight. *Proceedings 5th International Symposium on Landslides, Lausanne,* **1**, 379-382.

BBC Homepage, 22 March, 2001.

Bhandari, R. K. and Hutchinson, J. N., 1982. Coastal mudslides in the Oligocene clays of Bouldnor Cliff, Isle of Wight. *Landslides and Mudflows,* Reports of the Alma-Ata International Seminar (UNESCO and UNEP), (ed. A. Sheko), 176-199, October, 1981. Moscow: Centre of International projects, GKNT.

British Geological Survey, 1995. Wight. 1:250 000. Solid Geology, Second Edition. Edinburgh, Scotland: British Geological Survey.

Bromhead, E. N., Chandler, M. P. and Hutchinson, J. N., 1991. The recent history and geotechnics of landslides at Gore Cliff, Isle of Wight. *International Conference on Slope Stability Engineering-Developments and Applications* (ed. R. J. Chandler), 189-196. London: Thomas Telford.

Bromhead, E. N., Curtis, R. D. and Schofield, W., 1988. Observation and adjustment of a geodetic survey network for measurement of landslide movement. *Proceedings 5th International Symposium on Landslides, Lausanne,* **1**, 383-386.

Chandler, M. P., 1984. *The coastal landslides forming the Undercliff of the Isle of Wight.* Ph. D. Thesis (unpublished), Imperial College, University of London.

Chandler, M.P. and Hutchinson, J.N., 1984. Assessment of relative slide hazard within a large, pre-existing coastal landslide at Ventnor, Isle of Wight. *Proceedings 4th International Symposium on Landslides, Toronto,* **2**, 517-522.

Clark, A. R., Lee, M. E. and Moore, R., 1994. The development of a ground behaviour model for the assessment of landslide hazard in the Isle of Wight Undercliff and its role in supporting major development and infrastructure projects. *Proceedings 7th International Congress International Association of Engineering Geology,* Lisbon, **6,** 4901-4913.

Clark, A. R., Moore, R. and McInnes, R.G. 1995. Landslide response and management, Blackgang, Isle of Wight. *30th MAFF Conference of River and Coastal Engineers, Keele University, 5-7 July, 1995,* 6.3.1-6.3.24.

Clayton, C. R. I., 1978. *Chalk as a fill.* Ph. D. Thesis (unpublished), University of Surrey.

Clayton, C. R. I., 1983. The influence of diagenesis on some index propeerties of chalk in England. *Geotechnique,* **33,** 225-241.

Clayton, C. R. I. and Matthews, M. C., 1987. Deformation, diagenesis and the mechanical behaviour of chalk. From: *Deformation of Sediments and Sedimentary Rocks* (eds M. E. Jones and R. M. F. Preston), Geological Society Special Publication No. **29,** 55-62.

Collins, M. and Ansell, K., 2000. *Solent Science - A Review. Proceedings in Marine Science,* **1.** Amsterdam: Elsevier.

Conybeare, W. D. and Phillips, W., 1822. *Outlines of the geology of England and Wales.* London: William Phillips.

Cruden, D. M., Thomson, S. and Hoffman, B. A., 1991. Observation of graben geometry in landslides. *International Conference on Slope Stability Engineering-Developments and Applications* (ed. R. J. Chandler), 33-35. London: Thomas Telford.

Darwin-Fox, W., 1862. When and how was the Isle of Wight severed from the mainland? *The Geologist,* **5,** 452-454.

Denness, B., 1969. *Fissures and related studies in selected Cretaceous rocks of SE England.* Ph. D. Thesis (unpublished), Imperial College, University of London.

Devoy, R. J., 1987. The estuary of the Western Yar, Isle of Wight: sea-level changes in the Solent region. In: *Wessex and the Isle of Wight - Field Guide* (ed. K. E. Barber), 115-122. Cambridge: Quaternary Research Association.

Dike, E.F., 1972. *Ophiomorpha nodosa* Lundgren: environmental implications in the Lower Greensand of the Isle of Wight. *Proceedings of the Geologists' Association,* **83,** 165-177.

Dix, J. K. 2000. A geological and geophysical investigation of the submerged cliff at Bouldnor. In: *Coastal change, climate and instability* (eds R. G. McInnes, D. J. Tomalin and J. Jakeways), European Commission LIFE project LIFE-97 ENV/UK/000510 (1997-2000). Final Technical Report, **2,** Palaeo-environmental study areas, Area, P1, 6. Newport, Isle of Wight: Isle of Wight Council.

Dyer, K.R., 1975. The buried channels of the 'Solent River', southern England. *Proceedings of the Geologists' Association,* **86,** 239-245.

Edmunds, F. H. and Bisson, G., 1954. Geological report on road subsidence at Whitwell Road, Ventnor during 1954 (unpublished). 30th December, 1954.

Ellis, N., 1986. *Morphology, process and rates of denudation on the chalk shore platform of East Sussex.* Ph. D. Thesis (unpublished), Brighton Polytechnic.

European Union LIFE Project, 2000. *Coastal change, climate and instability* (Project Manager R. McInnes, Project Officer Jenny Jakeways). LIFE - 97

ENV/UK/000510 (1997-2000). Newport: Isle of Wight Council.

Fell R. and Hartford, D., 1997. Landslide Risk Management. In *Landslide Risk Assessment* (eds D.M. Cruden and R. Fell), 51-109. Rotterdam/Brookfield: Balkema.

Ferril, D. A. and Morris, A. P., 2001. Displacement gradient and deformation in normal fault systems. *Journal of Structural Geology,* **23**, 619-638.

Foundation and Exploration Services Ltd., 1995. Highway stabilisation at 'Woodlands', Undercliff Drive, St Lawrence, Isle of Wight. **1**, Factual Report on Ground Investigation (unpublished).

Gale, A. S., Huggett, J. M. and Gill, M., 1996. The stratigraphy and petrography of the Gault Clay Formation (Albian, Cretaceous) at Redcliff, Isle of Wight. *Proceedings of the Geologists' Association,* **107**, 287-298.

Geomorphological Services Ltd, 1991. Coastal Landslip Potential Assessment, Isle of Wight, Undercliff, Ventnor, 1:2500 scale maps (Crown Copyright). Newport Pagnell: Geomorphological Services Ltd.

Gibbard , P. L. and Allen, L. G., 1994. Drainage evolution in south and east England during the Pleistocene. *Terra Nova,* **6**, 444-452.

Hodges, W. and Woodruff, M. 2002. Borehole correlation difficulties in unstable coastal slopes between Cowes and Gurnard in the Isle of Wight, U.K. (this Conference).

Hosking, A. and Moore, R. 2001. Preparing for the Impacts of Climate Change. Report for SCOPAC, Isle of Wight.

H R Wallingford, 1991. Bonchurch, Isle of Wight. A random wave physical model investigation. Report EX 2347. Wallingford: Hydraulics Research Ltd.

Hutchinson, J. N. , 1965. *A reconnaissance of coastal landslides in the Isle of Wight.* Note No. EN 35/65. Watford: Building Research Station.

Hutchinson, J. N., 1970. A coastal mudflow on the London Clay cliffs at Beltinge, North Kent. *Geotechnique,* **20**, 412-438.

Hutchinson, J. N., 1980. Various forms of cliff instability arising from coast erosion in the U.K. *Fjellsprengningsteknikk-Bergmekanikk-Geoteknikk 1979*, 19.1-19.32. Trondheim: Tapir *for* Norsk Jord-og Fjellteknisk Forbund tillknyttet N. I. F.

Hutchinson, J. N., 1986. Cliffs and shores in cohesive materials: geotechnical and engineering geological aspects. In: *Proceedings Symposium on Cohesive Shores, Burlington, Ontario* (ed. M. G. Skafel), 1- 44. Ottawa, Canada: Associate Committee for Research on Shoreline Erosion and Sedimentation; and National Research Council Canada.

Hutchinson, J. N., 1987a. Some coastal landslides of the southern Isle of Wight. In: *Wessex and the Isle of Wight. Field Guide* (ed. K. E. Barber), 123-135. Cambridge: Quaternary Research Association.

Hutchinson, J. N.,1987b. Mechanisms producing large displacements on pre-existing shears. 1st Sino-British Geological Conference on Geotechnical Engineering and Hazard Assessment in Neotectonic Terrains. *Memoir of the Geological Society of China,* **9**, 175-200. Taipei, Taiwan: Geological Society of China.

Hutchinson, J. N., 1991. Theme Lecture. The landslides forming the South Wight Undercliff. *International Conference on Slope Stability Engineering-Developments and Applications* (ed. R. J. Chandler), 157-168. London: Thomas Telford.

Hutchinson, J. N., 1995. The significance of tectonically produced pre-existing shears. *Proceedings 11th European Conference on Soil Mechanics and Foundation Engineering, Copenhagen,* **4**, 4.59-4.67.

Hutchinson, J. N., 2001. Fourth Glossop Lecture. Reading the ground: morphology and geology in site appraisal. *Quarterly Journal Engineering Geology and Hydrogeology,* **34**, 7-50.

Hutchinson, J. N., in press. Chalk flows from the coastal cliffs of north-west Europe. In: *Catastrophic landslides: occurrence, mechanisms and effects,* Geological Society of America, *Reviews in Engineering Geology,* **15**.

Hutchinson, J. N., Bromhead, E. N. and Chandler, M. P., 1991a. Investigations of the landslides at St Catherine's Point, Isle of Wight. *International Conference on Slope Stability Engineering-Developments and Applications* (ed. R. J. Chandler), 169-179. London: Thomas Telford.

Hutchinson, J.N., Bromhead, E. N. and Chandler, M. P., 2002. Landslide movements affecting the lighthouse at St Catherine's Point, Isle of Wight (this Conference).

Hutchinson, J. N., Brunsden, D. and Lee, E. M., 1991b. The geomorphology of the landslide complex at Ventnor, Isle of Wight. *International Conference on Slope Stability Engineering-Developments and Applications* (ed. R. J. Chandler), 213-218. London: Thomas Telford.

Hutchinson, J. N. and Chandler, M. P., 1985. Collapse of well at "The Grove", South Grove Road, Ventnor, Isle of Wight. Report to the Chief Fire Officer, Isle of Wight County Council in connection with the inquest on Romaldas Ramunas Girenas deceased. London, 5th June, 1985.

Hutchinson, J. N. and Chandler, M.P., 1991. A preliminary landslide hazard zonation of the Undercliff of the Isle of Wight. *International Conference on Slope Stability Engineering - Developments and Applications* (ed. R. J. Chandler), 197-205. London: Thomas Telford.

Hutchinson, J. N., Chandler, M. P. and Bromhead, E. N., 1981. Cliff recession on the Isle of Wight SW coast. *Proceedings 10th International Conference on Soil Mechanics and Foundation Engineering, Stockholm,* **1**, 429-434.

Ibsen, M-L. and Brunsden, D., 1997. Mass movement and climatic variation on the south coast of Great Britain. In: *Rapid mass movement as a source of climatic evidence for Holocene.* (eds J. A. Matthews; D. Brunsden; B. Frenzel; B. Gläser, and M. M. Weiß,). Paläoklimaforschung, Palaeoclimate Research, **19**, Special Issue, ESF Project *"European Palaeoclimatic and Man 12", 171-182.*

Institute of Geological Sciences, 1976. Isle of Wight. Special Sheet. Drift edition. 1:50 000 Series. Southampton: Ordnance Survey, for the Institute of Geological Sciences.

Jones, D. K. C. and Lee, E. M., 1994. *Landsliding in Great Britain.* London:

Department of the Environment, H M S O.

Lee, E. M., Moore, R. and McInnes, R. 1998. Assessment of the probability of landslide reactivation: Isle of Wight Undercliff, UK. In: *Engineering Geology: a View from the Pacific Rim* (eds D. Moore and O. Hungr), 1315-1321 Rotterdam/Brookfield: A. A. Balkema.

Long, A.J. and Tooley, M.J., 1995. Holocene sea-level and crustal movements in Hampshire and Southeast England, United Kingdom. In: *Holocene Cycles: Climate, Sea Levels and Sedimentation* (ed. Frinkl Jr.), *Journal of Coastal Research*, Special Issue **17**, 299-310.

Majdalany, F., 1959. *The Red Rocks of Eddystone*. London: Longmans.

Maquaire, O. and Gigot, P. Reconnaissance par sismique réfraction de la décompression et de l'instabilité des falaises vives du Bessin (Normandie, France). *Geodinamica Acta*, Paris, **2**, 150-159.

McInnes, R.G., 1996. A review of coastal landslide management on the Isle of Wight. *Proceedings 7th International Symposium on Landslides, Trondheim*, **1**, 301-307.

McInnes, R.G., Tomalin, D. J. and Jakeways, J., 2000. *Coastal change, climate and instability*. European Commission LIFE project LIFE-97 ENV/UK/000510 (1997-2000). Newport, Isle of Wight: Isle of Wight Council.

McIntyre, G. and McInnes, R. G., 1991. A review of instability on the southern coasts of the Isle of Wight and the role of the local authority. *International Conference on Slope Stability Engineering - Developments & Applications* (ed. R. J. Chandler), 237-244. London: Thomas Telford.

Mew, F., 1934. *Fifty years back of the Wight*. Newport, Isle of Wight: Isle of Wight County Press.

Mimram, Y., 1975. Fabric deformation induced in Cretaceous chalks by tectonic stresses. *Tectonophysics*, **26**, 309-316.

Moore, R., Clark, A. R. and Lee, E. M., 1998. Coastal cliff behaviour and management: Blackgang, Isle of Wight. In: Geohazards *in Engineering Geology.* (eds J. G. Maund and M. Eddleston). Geological Society, London, Engineering Geology Special Publication, **15**, 49-59.

Moore, R., Lee, E. M. and Brunsden, D. 2000. Cowes stability study. Report for Isle of Wight Council, Newport. Swindon: Halcrow Group Ltd.

Moore, R., Lee, E. M. and Noton, N. H., 1991. The distribution, frequency and magnitude of ground movements at Ventnor, Isle of Wight. *International Conference on Slope Stability Engineering- Developments and Applications* (ed. R. J. Chandler), 231-236. London: Thomas Telford.

Mortimore, R. N. and Fielding, P. M., 1990. The relationship between texture, density and strength of chalk. *Chalk*, 109-132. London: Thomas Telford.

Mottistone, Lord, (General J. Seely), 1937. *Fear, and be slain*. (8th edn). London: Hodder and Stoughton.

Mudie, R., 1838. *Hampshire: its past and present condition, and future prospects; Vol III. Isle of Wight and the Channel Islands.* Winchester: D. E. Gilmour.

Nicholls, R. J., 1987. Evolution of the upper reaches of the Solent River and the formation of Poole and Christchurch Bays. In: *Wessex and the Isle of Wight. Field Guide* (ed. K. E. Barber), 99-114. Cambridge: Quaternary Research Association.

Owen, H. G., 1971. Middle Albian Stratigraphy in the Anglo-Paris Basin. *Bulletin of the British Museum (Natural History) - Geology, Supplement* **8,** London.

Phipps, Z., 1992. *Mottistone Manor, Isle of Wight.* GCE Dissertation, Cheltenham Ladies College.

Preece, R. C., 1980. The biostratigraphy and dating of a Postglacial slope deposit on Gore Cliff, near Blackgang, Isle of Wight. *Journal of Archaeological Science,* **7,** 255-262.

Preece, R. C., Kemp, R. A. and Hutchinson, J. N. 1995. A Late-glacial colluvial sequence at Watcombe Bottom, Isle of Wight, England. *Jl Quaternary Science,* **10,** 107-121.

Preece, R. C. and Scourse, J. D., 1987. Pleistocene sea-level history in the Bembridge area of the Isle of Wight. In: *Wessex and the Isle of Wight: Field Guide* (ed. K. E. Barber), 136-155. Cambridge: Quaternary Research Association.

Preece, R. C., Scourse, J. D., Houghton, S. D., Knudsen, K. L. and Penney, D. N., 1990. The Pleistocene sea-level and neotectonic history of the eastern Solent, southern England. *Philosophical Transactions of the Royal Society of London,* **328 B,** 425-477.

Price, F. G. H., 1879. *A Monograph of the Gault.* London: Taylor and Francis.

Price, D. and Townend, I., 2000. Hydrodynamic sediment process and morphological modelling. In: *Solent Science - A Review* (eds M. Collins and K. Ansell), 55-70. Amsterdam: Elsevier.

Reid, C., 1905. The island of Ictis. *Archaeologia,* **59,** 218-288.

Reid, C. and Strahan, A., 1889. The geology of the Isle of Wight (2nd edn). *Memoirs Geological Survey England and Wales.* London: H. M. S. O.

Rendel Geotechnics, 1993. Ventnor Wastewater Scheme, Isle of Wight, Interpretative Geotechnical Report on the Ground Investigation. Volume 1: Report (unpublished).

Rendel Geotechnics, 1994. Coastal Landslip Potential: St Lawrence, Isle of Wight, 1:2,500 scale maps (Crown Copyright). Birmingham: Rendel Geotechnics Ltd.

Rendel Geotechnics, 1995. Niton Extension Study, 1:2,500 scale maps (Crown Copyright). Birmingham: Rendel Geotechnics Ltd.

Scaife, R. G., 2000. Palaeo-environmental investigation of submerged sediment archives in the West Solent study area at Bouldnor and Yarmouth. In: *Coastal change, climate and instability,* Final technical report, **2,** Palaeo-environmental study areas, Area P1, 7.

Skempton, A. W. 1946. Earth pressure and the stability of slopes. In *The Principles and Application of Soil Mechanics,* 31-61. Westminster: The Institution of Civil Engineers.

Soil Mechanics Ltd., 2001. Ground investigation at Undercliff Drive, Niton, Isle of Wight. Report 101036 (unpublished).

Tomalin, D., 2000a. Geomorphological evolution of the Solent Seaway and the Severance of Wight: A Review. In: *Solent Science - A Review* (eds M. Collins and K. Ansell), 9-19. Amsterdam: Elsevier.

Tomalin, D., 2000b. Wisdom of hindsight: palaeo-environmental and archaeological evidence of long-term processual changes and coastline Sustainability. In *Solent Science - A Review* (eds M. Collins and K. Ansell), 71-83. Amsterdam: Elsevier.

Tomalin, D., 2000c. Discussion and evaluation: archaeology, palaeo-environmental evidence and coastal change. In: *Coastal change, climate and instability* (eds R. G. McInnes, D. J. Tomalin and J. Jakeways), European Commission LIFE project LIFE-97 ENV/UK/000510 (1997-2000). Newport, Isle of Wight: Isle of Wight Council.

Tomalin, D., 2000d. Recognition of major land loss from the rear scarp at Gore Cliff. In: *Coastal change, climate and instability* (eds R. G. McInnes, D. J. Tomalin and J. Jakeways), European Commission LIFE project LIFE-97 ENV/UK/000510 (1997-2000). **2**, P4, Section 6.3. Newport, Isle of Wight: Isle of Wight Council.

Tomalin, D., in press. "Wihtgarasbyrig" explored. *Proceedings of the Isle of Wight Natural History and Archaeological Society.*

Tomalin, D., and Scaife, R. G., 2000. Conclusions and the way ahead: towards the management of the European coastal palaeo-environmental resource. In: *Coastal change, climate and instability* (eds R. G. McInnes, D. J. Tomalin and J. Jakeways), **1**, Section 7.6.

Toms, A. H., 1955. Isle of Wight, Ventnor: landslips. Report of an investigation of the fundamental nature of the landslips, consequent on the development of deep fissures in the Whitwell Road (unpublished).

Varley, P. M., Warren, C. D., Avgherinos, P., 1996a. Castle Hill West landslip. In: *Engineering Geology of the Channel Tunnel,* (eds C. S. Harris, M. B. Hart, P. M. Varley, and C. D. Warren), 295-309. London: Thomas Telford.

Varley, P. M., Warren, C. D., Rankin, W. J. and Harris, C. S. 1996b. Site investigations. In: *Engineering geology of the Channel Tunnel,* (eds C. S. Harris, M. B. Hart, P. M. Varley, and C. D. Warren), 88-117. London: Thomas Telford.

Velegrakis, A. 2000. Geology, geomorphology and sediments of the Solent System. In: *Solent Science - A Review* (eds M. Collins and K. Ansell), 21-43. Amsterdam: Elsevier.

Velegrakis, A., Dix, J. K. and Collins, M. B. 1999. Late Quaternary evolution of the upper reaches of the Solent River, Southern England, based upon marine geological evidence. *Journal of the Geological Society, London,* **156,** 73-87.

Wade, S., Hossell, J., Hough, M. & Fenn, C. (eds). 1999. *The impacts of climate change in the South East: technical report.* Epsom: W. S. Atkins.

Webber, N. B., 1980. Hydrography and water circulation in the Solent. In *The Solent System - an assessment of present knowledge.* N. E. R. C. Publications, Series C, No. **22**, 25-35.

Whitaker, W., 1910. The water supply of Hampshire (including the Isle of Wight).

Memoirs of the Geological Survey England and Wales. London: H.M.S.O.

White, H. J. O., 1921. A short account of the geology of the Isle of Wight. *Memoirs of the Geological Survey of Great Britain England and Wales.* London: H.M.S.O.

Wong, H. N., 2001, Recent advances in slope engineering in Hong Kong. In: *Geotechnical Engineering: meeting society's needs.* (eds K. K. S. Ho and K. S. Li) **1**, 641-660. Rotterdam/Brookfield: A. A. Balkema.

Wooldridge, S.W, and Linton, D.L., 1955. *Structure, surface and drainage in South-East England.* London: George Philip and Son Limited.

Worsley, Sir Richard, 1781. *The history of the Isle of Wight.* London: A. Hamilton.

Zenkovich, V. P., 1967. *Processes of Coastal Development.* Edinburgh and London: Oliver & Boyd.

Session 1:

Instability – planning and management

It is essential to translate accurate and up-to-date instability information into an effective policy framework for the technically, economically and environmentally appropriate management of unstable areas. This represents the first stage in formulating management strategies.

Setbacks in Edmonton, Alberta, Canada

P. BARLOW, AMEC, Edmonton, Canada
P. CAZES, Grenoble Institute of Science and Technology (ISTG), France
D. CRUDEN, University of Alberta, Canada, and
D. LEWYCKY, Public Works, City of Edmonton, Canada

INTRODUCTION

Three documented case histories of landslide damage to residences in Edmonton, the Grierson Hill , Lesueur and Mill Creek Slides (Figure 1, Table 1), provide a century of observations. They prompted a study by Alberta Environment (Tedder, 1986) of slope stability along the North Saskatchewan River and its tributaries as far as Fort Saskatchewan 20 km to the North East. The study suggested a method of estimating setbacks from the crests of valleys beyond which sites would not be adversely affected by valley crest regressions for the lifetime of the structures, say 50 years (Cruden, Tedder, Thomson, 1989). Development within setbacks would require detailed geotechnical studies and, possibly, slope stabilization works.

The setback guidelines were useful for planning purposes and were extended by a later study for Alberta Municipal Affairs that reviewed slope failures within municipalities across the Province of Alberta (Cruden and De Lugt, 1991, De Lugt et al., 1993). None of the 10 failures documented would have affected housing beyond the setbacks from slope crests. The 3 Edmonton case histories were included in the Province-wide survey and their setbacks estimated at the times of slope failure.

The guidelines are shown schematically in Table 1. Contemporary maps and air photographs were used to determine the slope height, Hs, the inclination of the oversteepened slope, α, and the position of the structure with respect to the crest of the slope. More recent maps and photos which show positions of the crest of the valley slope can be used to determine average values of net river erosion (En). Erosion rates of terraces along the North Saskatchewan River average 0.3 m/year. Rates on tributary creeks may be half that value and average rates over extensive reachs may be as little as 0.08 m/year. Erosion rates of the Cretaceous bedrock might be an order of magnitude less than those in alluvial deposits

Ultimate slope angles, βu, are the angles of mature, abandoned slopes with the same geology and groundwater conditions as the oversteepened slope. Abandoned slopes are those not being eroded by a river at present. Mature, abandoned slopes have not been eroded by a river for several hundred years or more. These abandoned slopes can be identified through air photo interpretation or from field reconnaissance. Slope angles may be read from maps or measured on site. All the quantities necessary to determine the setback, SG, at a site were found by examination of existing maps and air photos or by unobtrusive site exploration.

Two recent events at Whitemud Road and Blackmud Creek (Figure 1, Table 1) have followed engineering studies of the sites. In this paper we briefly review and update knowledge of the first 3 events, then we consider the 2 current slope failures at greater length.

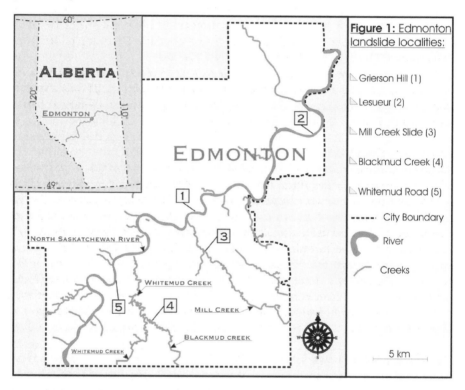

Figure 1: Edmonton landslide localities:

Grierson Hill (1)

Lesueur (2)

Mill Creek Slide (3)

Blackmud Creek (4)

Whitemud Road (5)

----- City Boundary

River

Creeks

5 km

GRIERSON HILL

Edmonton's largest and most damaging landslide, the Grierson Hill slide, developed following the floods and wet summers of 1900 and 1901. On September 27, 1901, the Edmonton Bulletin reported "the wet weather of the last two seasons has brought to a crisis the impending landslides along the hill face in the eastern end of town. The situation has now become one of grave importance to property owners along the stretch affected, a distance of some 500 yards. Along the hill front, the earth is splitting and sinking, the face of the hill sliding downward. These cracks run in different places, but always parallel to the river and about 20 feet back from the hill brow. An odd feature of the slides is that of all the earth that has changed places, it is impossible to see where any of it has gone. These slides are confined to the east end of town and to a single row of lots abutting on the high bank of the valley. The slides are confined to the part under which coal mining has been principally carried on" (Cruden and Thomson, 1993, pp. 84-85). The long history of this landslide has recently been updated by Martin et al., (1998).

LESUEUR

The Lesueur landslide is on the right and southern bank of a bend of the North Saskatchewan River, now within the eastern limits of Edmonton. In April 1963, snow melt revealed a small, distinct crack that arced across the lawn and terminated at the basement wall, 3 m from the northwest corner of the Lesueur house. The scarp grew to 1.8 m high by 20h00, September 3, 1963 and 6.7 m by 8h00, September 4. A corner of house's foundations was cantilevered over the main scarp (Thomson,1971) of the translational landslide. The main scarp was 50 m across, the slide's lateral margins diverging to 150 m at river level. Drilling located the surface of rupture 32 m below the slide crown and allowed an estimate of the volume of displaced material of 0.76 Mm^3. The displaced material of the landslide continues to move as a composite debris-slide debris-flow (Cruden et al, 1998).

MILL CREEK AND 76 AVENUE

Three houses were destroyed by the 1974 landslide into Mill Creek at 76 Avenue. Thomson and Tiedemann (1982, Figure 8) showed the main scarp of the landslide at the south wall of the western house, passing north of the central house and affecting the south west quarter of the eastern house. The uphill-facing scarp of the graben passed south of the central house, which was severely distorted by the landslide. All 3 houses were evacuated and demolished. The street, 76 Avenue, was distorted as was the sidewalk and the abandoned grade of the Edmonton, Yukon and Pacific Railway. The slide was attributed to the rise in groundwater levels associated with urban development on the Interior Plains (Hamilton and Tao, 1977). However, flows in Mill Creek had, no doubt, increased following land clearing and development upstream and presumably the outside of the meander had been eroded.

BLACKMUD CREEK AND 23RD AVENUE

The 23 Avenue slide occurred on property on the south side of the Avenue overlooking Blackmud Creek about 500 metres upstream of its entry into Whitemud Creek. 23 Avenue enters the valley of Whitemud Creek immediately west of the property (Figure 1).

Figure 2 showed fine-grained glaciolacustrine deposits over till which rested directly on shales, sandstones, coals and bentonites of the Upper Cretaceous Edmonton Group (Treen and Kallenbach, 1992). The slope had an inclination of 34° before the slide over a height of 33 metres. A terrace a metre or so above the usual level of the creek might be interpreted as the eroded remnants of a previous translational slide, which had left an untreed scarp along the crest of the valley. The toe of the surface of rupture of the slide emerged from the bank a metre or so above creek level, apparently following the bentonite seam encountered in the boreholes. A counter scarp indicated the down-slope margin of the active block that drove the translational slide. Mature trees of the passive block remained upright.

The property has been constructed on a promontory extending southwards into the valley of Blackmud Creek. Ground cracking was observed as snow melted in the spring of 1992. Movement extended despite the installation of 6 m long, cast-in-place concrete piles during the summer and extensive additional piling 18 m long in the late fall. Significant subsidence began on October 17. Additional piles with tie-back anchors were installed in November. Further movement had exposed 10 m or so of the pile wall by the summer of 1999. The house holders have recently recovered some of their damages from the city (Bielby, 2000).

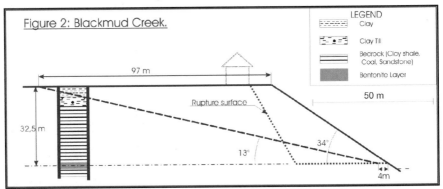

WHITEMUD ROAD

The Whitemud Road slide took place on Saturday, October 23, 1999; the residence at 4112 Whitemud Road collapsed into the graben of a translational slide which removed the back of the property and the backyards of two properties to the south and one to the north. The rupture surface of the slide daylighted on the east bank of the North Saskatchewan and displaced material fell from the steep slope into the river. Two neighbouring houses were demolished in March, 2000 and four residences are at least temporarily abandoned. Statements of claim have been filed by the owners of the destroyed residences.

Table 1 records the slope was 65 m high at an angle of 34°, the highest and steepest of the slopes in the city that has seen a bedrock landslide. Figure 3 indicates a bench up to 15 metres wide on the slope where seepage flows from sands over a clay till deposited on the weathered and eroded surface of the Upper Cretaceous bedrock. The bedrock below, bare of vegetation and colluvium, sloped at 42° into the river. The vegetated slope above the bench in the surficial materials had an angle of 38°. The toe of the surface of rupture was 45 m below the slope crest and 7 m below the top of the bedrock where a bentonite seam contained the surface of rupture (Barlow, 2000).

The displaced material did not move uniformly. An active block falling into an 18 m deep graben drove forward a central passive block. This left the lateral margins of the landslide unsupported; the failure of the northern margin was followed by flow on the southern margin stimulated by much groundwater seepage. Small rotational slides at the toe of the rupture surface retrogressed into the passive block. Generally, however, the upslope displaced material supported upright trees confirming the translational displacements on the rupture surface. Of 0.25 Mm^3 of displaced materials, 0.05 Mm^3 dropped in the river and 0.20 Mm^3 stayed on the slope.

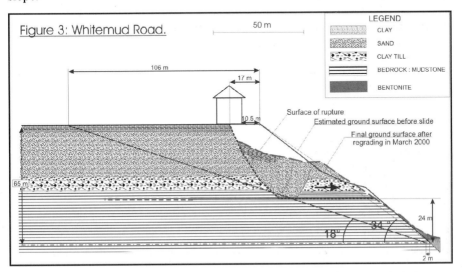

Figure 3: Whitemud Road.

DISCUSSION

The 5 slides have similar features. Each appears to be a single translational slide with a horizontal surface of rupture and a steep scarp, tracing a lunule on the crest of the valley wall. The rupture surfaces followed bedding in the Edmonton Group, each following different bentonite seams. In 3 cases, the slopes clearly fell into Hutchinson's Type 2 where erosion at the toe of the slope exceeded the rate of colluvial deposition. Some bedrock had likely been bared at each slope toe (Hutchinson, 1973). Possible anthropogenic triggers of enhanced erosion include bridge construction and storm sewer construction, urban development enhancing run off into Mill Creek and Blackmud Creek and the dumping of mining wastes over the bank opposite the Lesueur landslide. Each of these supplemented the active erosion that characterizes the outside of meander bends on Prairie creeks and rivers. Upslope anthropogenic influences associated with residential development were also considered to be a contributing factor for the slides at Whitemud Road and Blackmud Creek. While there is no continuous history of piezometric data at either site, it was considered that urban development increased groundwater levels and associated seepage discharge on the slopes in these areas (Thomson and Tiedemann, 1982; Bielby, 2000)

At Lesueur the landslide reactived the displaced material of an earlier slope movement. The geomorphology of the slopes at Whitemud Road and Mill and Blackmud Creeks suggested that these slopes had also failed previously so there is evidence of the cyclic behaviour characteristic of Type 2 slopes (Hutchinson, 1973, Figure 5).

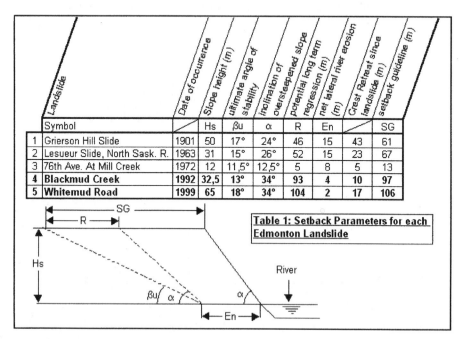

	Symbol		Hs	βu	α	R	En		SG
1	Grierson Hill Slide	1901	50	17°	24°	46	15	43	61
2	Lesueur Slide, North Sask. R.	1963	31	15°	26°	52	15	23	67
3	76th Ave. At Mill Creek	1972	12	11,5°	12,5°	5	8	5	13
4	Blackmud Creek	1992	32,5	13°	34°	93	4	10	97
5	Whitemud Road	1999	65	18°	34°	104	2	17	106

Table 1: Setback Parameters for each Edmonton Landslide

The landslide mechanism can be idealized in four stages (Figure 4). The main scarp develops first, following stress-relief joints. The displaced material moves as a rigid block along a bentonite seam, pushed by water pressures which fall as the surface of the rupture dilates (Figure 4.2). Movement leaves the head of the displaced material unsupported and, over a few months, a counter scarp develops which separates an active block (Figure 4.3). Calculations show (Cazes, 2001) a substantial increase in driving forces as this active block sinks into the widening graben. A new equilibrium is reached when the active block rests on the bentonite seam (Figure 4.4).

Grierson Hill has been stabilized (Martin et al., 1998) by drainage, recontouring and the construction of berms at the toe to prevent further river erosion and to buttress the slope. The first berm designed in the city has satisfactorily stabilized the slope north of the University of Alberta campus over the last 30 years (Thomson, 1970). Other slopes downstream have been stabilized for the Capital City Park (Thomson and Townsend, 1979)

The experience documented from these 5 landslides suggests that stabilization is necessary for a Type 2 slope where construction is planned within the setback guidelines. The guidelines can be estimated by the sum of the retrogressions of a Type 1, shallow landsliding, slope and a Type 3, abandoned, slope.

Figure 4 : Variation of the safety factor, F, with landslide stage.

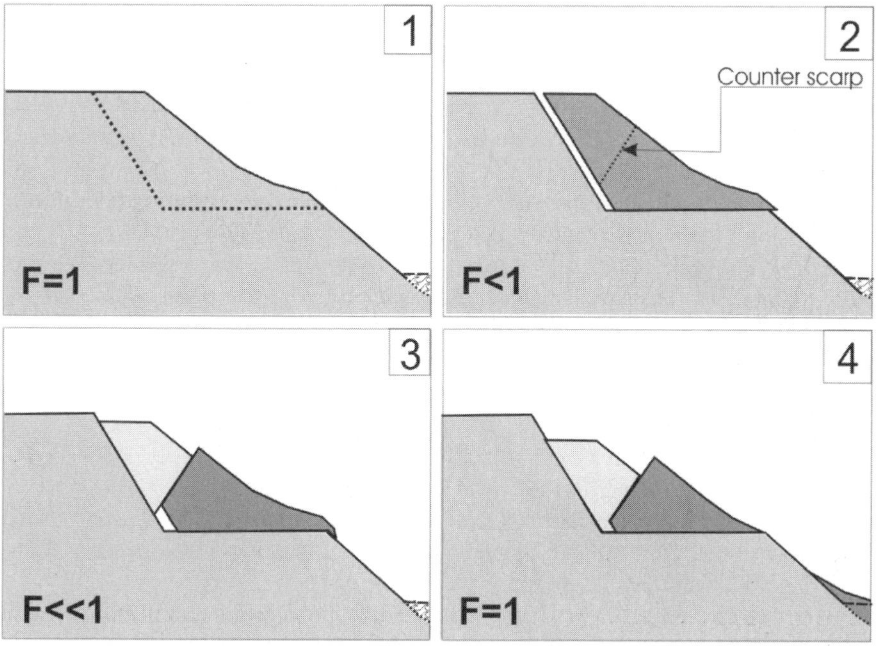

REFERENCES

Barlow, P., 2000. Whitemud Road Landslide Geotechnical Assessment, AMEC Earth & Environmental Limited, Edmonton.

Bielby, M.B., 2000. Reasons for judgment, Papadopoulos v. Edmonton (City of), Court of Queen's Bench of Alberta, Judicial District of Edmonton, 17th March, 24 p.

Cazes, P., 2001. Slope stability along the Saskatchewan River at Edmonton : setback analysis and method of construction, Grenoble Institute of Science and Technology (ISTG), Grenoble, Report of internship.

Cruden, D.M., Tedder, K.H., Thomson, S., 1989. Setbacks from the crests of slopes along the North Saskatchewan River Valley, Alberta, Canadian Geotechnical Journal, 26:64-74.

Cruden, D.M. Lugt, J.S., 1991. Setbacks from slope crests for structures, Alberta Municipal Affairs, Innovative Housing Grants Program, Edmonton, 56 p.

Cruden, D.M. Peterson, A.E., Thomson, S. Zabeti, P., 1998. The Lesueur Landslide: 35 years old but still active. Proceedings, 51st Canadian Geotechnical Conference, Vol. 1, 17-22.

Cruden, D.M. Thomson, S., 1993. Learning the ground rules, in Godfrey, J.D., Edmonton beneath our feet, Edmonton Geological Society, Edmonton, pp. 84-99.

De Lugt, J., Cruden, D.M., Thomson, S., 1993. A suggested method for estimating setbacks from the crest of slopes on the Interior Plains in Alberta, Canadian Geotechnical Journal, 30: 863-875.

Hamilton, J.J. and Tao, S.S., 1977. Impact of urban development on groundwater in glacial deposits. Proceedings, 30th Canadian Geotechnical Conference, Volume 2, pp. 1-35.

Hutchinson, J.N., 1973. The response of London Clay Cliffs to differing rates of toe erosion, Geologia applicata e idrologeologica, 8:221-239.

Martin, L.R., Lewycky, D.M. and Ruban, A.F., 1998. Longterm movement rates in a large translational landslide. Proceedings, 51st Canadian Geotechnical conference, volume 1, pp. 23-30.

Tedder, K., 1986. Slope stability in the North Saskatchewan River Valley, Report, Water Resources Management Services, Alberta Environment, Edmonton, 253pp and 15 Plans.

Thomson, S., 1970. Riverbank Stability study at the University of Alberta, Edmonton, Canadian Geotechnical Journal, 7: 157-168.

Thomson, S., 1971. The Lesueur landslide, a failure in Upper Cretaceous clay shale. Proceedings, 9th Annual Symposium on Engineering Geology and Soils Engineering, Boise, Idaho, pp. 257-288.

Thomson, S., Townsend, D.L., 1979. River erosion and bank stabilization, North Saskatchewan River, Edmonton, Alberta. Canadian Geotechnical Journal, 16: 567-576.

Thomson, S. and Tiedemann, C.E., 1982. A review of factors affecting landslides in urban areas, Bulletin Association of Engineering Geologists, 19:55-65.

Treen, C., and Kallenbach, B, 1992. Slope Stability Evaluation, 23 Avenue and Blackmud Creek West of 119 Street/23 Avenue intersection., HBT AGRA Limited, Edmonton.

Planning for management of landslides in communities in the United States – A cooperative project between the American Planning Association and United States Geological Survey

LYNN M. HIGHLAND, U.S. Geological Survey, Golden, CO, USA
SANJAY JEER, American Planning Association, Chicago, IL, USA

Landslide hazard mitigation at the land-use planning stage has been successfully implemented by a number of communities in the United States. Despite these successes, landslides hazards (whether human-induced or naturally-occurring) generally do not appear as a major concern during the land-use planning stage or at the site development stage, although some communities in landslide-prone areas have some form of local review. In many cases of course, the terrain is flat and the land is not subject to landsliding. However, there is some type of landslide activity in all 50 States, whether in the form of slumps, slides, coastal bluff erosion, river bluff collapse, rockfall, debris flow, lateral spreading, volcano edifice collapse, earth flows, or other types of landslide process.

In order to begin to encourage effective planning for landslide hazards by communities, the American Planning Association (APA) and the U.S. Geological Survey (USGS), are co-sponsoring a written planning guide entitled, "Landslide Hazards and Planning." This guidebook will help local planners incorporate landslide hazard planning into the comprehensive planning process in several key planning aspects. It will include steps to identify, map, and designate appropriate land uses for areas susceptible to landslides. It will also cover "best practices" for development, land-use compatibility, environmental assessments, site-plan reviews, land acquisition, post-disaster planning, and will augment land-use planning with geotechnical studies. Initially, the intention is to distribute this guide to APA's planning advisory service subscribers and then expand the distribution using the world-wide web.

We will first describe the nature and general extent of the landslide problem in the United States; how it is currently understood and how it is addressed as part of local planning efforts at the city and county level governments. The second part of this report will describe the concept and implementation strategy for the Planning Guide.

Instability – Planning and Management, Thomas Telford, London, 2002, 81–88

PRECIPITATION-INDUCED LANDSLIDES

Landslides are predominant in mountainous regions in the United States (Radbruch-Hall and others, 1982; Godt, 1997). The California coastal ranges are subject to landslides during the winter storm period (December through April) when precipitation from rainfall can be high. Wildfires, although a normal part of the ecological cycle, nevertheless denude slopes of vegetation and exacerbate the landslide problems in many parts of the western United States. The Sierra Nevada Mountains in California are also subject to landslides due to snowmelt, rain on snow events, and normal precipitation. The western parts of the states of Washington and Oregon experience high rates of precipitation and are comprised of many hilly and mountainous areas that have experienced burgeoning population growth and resultant disturbance of the land.

The mountainous areas of the states of Utah, New Mexico, and Nevada are subject to rockfall, debris flows, and slow-moving landslides that are triggered during spring and summer snowmelt and rainfall. Colorado's Rocky Mountains experience episodes of rockfall, debris flows and mudflows, which are triggered by stalled rainstorms, spring snowmelt, and rain-on-snow events. In all of these regions population growth and resultant new housing and highway construction have made previously unsuitable, steep building sites suddenly attractive for development considerations. Landslides occurring on the heels of wildfires have been commonplace in these areas in recent years, as there are still large tracts of open forests and grasslands that surround pockets of development.

Eastern and Midwestern states, especially areas located along major rivers such as the Mississippi, experience episodes of river bluff collapse. The city of Cincinnati, Ohio, and neighboring areas of the states of Ohio, Kentucky and Indiana, have experienced major, costly landslides. Mountainous regions of the states of New York, New Hampshire, and Vermont experience moderate to severe landsliding, and the Appalachian and Blue Ridge Mountains are subject to debris flows and landslides, especially during periods of extremely high rates of precipitation from hurricanes and large storms. Hilly and mountainous areas in the states of Virginia and West Virginia have experienced a large number of severe flooding and debris flow incidents. Coastal areas in the U.S. are subject to coastal retreat processes, rockfall, and other types of landslides (Radbruch-Hall and others, 1982, and Godt, 1997).

EARTHQUAKE-INDUCED LANDSLIDES

Seismically induced, (or earthquake-induced) landslides remain a concern in many parts of the U.S., particularly in the states of California, Alaska and Washington. Although recurrence rates for damaging earthquakes are lower in the eastern U.S. than for the west, the higher densities of population and of the built environment render these areas extremely vulnerable to costly earthquake damage. The eastern part of the U.S., due to its unique geology and soil characteristics, experiences lower attenuation of earthquake shaking, so that earthquakes affect a geographically broader area, rather than being locally confined, as is the case in California and other western regions of the U.S.

VOLCANO-RELATED LANDSLIDES

The state of Washington's Mount Rainier, a currently active volcano, is only one of several potentially hazardous volcanoes in the Cascade Mountain Range. The primary hazard from these volcanoes, in addition to lava flows and gas emanations, are eruption-related lahars (volcanic mudflows and debris flows). Hot eruptions cause rapid snow and glacier-melt on these volcanoes, precipitating fast-moving, potentially lethal, debris avalanches and lahars. The 1980 eruption of Mount St. Helens, in the Cascades, caused the largest historical landslide in the world. Another phenomena, glacial outbursts, occur when glaciers are rapidly heated by warm weather and trap meltwater, which accumulates under the weight of the glacier. This water may suddenly burst free of the glacier, coursing down stream channels and causing massive debris flows. Many of the Cascade Range volcanoes are host to glaciers, and experience these outbursts from time to time. Volcanic edifices, made up of material that accumulates around the summits of volcanoes through successive eruptions, are often unstable and prone to collapse, causing an additional type of landslide hazard. Communities in the vicinity of volcanoes must not only carefully plan new development that may be threatened by actual volcanic eruption, but they must plan for the accompanying hazards from landsliding.

Alaska and Hawaii, states not conterminous to the "lower 48" states, experience all types of landslides. The City of Honolulu, on the island of Oahu in Hawaii, has been subject to bouts of costly landsliding in its hilly areas. Alaska has experienced landslides in its mountainous regions, as well as several very large and damaging landslides in the city of Anchorage, during the devastating 1964 Good Friday earthquake (Hansen, 1965). Both regions have active volcanoes and both are subject to large earthquakes. Some of Hawaii's regions have recorded the highest rates of rainfall in the world, and Alaska experiences recurring freezing, thawing, snowmelt and high precipitation in coastal areas—climate that is conducive to landslide occurrence.

CURRENT CHALLENGES AND APPROACHES TO MANAGING LANDSLIDE HAZARDS IN THE UNITED STATES

When addressing development in landslide-prone areas, jurisdictions encounter several challenges. Many municipalities and counties are expanding geographically, with new development extending into hilly, and potentially unstable areas. One of the more difficult challenges some cities and counties face is the occurrence of landslide problems in areas of old development, both commercial and residential. Another problem is the fact that in the U.S. there is virtually no landslide insurance available for city and county governments, nor is landslide insurance available to private property owners for residential or commercial structures. In some cases, residents have received relief from earthquake-induced landslide damage by purchasing earthquake insurance, which is available in the U.S. Some insurance settlements have occurred in the past to those homeowners who have purchased federally subsidized flood insurance. This policy generally covers losses from mud and debris, but only when associated with flooding. Another drawback is that only residents in federally designated flood zones are even eligible to purchase this type of flood insurance, making it only available to a few property owners. The bulk of those suffering landslide devastation to their homes get little relief, except in cases of Presidentially declared disaster areas, when low-cost loans are sometimes available to landslide sufferers. Many landslide-related monetary losses are recouped through lawsuits, the unfortunate consequence of the two problems of no available insurance and sporadic, non-comprehensive urban landslide planning. As there is no federal government-level oversight of landslide hazard evaluations of private property; the mitigation of landslide hazards is left to local communities. Creative solutions such as Geologic Hazard Abatement Districts (GHADS) are sometimes used. GHADS are comprised of discrete localities, such as neighborhoods, in which residents organize and decide to self-insure for hazards such as landslides. As of this writing, GHAD's are enabled by state-level legislation in the state of California only (Olshansky, 1986). In California, a GHAD may be proposed by one of two means:

(1) a petition signed by owners of at least 10 percent of the real property in the district, or

(2) by resolution of a local legislative body.

As another remedy, a homeowner can file a lawsuit in court, in order to recover damages from municipal planning and building departments and/or developers. Often, a municipality or developer may disavow prior knowledge of a potential hazard, and/or settle with the plaintiff, rather than exposing site permit and/or building practices to the scrutiny of a public trial. Some kind of compromise is often preferable to an actual trial, which avoids leaving the decision of awarding cash compensation to the whims of a jury. In many cases it is difficult to determine who is at fault for landslide damage, as development in some areas may have precluded adequate knowledge of landslide processes. Much of the United States is not mapped for landslide susceptibility on a scale that is useful for builders and homebuyers, especially in areas of older development. Hiring private geotechnical engineering consultants for site evaluation (which may include expensive investigative methods such as drilling) is a viable solution, and is often the best way to completely evaluate vulnerability to landslides. (Some communities will retain the services of a consultant in their building departments even if there is no legal requirement to do so). Ultimately, local communities must decide for themselves whether to invest in planning for landslides, or to take a risk that landslides won't occur. As landslide hazard maps become more available and more sophisticated, communities can decide whether planning for landslides is cost effective.

CURRENT PLANNING PRACTICES FOR LANDSLIDE HAZARDS

Planning functions in the U.S. vary widely in scope, purpose, and intent. They also vary by level of government. For example, local governments prepare long-range, community-wide plans, and they adopt land-use controls, using tools such as zoning ordinances and land subdivision regulations. State governments authorize specific actions for local governments to undertake through enabling legislation and coordinating emergency response in times of disaster. The federal government plays a similar role for public lands through a variety of agencies which have stewardship of land for specific purposes, for example, the Bureau of Land Management, Forest Service, National Park Service, Military installations, and the cabinet-level Interior Department. Indirectly, the federal government also influences trust lands for Native Americans, whose reservations are managed autonomously by tribal governments. The American west, described as those regions located roughly west of 100 degrees West Longitude, is approximately two/thirds federal government- and Native American-owned. In this case, planning and land use matters are generally outside the scope of state and local government, although local and state governments have some input into planning for these regions, the extent varying greatly from state to state.

Of the communities that deal with landslide hazards as part of their local regulatory processes, it is common in most of them to implement one or two regulatory tools. For example, in developed areas, the focus is usually on zoning or building codes, although some attention is given to landslides at the subdivision or site planning stages. For larger land areas, the focus is on land-use maps augmented by an overlay zone or an environmental designation for the steepest slopes. Very few communities take into consideration all of the factors that cause landslides in a coordinated, comprehensive strategy, employing a variety of planning tools in addition to zoning, building codes, and plan designations. It is known from planning for other risks, such as fires, floods, and pollution, that taking landslide risks into account earlier in the planning process will not only minimize losses, but may avoid them altogether (American Planning Association, 2000).

Notable areas in the United States that are making progress in implementing landslide hazard planning in the urban planning process are the state of California, some cities and counties in Colorado; the city of Cincinnati and Hamilton County, Ohio, and the city of Pittsburgh, Pennsylvania. The city of Seattle and the surrounding Puget Sound area, in dealing with a multiplicity of geologic hazards, have made great strides in planning, through a unique cooperative project with the University of Washington, the U.S. Geological Survey, the Federal Emergency Management Agency (FEMA) and the private consulting sector. The cooperative project has resulted in enhanced landslide mapping, development of preliminary probabilistic landslide maps, and has increased comprehensive public education about landslide hazards.

Despite gains in understanding landslide processes, grading ordinances are sometimes still the major tool for mitigating the effects of potentially unstable land in many cities and counties; however, they still do not solve the perplexing problem of unstable land in old, established developments. As noted previously, cities states, and counties are autonomous in management of local affairs and planning regulations are a checkerboard of variation.

APA/USGS DEVELOPMENT OF A PLANNING GUIDE FOR URBAN PLANNERS

The American Planning Association is a non-profit public interest and research organization representing 35,000 practicing planners, officials, and citizens involved in urban and rural planning issues in the United States. Sixty-five percent of APA's members are employed by state and local government agencies. The U.S. Geological Survey, a U.S. federal government agency in the Department of Interior, provides scientifically integrated research in earth, water, biological, and map-related disciplines. While these two organizations differ in concept and objectives, they have some mission goals in common. The APA's research mission is similar to the USGS's Landslide Hazards Program in that both (a) provide information for land-use planning (b) partner with local land-management agencies, (c) assist local agencies in evaluating landslides and recommend strategies for mitigation; and (d) disseminate technical information to local communities about landslide hazards (Spiker and Gori, 2000). These common interests have brought the two organizations together to produce the guidebook.

The primary audience for this landslide planning guide will be urban and regional planning practitioners, public officials, and other public or private entities, whose actions and policies affect development patterns in landslide-prone areas. While this audience is the principal focus, others will be targeted, such as researchers and educators in academia, and local geologists and other engineering professionals interested in planning processes. Geologic Information Systems (GIS) professionals, and others involved in spatial technologies will be interested in GIS applications that will be associated with the guidebook. Begun in the year 2000, the project is organized into five tasks, which will be spread over a 24-month period.

Task 1 – Performance of Background Research – The APA research staff will survey current approaches to incorporating landslide factors in the planning process. This task will include the study of current zoning codes, the mailing of a questionnaire on the subject to all 1,700+ guidebook subscribers, and phone interviews of a selected sample of local government agencies. The APA research staff will also meet with public and private organizations representing geologists, earth scientists, emergency planners, and related environmental groups. GIS specialists will also be consulted as to how best to present this type of data.

Task 2 – APA will create a Technical Advisory Panel (TAP) - The TAP will include nationally well-known experts on this topic, APA project staff, USGS staff, and representatives from the private consulting sector. There will be a worldwide web-based information exchange center created for this purpose. APA project staff in collaboration with USGS staff and the TAP members will serve as presenters and organizers of conference sessions at various national and regional events. Hosting sessions on landslide hazards will be key to promoting best planning practices to the project's audience.

Task 3 – Preparation of Working Papers – APA will commission several (up to 10) working papers from authors and experts on various topics identified in the scope to address key issues. The theme of the working papers will focus on evaluating approaches of integrating landslide hazards into the planning process. The papers will include topics such as geology and landslide processes, zoning issues, mitigation, GIS and technology, legal issues, insurance, and engineering.

Task 4 – The presentation and discussion of case studies – The case studies will analyze current practices so as to refine the recommendations in the guidebook. The APA staff will select three jurisdictions for in-depth studies to document and analyze how landslide hazards are incorporated into the planning process.

Task 5 – Preparation and publication of the guidebook - An outline of the guidebook was developed at a USGS-sponsored symposium held in February 2000 in Chicago, Illinois. The guidebook will include case studies of real world examples of best practices; it will identify obstacles to implementation and suggest ways to overcome these obstacles. The report will also include sample language for incorporation of landslide concerns into planning documents, such as comprehensive or general plans, zoning ordinances, and subdivision regulations. Finally, the guidebook will provide useful links to online materials, such as working papers, data and mapping resources, GIS templates, and supporting software resources. As a part of a chapter in the guidebook, the sources section will include methods to incorporate existing USGS, FEMA, and other federal and state government sources. This section will be geared toward small jurisdictions, which may not have dedicated planning staff to be able to assemble necessary background information (base maps, land-use data, terrain and slope data, etc.).

THE FINAL VERSION OF THE GUIDEBOOK AND OTHER PRODUCTS
Following publication, APA will distribute the guidebook by direct mail to all 1,700+ subscribers of APA's Planning Advisory Service (PAS), a group consisting primarily of local, county, and regional planning agencies and planning consultants across the nation. APA will also print additional copies for distribution to non-PAS subscribers through its Book Service. The Book Service distributes 8,000 to 13,000 PAS Reports each year as individual orders or group orders, such as course

adoptions at graduate schools of planning. Sections of the guidebook will also be available on the web at: http://www/planning.org/landslides

REFERENCES

American Planning Association, 2000, Proposal to United States Geological Survey, Landslide Hazards and Planning, Chicago, Illinois, 14 p.

Baum, R.L., Harp, E.L., and W. Hultman, 2000, recent landslides on Coastal bluffs of Puget Sound between Shilshole Bay and Everett, Washington, U.S. Geological Survey Miscellaneous Field Studies Map, MF-2346.

Cruden, David M., and D. J. Varnes, 1996, Landslide types and processes, in Turner, A.K., and R. L. Schuster, eds., Landslides: Investigation and Mitigation, Transportation Research Board, Special report 247, National Academy Press, Washington, D.C., p. 36 – 71.

Coe, J.A., Michael, J.A., Crovelli, R.A. and W.Z. Savage, 2000, Preliminary map showing landslide densities, mean recurrence intervals and exceedence probabilities as determined from historic records, Seattle, Washington, U.S. Geological Survey Open-file report, 25 p.

Godt, Jonathan, 1997, Digital Compilation of Landslide Overview Map of the Conterminous United States, Open-file Report 97-289

Hansen, W.R., 1965, Effects of the earthquake of March 27, 1964, at Anchorage, Alaska, U.S. Geological Survey Professional Paper 543-A, p. A1-A68

Olshansky, R. B., 1986, Geologic Hazard Abatement Districts: California Geology, v. 39, no. 7, p. 158-159

Radbruch-Hall, D.H., R.B. Colton, W.E. Davies, I. Lucchita, B.A. Skip, and D.J. Varnes, 1982. Landslide overview map of the Conterminous United States. U.S. Geological Survey Professional Paper 1183, 25 p.

Spiker, E.C., and Paula L. Gori, 2000, National Landslide Hazards Mitigation Strategy: A Framework on Loss Reduction, U.S. Geological Survey Open-file Report 00-0450, 49 p.

Landslides, land-use planning and risk management: Switzerland as a case-study

DR O. J. LATELTIN, Federal Office for Water and Geology FOWG, Biel, Switzerland

ABSTRACT

In Switzerland, numerous slopes are affected by small movements related mostly to ancient landslide mechanisms of post-glacial age and to the progressive failure of rock slopes induced by weathering and water pressure in the fault systems. The rapid rockfalls and slides induced by the withdrawal of the glaciers, followed by long term slow movements, have carved a peculiar morphology which is discernible on large scale topographic maps (1:25'000). It is thus possible to assess that more than 8% of the Swiss territory has been affected by landslides.

Land-use planning and the resulting zoning laws are among the most effective tools for the prevention and mitigation of natural disasters. Since 1991, new federal regulations requires the Cantons to establish hazard maps and zoning for mass movements to restrict development on hazard-prone land. A three step procedure has been proposed for this risk management in Switzerland. An indispensable prerequisite for **hazard identification** is information about past events (maps of phenomena and register of events). In a second stage, **hazard assessment** implies the determination of intensity over time. Hazard mapping for the Local Plan represent four classes hazard: high danger (red zone), moderate danger (blue zone), low danger (yellow zone) and no danger (white zone).

In the third step (**Planning of measures**), a general acceptable risk has been fixed a priori. The codes of practices for land-use planning in areas prone to landslides can be summarized as follow:.in red zones building is strictly prohibited and in blue zones building is allowed with restrictions. In yellow zones, building is possible without restrictions. The regional authorities are participating actively to this hazard mapping to reduce the potential losses by a better land-use planning.

INTRODUCTION

Located in the Alps, Switzerland is a small "hazard prone" country (covering 41 300 km^2 with 6.7 million inhabitants) exposed to natural disasters, such as debris flows, earthquakes, floods, forest fires, hail storms, landslides, rockfalls, snow and ice avalanches and wind storms. Switzerland's geological structure is essentially the

Instability – Planning and Management, Thomas Telford, London, 2002, 89–96

result of a collision of the African and European plate over millions of years. 57% of its surface lies in the Alps, 30% in the Molasse Basin and 13% in the Jura . Rainfall is fairly abundant (50 cm in central Valais, 250 cm on Säntis at 2503m) and rather evenly distributed throughout the year. Towards the interior of the Alps, the timberline rises from 1700m to 2400m, the snowline from 2500m to 3200. Most of the population and all major cities are concentrated in the Molasse Basin.

By contrast, the shaping of our present landscape by water and glaciers has taken place over the past two million years. In the Swiss Alps numerous slopes are affected by small movements related mostly to ancient landslide mechanisms of post-glacial age and to the progressive failure of rock slopes induced by weathering and water pressure in the fault systems. The rapid rockfalls and slides induced by the withdrawal of the glaciers during the Preboreal period (10'000 to 9'000 BP), have carved a peculiar morphology which is discernible on large scale topographic maps. It is thus possible to assess that more than 8% of the Swiss territory (2500 km^2) has been affected by landslides. Residual movement (up to cm/year) are often ignored, due to the size of many unstable slopes or the lack of visible signs of activity.

Recently, the summer of 1987 registered more than 600 debris flows and considerable potential for debris flow formation still continues to exist particulary in the periglacial belt. The Randa rock avalanche of 1991 (30 million m^3 of fallen debris) cut off the villages of Zermatt, Täsch and Randa from the rest of the valley for two weeks. In 1994 a prehistoric landslide experienced a strong reactivation with historically unprecendented rates of displacement up to 6m/day, thus causing the destruction of a village of Falli-Hölli (41 houses). In August 1995, a debris flow (40'000 m^3) cut the highway in Villeneuve near Montreux, destroying some houses and wineyards. During the winter of 1996, two rock avalanches (volume:3 million m^3 of fallen rock) dammed a narrow alpine valley near Glarus, causing some important damage to a nearby located electrical powerplant. As a direct consequence of the strong precipitation phases in February and May 1999, more than 350 landslides and slope failures were triggered, mainly in the states of Aargau, Basel, Bern, Glarus, Lucerne, Nidwalden, Obwalden, Schwyz, Solothurn, St. Gallen, Uri, and Zurich.

REGULATIONS
Switzerland is a Federal country where 26 Cantons are sovereign in principle: the central authorities only have jurisdiction in those domains determined by the Federal Constitution and all other state powers automatically belong to the Cantons or to the communities. Each Canton has its own government, constituting laws and regulations within the framework defined by the relevant Federal laws. The prevention and management of natural disasters follow the same rules. Two new regulations, the Federal Law on Water Management and the Federal Forest Law came into force in 1991. Their purpose is, amongst others, to protect human lives, objects of value the environment from the damaging effects caused by water,

landslides, snow and ice avalanches and forest fires. The cantons are required now to establish registers of events and hazard maps of scale 1:5'000 (Local Plan) depicting endangered areas, and to take hazards into account for the purposes of land-use planning. For the elaboration of the registers of events and hazard maps, the Federal government is providing subsides to the Cantonal authorities up to 70 per cent of the costs.

RISK MANAGEMENT

In Switzerland, strong efforts have been made to apply the same strategy and similar approaches for dealing with all kind of natural hazards (now avalanches, landslides, floods, debris flows, rock falls) The identification of natural hazards, the evaluation of their impact on human beings and on their properties and on the environment and the general risk assessment are decisive steps towards the selection and dimensioning of adequate protective measures. Therefore, a three step procedure has been proposed for debris flows, landslides and rockfalls, as well as for floods and snow avalanches as it is shown in Figure 1.

---

1. WHAT MIGHT HAPPEN AND WHERE ?
(Hazard identification)
Documentation:
- Thematical maps (e.g., topography, geology)
- Aerial photographs
- Geodetic measurements
- **Register of events**
- **Map of phenomena**
⬇

2. HOW OFTEN AND HOW INTENSE CAN IT HAPPEN ?
(Hazard assessment)
Determination with:
- Simulation, modelling
- **Hazard map**
⬇

3. HOW CAN WE PROTECT OURSELVES ?
(Planning of measures)
Transposition for:
- **Risk assessment (cost/effectiveness, protection objective)**
- **Land-use planning (code of practice)**
- **Preventive and protective measures**
- **Warning systems, emergency planning**

Figure 1: Today's approach to risk management in Switzerland:
a three step procedure

FIRST STEP: HAZARD IDENTIFICATION
It is the aim of this first step to collect all information indicating an existing danger. Important indications often can be found in a register of events or in traces of earlier events like e.g. the headscarp of a landslide.

Some recommendations for the uniform classification, representation and documentation of natural phenomena indicating hazardous processes have been established by the federal administrations [1]. Therefore, the elaboration of **the Map of Phenomena** should be based on a uniform legend. According to the mapping scale (e.g., 1:50,000 for the Cantonal Master Plan, 1:5,000 for the Communal Local Plan), this legend offers a great number of symbols in a modular manner. Following the recommendations and this uniform legend, standard maps can be established, exhibiting the different phenomena indicating hazardous processes within an investigation area. Thanks to nation-wide development and practical application, maps from different parts of the country can be quite well compared.

A first version of recommendations for the elaboration of a coherent **Register of Events** including special forms for every phenomenon (snow avalanches, landslides, rockfalls, debris flows, and floods) and a digital database (StorMe) is in test operation with selected Cantons. On such a basis, all Cantons will have to work out their own registers in the near future.

SECOND STEP: HAZARD ASSESSMENT
Hazard means the probability of occurrence of a potentially damaging natural phenomena within a specific period of time and in a given area. Hazard assessment implies the determination of the magnitude or intensity of an event and it's frequency or return period (see figure 2).

	Probability of occurrence (in 50 years)	Return period (in years)
high	100 to 82 %	1 to 30
medium	82 to 40 %	30 to 100
low	40 to 15 %	100 to 300

Figure 2: In the table above, the probability of occurrence in 50 successive years is related to the return period by the binomial distribution assuming one or more independent occurrences in n (= 50) years.

The relation can be expressed as: $P_n = 1 - (1-1/T_r)^n$ where P_n is the probability of at least one occurrence in n successive years, and T_r is the return period in years for an event of a particular magnitude.

The main result of the second step in risk management in Switzerland is a hazard map depicting what areas are threatened by hazardous processes. Hazards maps established for the **Master Plan** (e.g., scale 1:50,000) display all hazard-prone zones at the Cantonal level. The classification is made in a simple way: endangered or not endangered perimeters. Based on a diagram combining intensity and probability, hazard mapping for the **Local Plan** (e.g., scale 1:5,000) represents four classes or grades of hazard: highly endangered (red), moderately endangered (blue), lowly endangered (yellow), and not endangered (white).

Some Federal recommendations for land-use planning in landslide-prone areas [2] (Lateltin 1997) and in flood-prone areas [3] have been proposed to the Cantonal authorities and to planners for the establishment of **hazard maps** using an intensity/probability-diagram. Similar recommendations have existed since 1984 for snow avalanches (Guidelines for the consideration of avalanche hazards in planning land-use activities, [4].

The detailed quantitative criteria for landslide intensity were chosen as follows:

Phenomenon	low intensity	medium intensity	high intensity
Rockfall	$E < 30$ kJ	30 kJ $< E < 300$ kJ	$E > 300$ kJ
Landslide	$V < 2$ cm/year	V: dm/year	$V > $ dm/day
Debris flow	_____	$D < 1$ m and $V_w < 1$ m/s	$D > 1$ m $V_w > 1$ m/s
E: kinetic energy H: horizontal displacement	V: mean annual velocity of landslide V_w : flow velocity		D: thickness of debris front

Figure 3: Criteria for intensity of rockfall, landslide and debris flow

Mass movements often correspond to gradual (landslides) or unique (rock avalanches) events. It is, therefore, difficult to assess the return period for a massive rock avalanche or to predict the reactivation time of a latent landslide. For repeated processes like periodic rockfalls, snow avalanches, floods or debris flows it is much easier to evaluate the corresponding intensity and return period, at least if the necessary observations on past events are available, e.g. in the form of a register of events.

THIRD STEP: PLANNING OF MEASURES

Risk management is an integral approach of human thinking and acting covering the anticipation and the assessment of risk, the systematic approach to limit the risk to an accepted level and to undertake the necessary measures. Rock avalanches and landslides are the result of the temporal and spatial overlapping of the two independent domains *potential hazard* and *potentially endangered objects.* The potential hazard is described in step 1 and 2 of the procedure by the probability of occurrence and the intensity/extent of the events. The spatial area in use corresponds to the probability of presence of any objects and the value of these objects.

Unfortunately, the methods for use in vulnerability and damage assessment leading to **risk assessment** are less developed than the methods for hazard assessment. The development of protection objectives is a simple way to make a risk assessment. The choice of protection objectives is made according to the values of the objects to be protected. Varying reference levels are applied according to object category (e.g. residential areas, infrastructure installation, agriculture areas). Should the existing degree of protection be lower than the protection objectives, this means there is a protection deficit. For reducing such deficits, land-use planning, structural protection measures (e.g. dams, nets, reforestation, drainage), forecasting and early warning systems or emergency planning (evacuation, closure of roads) should be done, according also to a cost/benefits analysis.

The hazard map is the technical base of land-use planning in areas threatened by landslides, rock avalanches as well as flooding and snow avalanches. It must be implemented in the following procedures:

- establishment and approval of plans (Master and Local Plan)
- planning, construction and use of buildings and installations
- building licence, construction and other forms of land-use
- award of subsidies

The **codes of practice** for land-use planning in endangered areas are the following:

In **red zones** substantial damage to buildings leading to their possible destruction has to be expected, putting people inside buildings into danger. Therefore building is essentially prohibited.

In **blue zones** slight to moderate damage to buildings has to be expected. People are endangered in the field but not inside of buildings. Therefore construction is allowed, provided that certain safety requirements are met. Parts of buildings exposed to hazardous processes have to be designed to withstand possible impacts. To minimise fatalities, communities are required to establish warning systems and evacuation plans. Unusually large assemblies of people in blue zones have to be avoided as far as possible.

In **yellow zones** slight damage to buildings may occur, but people are not endangered neither in the field nor in buildings. Therefore construction is allowed without restrictions but the owners have to be informed of existing dangers.

CONCLUDING REMARKS

The hazard map indicates which sectors are slightly or not appropriate for utilisation, according to existing natural hazard. The hazard map acts as a reference document for:

- the integration of such information in land-use planning (cantonal Master Plan or communal Local Plans, including the delimitation of hazard zones, construction prescripts, building licence)
- the planning of protective measures for objects and damage reduction by the owners.

Conflicts may happen when the hazard map is compared with land-use. Since land-use can not or hardly be modified, specific construction measures are required to reach the desired protection level. Hazard maps are also considered for the planning of protective measures and the installation of warning systems and emergency plans. The federal recommendations for land-use planning in alpine areas prone to landslides is a contribution to mitigate natural disasters by restricting development on unstable or endangered zones

REFERENCES

[1]   OFFICE FÉDÉRAL DE L'ENVIRONNEMENT, DES FORÊTS ET DU PAYSAGE / OFFICE FÉDÉRAL DE L'ÉCONOMIE DES EAUX , 1995: Légende modulable pour la cartographie des phénomènes, Recommendations. – Série dangers naturels, OCFIM, 3000 Berne.

[2]   OFFICE FÉDÉRAL DE L'ENVIRONNEMENT, DES FORÊTS ET DU PAYSAGE / OFFICE FÉDÉRAL DE L'ÉCONOMIE DES EAUX / OFFICE FÉDÉRAL DE L'AMÉNAGEMENT DU TERRITOIRE, 1997: Recommendations, Prise en compte des dangers dus aux mouvements de terrain dans le cadre des activités de l'aménagement du territoire. – Série dangers naturels, OCFIM, 3000 Berne.

[3]   OFFICE FÉDÉRAL DE L'ÉCONOMIE DES EAUX / OFFICE FÉDÉRAL DE L'AMÉNAGEMENT DU TERRITOIRE / OFFICE FÉDÉRAL DE L'ENVIRONNEMENT, DES FORÊTS ET DU PAYSAGE, 1997. Recommendations, prise en compte des dangers dus aux crues dans le cadre des activités de l'aménagement du territoire. – Série dangers naturels, OCFIM, 3000 Berne.

[4] OFFICE FÉDÉRAL DES FORÊTS / INSTITUT FÉDÉRAL POUR L'ÉTUDE DE LA NEIGE ET DES AVALANCHES, 1984 : Directives pour la prise en considération du danger d'avalanches lors de l'exercice d'activités touchant l'organisation du territoire. OCFIM, 3000 Berne.

The development and current state of land instability mapping

DR BRIAN R MARKER, Department for Transport, Local Government and the Regions, London, UK

INTRODUCTION

This paper briefly reviews some developments in land instability mapping over the past 30 years or so. It deals with broad scale mapping (normally 1:2500 scale or greater) rather than mapping of individual sites. Consideration is limited to landslides and to subsidence into underground cavities. Broader reviews of this subject are available in, for example, Commission on Engineering Geology Maps of IAEG (1976), McCall and Marker (1989) and Bobrowsky (2001). In addition, a major review of hazard mapping, in the widest sense of the term, is being prepared by Commission 1 of the International Association of Engineering Geology and the Environment with a view to publication in 2003.

Mapping of potentially unstable land grew from site investigation mapping during the 1960's and 1970's. It developed to provide an indication of ground conditions that might be expected in an area and as a context for interpretation of site investigation results. This remains the principal purpose of most hazard mapping. The intended audience consisted mainly of engineering geologists and, thus, the products were essentially technical in character. Most studies were undertaken to develop or demonstrate methods and the areas were often chosen following significant instability events, which raised consciousness amongst public agencies of the need for funding. There have been few national systematic mapping exercises such as that in the former Czechoslovakia (see McCall and Marker, 1989 page 218).

Much of the early work concentrated on landsliding, which, at the time of the 1976 IAEG review, was, by far, the most developed type of hazard mapping. In subsequent years work extended to a wider range of potential hazards, notably to subsidence, and maps became more diverse. Engineering geomorphology played a progressively more important part. This paper considers briefly:
  a) maps recording previous instability events;
  b) maps providing interpretations of potential hazards; and
  c) results providing information on hazards for people untrained in geosciences, such as many land use planners.

As yet, few examples of maps providing indications of relative risks arising from potential instability events have appeared in the literature.

MAPS RECORDING PREVIOUS INSTABILITY EVENTS

Maps showing the distribution of instability events and evidence of past instability are a starting point for most investigations. Some of these simply show spot records located at the central point of each event or occurrence without any attempt to define the actual limits (see, for instance, Durville and Hamoroux, 1995). However, some potentially unstable structures such as mine openings require spot representation (e.g. Freeman Fox Ltd 1988, map 5). Other investigations define actual geographical limits of all but the smallest events (for example, Owen and Walsby, 1989). Some include information on dates of instability events, where known, on the map (Fig. 1).

These are of obvious importance in compiling desk study information prior to site investigation as well as to practical site works since a high proportion of subsidence and slope instability events in development areas result from disturbance of previously unstable features. Such maps are based on surface (topographical, geological and geomorphological) and, where relevant, underground mapping, and analysis of results of site investigations and other documents.

In areas of former underground mining, analysis of mine plans or other documents can give a good picture of past activity. However coverage is often incomplete. In the UK, for instance, while preparation of mine plans for most mines became a requirement in 1856, not until 1872 was it required that the plans should be deposited. Prior to that date, and sometimes after it, mining was undertaken that is now only known if openings remain accessible, if workings collapse or if workings are encountered in site investigations. In addition, old plans may be presented at a variety of scales providing a challenge for accurate comparison with modern geographical representations (Holt and Marker, 1987). Where the coverage of plans is deficient then other documentary evidence may be valuable (see, for instance, Howard Humphreys and Partners, 1993).

In some forms of mining, extraction may take place at several levels giving rise to relatively complicated mine plan information (Fig 2). Such maps can be difficult to read when plotted on a single sheet and are unwieldy where large sets of overlays are employed. However GIS now provides an ideal means of overcoming these difficulties.

Understanding of the extent and mechanisms of ground movement can be improved by plotting data on damage to buildings and constructions as was done for example in respect of mines and dissolution structures in chalk in Norwich (Howard Humphreys and Partners, 1993) or landsliding at Ventnor, Isle of Wight (Lee et al. 1991).

Most record maps are accompanied by information on instability events, shafts and mined ground, and of related structures and phenomena. In the past, these usually consisted of card indexes or ledgers in which each location has a distinctive registration number and National Grid references. Databases have now made access to, and searching for, information much easier. Grid referenced is ideal for incorporation into databases and GIS. However there is a tendency for data to be collected after a serious instability event has caused public concern, only to fall into disuse without updating as memory fades. This wastes the value of the original investment in the work particularly where card index systems are difficult of access or vulnerable because no duplicate copies exist.

HAZARD MAPS
While a knowledge of past events and of features that can become unstable is valuable, potential ground movements are not necessarily limited to areas in which past failures have occurred. It is, therefore, important to be able to assess the implications for the wider area. At a simple level it is possible to plot numbers of features, such as mine openings or landslides, that might be associated with instability events per unit area (see, for example, Fig. 3 for sink holes, or Doornkamp and Lee, 1992 in respect of mineshafts).

Another option is to map factors that are likely to influence instability events and to compare these with the known distribution of past occurrences. A map of slope steepness in the Bradford area, for instance, shows the association of landsliding with steep slopes in certain geological formations (Fig. 4). This clearly illustrates the need for caution on steep slopes in these neighbourhoods. It has been recognised in planning guidance that slope angle maps can be a useful first step where more detailed studies cannot be implemented quickly (DETR, 1996).

Methods for extrapolation have also been developed. For instance, a study of coal mining instability in the Islwyn area of south Wales was based on analysis of mine plans, site investigation reports and historical documents. Causes and depths of 396 subsidence events were reviewed. It was established that most took place where workings were below cover less than 6 times height of working. The seams that most subsidence events were associated with were also identified. Therefore a method was established to define an envelope of ground above the seam outcrop where subsidence, if any, is most likely to take place. Uncertainty about the positions of seam outcrops also had to be taken into account (Fig. 5).

Numerical weighting may be used to express relative levels of hazards. For instance, Edmonds et al (1987) considered dissolution cavities in the chalk around Henley on Thames, Oxfordshire. Factors governing the spatial distribution of natural solution features in chalk were identified as:
- Lithostratigraphy of the chalk G1
- Nature of post Cretaceous deposits G2

- Groundwater table H1
- Topographical relief and drainage H2
- Former drainage paths GM1
- Glacial deposits (the presence which may be conducive to dissolution) GM2

Each was given weightings related to their likely significance and were combined into a formula: Subsidence hazard rating = (G1 + G2+ H1 + GM1 + GM2)
The results (Fig.6) compared well with known patterns of distribution.

All of these types of maps are useful to geoscientists but require additional interpretation if they are to guide others in decision procedures, but care is needed to ensure that suitable specialists are involved when appropriate.

MAPS FOR LAND USE PLANNING

Many attempts have been made to express information on ground conditions for a wider audiences but many results have proved to be too technical (Marker, 1998). Geoscientists often take matters that are unfamiliar to other types of professionals for granted, or fail make allowance for the ways in which information is actually used in preparing development plans, development control and building control. Also, few land-use planners have any previous backgound in the earth sciences (Worth, 1987).

A programme of research into applied geological mapping was undertaken by a number of contractors for the former Department of the Environment between 1980 and 1996. The series developed from thematic geological maps, through engineering geology maps, to results that were much better fitted to the requirements of planners and developers (Smith and Ellison, 2000). In general, the utilisation of results by the intended audiences remained disappointing, except where these were built into formal land use planning and building control procedures and local authority officers were closely involved in the research and preparation of the guidance.

However the potential of linking hazard maps to land use planning objectives was more directly addressed in specific land instability studies commissioned by the same Department (see Brook, this conference). Amongst these was the Islwyn study, mentioned earlier, in which the map was linked to an advice note on procedures for planning applications that was issued by the local planning authority. A study of landsliding in the Ventnor area provided a planning guidance map which gave advice on responses that might be adopted during preparation of the development plan and in development control on the face of the map (Fig. 7 shows a simplified extract redrawn from the map). The work was later developed and extended to other parts of the Isle of Wight Undercliff (Clark et al. 1994; Moore et al. 1995). Work on dissolution associated with gypsum in the Ripon area (Thompson et al, 1996, 1998a) led to the definition of 3 categories of ground which were described in terms of development planning and development control procedures. Advice to planners and building control officers on the content that might be expected in a site investigation report and on the need for a competent person to undertake that investigation was also included. Some good practices coming from the series of studies are set out in more detail in Thompson et al. 1998b.

A current project, jointly sponsored by NERC and DTLR, is now examining the development of environmental information systems for sustainable urban planning and management. Scientists, including geoscientists from the British Geological Survey (BGS), have developed summaries of planning procedures used by local authorities and which need to be informed by environmental information. The next step is to develop interfaces that relate environmental data to planning procedures. The work is expected to result in a prototype software system by mid 2004. In addition, BGS has indicated that future map revision will include some appropriate thematic mapping thus helping to overcome the lack of systematic coverage although over a long period (Walton and Lee, 2001).

CONCLUSIONS
The past few decades have seen a progression from maps recording locations of instability events, to a variety of types of hazard maps, and maps that directly address the requirements of land use planners. The most successful planning maps have been those that have involved planners and building control officers in research and design. However it is important that maps for that purpose are used properly and with the regard to the need for appropriate expert advice. Few attempts to map risks associated with land instability have been published as yet. There is scope for the use of GIS to manipulate socio-economic data alongside hazard assessments to assist in that process and to help guide prioritisation of remediation.

Problems of expense and handling associated with large sets of maps can now be overcome using GIS. This is already employed extensively by academics and consultants but, while many planning authorities are beginning to use GIS few employ it in day-to-day use on matters such as land instability. In many cases, investment would be needed to develop the data sets needed for GIS layers, even where existing paper-based sets of information exist. However, growing use of GIS in local authorities does provide an opportunity to emphasise the potential for assessment of land instability, and other potential problems. Current work on an environmental information system for planners might assist in that process.

The BGS is beginning to address the lack of systematic coverage of thematic maps, but that will take a long time. The probable pattern of hazard mapping in the UK in the coming years will probably continue to be as a response, after the event, to specific occurrences of land instability.

The different training backgrounds of geoscientists and of land-use planners will continue to impede communication unless each are given some exposure to the discipline of the other during respective courses. In many cases, it would be enough for planners to appreciate the significance of land instability, to know where to go for advice, and to be able to frame the right questions. For the geoscientist, it is awareness of the structure and principles of the planning system, the stages at which information is needed, the terms and forms in which it should be presented and the levels of detail required.

Views expressed in this paper are those of the author and not necessarily those of the Department for Transport, Local Government and the Regions.

© British Crown Copyright. Reproduced by permission of the Controller of Her Majesty's Stationery Office.

REFERENCES

Adams, F T and Lovell, C W 1984 Mapping and prediction of limestone bedrock problems *Trans. Res. Rec.* 978, 1-5

Bobrowsky, P T [Ed] 2001 *Geoenvironmental mapping – method, theory and practice* Balkema (Rotterdam)

Clark, A R; Lee, E M and Moore, R 1994 The development of a ground behaviour model for the area of a landslide hazard in the Isle of Wight Undercliff and its role in supporting major development projects. *Proc 7th Congress IAEG, Lisbon*, VI, 4901-13.

Commission on Engineering Geology Maps of IAEG 1976 *Engineering geology maps.* Earth Sciences 15. UNESCO Press (Paris)

Culshaw, M G; Bell, F G; Cripps, J C and O'Hara, M *Planning and engineering geology* Geological Society Special Publication 4

Department of the Environment, Transport and the Regions (DETR) 1996 *Development on unstable land – landslides and planning* Planning Policy Guidance Note 14, Annex 1 HMSO (London)

Doornkamp, J C and Lee, E M 1992 *Geology and land use planning, St Helens, Merseyside - summary report* Rendel Geotechnics (Birmingham) 64pp

Durville, J L and Hamoroux, M 1995 Les risques dus aux carrieres souterraines abandonees et leur prevention - strategies et methodes de prevention. *Bull. Internat. Assoc. Eng.Geol.* 51, 114-128

Edmonds, C N; Green, C P and Hollingworth, I E 1987 Subsidence hazard prediction for limestone terrains as applied to the English Cretaceous Chalk. In: Culshaw et al [Eds] (separately referenced), 290-291

Freeman Fox Ltd 1988 *Investigations of ground characteristics in the area around Chacewater and St Day in the County of Cornwall.* Freeman Fox Ltd (London)

Howard Humphreys and Partners 1987 *Environmental geology study in the Bristol area.* Howard Humphreys and Partners (Leatherhead) 3 Vols.

Howard Humphreys and Partners 1993 *Subsidence in Norwich* HMSO (London)

Holt, D N and Marker, B R 1987 Benefits of engineering geology for land use planning in areas of past metalliferous mining In: Culshaw et al.,[Eds] (separately referenced) 75-80

Lee, E M; Doornkamp, J C; Brunsden, D C and Noton, N H 1991 *Ground movement in Ventnor, Isle of Wight* Geomorphological Services Ltd (Newport Pagnell)

McCall, G J H and Marker, B R 1989 Environmental geology mapping In: McCall, G J H and Marker, B R [Eds] *Earth science mapping for planning, development and conservation* Graham and Trotman (London)

Marker B R 1998 Incorporating information on geohazards into the planning process. In: Maund J C and Eddleston M (Eds*)* *Geohazards in engineering geology.* Geological Society Lond. Special Pub. 15 385-9

Moore, R; Lee, E M and Clark, A R 1995 *The Undercliff of the Isle of Wight – a review of ground behaviour* South Wight Borough Council (Ventnor) 68pp

Ove Arup and Partners 1995 *Islwyn shallow mining – summary report* Ove Arup and Partners (Cardiff) 35pp plus appendices

Owen, J F and Walsby, J C 1989 *A register of Nottingham's caves* British Geological Survey Technical Report WA/89/27 3 Vols.

Smith A and Ellison R A 2000 Applied geology maps for planning and development - a review of examples from England and Wales 1983-1996. *Quart.Jl.Eng.Geol.* 32, supplement S1-S44.

Thompson, A; Hine, P D; Greig, J R; and Peach, D W 1996 *Assessment of subsidence arising from gypsum dissolution (with particular reference to Ripon, North Yorkshire)* Symonds Travers Morgan (East Grinstead)2 Vols   95 and 285pp

Thompson, A; Hine, P; Peach, D; Frost, L and Brook, D 1998a Subsidence assessment as a basis for planning guidance in Ripon. In: Maund and Eddleston (Eds) *Geohazards in engineering geology.* Geological Society Lond. Special Pub. 15, 415-426.

Thompson, A; Hine, P D; Poole, J; and Greig, J R 1998b *Environmental geology in land use planning – a guide to good practice* Symonds Group (East Grinstead)

Walton, G and Lee, M K 2001 *Geology for our diverse economy* British Geological Survey (Keyworth)

Waters, C N; Northmore, K; Prince, G; and Marker, B R (Eds.) 1996 *A geological background for planning and development in the City of Bradford Metropolitan District. Vol. 1 A guide to the use of earth science information in planning and development.* British Geological Survey Technical Report WA/96/1 British Geological Survey (Keyworth)

Worth, D H 1987 Planning for engineering geologists In: Culshaw et al. [Eds] (separately referenced) 39-46

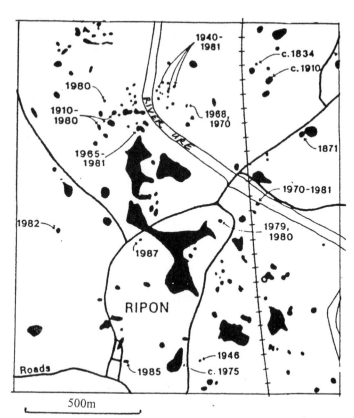

Figure 1: Subsidence structures and records of events associated with dissolution of gypsum, Ripon, North Yorkshire (based on Thompson et al, 1996)

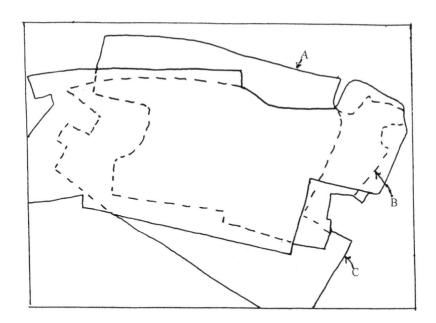

Figure 2: Limits of coal mining in three seams in the Bristol coalfield taken from mine plans (redrawn and simplified from Howard Humphreys and Partners, 1987) Original map at 1:10000 scale.

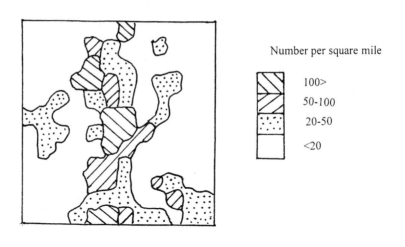

Figure 3: Numbers of sinkholes per square mile in part of Lawrence County, Indiana (based on a figure in Adams and Lovell, 1984). Original scale 1:250000.

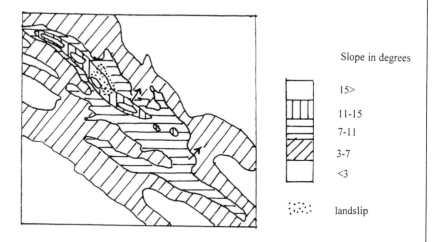

Figure 4: Generalised steepness of slopes in part of the Metropolitan Borough of Bradford (simplified from Waters et al. 1996). Original map at 1:50000 scale.

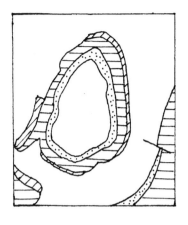
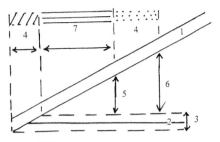

1 superficial deposits
2 seam outcrop
3 conjectural seam (elevation uncertain)
4 abandoned shallow mines possible (seam outcrop uncertain)
5 H/T migration ratio (H = thickness of rock above seam;
    T = thickness of coal extracted)
6 upper limit for H/T based on 2m high roadway
7 abandoned shallow mines known or anticipated

Figure 5: Development advice map in respect of ground mined for coal for part of Ebbw Vale, South Wales (redrawn and simplified from Ove Arup and Partners, 1995). Original scale 1:10000

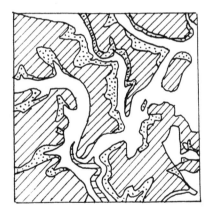

Hazard

	Very high	>600
	High	401-600
	Moderately high	301-400
	Moderate	201-300
	Moderately low	137-200
	Low	90-136
	Very low	55-89
	Generally absent	<55

(the extract shows only some of the categories identified in the legend)

Figure 6: Subsidence hazard assessment for natural solution features in chalk near Henley-on-Thames, Oxfordshire (based on a figure in Edmonds et al, 1987). Original scale 1:25000.

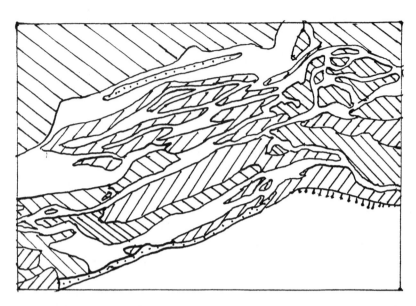

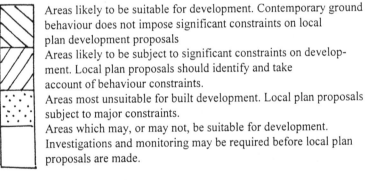

Areas likely to be suitable for development. Contemporary ground behaviour does not impose significant constraints on local plan development proposals

Areas likely to be subject to significant constraints on development. Local plan proposals should identify and take account of behaviour constraints.

Areas most unsuitable for built development. Local plan proposals subject to major constraints.

Areas which may, or may not, be suitable for development. Investigations and monitoring may be required before local plan proposals are made.

Defended coastline.

Figure 7: Classification of landslide potential zones in terms of land use planning and development responses for an area near Ventnor, Isle of Wight (simplified from a map in Lee et al, 1991). Original scale 1:2500

Cowes to Gurnard coastal stability: providing the tools and information for effective planning and management of unstable land

Dr R. MOORE, Halcrow Group Ltd, Birmingham, UK, and
Mr R.G. M^cINNES, Isle of Wight Council, Newport, UK

INTRODUCTION

The coastal slopes between Cowes and Egypt Point have, over the last 200 years, been extensively developed for residential, leisure and retail purposes. Initially, development was focused at Cowes on the more accessible gently sloping ground. As the demand and opportunities for development sites increased, development spread further west towards Gurnard. Other sites have been redeveloped with multi-storey flats or detached homes. The spread of development has, in places, occurred on steeper ground of marginal stability. This has led to an apparent increase in the number of reported problems of ground instability. The problems have been heightened in recent years as a result of several slope failures caused by construction activities.

In February 2000, the Centre for the Coastal Environment within the Isle of Wight Council commissioned Halcrow Group Limited to carry out a coastal slope stability study at Cowes and Gurnard on the north coast of the Isle of Wight. The main objective of the study was to provide strategic guidance and information on ground stability conditions between Cowes and Gurnard. Specifically, the study brief required the preparation of a series of geomorphological, ground behaviour and planning guidance maps. The maps are intended to assist decision-making by informing the planning process as well as provide a basis for assessing the requirements for stability investigations and reports in support of future development proposals in the study area.

The paper presents the main findings of the study and how the information will be used in future planning and management decisions in the area. In particular, it is recognised there is a need to ensure high standards of ground investigation, engineering design and construction to mitigate potential ground movement related problems.

STUDY AREA DESCRIPTION

The north-facing coastal slopes extending from Cowes to Gurnard (Plate 1) are the

Instability – Planning and Management, Thomas Telford, London, 2002, 109–116

most northerly landmass of the Isle of Wight. The coastal slopes form a prominent headland separating the Medina River and Estuary from the western Solent. The headland is characterised by a plateau forming the higher ground above gently sloping coastal cliffs of varying height up to 35m. The limit of the study area (about 100ha) extends along the coast from Market Hill, Cowes, west to Gurnard Marshes, and by up to 0.6 km inland of the shoreline.

Plate 1 – Study Area

Courtesy: Brian Manby

APPROACH

The conflict between development and unstable land is not unique to the area. Since 1984, central and local government has carried out extensive research to assess the significance and consequences of unstable land in Great Britain. Lee *et al.* (2000) provide a summary of the research, from which several key conclusions were drawn:

- There is a legacy of ancient landslides formed during past climatic conditions, i.e. many problems of ground instability are related to inadvertent reactivation of pre-existing landslides or failed ground.
- Coastal cliff recession is an intermittent process, with periods of little or no erosion separated by rapid and occasionally dramatic landslides, which may remove large sections of cliff in a single event.
- The impact of human activity which can have a significant effect on cliff stability, both at the site and on adjacent slopes.

As part of this research, the Department of the Environment (now the Department of the Environment, Transport and the Regions), the former South Wight Borough

Council (now the Isle of Wight Council), commissioned a series of studies[1] to develop an approach for landslip potential assessment. The main objective of these studies was to identify, collect and test appropriate levels of earth science information in support of planning and development decisions on unstable land. This research provided the background to the preparation of planning policy guidance (PPG14[2]), and associated guidance for the investigation and management of landslides in Great Britain aimed at planners and developers (Clark *et al.* 1996).

Central to the approach developed by these studies is the need to:
- Determine the nature and extent of unstable ground/landslide problems;
- Understand the past behaviour of unstable areas;
- Formulate a range of management strategies to reduce the impact of future ground instability.

The landslip potential approach, which uses detailed desk study and field mapping techniques was adopted for the Cowes to Gurnard study and provides a framework for understanding general planning and development control principles. It also assists in identifying key uncertainties and areas where subsurface investigation is most needed.

HISTORY AND IMPACT OF PAST GROUND MOVEMENT
There is little quantitative information on past ground movement within the study area. Aerial photographs and reports provide the main sources of evidence. Various sources of information have been used to assess the impacts of ground movement, including:
- Isle of Wight planning and building control records.
- Local knowledge: accounts of local residents, engineers and surveyors.
- Systematic survey of damage due to ground movement, carried out by Halcrow

The systematic survey of damage due to ground movement was carried out in accordance with the approach used by Moore *et al.* 1995). The survey was restricted to observations of external damage to buildings, walls and roads that could be directly linked to ground movement (i.e. the convergence of ground cracks/ settlement with structures showing visible signs of damage). Damage caused by ground movement was classified according to a five-fold severity scale from negligible to severe based on increased levels of damage and by inference, costs of repair.

The systematic damage survey was analysed to provide an indication of the scale of

---

[1]Geomorphological Services Ltd (1986-1987); Halcrow, Sir William & Partners (1988); Lee and Moore (1991); Moore, Lee and Clark (1995)

[2] Department of the Environment 1990, 1996

damage due to ground movement affecting pre- and post-1900 development. The impacts of moderate and serious damage to buildings, walls and roads indicated slightly higher incidence for pre 1900 development, which might be expected given the greater age and density of development at Cowes. The impacts on buildings were considered separately, which indicated a higher incidence of moderate to serious damage to post 1900 development. This probably reflects variation in the design of post 1900 buildings, which tend to be less substantial than earlier 'Victorian' buildings, and the spread of development onto marginally stable slopes. The distribution of damage showed marked concentrations of moderate to severe damage. These concentrations appear linked to landslide areas and there were notable linear distributions of damage above and at the base of the coastal slopes.

COASTAL SLOPE DEVELOPMENT
Geology
The solid strata to be found in the study area comprise Oligocene clays, marls and limestones. They include the Osborne Member and Bembridge Formations, comprising Bembridge Limestone and Bembridge Marls, overlain by superficial Plateau Gravel. Other superficial deposits in the area include alluvial deposits associated with stream valleys and colluvium or landslide debris associated with coastal landslides derived in recent geological time (i.e. Holocene).

There is uncertainty regarding the dip of strata across the study area. Bird (1997) indicates a general dip of strata for the northern part of the Isle of Wight to the south. The British Geological Survey County Series Map (1:10,560 scale) of the Cowes area indicates a southerly (landward) dip of strata up to 10° along Queen's Road. The available borehole data indicates extensive disturbance of strata (i.e. Bembridge Limestone), possibly due to coastal instability or cambering, from which reliable estimates of bedding inclination is not possible. From the available data it is only possible to conclude there is likely to be a southern component of dip for *in situ* strata.

Geomorphology
A geomorphological map of the study area was produced (Figure 1), which summarises the surface morphology of the coastal slopes and surrounding features. The geomorphological map is the product of extensive field mapping[3] supported by an interpretation of aerial photography and available geological information of the area. The following main features are identified on the Geomorphology Map:
- Plateau
- Estuary Slopes of the Medina
- Valley-side slopes

---

[3] Geomorphological field mapping comprised a detailed tape and clinometer (slope angle) survey of accessible areas, including private land where permission was granted.

- Degraded Coastal Slopes
- Coastal Mudslides
- Deep-Seated Coastal Landslides

The map shows the relative positions of the main geomorphological units that occur in the area, and identifies the nature and extent of individual landslide units. The spatial pattern of surface features, such as broad benches, steep slopes and cliffs, give vital clues about the extent and behaviour of landslides and their probable mechanisms of failure. In this way, the geomorphology map identifies a number of different landslide forms and their inter-relationships. Subsurface investigation is, however, needed to confidently model landslide mechanisms throughout the study area.

GROUND BEHAVIOUR ASSESSMENT
The ground behaviour assessment is based on the following key information:
- The stability review, which summarises past records of ground movement and the damage survey carried out as part of this study;
- An understanding of coastal slope development, which includes the geology and structure of the study area, the geomorphological map prepared for this study, and consideration of the causes of ground movement.

The approach used in the production of the ground behaviour map (Figure 1) involved the assessment of landslide activity within contrasting geomorphological units. The map identifies four distinct 'cliff behaviour units' or coastal landslides. Other ground behaviour units are also identified, which include the valley-side slopes, estuary slopes and plateau. The contemporary processes and impacts of ground movement in these areas can be expected to be somewhat different, as explained in the legend on the maps.

A cliff-top settlement zone has been defined to account for the potential settlement and recession of the cliff-top. The landward boundary is arbitrarily defined in places, 50m parallel the cliff edge, in the absence of an obvious geomorphological boundary.

The nature of ground movements can be divided into two distinct groups, namely:
- Sub-surface movements associated with the progressive creep of deep-seated landslides;
- Surface or superficial slope movements arising from the erosion or failure of steep slopes, the differential movement and settlement of clay slopes, and compression or ground heave.

Contemporary problems arising from ground movement tend to result almost entirely from superficial movements. The nature and significance of superficial ground movements varies across the study area, with different coastal slopes and landslide systems characterised by different problems. The relative risks can be determined by comparing the types of ground movement hazard with the pattern of existing development and services.

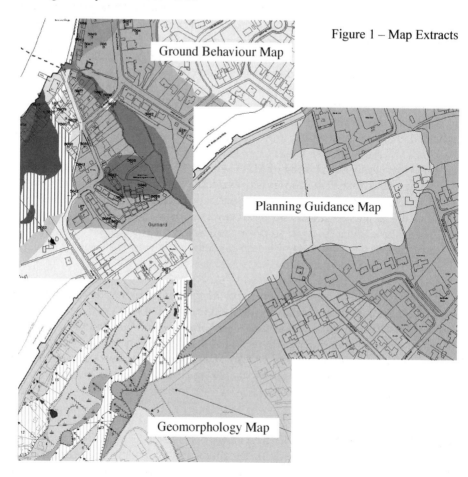

Figure 1 – Map Extracts

Ground Behaviour Map

Planning Guidance Map

Geomorphology Map

There appears to be a strong relationship between the types of ground movement that can be expected in different geomorphological units across the study area. Each geomorphological unit or landslide feature has its own characteristic range of stress conditions, which will affect buildings, walls and roads, producing a characteristic type and distribution of damage. For example, the concentration of severe damage at Gurnard Cliff has resulted from coastal mudslides, comprising undermining and recession of the cliff top, active settlement of the cliffs and translational movement of debris to the foreshore. Outward displacement and heave of mudslide lobes at the base of the coastal cliffs has promoted the destruction of coastal defences along this section.

PLANNING GUIDANCE

The Planning Guidance Map (Figure 1) is based on the assessment of ground behaviour and the variability in stability conditions across the study area. Guidance is provided on development plan policy and the control of development in areas subject to land instability. The map categorises the area according to the degree of impact which slope stability considerations might have on development proposals. Five categories have been distinguished, ranging from 'area suitable for development' to 'area unsuitable for development'.

In actively unstable areas, ground movement will affect new development; such areas should be avoided. Less unstable areas may be successfully developed, provided that the developer carries out appropriate mitigation and stabilisation measures. In areas subject to land instability, developers and homeowners must accept a higher level of commercial risk than would be expected in normal circumstances (provided that the risk is not associated with a significant safety risk). It is important in such cases, that prospective purchasers are made aware of the potential risks, along with their legal responsibilities with regard to safeguarding their property and neighbouring land from instability

These broad conclusions provide the framework for the development of planning procedures for the area that take account of the information now available on land instability problems. The overall objective of the guidance is based on 'Planning Policy Guidance: Development on Unstable Land' (PPG14)[4], which is:

> *"to ensure that development is suitable and that the possible physical constraints on the use of land are properly accounted for at all stages of planning. Although in some cases the appropriate response might be to prevent the development of land that is unsuitable, the principal objective of the guidance is to encourage the full and effective use of land in an acceptable and appropriate manner."*

COASTAL SLOPE MANAGEMENT

The role of human activity in initiating or reactivating many slope problems should not be underestimated. As experience in Ventnor has shown, in such circumstances many problems can be reduced if there is a programme of active landslide management (Moore *et al.* 1995).

A similar management strategy for the coastal slopes between Cowes and Gurnard is being implemented. The main objectives of such a strategy are to reduce the likelihood of future movement by controlling the factors that cause ground movement, and to limit the impact of future movement through the adoption of appropriate planning and building controls.

---

[4] Department of the Environment (1990, 1996)

CONCLUSIONS
The study has involved a review of available information coupled with geomorphological field surveys that has enabled an assessment of the scale and magnitude of ground instability across the study area. The broad conclusion of the study is that past incidents of ground movement and the contemporary distribution of serious and severe damage to buildings, walls and roads is mostly related to the four discrete landslides. Ongoing ground movements in these areas can be expected to cause similar occurrences and distribution of damage in the future. It is notable that the damage survey identified a higher incidence of serious damage to post-1900 buildings. In order to safeguard existing and new development in these areas appropriate detailed investigations should be carried out for the purpose of identifying measures to mitigate potential ground movement related problems.

In some areas, such as Gurnard Cliff, development would be inappropriate due to land instability constraints, among other factors. In other areas, new development and redevelopment of sites may be feasible provided the advice is taken into account and appropriate investigations and precautionary measures are implemented.

ACKNOWLEDGEMENTS
The contributions of Prof. D Brunsden, Mr EM Lee, Halcrow staff, Council officers, consultees and local residents during the study are gratefully acknowledged.

REFERENCES
Bird E (1997) *The Shaping of the Isle of Wight with an Excursion Guide.* Ex Libris Press.
Clark AR, Lee EM and Moore R (1996) *Landslide Investigation and Management in Great Britain: A Guide for Planners and Developers.* HMSO London.
Department of the Environment (1990) *Planning Policy Guidance Note 14: Development on Unstable* (PPG14). HMSO London.
Department of the Environment (1996) *Planning Policy Guidance Note 14 Annex 1: Development on Unstable Land: Landslides and Planning.* HMSO London.
Geomorphological Services Ltd. (1986-1987) *Review of Research into Landsliding in Great Britain.* Reports to Department of the Environment.
Halcrow, Sir William & Partners (1988) *Rhondda Landslip Potential Assessment.* Reports to Department of the Environment and Welsh Office.
Lee EM and Moore R (1991) *Coastal Landslip Potential Assessment: Isle of Wight Undercliff, Ventnor.* Report to Department of the Environment.
Lee EM, Jones DKC and Brunsden D (2000) The Landslide Environment of Great Britain. In: Bromhead EN et al. (eds) *Landslides in Research, Theory and Practice.* Volume 2;911-916.
Moore R, Lee EM and Clark AR (1995) *The Undercliff of the Isle of Wight: A Review of Ground Behaviour.* South Wight Borough Council, Cross Publishing.

Post-emergency actions and management of the "La Josefina" landslide (Ecuador)

RAÚL SARRA PISTONE, Dep. Director Geotechnical Service, COBA, Lisbon, Portugal

INTRODUCTION

In March 1993, a paramount landslide occurred at the "La Josefina" site on the Paute River in the Andes mountains of Ecuador. As a consequence a huge dam was created, obstructing the river valley, impounding almost 200 million m^3 of water (Figure 1).

When the level of the emergency spillway was reached, an extraordinary discharge of almost 11.000 m3/s destroyed most of the infrastructure in the downstream valley, unstabilizing huge land masses from the valley slopes, strongly modifying the valley and the river bed morphology.

The population directly affected by the disaster was estimated at 6000 persons. If the lateral effects (e.g. disruption of the transport system) are also considered, then the number of people affected could be approximately 580.000.

Damage to infrastructure was very important: the whole road system was destroyed, the Paute Hydroelectric Complex, which produces almost 60 % of the energy in the country, was severely damaged in access roads and bridges, transmission lines, protection works and some hydromechanical structures of the equipment.

The Government of Ecuador signed an agreement with the European Union for the "Rehabilitation of the Paute Valley", 80 % financed with European resources. For executive management the "Unidad de Gestión" was created, which operated for 3 1/2 years, until May 2000, having achieved the main expected goals.

Within the framework of the agreement, studies, design and construction of emergency and post-emergency works were performed for landslide stabilization along the valley, particularly for the "la Josefina" site, for the stabilization of the riverbed, basic conditions for the sustainability of interventions, for rehabilitation of the road and bridge system, soil recovery for agricultural purposes, etc. Such interventions allow the revival of the social and economic activities in the valley and the clear perception that after such a traumatic experience, coordinated actions could be of great help to overcome the disgrace.

Figure 1. Aerial view of "La Josefina" landslide

STABILISATION PROJECT

The landslide stability was studied in its three main components (S.Pistone, Torres Ochoa, S.e Pinto, 99):

A.   Stability of the remaining slopes of the M. Tamuga,
B.   Stability of the sliding mass
C.   Stability of the front slope

The front slope has been considered priority owing to its precarious stability and the eventual consequences of a new landslide. The slope was formed by the huge discharge of the natural reservoir. Shortly after the C° Tamuga landslide, a spillway channel started to be constructed with a depth of 17 m.

The sliding mass is heterogeneous, being formed of big granular dioritic blocks and andesitic rocks and colluvial deposits associated to an ancient sliding which is practically coincident with this one.

The front slope was considered the most critical part of the sliding mass, however its stability was constrained by the overall stability. The monitoring of the sliding mass and the available geodetic measurements had not shown signs of generalised instability (PRECUPA, 96, 97).

The solution envisaged the modification of the slope geometry and the surface drainage systematisation. This type of intervention was considered to be technically efficient to stabilise the slope and feasible under the economic point of view to be executed within the emergency program. The solution was discussed in a public session at the town of Cuenca.

The different calculation scenarios considered in the design, were associated with the different configurations of the saturated area within the sliding mass. The selected cases were analysed for seismic actions, bearing in mind the parameters recommended for that area (PRECUPA, 97, 98).

GEOTECHNICAL STUDIES
Given the dimension of the slope and the constraints to the interventions, the traditional site investigation methods could not be applied. Therefore, it has been considered useful to define the characteristics of the materials by means of *index properties*.

The index properties can be determined through relatively simple tests in most of the country laboratories provided with conventional equipment. From samples collected in four representative sites the following classification tests were undertaken: Slake durability tests – SDT, Los Angeles tests (LA), Rock fragment strength (Pa), Uniaxial compression tests (UCS), Point load tests (PLT).

The rockfills mechanical behaviour strongly depends on the stress conditions, the voids index and the strength of the rock fragments (Veiga Pinto, et al, 1998). Assuming the high compaction of the lower layer of the slide mass, the stress state and the rocks strength would control the behaviour of the strength parameters of the materials finally selected for the stability analysis as shown on table 1.

Table 1. Calculation parameters

	Layer $(\phi - c)$		
Case	Surface	Transition	Deep
I	42°- 0	42°- 15 kPa	47°-0
II	42°-0	45°-0	45°-0

STABILITY ANALYSIS
After adjusting the geologic-geotechnical model, stability analyses were executed focusing on different scenarios with different mass geotechnical characteristics and saturation levels.

The most critical cross sections were analysed. Considering that high confining stresses could affect the strength of the deep layer, the final analysis was performed on case II, which uses identical strength parameters for the transition and deep layers. It has also been simulated the fluvial erosion of the slope toe, having shown that the stability design strongly depended on the erosion control of the riverbed.

In order to control the longitudinal profile of the Paute river, with great erosion potential, it was necessary to construct transversal structures (weirs) built using large size rockfill cemented with concrete (Abril, 2001).

FRONT SLOPE GEOMETRY AND MONITORING

The slope geometry should accomplish the primary goal of assuring its stability, minimising the volume of material to be removed. Therefore, it has been chosen a variable curvature profile with smaller gradients at the slope base and higher gradients at the upper zone, from 1:2 to 1:1.3 (V:H). This geometry was adapted to the characteristics of the existing materials during the construction. The total volume of material to be removed according to the stabilisation design was of about 1.9 million of cubic meters (Figure 2)

Figure 2. The half-upstream front slope after stabilisation works

At the end of the Program it was possible to implement a geodetic control system of the Tamuga slope displacements. A benchmark net was installed in the remaining slope and surveyed using a geodetic base located on the opposite bank of the river.

Piezometers and inclinometers should be installed to control the stability at medium and long term. Their installation was cancelled owing to the very high local costs associated with drilling and equipment supply. They would be placed on the second phase of the Program.

IMPACT OF THE DISASTER ON THE PAUTE VALLEY

People who lived on this fertile valley had lost their productive structure, essentially the agro-industrial structure, being isolated from the rest of the country, with the consequent paralysing of their economy and the massive emigration (Figure 3).

Within the collaboration agreement established between the European Union and Ecuadorian Government, that resulted in the creation of the Unidad de Gestión (UG) – the managing entity – assisted by a team of European experts of the consortium formed by the engineering consultants ENB (Greece), COBA (Portugal) and STU-CKY (France), various studies and geotechnical works associated with the stabilisation of landslides at the Paute river area were undertaken, in particular the Tamuga landslide at La Josefina site.

Figure 3. Intervention Area of the Programme

Indicators and associated costs with the disaster

According to the "Initial Survey Report" prepared by the UG Co-directors in January 1997, the damages resulting from the disaster were the following (Table 2):

Effects on the Population. In conformity with the Evaluation Report of the CONPROE dated July 1993, the Tamuga landslide caused at least 100 casualties and more than 5.600 people were directly affected by the disaster. Considering also the problems that affected the transport network between Cuenca, Azogues and the eastern cantons, the indirectly affected population was estimated at 582.700.

Damages to the Paute Hydroelectric Development: INECEL estimated their damages at some 4.300 million sucres (approximately 2.25 million US dollars), resulting from the destruction of roads, access bridges to the power plant, towers and power transmission lines, protection works, etc.

Table 2 – Estimation of Direct Economic Losses

TYPE OF LOSS	TOTAL	
	SUCRES (Million)	USD (thousands)
Private Sector	79,833,000	41,863
Production Sector	13,702,000	7,185
Public Sector	146,227,000	76,679
TOTAL	239,762,000	125,727

* rate of exchange of June 1993: 1 USD$ = 1907,01 Sucres

The Project Management

The rehabilitation of the Paute River Valley is a process currently underway, in which various public institutions, Ecuadorian and international, and public enterprises have participated. The numerous works carried out by the UG have largely contributed to the reconstruction of the valley. The following main management principles were applied in the Programme (Bronzoni et al, 2001):

Co-direction Principle: The cooperation projects financed by the European Commission in Latin America (ALA projects) apply a co-direction system based on two co-directors – a national and a European - who are jointly responsible for the technical, administrative and financial management of the project.

Financial, Administrative and Technical Autonomy. The technical and financial cooperation projects financed by the European Commission in Latin America have technical, administrative and financial autonomy.

UG technical team. The co-direction of the "Rehabilitation of the Paute Area" Project dedicated some months to the selection of technical and administrative personnel, through careful evaluation of the applications, interviews and technical examinations. This initial effort permitted the constitution of a cohesive team that entirely assumed the aims of the project.

Contracting Procedures. The UG had applied the normal procedures for contracting works, services and supplies used in this type of the European Commission financed projects, suiting them as far as possible to the Public Contracting Law and other Ecuadorian legal standards.

Management of Public Resources: Strict Management of Funds. At the end of its activities, the UG was awarded a prize by the "Cámara de Industrias de Cuenca", a prize that is annually awarded to the best enterprises of the private sector.

Table 3. Operation actions carried out within the Contract "REHABILITATION OF THE PAUTE REGION", during the period September 96 – July 00 (Bronzoni, et al, 2001)

ITEMS	COST - EUROS
A. EMERGENCY WORKS	
Culverts	363.834
Stabilization of Cerro Tamuga	1.313.605
Bridges	1.118.707
Roads	94.943
Other Small Works	238.367
TOTAL	3.129.456
B. LAND RECOVERY	
Land Rehabilitation	775.122
Irrigation	916.122
Support to production activities	298.246
Forestation	468.194
SUBTOTAL	2.457.684
C. STABILIZATION OF THE PAUTE RIVERBED	
Weirs	2.616.415
Others	60.451
Retaining walls	118.972
Other Small Works	130.064
SUBTOTAL	2.925.902
HOUSING	
Disperse	945.509
Concentrated	932.540
Associated works	78.396
SUBTOTAL	1.956.445
E. SECTORS AND SERVICES	544.440
F. SOCIAL EQUIPMENT	215.522
TOTAL (A+B+C+D+E+F)	**11.229.449**
TOTAL ESTIMATE EQUIVALENT IN DOLARS EXCHANGE RATE USED: 1 EURO = 1.05 US$	11.790.921

Emergency and Post-emergency

The "Rehabilitation of the Paute Area" Project, conceived as an emergency action in the months following the disaster, began its operation in September 1996, three years and three months after the disaster.

The initial diagnosis indicated that it was no longer an emergency situation but instead a post-emergency situation, in a social, institutional and management context, requiring definitive reconstruction. On the other hand, it was clear that the three years of isolation that affected the Paute Valley after the disaster had aggravated problems that were already serious and usual before the disaster, as unemployment and emigration.

The co-direction of the Programme decided to begin the project, strictly applying the methodology foreseen for cooperation projects financed by the European Commission in Latin America. An initial phase of programming was accomplished in six months, and included the structuring of the technical and functional organization chart of the UG and the preparation of the Project's basic documentation: Initial Diagnosis, Overall Operation Plan and First Annual Operation Plan.

As a final result of this project management policy, approximately half of the total budget was applied in great infrastructure works (bridges, roads, stabilization of the Tamuga slope, construction of weirs for the stabilization of the Paute Riverbed). The rest of the budget was applied to finance social works and activities: gravity and sprinkler irrigation networks, production co-operative societies, community storage facilities for agriculture and cattle breeding materials, improvement in agriculture practices for communitarian groups. Forestation, conventional and photovoltaic electrification systems, housing, communal housing, water supply systems and waste water treatment (Table 3).

ACKNOWLEGEMENTS
To the technical team of the UG and the European Technical Assistance for its important collaboration in the successful achievement of the Programme goals.

REFERENCES
Abril, B., 2001. Plan de estabilización fluviomorfológica del Río Paute para mitigar el riesgo hidrogeológico de la zona de la Josefina. I Int.Symp. of Mass Movements. Cuenca, Ecuador.
Bronzoni, G., Paredes Roldán, E., Sarra Pistone, R. 2001. El Deslizamiento de La Josefina (Ecuador). Obras emergentes. III Simposio Panamericano de Deslizamientos. Cartagena. Colombia.
Informe de Diagnóstico Inicial, 1997. Programa "Rehabilitación de la Zona del Paute". Codirección de la Unidad de Gestión. *Informe inédito*
Informe Final, 2000. Programa ECU/B7-3010/94/44 de Rehabilitación del Valle del Río Paute. Codirección de la Unidad de Gestión. *Informe inédito.*
PRECUPA, (1997). *Informe de la Segunda Etapa.* Cuenca. Ecuador.
PRECUPA, (1998). *Informe Final.* Cuenca. Ecuador.
Sarra-Pistone, R., Torres Ochoa, M., Seco e Pinto, P., (1999). Deslizamiento "La Josefina", Ecuador. *XI Congreso Panamericano de Mecánica de Suelos e Ingeniería Geotécnica. ISSMGE.* Foz do Iguazú, Brasil.
Veiga Pinto, A., M. Prates, 1998. Aterros em vias rodoviarias. Projecto, Construção e comportamento. *Seminario. LNEC.* Lisboa.

Session 2:

Unstable land – problems and opportunities, legal and planning issues

Instability can impose physical constraints on development but also provide opportunities for appropriate development (including future public open space), for which land-use planning is the key. Land-use planning is essential to effective, economic and appropriate management of unstable land. Legal issues relating to the management of unstable land have had an increasingly important public profile over recent years.

Land-use planning as an element in risk management

DAVID BROOK, Department for Transport, Local Government & the Regions, London, UK

INTRODUCTION

The English planning system is concerned with development and the use of land. The Town and Country Planning Act 1990, as amended, aims to control development in the public interest through a plan-led system. Forward planning is through a hierarchy of Government policy guidance, regional planning guidance, and development plans. Proposals for development that accords with the development plan should be approved unless other material considerations indicate otherwise.

The range of considerations that can be material to planning decisions, both in preparing development plans and in determining applications for planning permission is wide. While they are not precisely defined, the courts have ruled that they can include anything related to planning, ie with development and the use of land.

The characteristics of land are fundamental to decisions about how it is used. A number of natural and artificial hazards can pose significant risks to people, property and the environment. Land instability, land affected by contamination and land liable to flooding can all cause significant problems to new development both during and after construction. The impacts of new development on these hazards can also adversely affect existing development. The planning system needs, therefore, to take account of these physical characteristics in both forward planning and the control of development.

Planning can help in the management of risks from physical hazards but it must work alongside other control systems that deal with health and safety and the control of pollution to do so effectively. There will also be a need in some cases for active intervention to mitigate the risks where necessary.

NATIONAL ASSESSMENTS

Experience in the early 1980s showed that, in many cases, planning decisions took little, if any, account of the nature of the land and concentrated on the social and environmental considerations. The physical considerations were often regarded as being the responsibility of either other control systems or of the developer. Few planners at that time knew the extent of physical problems within their areas and how they could consider them in preparing development plans or determining applications for planning permission.

To assist in the consideration of land characteristics in planning, the then Department of the Environment (DOE) instituted a research programme involving a series of national reviews of the physical problems that could seriously affect land and their relevance to development and planning. These reviews independently examined the constraints imposed by landslides (Geomorphological Services Ltd, 1986/1987; Jones & Lee, 1994), mining instability (Arup Geotechnics, 1991, 1992). natural underground cavities (Applied Geology Ltd 1993), adverse foundation conditions (Wimpey Environmental, & NHBC, 1995), erosion, deposition and flooding (Rendel Geotechnics, 1995) and natural contamination (Appleton, 1995).

The individual national reviews had common objectives:
- To review the geographical distribution of the potential constraint;
- To review the effects of the constraint;
- To summarise causes and mechanisms;
- To review methods of investigation, monitoring and remediation;
- To assess the relevance of the constraint to planning and development; and
- To identify any need for further research.

The reviews demonstrated the widespread nature of potential physical constraints, which could largely be overcome by appropriate engineering but at a cost. However, there remained large uncertainties regarding the true extent of most constraints and some uncertainty as to whether the planning system was the appropriate mechanism for considering some of them.

CASE STUDIES TO DEVELOP TECHNIQUES
In parallel with and following on from the national reviews further studies were undertaken to develop techniques that would assist planners in their consideration of physical ground characteristics. These looked at specific problems in particular areas but aimed to develop techniques that were more generally applicable.

Landslides were examined in South Wales (Halcrow, 1993) and the Isle of Wight (Geomorphological Services Ltd, 1991), mining subsidence in South Wales (Ove Arup & Partners, 1985, 1995) mining and other forms of subsidence in Norwich (Howard Humphreys & Partners, 1993) and natural gypsum subsidence in Ripon ((Symonds Travers Morgan, 1996). Potential problems due to methane (Wardell Armstrong, 1996) and radon (Appleton & others, 2000) emissions were examined on a wider basis, rather than being restricted to specific areas.

THE DEVELOPMENT OF PLANNING GUIDANCE

While coal mining subsidence had been specifically identified as a potential constraint that needed to be considered in planning (MHLG, 1961), there had been little reference to other physical characteristics of land in the wide range of planning guidance that was available. An important exception was that of land contaminated by industrial processes (DOE, 1987). In 1990, however, the then DOE (1990) issued Planning policy guidance note (PPG) 14. This specifically identified the stability of the ground as a material planning consideration, in so far as it affects land use and development.

Planning policy guidance note 14 – Development on unstable land

PPG 14 emphasised the need to consider land stability at all stages of the planning process, while recognising the prime responsibility of the land-owner/developer. Development plans should identify the existence of physical conditions leading to potential instability and outline the policies proposed to deal with them through the planning system. The principal requirement was the need for prospective developers to carry out such investigations as are necessary to satisfy the local planning authority that instability will not pose unacceptable risks to a particular development. PPG 14 recognised that, in some cases, information on the stability of the ground would be fundamental to the principle of development. In other cases, the principle of development could be resolved but further information was required to resolve specific details. A planning permission subject to conditions requiring appropriate investigation and remediation would be an appropriate response.

While the Building Regulations also covered the need to design and construct to avoid risks from subsidence and heave, these were generally not considered to include mining subsidence or landsliding. The Building Regulations were amended in 1991 to give wider consideration to subsidence and landslip than had previously been the case. In addition, the planning and building control systems were seen as being complementary. The planning system determined whether a building should proceed, in the light of the physical characteristics of the land, including any necessary land remediation, and other considerations. The building control system determined whether the detailed design and construction of buildings and their foundations would allow them to be constructed and used safely.

Mineral planning guidance note 12 – Treatment of disused mine openings...

The particular problems arising from disused mine openings were recognised as a special case because of the generally higher risk to life. Minerals planning guidance note (MPG) 12 (DOE, 1994a) advised on the planning considerations relevant to their treatment for public safety reasons, to safeguard conservation interests and to facilitate development. The need to consider the likelihood of mine openings being present when determining planning applications in appropriate areas was emphasised and good practice guidance on mining data systems was given

PPG 14 Annex 1 – Landslides and planning
The first annex to PPG 14 (DOE 1996) advised on how local planning authorities can identify areas of landsliding and how they should address the potential problems arising through the planning system. In particular, it outlined good practice on an incremental method of landslide hazard assessment, with examples from the case studies in South Wales and the Isle of Wight, and gave guidance on the preparation, content and format of slope stability reports to accompany planning applications.

PPG 14 Annex 2 – Subsidence and planning
The second annex to PPG 14 (DTLR, 2002a) advised on the consideration of subsidence when considering proposals for mining/underground construction, remedial action after subsidence has occurred and development on land liable to subside. It included illustrative examples from the case studies of techniques and good practice advice on mitigation of subsidence and treatment of mine openings, on mining data systems and on the content of stability reports. PPG 14 Annex 2 specifically advised that subsidence (and heave) due to shrinking and swelling clay is adequately controlled under the Building Regulations 2000 and is unlikely to be significant in planning terms.

MPG 5 – Stability in surface mineral workings and tips
MPG 5 (DETR, 2000) recognised the potential significance of instability in activities involving major earthmoving operations. It advised on the consideration of instability in relation to both quarrying and tipping operations and other development in or near to mineral workings. It recommended the submission, where appropriate, of a design report prepared by a competent person and gave good practice guidance on the design, inspection and assessment of excavated slopes and tips and related structures. It thus complements the requirements of the Quarries Regulations 1999, which aim to protect health and safety.

PPG 25 – Development and flood risk
Following a series of inland floods from Easter 1998 to the autumn of 2000, planning guidance on flooding (DOE, 1992) was revised. The new PPG 25 (DTLR, 2001) recommended the application of the precautionary principle to recognise the uncertainties in flood-risk estimation and the likely effects of climate change. It established a risk-based search sequence to avoid risk where possible and manage it elsewhere and recommended a minimum standard of defence where new development has to be located in areas vulnerable to flooding.

Land affected by contamination
To support the ongoing programme of urban regeneration, the then DOE (1987) issued guidance on the consideration of contaminated land in planning. It advised on the need for thorough investigation of previous land uses and the carrying out of any necessary remediation to enable the most suitable use to be selected. This

advice was extended in PPG 23 on planning and pollution control (DOE, 1994b). The Department for Transport, Local Government and the Regions (DTLR) has recently consulted on revised advice on land affected by contamination (DTLR, 2002b).   This complements the introduction of the statutory contaminated land regime under Part 2A of the Environmental Protection Act 1990.   The aim of both systems is to ensure that contamination is properly identified and remediated to a condition that is suitable for use by breaking the source-pollutant-receptor linkage that will have been established through appropriate risk assessment.

RISK MANAGEMENT STRATEGIES

The management of risks arising from land characteristics has always tended to be reactive.   The occurrence of a significant event requires an immediate response but also gives rise to proactive thinking about avoiding such events in future.   The result is that there are a number of well recognised strategies that may be applied depending on the particular circumstances.

Since hazard prediction is at best an uncertain science there will always be a need for an emergency response and crisis management strategy.   This will deal with the immediate after-effects of a hazard event, including the relief of the threatened population and repair and reinstatement of structures.   Significant urban areas and populations are already at risk from a variety of hazards and many areas have contingency plans in place to deal with events as they occur.   These should be tested on a regular basis.

Planning for losses will also be needed.   There is provision for the repair of damage caused by coal mining subsidence under the Coal Mining Subsidence Act 1991. For most hazards, however, loss cover is largely through the private sector insurance industry in the United Kingdom provided the standards of insurability are met. These basically require a large number of exposures, fortuitous occurrence, a loss that is definite in time and amount no significant concentration in vulnerable areas and reasonable premiums in relation to the potential financial loss.   Where these conditions are not met and the probability of occurrence becomes too high, there may be difficulty in obtaining insurance cover.   For example, it is rarely possible in Britain to obtain cover against coastal erosion and the insurance industry have recently indicated that their continued provision of flood cover may not be possible without significant action by Government to reduce their exposure in certain areas.

Modifying the hazard generally involves some form of engineering treatment to reduce the likelihood of hazard occurrence.   This may be carried out before development takes place or with existing development in place.   However, the costs are such that it is unlikely to be undertaken until there is a definite chance that an event will occur (or it is already occurring).   The stabilisation of landslides, the infilling of abandoned mines and the provision of flood defences are clear examples of this strategy.

Controlling the effects of a hazard can be achieved by avoiding development in the most vulnerable areas, by incorporating preventive measures in the design of new buildings and structures or by modifying existing structures to incorporate precautionary measures. In Britain this strategy is largely applied through the land-use planning system and the Building Regulations 2000.

CONCLUSIONS

A wide variety of hazards may arise from the characteristics of land, both natural and artificial. The United Kingdom has developed an active forward strategy to minimise the risks arising based on the results of research to identify the scale and nature of problems, and to develop techniques enabling them to be taken into account in land-use planning. Advice on how to deal with the problems and good practice in assessment and mitigation is now widely available to land-use planners. This will assist in reducing the risks to people, property and the environment. The prime responsibility for identifying and dealing with the hazards remains with the developer and land-owner

Even so, the planning system only deals with new development. There are limitations to what can be achieved through the planning system since it can only consider valid planning considerations. Other issues, such as those related to the control of pollution and health and safety need to be addressed through other legislation.

It also has to be recognised that there is much existing development already in vulnerable areas. While this has not been quantified for instability issues, we know that 10% of the English population live in areas vulnerable to flooding. Thus, however good the consideration of hazards through the planning system, the effect on overall flood risk is likely to be marginal. The release of appropriate sites to meet the Government's declared policy aims of locating 60% of new housing on previously developed sites and by re-use of existing buildings relies on an adequate consideration of the hazards to which they might be subject. Proper consideration of flooding, subsidence and contamination in particular can enable appropriate redevelopment to take place. There will remain some hazards to existing development that may need to be addressed by direct intervention to mitigate the risks.

The views expressed in this paper are those of the author and do not necessarily represent those of the Department for Transport, Local Government and the Regions.

REFERENCES
Appleton, J. D. 1995.   Radon, methane, carbon dioxide, oil seeps and potentially harmful elements from natural sources and mining areas: relevance to planning and development in Great Britain -- Summary Report.   *BGS Technical Report WP/95/4*, 40pp.
Appleton, J. D., Miles, J.H.C., Scivyer, C.R. & P.H. Smith, 2000.   Dealing with radon emissions in respect of new development – summary report and recommended framework for planning guidance.   *British Geological SurveyResearch Report RR/00/07*, 26pp.
Applied Geology Ltd, 1993.   *Review of instability due to natural underground cavities in Great Britain.*   Report to Department of the Environment, Royal Leamington Spa, Applied Geology Ltd, 14 vols.
Arup Geotechnics, 1991.   *Review of mining instability in Great Britain.*   Report to Department of the Environment, Newcastle-upon-Tyne, Arup Geotechnics, 25 vols.
Arup Geotechnics, 1992.   *Mining instability in Great Britain – summary report.*   London, Department of the Environment, 22pp.
DETR, 2000.   *Minerals planning guidance note 5: Stability in surface mineral workings and tips.*, Norwich, TSO, 40pp.
DOE, 1987.   *Department of the Environment Circular 21/87: Development of contaminated land,* London, HMSO, 10pp.
DOE, 1990.   *Planning policy guidance note 14: Development on unstable land.*   London, HMSO, 25pp
DOE, 1992.   *Department of the Environment Circular 30/92: Development and flood risk.*   London, HMSO, 11pp.
DOE, 1994a.   *Minerals planning guidance note 12: Treatment of disused mine openings and availability of information on mined ground.*   London, HMSO, 31pp.
DOE, 1994b.   *Planning policy guidance note 23: Planning and pollution control.*   London, HMSO, 39pp.
DOE, 1996.   *Planning policy guidance note 14: Development on unstable land: Annex 1: Landslides and planning.*   London, HMSO, 17pp.
DTLR, 2001.   *Planning policy guidance note 25: Development and flood risk.*   Norwich, TSO, 60pp.
DTLR, 2002a.   *Planning policy guidance note 14: Development on unstable land: Annex 2: Subsidence and planning.*   Norwich, TSO.
DTLR, 2002b.   *Consultation draft of planning advice on land affected by contamination*   London, Department for Transport, Local Government & the Regions.
Geomorphological Services Ltd, 1986/87.   *Review of research into landsliding in Great Britain.*   Report to Department of the Environment, 15 vols.
Geomorphological Services Ltd, 1991.   *Ground movements in Ventnor, Isle of Wight – a summary of the study of landslide problems in Ventnor carried*

out for the Department of the Environment. Milton Keynes, Geomorphological Services Ltd, 68pp.

Halcrow, Sir William, & Partners, 1993. *Rhondda landslip potential assessment – summary report.* Cardiff, Sir William Halcrow & Partners, 67pp.

Howard Humphreys & Partners Ltd, 1993. *Subsidence in Norwich – Report of the study on the causes and mechanisms of land subsidence, Norwich, carried out for the Department of the Environment.* London, HMSO, 99pp.

Jones, D.K.C. & E. M. Lee, 1994. *Landsliding in Great Britain.* London, HMSO, 361pp.

MHLG, 1961. *Ministry of Housing & Local Government Circular 44/61: Surface development in coal-mining areas.* London, HMSO, 2pp.

Ove Arup & Partners, 1985. *Mining subsidence – South Wales desk study.* Report to Department of the Environment, 5 vols.

Ove Arup & Partners, 1995. *Islwyn shallow mining – final report.* Cardiff, Ove Arup & Partners.

Rendel Geotechnics, 1995. *Erosion, deposition and flooding in Great Britain – a summary report.* London, Rendel Geotechnics, 81pp.

Symonds Travers Morgan, 1996. *Assessment of subsidence arising from gypsum dissolution.* East Grinstead, Symonds Travers Morgan, 2 vols.

Wardell Armstrong, 1996. *Methane and other gases from disused coal mines: the planning response.* London, TSO, 2 vols.

Wimpey Environmental & NHBC, 1995. *Foundation conditions in Great Britain: a guide for planners and developers.* Hayes, Wimpey Environmental, 2 vols.

evelopment setbacks from river valley/ravine crests in ewly developing areas

rEFAN FEKNER, BES, MA, ACP, MCIP, Senior Planner, Planning & Policy rvices Branch, City of Edmonton Planning and Development Department, Edmonton, berta, Canada

ACKGROUND AND ISSUES

ne crests of Edmonton's river valley and ravines often provide spectacular views and are desirable sites for residential development. Unfortunately, these sites may be bject to slope movements that may damage structures or services to them.

ne retreat of slopes from their crests is an ongoing and permanent process. The present alley of the North Saskatchewan River in the Edmonton region was excavated less than 2,000 years ago. The river has cut downwards (downcutting) and across (lateral osion) to a depth of more than 54 metres. The environmental factors primarily sponsible for slope instability are the river cutting the 'toe' of the slope and the retreat slopes towards a more gentler slope angle, called the "angle of stability." Urban evelopment, on or in close proximity to river valley or ravine banks, further contributes slope instability and generally increases the rate of slope retreat and the incidence of ope failure. In one study, man-induced landslides accounted for three-quarters of the 7 landslides sampled in Edmonton. The predominant man-induced contributor to slope stability and failure is the introduction of water, through the increase in groundwater vels and by overwatering in the vicinity of the edge of the river valley.

ince 1985, the City of Edmonton has had a sound and valid citywide policy (Top-of-e-Bank Roadway Policy) governing development within a narrow strip of land along ne edge of the river valley and ravine system. The Policy has used the top-of-the-bank TOB) roadway as the primary means of separating privately owned urban development om the publicly owned river valley and ravine system. A 7.5 metre setback from the rest establishes a minimum separation space and where a top-of-the-bank roadway is ot applied, the 7.5 metre setback is used for public access, typically a top-of-the-bank alkway. Unfortunately, top-of-the-bank roadways have been implemented in only one fth of the situations, resulting in urban development backing onto the public upland rea adjoining the crests of the North Saskatchewan River valley and ravine system.

Some 118 slope failures and nearly 400 encroachments onto public land have be documented by civic staff, identifying the loss of "environmental protection" previou afforded by the top-of-the-bank roadway. Furthermore, the City's acquisition of t slope, through the designation of "environmental reserve" at the time of subdivision, I had the effect of plac'ng higher maintenance costs, liability and risk upon t municipality. The City of Edmonton annually spends over $1,000,000 on slide repa Additional costs may arise from ongoing litigation by private property owners. In c recent court case that is currently under appeal, the City was found partially liable damages despite the negligence of the property owner.

Since 1985, the TOB Roadway Policy has been seriously eroded and the risk of slc instability and failure has been misconstrued (by those wishing to avoid its ma elements such as the TOB roadway) as infrequent, uncommon and largely avoidable. view of these concerns, the Planning and Development Department was given t mandate by the old Municipal Planning Commission to review the 1985 Poli including issues of "liability, maintenance, and the philosophy of public access a exposure to the river valley system."

The review of the 1985 Top-of-the-Bank Roadway Policy has identified a number of k deficiencies or problem areas:

- Environmental reserve is often taken for the slope only and thus does not include t potentially unstable upland area behind the crest of the river valley or ravine. Tl leads to a general misconception that the top-of-the-bank line (crest) h significance in determining the line between stable and unstable land (it doesn' The result has been the use of many different lines demarcating different land u servicing and development objectives.
- The environmental hazards assessment (typically referred to as geotechnical repor has employed a wide variety of terminologies and standards. The lack of a clear a uniform standard is not only confusing to the potential home buyer but is large unnecessary as it assumes different levels of environmental protection. The curre professional engineering practice is to use these various standards in conjuncti with different land use and development restrictions imposed through restricti covenants on the title of properties, to manage environmental risk.
- Restrictive covenants incorporate the recommendations of geotechnical enginee notify property owners of the environmental hazard risk and indemnify the Ci from future legal challenges. However, they are being poorly implemente Empirical evidence suggests that not only do residents not follow them but that t City has a great deal of difficulty in enforcing them. The reasons for po enforcement include the lack of administrative and political will, insufficie dedicated staff resources and budget, legal obstacles and public resistance.

Exceptions to the top-of-the-bank roadway, which promote a top-of-the-bank public walkway or public upland area, have become the 'rule.' This has created urban development backing onto the top-of-the-bank (crest) that in turn has contributed to increased slope instability and failure, encroachments onto public lands, and greater liability, risk and maintenance costs to the City of Edmonton.

Public access has not been optimized with the loss of the top-of-the-bank roadway. In some development situations where the Top-of-the-Bank Roadway Policy applies, public access has not being provided at all (17%) or has not been developed in a continuous manner along the entire length of the river valley or ravine edge.

)CAL PLANNING EXPERIENCE

ınners do not work in a political vacuum. The prevailing attitudes and interests of ty Council, affected property owners, the media, the public and the various interest ups must be considered in crafting new policy and regulations. In Edmonton, werful associations, the Urban Development Industry and the Greater Edmonton me Builders Association represent some of the dominant partners in the development ustry. Consultation with these associations, affected property owners and interest ups is a standard practice and some degree of concurrence is expected by the civic ministration and by City Council for the proposed new measures.

ıe essential struggle: moving the development boundary

ıe current development boundary, established by the 1985 Top-of-the-Bank Roadway licy, has been considerably eroded. The risks associated with development near the est of the river valley and ravine system have been diminished or largely overlooked in ıd use decision making. City Council, in its development decisions, has perceived ıblic access as the primary issue. Developers have generally been successful in oiding the requirement for a top-of-the-bank (TOB) roadway, thereby maximizing the ımber of lots backing onto the crest. This is done for strong economic reasons – more ts, reduced servicing costs with double sided development on a road, and more luable, 'viewpoint' lots. By avoiding the requirement for the TOB roadway, the velopment boundary has been moved closer to the crest. This has the effect of placing gher maintenance costs, enforcement responsibilities, liability and risk upon the unicipality.

tential home buyers of viewpoint property are also at risk from slope instability and ilure. Private encroachments onto public open space are a common problem. The unicipality poorly enforces restrictions on underground sprinklers and other water atures, imposed by restrictive covenants. Problems with enforcement include a lack of lministrative and political will insufficient dedicated staff resources and budget, legal stacles and public resistance. Often, residents are not aware of or prepared to adhere such restrictions until too late when instability problems become manifest. Home owners insurance is not available for slope failure. When slope failure occurs, reside want the City, which is a neighbour by virtue of the environmental reserve designati of the slope, to either fix the problem or to compensate them for their loss. In genei the policy and regulatory tools which are available for managing the environmen hazards risks associated with urban development backing onto the edge of the riv valley and ravine system are weak and subject to enforcement problems.

The job of pushing the development boundary back from the crest of the river valley a ravine system so that development is not exposed to an unacceptable risk began with t construction of a solid technical foundation and public education. The planning proc involved the creation of a multidisciplinary, interdepartmental working committee le by a senior planner; a review of current policies and practices; research on best practi from other public agencies, universities and professional organizations; and t preparation of public information strategies and materials. Background reports, a dr new City policy and accompanying amendments to several statutory and regulatc documents have resulted from this effort. The draft new City policy has benefited fr the excellent pioneering work conducted by the University of Alberta on appropri development setbacks from crests, based on studies of local slope failures a conditions. After internal civic circulation, a second phase of work has begun witl program of stakeholder consultation with development industry representativ professional associations and companies, and community and environmen organizations. The final stages of this work will involve City Council's discussion (a possible public hearing) and approval of the new City policy and accompanyi amendments, and the implementation of this package, including revised business a operating practices. The challenge is to raise awareness of environmental risks and find workable solutions for developers, builders, home buyers and the municipality.

FINDING SOLUTIONS
Finding solutions begins with clearly defined policy objectives. Environmen protection and public access are the major public objectives related to current top-of-tl bank policies and practices. "Environmental protection" means that the riv valley/ravine system is protected from urban development that may compromise integrity and long term stability. "Environmental protection" also means that urb development is safe and secure from environmental hazards that may result in loss persons and property. "Public access" means the provision of optimal access for lo residents and the general public to a continuous circulation system along the ent length of the upland area abutting the river valley and ravine system. Achieving the two objectives will ensure that the North Saskatchewan River Valley and Ravine Syste is preserved as a significant natural amenity feature and recreational opportunity for t citizens of Edmonton.

The basic approach taken in finding solutions to the previously outlined deficiencies or problems is that of risk management. See Figure 1. This approach emphasizes "environmental protection" by clearly establishing the line (called the Urban Development Line) between what land is reasonably safe on which to build (urban development) and what land is unsafe to build on, necessitating the taking of "environmental reserve" by the municipality to guarantee "environmental protection." "Environmental reserve" is very generally defined in superceding provincial legislation but given a more precise definition in the draft new City policy. The Urban Development Line would generally be identified at the statutory planning stage, or, in any event, at the next earliest stage in the planning and development process.

The policy tool kit: the estimated long term line of stability, a minimum public upland area setback, a caveat on the property title and the reintroduction of the top-of-the-bank roadway

For the majority of development situations, where potential slope instability and failure remain the significant environmental hazard to be managed, the estimated long term line of stability will become the Urban Development Line. See Figure 2. The estimated long term line of stability will be established through an environmental hazards assessment. Such a line of stability will incorporate the combination of the estimated long term line of stability and future instability factors. The estimated long term line of stability will set a higher standard for a structure than for a TOB roadway. The future instability factors will include estimated toe erosion over a given period; anticipated increases in the water table within a developed urban drainage basin; any anticipated or planned removal of vegetation; any planned placement of fill for lot grading purposes; and any anticipated redirection of the surface water runoff over the slope. It will provide consultants with a more conservative figure for establishing development setbacks. It will increase the separation space between privately owned urban development and the publicly owned river valley and ravine system. The Urban Development Line, identified in a caveat placed on the property title will provide residents with an opportunity to be aware of the risks attached to living along the crest of a slope. The additional separation space will also give residents time to react to various slope failures, instead of necessitating immediate remedial action. The Urban Development Line represents a clear and uniform development standard as well as a more reliable definition of what constitutes unsafe land that should be taken as environmental reserve at the time of subdivision.

The Urban Development Line also comes with a minimum public upland area setback to provide for public access and environmental protection. The setback begins at 7.5 metres and increases depending on slope steepness and depth. The Urban Development Line can be no less than the minimum public upland area setback.

Finally, the City will require the owner of any property adjacent to the Urb
Development Line to enter into an agreement, to be registered on title as a caveat. T
caveat will apprise the owner of the environmental risks associated with development
proximity to the river valley/ravine crest; will specify any additional developme
controls or restrictions (restrictive covenants) and will contain an explicit statement t
land will not remain stable in perpetuity.

Figure 1

The Concept of the Urban Development Line (UDL)

The urban development line distinguishes what lands are safe to build on for urban
development and what lands pose too serious a risk for such development to occur
(designated environmental reserve). Environmental reserve is taken by the City at the
subdivision stage of development and usually requires that land be left in a natural state
or as parkland. The urban development line is delineated through an environmental
hazards assessment which looks at a number of environmental risks such as slope
failure and underground instability. The urban development line applies one common
standard or formula, the estimated long term line of stability, which employs predetermined
existing and future instability factors.

Source: City of Edmonton Planning and Development Department, August 2001.

Figure 2
Major Components of the New City Policy:
Development Setbacks from River Valley/Ravine Crests
in Newly Developing Areas

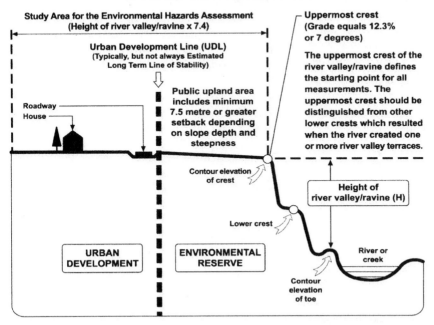

The Urban Development Line (UDL) divides urban development from environmental reserve. The UDL is established through an environmental hazards assessment (usually a geotechnical study). The study area for the hazards assessment is larger than the actual UDL in order to ensure that future urban development is "captured", preventing such development from occurring on the wrong side of the UDL (i.e. within the environmental reserve). Environmental reserve is measured back from the uppermost crest, a minimum of 7.5 metres or greater, depending on slope steepness and depth, and as further determined by the environmental hazards assessment. The majority of development fronts onto the top-of-the-bank (TOB) road, except for shallow ravines or gentle slopes where a TOB walkway may replace a TOB road.

Source: City of Edmonton Planning and Development Department, August 2001.

long with the risk management approach, the reintroduction of the top-of-the-bank roadway is being proposed for newly developing areas. Why? The argument can forcefully and convincingly be made that the top-of-the-bank roadway provides superior environmental protection, superior public access and superior protection from priv encroachments onto the public upland area adjoining the edge of the river valley and ravine system.

The top-of-the-bank rcadway offers superior environmental protection to resident properties, by providing additional separation space between urban development and valley/ravine edge. It lessens the concern that residents are contributing to increas slope instability through inappropriate water management practices (i.e. use of wa sprinklers). Also, a top-of-the-bank roadway serves to intercept overland surface wa flows. Finally, it allows the City access to lands adjacent to the river valley edge order to repair or stabilize unstable slopes.

As for public access, the top-of-the-bank roadway provides a year round, continuous a fairly linear facility for cyclists, pedestrians and motorists. A top-of-the-bank walkw will only be provided as an option to a TOB roadway where a shallow ravine or gen slope exists. The environmental risk is reduced and public access by a walkway considered adequate for the anticipated traffic. Where a top-of-the-bank walkway is be provided, provision should also be made to allow for public access and viewpoints penetrate the long wall of urban development at regular intervals.

Finally, the top-of-the-bank roadway eliminates the opportunity for priva encroachments unto the public upland area along the edge of the river valley and ravi system.

DRAFT CITY POLICY AND ACCOMPANYING AMENDMENTS AND RECOMMENDATIONS

The draft City policy is currently in circulation and discussions are being held w various representatives from the development industry, professional, community a environmental groups, and property owners. The policy is accompanied by amendme to the City's Municipal Development Plan Bylaw 11777, the North Saskatchewan Riv Valley Area Redevelopment Plan Bylaw 7188 and the new Zoning Bylaw 128(Finally, the report to City Council will contain recommendations affecting business a operating practices. This package of a new policy, accompanying amendments a implementation recommendations will likely be before City Council for their appro before the end of the year. The new City policy will bring together engineering, le; and planning solutions to a problem, slope instability and failure, that currently diminished or overlooked in land use decision making. It will join other City polic and development practices intended to ensure that the North Saskatchewan River Val and Ravine System is preserved as a significant natural amenity feature and recreatio opportunity for the citizens of Edmonton.

Environmental hazards and landuse planning for sustainable development: The Douala unstable coastal region, Cameroon

C. M. LAMBI, & L. F. FOMBE, Department of Geography, Faculty of Social and Management Sciences, University of Buea, Cameroon

INTRODUCTION

Man's activities on land and the effects of natural hazards seriously threaten biodiversity of coastal areas. These fragile zones are altered through industrial activity, poor construction methods, flooding, silting and landslides. The quest for better living conditions entails a strong desire to overcome the threats and ugly consequences of these hazards in Cameroonian cities which usually result from natural hazards. To overcome such adversity pre-empts a comprehensive regionally based planning strategy.

The Douala region is characterised by its diversified physical outlook. It stretches from a low lying and partially submerged suburb of Bonaberi marked by flash floods and industrial pollution to sporadic landslides in Limbe. These recurrent forces have endangered the lives of more than half a million people. Limited land, the structural adjustment of Cameroon's economy following the economic crisis in the past two decades and our poor urban redevelopment schemes have forced the desperate urban poor to encroach into wetlands where housing densities are remarkably.

This paper seeks to examine the underlying causes of intermittent and sporadic occurrences of floods and which continuously threaten the population. It also assesses the economic and ecological opportunities offered by these hazard-prone zones. In addition, it proposes long-term planning strategies that could be devolved as a balance between expanding population and the natural environment.

ENVIRONMENTAL SETTING OF THE REGION

The physical geography of Bonaberi epitomises a homogeneously low-lying region, characterised by mangroves and interspersed with creeks and lagoons of the River Wouri estuary. More than 75% of the land has an average altitude of 2 metres above sea level. The numerous tributaries of the River Wouri form an interface between Mabanda residential zone and Bojongo village, covering an area of approximately 20km². Its gentle gradient does not allow any coarse particles to reach the sea.

Instability – Planning and Management, Thomas Telford, London, 2002, 143–149

There is a thick mangrove swamp forest presently undergoing extinction resulting from urbanisation which covers the stretch of maritime land. This vegetation spreads over a landmass of 80km in length. The swamps give rise to the luxuriant growth of palms, some of which thrive in rivulets. The soils of the region are sandy. Their loose or unconsolidated property coupled with marine and river erosion keeps them mobile.

In the Littoral region Rainfall is heavy, recording about 1400mm yearly. The rainy season lasts for up to 10 months during which torrential downpours invariably create problems of run-off in the sandy low-lying areas with gradients below 1%.

Limited urban space for housing, coupled with a timid planning policy in Bonaberi have encouraged the population to build in waterlogged zones and on hill slopes.. Many land speculators acquired and moved into the waterlogged zones and wetlands in spite of the extremely hostile nature of these environments. In relatively elevated areas, people have built both on and at the foot of sandy slopes. Furthermore, the juxtaposition of industrial, residential and agricultural land is a basic feature here.

THE ENVIRONMENTAL THREATS

Floods, occasional sea incursions and landslides are natural occurrences of the Douala coastal region. Poor construction, deforestation and poor farming methods have endangered and modified this fragile ecosystem. These combined human and natural actions constitute very serious environmental threats.

Stagnant pools of water from heavy rain or the sea is a characteristic feature due to the level topography and absence of adequate drainage facilities. Coupled with filth, these zones have developed into permanent mosquito and other disease vector incubation grounds. Wastes from homes are also carelessly dumped with spills ending up in rills and the narrow gutters. Industrial waste such as millet chaff, wood, saw dust, discarded chemicals and metals end up in the various channels of River Wouri. Such materials increase the frictional drag of the River, given the low gradient. During downpours, the drag becomes great leading to overland flow. Flood waves travel down the River Wouri from its upper sections in the hinterlands for over 100km. It sometimes takes 4hrs. to reach the Bonaberi suburb where the speed reduces substantially by 75%. This slow rate of flow can last for 10hrs depending on the length of rainfall and the period of high tide. Both river water and high tides cause enormous floods which sometimes attain alarming heights of 1.0-3.0 metres (table 1). The enormous floods of August and September 2000 remain fresh in the memories of the inhabitants of the Douala metropolis.

Table 1: Flood heights and extent of damage for selected residential areas

Flood height	Flood duration (Hrs.)	Structures/property affected (categories)	Highly affected zones
Less than 2 m	2	A: Farms, Roads, Pipes	Ngwelle Mabanda Bojongo Ndobo
2 to 3m	3-4	B: Category A, Households/Foundations & Electric poles	
Above 3m	>5	C: Categories A, B, & Wooden/metal structures	

Source: Field survey, September 2001.

The intermittent invasion of land by the sea occurs almost weekly, and this also constitutes a serious environmental hazard. Returning industrial and domestic impurities contaminate the wells and affect both flora and fauna. The exploitation of quarries for sand, and constant dredging of the Wouri channel which divides the mangrove into two accelerates erosion. (Kuete, 1988:105).

Foundations of most houses are less than one metre above the ground and floors of most houses are underlain by semi-decomposed material (sawdust, and domestic trash). This condition enhances the fragility of the houses in a zone where the soil is highly inundated. Some 80% of temporary buildings within the Bonaberi shanty areas are constructed with wooden material. The frequent sea incursions (twice in 24 hours) become devastating (Kuete; 1988:95-7). Water level can rise to 2m in 5 minutes. The salty water with its high corrosive property affects metal structures and roads. Railway tracks in areas like Ndobo are completely submerged and traffic flow is halted during such moments. Wooden materials decompose faster under the high humidity (moisture) thereby reducing their life span

Given the sandy and loose nature of the soil, pit toilet and wells, drainage systems can not be constructed to adhere to safety levels. Pit toilets are shallow for the safe disposal of faeces. Contents of such toilets are often washed back into land during flash floods thereby contaminating the wells. The proximity of most wells to toilets and the low-lying topography enhances pollution of wells rendering them unsafe for domestic use. (fig.1). Deaths resulting from typhoid and dysentery are common. Pit toilets can not be easily reached for emptying by the services of hygiene and sanitation due to narrow streets to admit specialised vehicles. Even so, most owners do not concrete the shallow toilets nor are they deep enough. Depths of 10 feet are not even attained in most cases.

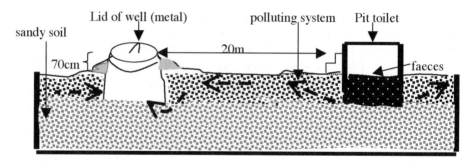

Fig. 1: The possibility of domestic water pollution by a pit latrine at close proximity.

Farming for market gardening products is widespread. This uncontrolled activity in the area further blocks the drainage channels because most ridges are formed against the dip of the slope in many areas. Consequently, overland flow is unconsciously directed to assume an opposite direction to the mouth of the river or the sea. The growth and propagation of algae and other forms of bacteria are not uncommon in the stagnant pools that eventually develop thereby enhancing diseases.

The above threats, therefore, have serious repercussions on the lives of the people living in the area. This has become almost an endemic problem for the urban poor just recovering from an economic depression. The desperate population thus finds itself in a vicious web unable to mastermind the threats.

MANAGEMENT APPROACHES FOR BIODIVERSITY AND SUSTAINABILITY

The activities of mankind are influenced by major movements of parts of the earth's crust Sparks (1986. 2),. Floods constitute one of those major hazards, which combine with pollution to threaten today's increasing urban population of most third world countries. These disasters need major long-term planning strategies to enhance and sustain the ecosystem. The solutions to floods, pollution and dredging in the Douala region can only emanate from the population. Unfortunately, the majority of these urbanites such as those in Bonaberi are primarily involved in informal and small income-generating activities that can not raise enough capital to respond punctually to these hazards. Bonaberi's population of half a million is growing at an annual rate of 12%. This indicates that by the year 2025, every thing being equal, this zone shall harbour over a million people with a density of 30.000 persons/km ^{2}.

The low lying topography of Bonaberi, its cool sea breeze, rich marine life, mangrove vegetation and human resources are essential prerequisites for economic development if astutely managed. These resources can create employment through tourism, enhanced fishing, while maintaining its ecological state, contrary to public adversity that it is an unmanageable territory.

Existing land use and management strategies:
In the past two decades, the population of this hazard prone zone has made unsuccessful attempts albeit public intervention to fight floods and the intermittent sea incursions. They are limited by their capacity to raise funds and to mobilise human effort in a zone where land ownership is fraud with much scepticism. Consequently, the technique applied so far to protect homes and property to sustain city life has been outstandingly very timid.

Attempts made to check sea incursions which are recurrent involves the building of protective solid walls across door posts or around buildings and the reclamation of water logged creeks through filling with earth and rubble. The people also use pieces of discarded plank from sawmill factories, stones or cement blocks to serve as bridges. In search of housing facilities, the mangrove and other water logged vegetative plants like raffia and brush wood are systematically felled depriving reptiles, amphibians and animals of their natural habitat. As the effects of high tides in the Bonaberi wetlands have been minimized by the ubiquitous offshore mangrove forests, they should be protected from exploiters. Courses of rivulets in some parts of the zone become narrower due to reclamation of the land for building. Most rivulets hardly attain 3 metres in width. Their importance in channel flow is ignored due to water stagnation and the lack of finances to undertake huge drainage schemes.

Given the appalling landscape situation of the western part of Bonaberi for example, 50 % of the resident population claim to erect temporary buildings with the hope of finding a permanent and more secured land elsewhere when living conditions improve. Field survey reveals that 90% of this category has come to live permanently in this suburb.

Recommended planning measures:
The techniques used by the local population to fight environmental hazards (flood, silting and pollution) have ipso facto proven ineffective due to the progressive deterioration of physical structures and the health of the people. The plausible measures that can be implemented to combat these threats are considered vis-à-vis the changing water levels, conservation methods, drainage, agriculture and industrial pollution.

Changing water levels: The rise in sea level and river overflow due to heavy rains entails the widening and deepening of channels. Regular draining exercises before, during and after the rainy season should be instituted. In areas like Ndobo where high flood levels are often recorded, stream channels can either be deepened by one metre or widened to contain the excess flow. This will not only restore the streams' track but also ensure that the occurrences of sea incursions do not affect human investment. With this achieved, some 25metres on both banks of streams in seriously inundated areas be set aside for raffia and aquatic life to thrive. On building sites, foundations should be made of concrete material and should be raised by 1.0-3.0 metres above water table so that an appropriate drainage and sanitary system can be created for easy evacuation of household effluents.

Conservation methods: Reclaimed areas especially along streets can be planted with eucalyptus as a measure to check against deforestation and water logging since this tree takes up and evaporates much ground water. The process of forest regeneration should be obligatory through legislation to enhance environmental protection so as to avoid "urban deserts."

Urban agriculture and drainage: Considering the low economic status of the population, horticulture to sustain urban life is imperative. A survey in Bonaberi indicates that 80% of the females though involved in other activities, do market gardening in the vicinity and especially on the uninhabited marshy zones. Crops such as huckleberry, green/bitter leaves are cultivated to feed the urban population. Unfortunately, the activity exacerbates the drainage problems and most of the water used is polluted by toxic chemicals from industries or may contain high levels of marine salt. This activity can only be undertaken in the dry months of the year when the water table drops. Agriculture should be improved through better channelling of water (creation of drains), and the use of sites which are unfit for housing, thus providing sustainable employment for the majority of the female population interested in the activity.

Combating pollution: Effluents from industries and homes (fig.1), constitute a major hazard. In 1985 when Bonaberi was slated as an industrial zone to relief the congested centrally located Bassa region, there were barely 27 industries. Today, more than 50 factories such as SCI, NOSUCA and Cimencam are located in the south west and north east edges of the zone. To keep the population at bay, an enforceable measure should limit the residential zone to between 25 and 50 metres from industrial edge. This industrial edge has to be defined and scrupulously respected by the population to avoid industrial pollution. Forestland could constitute a 'green belt' between the residential and industrial territory. The elimination of water and atmospheric contaminants could be achieved through an industrial code of conduct for the treatment and disposal of industrial wastes. It is recommended that pit toilets on the sandy soils should be lined with plastic sheeting as suggested by Kabananukye (1994). This process could prevent compounds volatilised in the soil and gases as ammonium and nitrogen from contaminating ground water and plants.

CONCLUSION
Natural phenomena such as floods are, so far, beyond man's control. But his daily actions gradually contribute to global atmospheric and ecological changes, which threaten his very survival. In this major Cameroonian metropolis, the intervention of public authorities in urban development has been slow and less punctual. There is need for a timely intervention to protect marginal and fragile environments so that they can continuously accommodate the growing population of most third world cities. A progressive resettlement scheme away from endangered lands should constitute a long-term planning measure in the Douala Bonaberi region.

REFERENCES
Goudie, A., (1990). The Human Impact on the Natural Environment, 3rd edition, Basil Blackwell ltd., Oxford.

Kabananukye, I.B.K., (1994). Sanitation and garbage: Environmental Management in Kampala City, Makerere Univ., Kampala.

Keller, A. E., (1979). Environmental Geology, 2nd. Edition, A Bell & Howell Company, London.

Knapp, B. J., (1992). Systematic Geography, Collins Educational, London.

Kuete, M., (1998). Le Milieu de l'Ecosysteme Mangrove: Le Cas des Bouches du Cameroun, in JASS, vol. 1, No. 1, pp91-109.

Ward, R. C., (1978). Floods, A Geographical Perspective, Hasted.

Environmental hazards and landuse planning for sustainable development: The Limbe unstable coastal region, Cameroon

M. LAMBI, S.S. KOMETA, L.F. FOMBE, Department of Geography, University of Buea, Cameroon

INTRODUCTION

The strange blend of catastrophic events in the month of June 2001 in the coastal town of Limbe was extremely bewildering. Floods, mudflows, landslides, powerful marine waves stormed Limbe at the same time after a heavy and prolonged deluge. The year 2001, will thus remain imprinted in the memories of most Limbe inhabitants. By an unprecedented climatological anomaly, the rainy season of 2001 exhibited an ugly face which caused environmental instability that had never been seen in the past 20 years in this coastal city. These floods were the worst. On hand to remind the residents of this city that they live in an extremely high risk area were the alarming flood waters within the lowlying coastal areas, the strong and unprecedented marine erosion caused by oceanic waves that pounded like giant hammers on the reclaimed fragile coastal installations, and finally the alarming and numerous rapid landslides and mudflows which affected the anthropogenetically altered hill slopes. While the latter swept houses and property and took a death toll of 23 from the 24th June 2001 rainstorm, the excessive floods swept down houses, flooded homes and destroyed property and consequently created hundreds of environmental refugees. In much the same way, some of these seaside residents lost their homes as part of buildings were gracefully drifted into the sea by the sledge hammering effects of the oceanic waves. The sea waves and the floods from the heavy deluge ran wild. Limbe was caught in a frenzy. An if these tragedies were not enough, the high sea level rise which could be likened to the effect of tsunamis thus came to complete the already dismal situation of the suffering victims. It is needless to emphasize that floods of such magnitude rendered many inhabitants of this risk prone areas homeless. The simultaneous occurrence of these environmental hazards seem to give the impression that nature had conspired with all its elements to make livelihood difficult for an already threatened and traumatized urban poor who are constrained by their delicate economic status to settle in the vulnerable zones.

Sustainability – Planning and Management, Thomas Telford, London, 2002, 151–159

THE PROBLEM

For the coastal town of Limbe, several attractions brought people from the Cameroor hinterland in search of better economic opportunities. Among these attractions is mistaken notion the Limbe is Cameroon's "OPEC city." It was also mistakenly belie that the presence of the petroleum refinery and as a redistribution centre for petrole products for Cameroon meant that Limbe is a wealthy city with more econol opportunities. The consequences of these expectations were an unusual influx of pee into a city which neither has an adequate housing scheme nor minimum urban utilit Worse still, suitable land for settlement is extremely limited. In spite of these proble the population has squeezed itself into the poor urban shanties, most of which are loca either in the seasonally or permanently flooded (LUC N°1419) after some haphazard poor reclamation attempt.

Limbe with a small surface area confined between the surrounding volcanic hills and Atlantic Ocean was the home of 51.600 inhabitants in 1976 giving a population den: of 93.8 per sq km. The 1987 census, however, showed that the population had gro astronomically to 87.600 inhabitants giving a density of 147 person per sq km, bar eleven years later. This low-lying coastal plain does not only carry the congested bu up area, but also part of the Cameroon Development Corporation oil palm plantatic Thus as a dynamic economic urban centre in Cameroon's so-called "OPEC City" m: people were forced to encroach on marginal lands – the steep forested hill slopes of Mabetta–Towe hills known for catastrophic landslides, the low-lying season: inundated areas, and some permanently wetland zones (Fig.1).

Secondly, in terms of drainage problems and the hydraulic structures in place, increasing urbanisation in Limbe has had a negative impact on the environment. Inde the available urban hydraulic structures are poor. Some parts of Limbe suffer from poorly conceived scheme of drainage facilities while other areas of the same city hav total dearth of these structures. Here, it is has been observed that the extremely monsoonal climate, the natural waterways and the existing hydraulic structures can successfully accommodate the volume of surface run-off because of human interferen through increasing urbanisation. The average annual rainfall is 4050mm or thereab The rainfall ranges from a minimum value of 400mm a month from December February and attains a monthly maximum value of 700mm in the months of July . August. Such heavy rainfall should leave no room to chance or haphazard planning terms of evacuation routes (storm drains) for flood waters.

Furthermore, there are either limited and inadequate attempts or none whatsoever at l reclamation in the vulnerable flood prone areas prior to the construction of hous Consequently, floods remain yearly visitors at the doorsteps of homes located in

nerable zones. So the urban poor have knowingly and unfortunately elected to live
h these wetland problems.

E ENVIRONMENTAL SETTING

: site of Limbe remains pregnant with overwhelming environmental problems.
ierally, topographic lay out of Limbe shows a generalized basin-like structure with a
nber of topographic elevations. These are hillocks within the low-lying zone which
sist of Mbende hills, cassava Farm hill which connects with the Mabetta Hills,
:onut Island, and the Botanic Garden Hills.   Some of these hillocks rise to 130m
·ve sea level while the surrounding chains rise to spot heights of 295m in the Mabetta
ls and 362m spot heights in the Towe Hills.

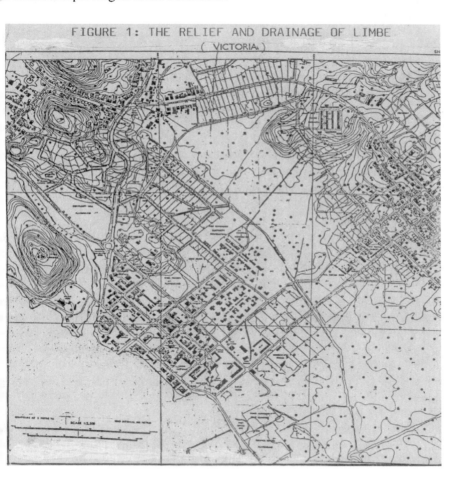

FIGURE 1: THE RELIEF AND DRAINAGE OF LIMBE
(VICTORIA)

As an ideal area for settlement, Bota has Limbe Camp, Camp Sic (real estate) and Bota GRA. On the eastern side is Gardens and Half Mile. Within the ensconced l lying zone, many parts of which lie 1–2m above sea level and some volcanic scoriace cones have been intensively colonized by settlements notwithstanding the risks which fragile, loose scoria and ash may present when disturbed by natural or artificial fact Against this natural hazard, man remains helpless.

While the relatively elevated topographic sites within the foothill areas of the hillc may be ideal locations for settlements because of their freedom from floods, the fra and loose nature of the scoriaceous cones make them vulnerable if intense seis activity were to occur. And Limbe lies on the Cameroon Volcanic Line characterisec seismic activity.

LIMBE FLOOD DISASTER ZONE
Without the present high degree of urbanisation way back in the 1960s, certain part Limbe, were however, known as disaster areas. Foremost of these is the Clerks Quart the Chruchs Street zone, Motowo Quarter, extensive parts of Unity Quarter New Lay and the Jengele Quarter otherwise known as "Crab Quarter".   These are remark wetlands which under the German and British administrations were left to nature. these were not suitable for human settlement, they were recognized non-habitable zo In the circumstances, these seasonally incendated areas posed no problem or challeng the inhabitants and municipal authorities of the then Victoria which was later re-bapt as Limbe following the Cameroonization of towns, divisions and the provinces of country.

This disaster zone has not been inhabited for long time. Most of those who ignora thought that they could settle and contain the yearly floods or put some devices in p to mitigate the effects of the floods abandoned the area when they realized that magnitudes of the floods were increasing with greater urbanization. Secondly, they a realized that their houses were constantly sinking. Overwhelming evidence in respect is prevalent everywhere. Houses thus affected by sinking have some of t windows now either at ground level or may be some 20–30cm above ground level. where do we put windows at ground level or less than 50cm above the surface. The dc are much shorter than the traditional 2m which is an ideal door height. Moreover, flood waters surrounding the houses are higher than some floors. Consequently, s floors are simply inundated during the entire rainy season. This is why the said disa zone is totally inhospitable for settlement.

The presence of causeways or topographic traps meant that not all the rivulets, stre and springs which flowed undisturbed by man-made structures lost some of t

ıways and were forced to stagnate forming rainy season semi permanent /transient
ɛs some of which have constant supply routes. The transient water-bodies have varied
ıensions as shown by table 1.

ıle 1: Locations, Dimensions and Height of Flood Peak.

Area/Name	Dimensions at peak	Flood peak height
ıurch Street	80 – 250m	1.0 – 1.3m
ıe Disaster zone	100 – 250m	1.75m
erks Quarters	80 – 300m	1.0 – 1.5m
otowo/Jengele	Very extensive	1.5 – 2.0cm
awo Quarter	Very extensive	1.5 – 2m

Source: Field Survey June –August 2001

ɛse exceptionally high flood peaks or levels in the flood-prone zones permitted the use
ɪ lying boats by the marines for rescue operations.

USE OF FLOODS

ırt from the effect of increasing urbanisation, deforestations on the Mabetta Hills,
ɪve and Livanda ridges, human encroachment on the river channel, the heavy and
ɪonged deluge in the months of June and July 2001, and poor drainage were the major
ɪors that accelerated and exacerbated the Limbe floods.

ɛ 1980s, the oil boom period was an era of prosperity during which an engineering
ɪpany built roads in Limbe. Many of these, however, were causeways across areas
ɪ were flooded each year. Church Street for example and the roads around Clerks
ırters are all causeways built from scoriaceous materials. These relatively elevated
ɪseways with thicknesses of 30–50cm of volcanic infills thus behaved in much the
ɪe way as topographic traps for periodic run off and natural water pathways which
ɪied the surface discharge into the sea.

ɪough drains were provided during the road construction, they were, indeed, too small
ɪhe first place and totally inadequate to carry the torrential run off for this coastal
ɪn. As some drains were quickly choked by debris infill, their average percentage
ɪctiveness was only 56.77 as shown in Table 2.

ɪeover, population growth came along with more accumulation of solid wastes. Like
ɪy developing cities, people with limited environmental education occasionally and
ɪdestinely dump (LUC N°1056/2) their refuse in drains. This illegal dumping of
ɪbage in the Jengele steam by some inhabitants has turned it to a pseudo-dump site

thus clogging water passage so that it flows with difficulty and sluggishly into the
This situation has been worsened by sedimentation on the stream beds which has b
responsible for raising the stream channels. These raised stream beds trigger overs}
during heavy downpours causing floods. The current problem of flooding therefor
largely human induced through increasing urbanisation, inadequate provision of dr;
and the poor disposal of solid waste.

Table 2: Dimensional Values & Effectiveness of Drains in the Limbe Municipality

City Location	Width of drain (cm)	Depth of drain (cm)	Thickness of Debris infill (cm)	Percentag Effectivene
Church Street (Rd)	52	62	55	11
Church Street (Ld)	52	62	58	6.5
Garden's street from Half mile	62	60	60	0
Garden Main street	66	104	10	90
Clerks Quarters Main street (Rd)	66	104	8	92
Clerks Quarters Main street (Ld)	66	104	8	92
Clerks Quarters Lane 1	70	60	7	88
Clerks Quarters Lane 2	70	60	9	85
Manga William Street (Left drain)	46	61	20	67
Manga William Street (Right drain)	50	53	20	62
Down Beach Main Administrative Road	66	104	15-20	83
Mean = $\sum x/11$	59.27	70.9	26.82	56.77

KEY: Rd (right drain), Ld (Left drain)

CATASTROPHIC LANSDLIDES
The balance sheet of the June 2001 Limbe landslide, unfolds the catastrophic natur(
this mass wasting phenomenon which caught its residents by surprise. The death to}
23, the total destruction of several homes which were drifted and buried down slope,

vy loss of property and the creation of hundreds of the Mabetta environmental
igees go to tell a dismal story.

th population growth came an increased pressure on land for settlement and
iculture which led to human encroachment on the steep hill slopes. In this, way, the
betta–Towe-Livanda hills fell prey to urban construction and peasant farming.

irough deep weathering of volcanic ash and scoria on the above massifs followed by
ious forms of human activity on the highly devegetated steep hill slopes had far
ching consequences. The felling down of its vegetation reduced the infiltration
acity of the soil. At the same time, the deforestation augered well for accelerated
sion so that the increased run off was a contributory factor for the June 2001 wet
son flooding in the low lying parts of Limbe.

e absence of a good vegetal cover implies that the binding property of soil particles on
steep and thoroughly weathered Mabetta–Towe Hills was reduced. The luxuriant
iical vegetation cover intercepts the heavy precipitation and the process of infiltration
:s not leave much water for overland flow. On the Moliwe –Bonadikombo massif for
imple, which enjoys a freedom from all forms of human interference (no homes,
ns, and no deforestation or timber exploitation), there are no landslides or other forms
slope failure. Although man sometimes opens up small crop farms inside part of this
est Reserve, this type of anthropic practice has negative effect on slope stability.
iere man has evidently pushed his luck too far in Limbe by extending settlements into
unstable hills, he has provided part of the trigger mechanisms for the occurrence of
dslides. The heavy deluge of June 2001 caused a super saturation of the regolith on
60°-70° slopes of these volcanic hills. The presence of oversteepened slopes created
vertical cuts for building, and the concentration of run off on such saturated and
icate slopes became good panaceas for the massive slope failure, which consisted of
eral landslides and mudflows.

OSPECTS AND OPPORTUNITIES

ian hydrology and geomorphology have never been taken seriously in Cameroonian
es. Even the major metropolis of Douala and Yaounde which should have received
ter attention leave much to be desired as they show obvious missing urban links which
ray our poor planning strategies or the lack of it. Indeed, the settled areas that have
wn into cities do not only require careful policy planning over time, but also require a
orous implementation of the strategic master plans. Perhaps by design or in error,
ne of the homes in the problematic areas actually had building permits.

The steep slopes of the scoriaceous cones and chains within the Limbe urban area are
fragile for settlement and other forms of human activity. Some of these are old volca
of pyroclastic mixed with ash which have undergone profound chemical weathe
which when lubricated and saturated by the heavy precipitation are often liabl
slumping, mudflows and landslides. Such highly fluid material steep slopes l
cohesion and flows out with the least disturbance which may range from earth tren
(seismic activity), overloading and the stripping of the overlying vegetation. But in
case of the recent multiplicity of landslides the excess and persistent rainfall w
caused over saturation and overloading was the motive cause of the slope failure. Th
delicate hilly environments and their steep slopes should be avoided since the bala
sheet for the last landslide was a rude shock to Cameroonians. Instead of hu
occupancy, the fragile hilly areas should be declared natural forest reserves in which t
resettlement and farming would be strictly prohibited. The flooded coastal lowland
which water at high tide rises above sea level warrant coastal protection. The importa
of sea defences along the Limbe (Victoria) coastline came from the Germans who t
stone embankments on those headlands on which their administrative offices w
located. The coastal urban dwellers need to feel secure that the voracious sea waves
not encroach on their real estate properties (Theobald, 1995). A proper coastal protec
scheme should be put in place since nothing else has been done after the period of
German occupation of Cameroon.

The problem of poor drainage for this coastal city should be given priority so that
drains be adequately revamped to carry run off of a monsoonal character. This calls f
total redesigning of the urban drains so that they are bigger and deeper. Furthermore
problem of proper waste disposal be addressed through sensitisation and environme
education. In the periodically flooded areas, proper reclamation should be don
rehabilitate settlement. Such rehabilitation process must make provision for an effic
drainage system in order to avoid pockets of standing water. The Motowo-Jen
quarter, a permanent wetland zone, should be left to nature since all indigenous dev
to combat floods have met with repeated failures.

The municipal authorities should embark on a resettlement scheme to avoid
anomalous concentration of huge populations into restricted and hostile ur
environments where they are exposed either to landslides, mudflows or floods in
wetlands. For the purposes of relocation, the Cameroon Development corpora
(CDC) a giant Agro-industrial concern should cede land to the municipality. Th
catastrophic phenomena are obvious indicators that the time has come when man mus
given priority over plants or crops. In this way our urban development would be pe
centred or socially oriented rather than the present economic/financial orientation w
man is merely used as a tool of production for profit maximisation. It is only throug

per resettlement scheme that we can avoid the repeated requiems in the future from
ironmental disasters in Limbe.

)NCLUSION
s the intensity of the natural processes at work and the human interference that stand
as the root causes of the recurrent Limbe disasters (Lambi, 1987; 2001).

:FERENCES
mbi, C.M. (1997): Human Interference and Environmental Instability. The Case of the
Limbe Landslide. Cameroon Geographical Review, the University of Yaounde,
Volume 10, No7, pp44-52.
mbi, C.M. (2001): Environmental Issues: Problems and Prospects. (Editor) Unique
Printers, Bamenda, Cameroon, pp133-146.
JC, File N°1419: Flood in Limbe, Limbe Urban Council.
JC, File N° 1056/Vol.2: Refuse Evacuation. Limbe Urban Council.
eobold, M. (1995): Strengthening the sea defences. Guyana.
In Courier, N° 150, March/April, pp 50-51.

Mapping of landslide-prone areas in the Saguenay region, Québec, Canada

D. ROBITAILLE, D. DEMERS, J. POTVIN, F. PELLERIN, Géotechnique et géologie, Ministère des Transports du Québec, Québec, Canada

ABSTRACT
The report shows an example of a recent mapping program performed in the Saguenay area, where more than 1 000 landslides occurred within a 36-hour period during the torrential rains of July 1996. The paper briefly describes the more common types of landslides in the clayey soils of the Quebec province. It also explains the different steps in producing various management maps. The accuracy of the digital tools allows the production of maps on a 1:2000 scale with a sufficient reliability, which facilitates their use.

INTRODUCTION
Following the St-Jean-Vianney disaster in 1971, where an earthflow caused the death of 31 people, as well as causing the destruction of 20 residences and the relocation of 200 others (Tavenas et al., 1971; Bergeron, 1983), the Quebec provincial government launched a program to map landslide-prone areas in a part of its territory. This program, which was spread over a ten-year period, resulted in the production of zoning maps on a scale of 1:20000 (Lebuis et al., 1983). From about the mid 1980s, these maps were integrated into the land-use plans prepared by the territorial administrative groups called "regional county municipalities" (RCMs). The mapped zones were to be managed through a normative framework suggested by the government but applied by the local municipalities. The standards applied to various activities within the zones identified on the maps, some of which were prohibited, while others were permitted under certain restrictions.

At the beginning of the 1980s, following a decentralisation policy, the mapping program was discontinued and technical support from the government was considerably reduced. During the implementation of the first versions of these land-use plans, problems relating to the areas prone to landslides received less attention. Nevertheless, over time the application of a normative framework provoked increased concerns while various deficiencies appeared. These included a scarcity of precision on the maps or their unavailability in some cases, the lack of rigour in the zoning criteria, as well as the lack of technical support, awareness and knowledge on the part of the various participants involved.

Instability – Planning and Management, Thomas Telford, London, 2002, 161–168

In 1996, Quebec's Saguenay region received torrential rainfalls that provoked more than one thousand landslides in less than 36 hours, causing the death of two children buried under the debris of a mudflow (Demers et al., 1999). Following this event, a new mapping program of the landslide-prone areas was started in this region and an interdepartmental committee was created to review Quebec's approach regarding landslide risk management.

The map-making underway in the Saguenay – Lac-Saint-Jean region became a pilot project allowing the application of results from the interdepartmental committee's consultations and the new computer and technological tools available, while at the same time developing an approach that is more focused on the needs of the users of future maps.

This report summarises the local geological context, covers the methodology used and sums up the practical solutions that had to be applied to the map-making problems so as to make them more accessible.

GEOLOGICAL CONTEXT

The majority of the population in the Saguenay region has constructed on clayey soils, deposited during the water surge following the melting of the Wisconsin glaciers about 8,000 years ago. When the waters were receding, the development of marine terraces and a drainage network created slopes reaching 40 metres in height, most of which are located on the banks of rivers and streams. The three most frequent types of landslides that affect these clayey slopes can be grouped into two groups: the first involves small surface area landslides that include superficial and rotational landslides, while the second type includes all the large-scale landslides, which we commonly call "earthflows".

The superficial slides occur in steep slope and affect the thin layer of the superficial altered clay. The debris of these slides spreads out quickly at the foot of the slope. These slides are very frequent and hundreds of them occur every year. Because their failure plan is not very deep, the surface area and volume involved are fairly small in comparison to the other two types of slides. Nevertheless, the impact of the debris can cause a great deal of damage.

Rotational landslides generally occur in slopes situated on the banks of rivers or streams where erosion is present. Their deeper failure plan provokes land loss at the summit of the slope. These landslides are very frequent along riverbanks, with more than a hundred occurring every year. A number of these are also triggered by man's poor practices, chiefly by inadequate grading work or bad drainage.

Earthflows of natural origin occur in sensitive clays ($I_L>1.2$ and $S_{ur}<1$ kPa), generally along riverbanks where an initial deep rotational landslide has occurred, which is a starting point for the retrogressive progress (figure 1). The clay of the Saguenay region is characterised by its very high sensitivity ($S_t=S_u/S_{ur}$), measures of which often exceed 1000. In some areas, the clay has extremely low remoulded shear strength and behaves much like a liquid that is more or less viscous. In some cases, almost all of the clay liquefies during a landslide and the crater can be completely emptied, as was the case in Saint-Jean-Vianney (Potvin et al., 2001). The retrogression distance at the summit of the slope can vary anywhere from several metres to more than a kilometre, and affect a surface area of several hectares. These landslides generally occur within a matter of minutes. Despite their infrequency (1/year), they constitute a major risk due to their magnitude and their destructive nature.

Figure 1 : Earthflow that occured in 1983 on the shores of Lake Saint-Jean. The slope is 20 meters high.

DESCRIPTION OF THE MAPPING METHOD
Map production of the areas prone to landslides is carried out by superimposing different layers of information. During the first stage, we proceed to identify the nature of the bedrock and its topography, as well as the nature of the soil at a depth that is at least equal to the height of the highest slope in the region. This operation starts with the interpretation of aerial photos, compilation of drilling and borehole data, as well as field observations. This first map is called the **geological map** or **basic map**.

The geomorphologic elements are found in the second layer of information. An inventory is made of the scars left behind by major landslides and the geographical position of all other minor landslides, which are collected from a variety of aerial photographic coverage, archives or eyewitnesses. Areas of erosion, the leading natural factor in the weakening of a slope, are also identified using aerial photos and field observations. Geomorphologic elements that can play a stabilising role such as alluvial plains are also compiled, as are other elements that represent obstacles to the development of large landslides, such as secondary gullies.

All slopes measuring 5 metres and more, with an inclination of 14° or more, are mapped and classified into two categories. The "**steep slopes**", with a inclination equal to or greater than 22°, are slopes where a landslide can occur naturally, as has been demonstrated in the detailed description of the 350 landslides that took place in the region in 1996 (Perret et Bégin, 1997). The slopes with a inclination between 14° and 22°, called "**moderate slopes**", are also mapped because they can be affected by failure if man's harmful activities such as excavation at the foot of a slope or landfill at the slope summit are not controlled. Relief maps with contour lines spaced out from 2 to 3 metres (scale 1:10000) are usually used in the zoning specifications. When more precise topographical surveys are necessary, such as in urban areas or very rugged terrain, we can use 1:2000 scale topographical maps where available, or airborne laser surveys if need be.

Still on the same layer of information, the intensity of gullying is indicated by symbols and is divided into four categories: gullies that are 5 to 10 m high, those that are 10 to 20 m high, those that are 20 to 40 m high and those that are higher than 40 m. This allows for a quick visualisation of the slope height and is a means to simplifying the subsequent map-making process. Correctives measures, like riprap, are also indicated. The ensemble of this data gives a landslide **susceptibility map**.

The following step, which leads to the production of a **hazard map**, requires the interpretation of all the information previously collected. For each type of landslide, a degree of susceptibility is associated, which depends on the geological, geomorphological and geotechnical characteristics of the site and the soil. In our case, five hazard classes have been identified, which are based on the precursory work of Lebuis et al. (1983).

The first three classes permit us to classify the zones by the danger that small surface area landslides represent. The first class, called the high susceptibility zone (**HSZ**), groups together slopes with a steep inclination (≥22°) and which present major signs of instability or being subject to erosion. In these zones, the stability of the slope degrades continuously and the occurrence of a landslide is strongly possible without, however, the possibility of foreseeing the event. The second class, called the moderate susceptibility zone (**MSZ**) also includes all the slopes with a steep inclination (≥22°), but which show neither signs of instability nor active erosion. In these zones, the stability of the slope does not degrade with time, but may vary according to meteorological conditions. A combination of specific factors (e.g.: abundant rains and rapid spring thaw) or a major event (e.g.: rains in July of 1996 in the region) can be the triggering factors. The occurrence of a landslide is therefore linked to the probability of such conditions occurring. Finally, the last class, called the low susceptibility zone (**LSZ**) groups together slopes of a moderate inclination ($14° \leq \beta < 22°$), where no landslide of natural origin should occur, unless provoked by an exceptional seismic activity of an extreme magnitude. Although they are stable in the long term, development on these slopes or in their proximity could provoke failures. That is why particular precautions are necessary when using these lands. For these three classes, the hazard zone includes the slope itself and strips of land at the summit and at the base. These strips can be affected by landslides and are sensitive to any human activity that could constitute a triggering factor.

The last two classes identify the areas where large-scale «earthflow» landslides could occur if the necessary conditions were to arise. In general, the land involved is more or less flat and is located in the back of previously identified slopes.

The earthflow-exposed zone (**EEZ**) includes a strip of land located at the summit of the slope, exclusively behind the **HSZ**. The presence of sensitive clay is characteristic of this zone. A deep rotational landslide in the slope could initiate retrogressive movement. At this site, evaluation of the probable retrogression distance is based on the comparisons of adjacent earthflow scars (Lebuis et al., 1983).

Lastly, the hypothetical earthflow exposed zone (**HEEZ**) includes a strip of land situated at the summit of the slope in the back of the steep and moderate slope terrain. It is characterised by the presence of sensitive clay. A major retrogressive slide could not occur here unless exceptionally unfavourable conditions were to arise. The evaluation of the probable retrogression is based on the size of the largest earthflow scars in the sector concerned. The identification of this zone, where development constraints are minimal, provides, nevertheless, a better understanding of this territory for the management of any major emergencies or disasters.

The use of aerial orthophotography as a planimetrical tool is useful in the regional evaluation of possible risks affecting the territory based on the hazard map. Furthermore, it is an indispensable tool for the management of emergency situations.

The last step is used in the production of what is called the "**development constraint map**", which is basically intended for land-use managers (figure 2). It is designed especially to assist with the application of municipal regulations concerning the landslide-prone areas. The zoning, indicated by an alphanumerical nomenclature, is associated with a series of constraints to be respected by land users. It uses a less alarmist key that the hazard map and it is better adapted to neophytes.

In land where the failures are controlled by the presence of clayey soil, three classes of constraints are identified. Class **A** includes all zones susceptible to small surface landslides. This first category is subdivided in two sub-classes according to the geometry of the slope. All steep slopes ($\geq 22°$), whether high or moderate hazard, are grouped together under the heading of **A1** and are regulated more severely because of the danger of natural occurring landslides. Moderate slopes are grouped together under class **A2**, less restricting, since only man-made activities are susceptible to triggering landslides. For the **A1** and **A2** categories, the zones regulated include the slope itself and the strip of land situated at the summit and at the base of the slope. At the summit, the width of this strip is equal to two times the height of the slope (2H), to a maximum of 40 metres. At the base, the width of this strip is equal to two times the height of the slope. Taking into consideration the height of the slopes in the region, the width of these strips of land can vary from 10 to 80 metres. The dimensions of these strips have been fixed empirically based on the inventory of landslides in the southern part of the Quebec territory.

Representation of the **A1** and **A2** types on the map is a challenge in terms of precision because of the small surface area that they occupy. Using the 1:20000 or even 1:10000 scale, it is impossible to have reliable and usable limits for a parcel of land for such use as issuing a building permit for example. These scales are used more for planning land use in general. The use of aerial numerical orthophotography allows us to obtain a very good representation of planimetrical elements on a 1:2000 scale based on photos using a 1:40000 scale, with a deviation of not more than 4 metres. If necessary, an increased precision can be obtained from the photos on a larger scale (1:15000 for example). However, in order to compensate for all the errors possible (imprecision of the methods or tools available) in determining the width of the controlled strips, we use the maximum value in the slope height category such as mentioned in the steps of the susceptibility map. As an example, for a slope that is 6 metres high, the width of the controlled strip identified on the map will be 20 metres rather than 12 metres (table 1).

Table 1: Effective width of controlled strips used in mapping.

Slope height categories (m)	Controlled strips width (m)	
	Top of the slope	Bottom of the slope
$5 \leq H \leq 10$	20	20
$10 < H \leq 20$	40	40
$20 < H \leq 40$	40	80
>40	40	2H

Class **B** is associated with large-scale landslides. Its representation on a map poses less problems of scale since it always involves zones that attain 100 metres or more. Classes **B1** and **B2** correspond respectively to classes **EEZ** and **HEEZ** on the hazard maps. In general, the limits of these zones are slightly enlarged to include the corresponding hazard zones while following the elements that are easy to spot on the map (lot lines, roads, etc.). It is thus easier for the land-use manager to apply these limits in the field. The "B" zones are still situated behind zones **A1** or **A2**, at a distance from the slope where most of such activities as landfilling have no direct effect on slope stability. Nevertheless, constraints similar to those of the zones with steep slopes (**A1**) apply for class **B1** with regards to activities where human life can be at risk (habitable buildings or prolonged stay sites). For class **B2**, only certain types of buildings (public buildings, factories, etc.) must meet with various standards.

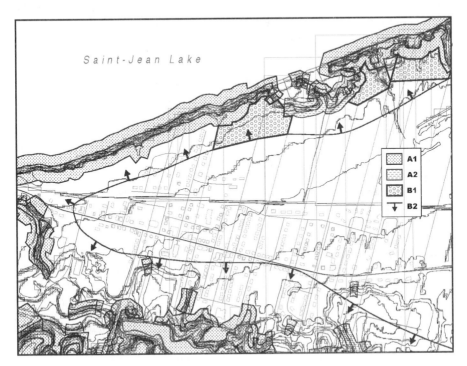

Figure 2: Development constraint map intended for land-use managers.

The method developed allows us to obtain maps of land development constraints, the use of which is intended primarily for the prevention of risks associated with landslides. They can be used easily by the land-use managers with regards to parcels of land (scale of 1:5000 in a rural setting and 1:2000 in an urban area), which meet the essential requirements of municipal needs.

Detailed description of the new normative framework developed by the interdepartmental committee will be the subject of a future publication.

CONCLUSION

The Quebec experience over the last 25 years, such as revealed by the interdepartmental committee, demonstrates that map production is essential in ensuring better risk management with regards to landslides. However, the production of zoning maps does not constitute an assurance of good management. Hazard maps must be accompanied by development constraint maps, which represent a popularisation for the neophytes. These development constraint maps must be produced with usable scales and not only at the regional level, but also at the level of a parcel of land. Using computerised and high technology products, we are now able to obtain satisfactory resolutions. However, it is still necessary to use map-making devices in order to determine safety perimeters in zoning so as to compensate for inaccuracies inherent in the regional approach. Moreover, the key must be easy to understand and ideally, alarmist terminology should not be used. Technical support for the various participants and awareness programs will have to accompany any emission of maps. Finally, the interdepartmental committee's analysis indicates that it is impossible to ensure map uniformity without the presence of a central organisation that would also assume the responsibility for any criteria used in zoning.

BIBLIOGRAPHY

Bergeron R., 1983. *Introduction historique au phénomène des coulées argileuses québécoises.* Ministère de l'Environnement du Québec, internal document, 26 p.

Demers D., Potvin J. and Robitaille D., 1999. *Gestion des risques de glissement de terrain liés aux pluies des 19 et 20 juillet 1996 au Saguenay – Lac-Saint-Jean.* Report submitted to the Bureau de reconstruction et de relance du Saguenay – Lac-Saint-Jean.

Lebuis J., Robert J.M. and Rissmann P., 1983. *Regional mapping of landslide hazard in Quebec.* Symposium on soft clay slopes. Linköping, Sweden, Report No. 17, Swedish Geotechnical Institute, p. 205-262.

Perret D. and Bégin C., 1997. *Inventaire des glissements de terrain associés aux fortes pluis de la mi-juillet 1996, région du Saguenay / Lac-Saint-Jean.* Institut national de la recherche scientifique (INRS-Géoressource). Report for the Bureau de reconstruction et de relance du Saguenay – Lac-Saint-Jean.

Potvin J., Pellerin F., Demers D., Robitaille D., La Rochelle P. and Chagnon J.Y. 2001. *Revue et investigation supplémentaire du site du glissement de Saint-Jean-Vianney.* Proceedings of the 54th Canadian Geotechnical Conference, Calgary, vol. 2, p. 792-800.

Tavenas F., Chagnon J.Y. and La Rochelle P., 1971. *The Saint-Jean-Vianney landslide: observations and eyewitness accounts.* Canadian Geotechnical Journal, vol. 8(3), p. 463-478.

The Franklands Village Landslip, West Sussex

M. W. STEVENSON, Southern Testing Laboratories Ltd., East Grinstead, West
Sussex, UK, and
D. J. PETLEY, School of Engineering, University of Warwick, Coventry, UK

SUMMARY
A catastrophic landslip developed in December 1993/January 1994 within a housing
development in Reed Pond Walk, Franklands Village, near Haywards Heath, West
Sussex, UK. The landslip covered a plan area of approximately 4600 m^2 and resulted
in the demolition of 14 flats and houses, the destruction of the road and the fracture
and loss of main services.

This paper describes the development of the landslip and discusses the investigative
methods and remedial measures undertaken. Reference is also made to the measures
taken to ensure the safety of the public during the emergency which developed, and
the need to respond to world-wide media interest.

SITE HISTORY
Old Ordnance Survey maps of the area show the site to be wooded and undeveloped
in 1875, but by 1899, a "gravel pit" had been excavated up slope, to the immediate
south east of the landslip which developed in 1993/1994. The houses in Reed Pond
Walk were constructed in the 1930's as part of a scheme to provide work for the
unemployed and the provision of low cost rental accommodation. By 1993, the
dwellings were owned by Franklands Village Housing Association with the
exception of two privately owned houses (Nos. 110 and 112). Vacant land on the
higher ground to the south east which included the former pit area was developed by
another party in the late 1980's and completed in 1990.

TOPOGRAPHY
The position and setting of the site is indicated in Fig. 1. The original development
comprising two storey flats and detached houses is set on terraced valley slopes
which fall towards a stream in the floor of the valley in the north west. Overall
slopes are of the order of 13° increasing to 16° in the lower wooded area to the
north. Terracing of the valley slope to facilitate development resulted in low,
localised steepening up to 30 to 45°. The crest of the valley lies to the rear (east) of
the houses with the difference in level from the crest to the toe of the slope being
approximately 22 metres.

In 1985/86, the stream had been partially culverted in the southern part of its length in works associated with slope stabilising measures on the western (east facing) valley slopes. On the properties ultimately affected by the 1993/94 landslip, the gardens were generally well tended and laid to lawn with dividing hedges. There were occasional small trees with some mature oak at the toe and in the more wooded undeveloped area to the north and the southeast.

GEOLOGY

The 1:10560 Geological Map of the area shows the solid geology to be within the Hastings Beds of the Lower Cretaceous Wealden Series. The Hastings Beds were uplifted by Tertiary folding which has lead to their exposure in the anticlinal Wealden dome. This uplift has resulted in geological faulting which has strongly influenced the extent of the subsequent landslip.

The solid geological structure is:

Upper Tunbridge Wells Sand.		Silts thinly bedded, sandstone and mudstone.
Grinstead Clay	Upper Grinstead Clay	Clay-shales and mudstone.
	Cuckfield Stone	Fine calcareous sandstone and mudstone.
	Lower Grinstead Clay.	Clay shales and mudstones
Lower Tunbridge Wells Sand	Ardingly Sandstone	Massive fine sandstone.

This is a succession typical of the High Weald where the thicker sandy parts of the sequence were deposited on broad alluvial flats cut by meandering distributing channels while the clays were deposited in shallow lobes which were at times locally traversed by distributary channels bordered by sandy levee (Gallois and Worssam, 1993). The solid geology has been subjected to periglacial conditions during the periods of the last ice ages and their geotechnical properties have been significantly influenced by solifluction, landslip, cambering, valley bulging and cryoturbation. In the Weald, the solid succession is commonly obscured by the resulting downwash and/or displaced by cambering and landslip.

The geological map indicates a shallow southerly strata dip, but subsequent investigations revealed a north-westerly dip of 8 to 10 degrees. The succession is cut by a geological fault (The Sandrocks Fault) on the lower slopes downthrowing the Upper Tunbridge Wells Sands to the northwest. The fault trends north-east to south-west. To the north of the site the Sandrocks Fault is cut by the Abbots Leigh Fault which trends north west to south east and again downthrows the Upper Tunbridge Wells Sand, but this time to the north east. A further unnamed fault was encountered to the immediate south of the site, trending east west and effectively delineating the area at risk from landslip. These geological faults are indicated on Fig. 1

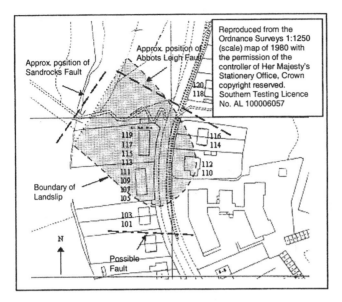

Figure 1: Site plan

SOIL CONDITIONS ENCOUNTERED

Initial site investigations (performed in early 1993) were restricted by the need to gain access within well-maintained gardens and to minimise disturbances to residents. Boreholes were undertaken using a specialist, breakdown, restricted access shell and auger rig while the trial pits were excavated by hand methods. Following the landslip, further investigations were made. Locations of all the trial pits and boreholes are shown on Figure 2.

The succession of soils encountered in the boreholes and trial pits reflected the anticipated geological succession and faulting. Some made ground was found, being less than 1 metre in thickness and associated with the terracing and development of the road and properties constructed on the site.

Head deposits were encountered primarily in the northern part of the site where up to 2.6 metres was found. This appears to represent a head infilled valley which trends towards the floor of the valley, possibly associated with the geological faulting on the site. The head thins towards the southwest. A thin covering of head was associated with the Tunbridge Wells Sands to the extreme north, east and south of the site.

Within the fault bounded area, the full strata sequence was encountered to the east of Reed Pond Walk, where the basal members of the Upper Tunbridge Wells Sands were exposed. Downslope the Grinstead Clay outcrop was masked by soliflucted and previously landslipped clay. The Cuckfield Stone is a useful marker horizon and was found to outcrop beneath the slipped clay on the lower valley slopes in the rear gardens of flats numbers 105 to 111. The limestone was not recorded in the post slip investigation in borehole A, possibly because of lateral variation or even as a result of landslipping. The underlying Ardingly Sandstone was proven in the valley floor.

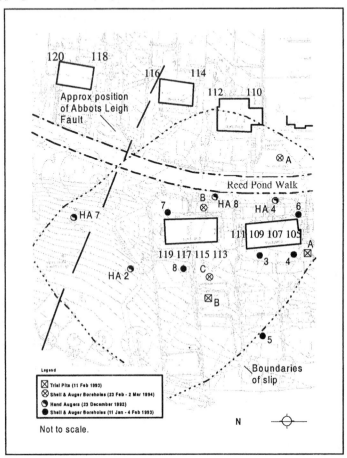

Figure 2: Locations of pits and boreholes

DESCRIPTION OF THE LANDSLIP

As stated earlier, the main development of houses in Reed Pond Walk was made in the 1930's. Some further development, consisting of 2-storey flats, occurred in the former pit area to the south east in the 1980's and 1990's.

The first indications of potential problems occurred on 20 March 1991, when problems with drains and some cracking were reported from the occupants of No. 119 Reed Pond Walk.

On 8 December 1992, following rainfall, cracking was reported in Numbers 110/112. On the following day (9 December 1992), cracks were reported in Numbers 105 to 119. A site visit by personnel from Southern Testing Laboratories Ltd. identified the first clear evidence of movement. In the rear gardens of flats numbers 113 – 119, a tension gash in the form of a 180 mm scarp was found, trending southwest-northeast, and a bulge downslope was also identified, reflecting a shallow localised slip. In the footpath to the south of block 105 – 111, cracking showed vertical movement of 65 mm and horizontal, north-westerly, movement of about 30 mm. Similar movement was evident on the damp course of the adjoining properties, where the front of the property appeared to have settled relative to the rear of the property. Upslope of Reed Pond Walk, tension gashes were also visible in the garden paths. A trial hole excavated through one of the tension gashes, showed an open gash (10 mm wide) bearing 338 to 194 degrees. This movement agreed with the apparent forward rotation of the adjoining property (no. 112). This movement appeared not to increase through the drier months of 1993, but further movement was observed in October 1993.

As a consequence of the movements observed in late 1992, a site investigation was undertaken by Southern Testing Laboratories Ltd to identify the extent and nature of the landslip. The investigation was completed in April 1993. The evidence at that time suggested relatively shallow rotational movement on the lower slopes with a lateral "wedge" slip upslope of Reed Pond Walk. Movements were relatively small. Analyses indicated that further movements could be expected following periods of heavy rainfall. A series of remedial measures were proposed at that time, which would have necessitated the temporary removal of the occupants of two houses and two blocks of flats and the rehousing of the occupants. No remedial measures had been performed however before the onset of further substantial movements later in 1993.

The first signs of these movements occurred on 18 October 1993 when cracking was found in the property at No.110/112 Reed Pond Walk: these cracks were 25 mm wide vertically and laterally. Some slight cracking was also found in the lower flats (Nos. 105 – 119). Additionally, a movement of 40 mm was identified in the gas pipe in the road.

The movements then developed rapidly. In early December 1993, the occupants of properties Nos. 105 – 119 were evacuated. By 23 December 1993, distortion of 100mm was apparent in Nos. 110 and 112 and the joints in the road had opened by about 150 mm. There were also signs of the development of a graben feature in front of the recently completed flats with the formation of a backscar some 100mm high.

On 24 December 1993, there were clear signs of the development of a large failure and the decision was taken to evacuate the occupants of the houses at 101, 103, 114 and 116 Reed Pond Walk.

By 26 December 1993, horizontal movements in the joints in the road had reached 300 mm. The graben was more apparent , trees were exhibiting clear backwards tilt, and the road was sinking by 300mm. There was a developing bulge at the bridge/culvert with upwards movements of some 100mm, and the path alongside the flats Nos. 105 – 111 to the south had dropped by about 100 mm. The upper house (Nos. 110 and 112) was ᴄlose to collapse.

On 28 December 1993, the house (110 and 112) was demolished.

By 30 December 1993, the graben was clearly identified with a back scar some 400mm high in front of the new flats.

By 1 January a back scar with about 1m vertical movement had formed in the wooded area. The sewer in Reed Pond Walk was broken and replaced by a flexible surface pipe.

Over the next 3 weeks, the movements developed into a substantial failure. Surface movement was towards the north west with lateral movements of some 4 m and lateral downwards movements of 1.2 m in the upper valley slopes, and lateral movements of 3 m and an upward movement of between 0.2 and 1.3 m in the toe of the slopes. The back scar in front of the new flats reached a maximum of 4 m in height with a subsidence graben 6 m wide in front of the scar. These severe movements fractured and severely damaged all services and access roads and paths. Flats Nos. 105 – 119 were demolished on 10 January 1994.

The movements comprised the forward movement of a large sliding block, together with a deeper seated rotational movement whose toe bulged upwards at stream level. The dimensions of the slip were strongly influenced by the geology, with geological faulting bounding the slip to the north and in the north west corner where the toe bulge moved upslope to the north of the footpath in the wooded area. In the lower valley slopes, to the south west (flats nos. 105 – 111), the outcrop of the Cuckfield Stone, and the position of the original culvert, appear to have controlled the shape of the southern slip boundary.

A photographic record of the landslip has been given by Stevenson, 1994.

EMERGENCY REMEDIAL MEASURES
The onset of the large scale movements in December 1993 coincided with the Christmas and New Year holiday period. As stated previously, the occupants of the main blocks of flats positioned within the slip mass had been moved to alternative accommodation in early December with the evacuation of a further four houses immediately prior to Christmas. Substantial distress occurred in the road making it impassable and re-routing of emergency access for the emergency services (fire, ambulance, etc.) was arranged. A temporary sewer was routed over the surface of

the slip. These works were undertaken by the Housing Association, their Engineers (K. A. Marshall and Partners) and the Statutory Authorities. Timely and safe actions ensured that there were no losses of life or injuries during this period.

Following discussions and agreement with the insurers and other interested parties, emergency remedial measures were installed as follows:

1. Diversion of surface water inflow.
2. Blocking and diversion of all drainage off-site.
3. Demolition of three housing blocks (nos. 105 – 119, 110 and 112)
4. Construction of temporary hardcore access roads for plant.
5. Installation of three 2 to 4 m deep counterfort drains using a designed filter medium and a geofabric wrapped slotted pipe, discharging into the stream.
6. Construction of a shallow drain at an observed spring line.
7. Construction of a toe bund constructed from hardcore spoil obtained from the excavations for the drains and placed on a sand blanket.
8. Installation of monitoring piezometers and slip indicators.

The cost of these emergency measures was approximately £50,000.

These emergency measures were installed during January and February 1994 and successfully arrested the movements on the site.

CAUSES OF THE FAILURE

The causes of the landslip have been discussed by Walbancke et al., (2000). Briefly, as a part of the development carried out in 1990/1991, a storm water balancing tank was constructed for surface water runoff. The tank was 46 m long, 1050 mm internal diameter and constructed with concrete pipes. The pipe trench was about 3.3 m deep and a granular bedding material was placed around approximately two thirds of the pipe periphery. There was no provision for draining the granular bedding material. The granular material formed a focus for the accumulation of water and acted as a reservoir feeding the downslope flow and increasing the pore water pressures throughout the slope but particularly in the lower parts.

PERMANENT REMEDIAL MEASURES

Following the failure a detailed site investigation was performed during March and April 1994. The aims of the investigation were to identify the extent and causes of the landslip, and to obtain information for the design of permanent remedial measures.

The permanent remedial measures comprised:
1. Culverting of the stream and extending the toe level over the full length of the slip area.
2. Culverting other site levels and raising them by 4 m, with the slope then regraded to 10°.
3. Installation of additional counterfort drains.
4. Prevention of discharge from the granular fill around the storm water balancing tank.

These measures were completed in August 1994, and allowed the reconstruction of the road and rebuilding of properties on the site.

MEDIA INTEREST
As the main slip developed in late 1993, local residents contacted a local radio station. The slip coincided with the occurrence of widespread flooding in southern England and quickly attracted the attention of major television networks. The questions raised by the media ranged from the simple (e.g. is it the biggest ever), to the more serious (e.g. is it related to climate change). In view of potential liabilities, great care was needed in dealing with the media to ensure accuracy with regard to safety etc., but not to place blame. The slide was reported on national TV news in the UK, on one occasion even being placed before an item concerning President Clinton.

Ultimately the landslide was reported world wide including Australia and North America, and resulted in an influx of visitors who were difficult to keep out of the slip area. One entrepreneurial local resident set up a stall to sell souvenirs!

ACKNOWLEDGEMENTS
The authors wish to acknowledge the contributions of the following in connection with the Franklands Village Landslide:
Franklands Village Housing Association, and in particular the late Mr. P. Reeves;
Mr. Ken Marshall of K. A. Marshall and Partners;
Mr. R. Coe and Dr. H. J. Walbancke of Binnie Black and Veatch;
Mr. M. Scott and Mr. D. Vooght of Southern Testing Laboratories Ltd..
We are also grateful to Mr. D. Spearman of Southern Testing Laboratories Ltd. for assistance in the preparation of the figures.

REFERENCES
Stevenson, M. W., 1994. The Franklands Village Landslip. *Quarterly Journal of Engineering Geology,* 27, 4, 289 – 292.
Walbancke, H.J., Stevenson, M.W. and Coe, R.H., 2000. The effect of a surface water attenuation facility on the stability of a slope. Landslides in Theory and Practice, ed. by Bromhead, E., Dixon, N. and Ibsen, M-L,. Thomas Telford, London. 3, 1527 – 1532.

Instability relevance on land use planning in Coimbra municipality (Portugal)

A. O. TAVARES, BSc, Assistant Professor, University of Coimbra, Portugal
A. F. SOARES, DSc, Full Professor, University of Coimbra, Portugal

ANTHROPHIC OCCUPATION

The study area, located in central Portugal, is of about $316Km^2$ and has been characterized during the last decades by important development of urban and outer urban occupation, centered in Coimbra city, and by the crossing of road infrastructures.

The comparative analysis between the anthrophic occupation areas in the early 1980s and the results of the land-use planning, defined in 1993, shows an increase from $42Km^2$ to $95Km^2$. All the areas with residential use or in reserve as well the industrial plants and current or projected infrastructures and equipments were taken into account for this evaluation.

This evolution caused a marked change in the relation with the physical factors, in the two temporal sets, as illustrated by:
-   The progressive development and building in areas with severe slopes, with the major relative amount in areas characterised by 8-16% and 16-26% slope steepness;
-   The new relevance on lithological units occupation, namely on Jurassic marl-limestone alternations and on Cretaceous outcropping (sandstones, mudstones and limestones), in opposition to the traditional units occupation (Triassic conglomerate and sandstone and Lower Jurassic dolomitic-limestones) which present a declining relevance;
-   The progressive remoteness between the anthrophic areas and the permanent water streams (stream order >=3, based on Horton's numbering), which express the increasing in slopeness occupation and indirectly the actual conditioning policy in agricultural and natural areas;
-   The translation into areas characterised by growing levels of instability associated with the geomorphic processes, according to the methodology explained below.

PHYSICAL CHARACTERISATION

The global physical analysis set off the contrast and the variation on the lithological units, structural and morphological features, drainage setting and hydrological nature.

Twenty-five lithological units were identified, divided into substratum and superficial representation, and mechanically characterised. Table 1 contains the synthetic compilation of all the units and the outcropping areas.

Table 1. Lithological units representation

Lithological domains			Lithological units	Km^2	Lithological domains			Lithological units	Km^2
Phyllites, greywackes, shales and quartzites	Precambrian Ordov./Silurian	Substratum	1	50,1	Limestones and marl-limestone alternations	Lower and Middle Jurassic	Substratum	10	24,5
			2					11	
								12	
			3					13	
								14	
								15	
								16	
Conglomerates, sandstones and mudstones	Carboniferous Upper Triassic		4	36,9	Sandstones, mudstones, and limestones	Cretaceous Tertiary		17	68,1
			5					18	
			6					19	
			7					20	
Dolomitic limestones and claystones	Upper Triassic Lower Jurassic		8	40,6	Sandy-conglomerate deposits, tuffs and travertines, peats	Plio-Quaternary	Superficial	21	96,3
								22	
								23	
			9					24	
								25 (alluvium)	(64)

The studied area is represented by a thrust-system, uplifted to the east, and gentle dipped and depressed to the west, by N-S faults. Other faults with strikes NE-SW to ENE-WSW, NW-SW to NNW-SSE and E-W generate blocks with a noticeable thickening of the lithological units. The strike sets have a strong topographic expression, which impose different values on drainage density and affect the infiltration rates. The structural feature also controls the properties of the rock materials with strength degradation, fracture intercept closeness and deep weathering penetration.

The east uplift of the metamorphic units, with slopes steepness often > 26%, has historically produced the repulsion of the human occupation. The anthrophic areas have been specially developed on the gentle dipping monocline defined on Upper Triassic conglomerates and sandstones or on the Lower Jurassic dolomitic limestones.
Another physical aspect that has been conditioning the human fixation and the urban growing is Mondego river (average annual flood $2500 \times 10^6 m^3$) characterized by

large sediment loads and flood hazards which compelled to engineering floor-protection works with channel improvements and multipurpose storage reservoirs in the 70s. In the urban area of Coimbra the Mondego river channel changes from an upstream incised dip valley to a downstream large valley floor (maximum wide over 4800m, and deposits thickness over 40m) (Tavares, 1999).

Instability processes

Instability processes related to the mass movements have been historically observed and described, according to Varnes (1978) and Cruden & Varnes (1996) classifications. These processes have achieved special relevance with the recent anthropogenic occupation. This relevance is related to the increasing of the instability manifestation numbers, the displaced mass materials and the economic and social losses.

Evidence of falls and translational slides is recognized in most of the substratum units but especially on metamorphic rocks (close to major faults and folds or in the relativity narrow zone of magmatic intrusions) and dolomitic limestones (within pelitic lenses, major faults and sinkholes filled by Red Sands with moderate plasticity). The triggering factors of these movements are chiefly the slope cutting due to road and building excavation.

Rotational slides in different states of activity (WP/WLI UNESCO, 1993a; WP/WLI UNESCO, 1993b), are essentially recognized by involving the marl-limestone alternations (units 10, 12 and 13) and by claystone-dolomitic limestone alternations (unit 8), usually associated with faults. The slope angles of the affected areas have a large range values (8-16%; 16-26%, >26%), with a frequently NW-NE aspect. The causal factor is mainly the human action with cuts in potential unstable slopes for road and house construction, vegetation cover alteration, and drainage or infiltration capacity change. The recent evidence has been mapping in outer urban areas, to the north and southwest of Coimbra city, due to the increasing occupation pressure in the last decades. Rotational slides, inactive relict to active (op. cit.), are also locally mapped and observed on Cretaceous sandy-mudstones (unit 19), which present moderate to highly plastic namely in conjunction with fault pattern. Bank river erosion and artificial cuts are recognized as triggering factors.

Lateral spreads are observed on fractured dolomitic limestones ($L_{45}F_{45}$, ISRM 1981) alternate with moderately plastic clays (thickness layers between 1-180cm) and on limestones ($L_{45}F_{45}$, op. cit.) alternate with moderately plastic marls – units 10, 12 and 13 - (thickness layers 10 to 250cm). These movements are related to average slope angle ranging between 8-20%, NW-NE slope aspects, and bare or sparse vegetation cover. The main triggering factor of these movements is the rainfall.

The debris and earth flows occurrence map suggest that there are three main lithological domains involved which have typical field drainage and morphological slopes: (1) the metamorphic units when moderately to highly weathered (W_{34}) and associated with slopes over 26% and gully erosion; (2) the interbedded sandy-mudstones of the Upper Triassic (unit 7) with low vertical permeability and correlated with slopes over 16% and frequently with NW-NE slope aspects; (3) the Plio-Quaternary sandy-conglomerate deposits specially on Red Sands – unit 21 - (with moderate plasticity and low compactness), on colluvial/slope deposits (unit 24) and on anthropogenic fills, usually defined on an open slope and enters a gully or initiates in a gully (according Fannin & Rollerson, 1993, methodology). The main triggering factors are the rainfall (specially when seasonal water courses were changed), the anthophic slope cuts and the plant cover destruction.

The gully erosion has special incidence on the Jurassic tiny marl-limestone alternations (unit 12) and on calcareous marlstones (unit 14), which present slopes under 16% and have special development within S and W slope aspects. These processes are also observed on the cretaceous White Sandstones (unit 17) with low superficial compactness and on sandy-mudstones (unit 19) with strongly vertical permeability variations.

Karst subsidences and sinkholes, frequently filled with Plio-Quaternary materials, have been identified on Jurassic units as: dolomitic limestones (unit 9), marls and marly limestones (unit 10) and calcarenites to calcirudites (unit 16). Karst forms have also been recognized on the superficial tuffs and travertines (unit 23). Associated with these forms, falls and translational movements related with old quarries and recent anthophic slope cuts have been observed.

The evidence of these instabilities illustrates the progressive disequilibrium between surface and forms, concerning human activities. The urbanization, the road infrastructures, the vegetation destruction, and the change on soil proprieties, drainage network and infiltration rates generate an active disturbance period (Toy & Hadley, 1987). This state of disturbance had special relevance in the 2000-2001 winter.

During this period a general state of activity reactivation and the increase in the rate of movements of several unstable areas were produced. The main triggering factor was the rainfall with the increase of the monthly data, as shown in Figure 1. Since the last days of December major translational, rotational and complex slides became active or were reactivated and lateral spreads or flows took place. The long-lasting rainfall produced cumulative values that surpassed the average annual rainfall in January and a variation of 162% (862 to 1396mm) in the cumulative rainfall records in April when compared with the average regime (Figure 2).

Some examples of instability could be pointed out to show the significance of the cumulative rainfall during this long-lasting period. The most destructive mass movement occurred in the urban area of Coimbra city where the rainfall associated with wind gusts produced a complex movement involving a rotational slide followed by an earth flow. An anthrophic cut on a slope deposit, covered with high trees, and overloaded by a top earth fill with a basal spring, were the previous conditions of the movement which caused severe structural damages to an eighteen floor building, the destruction of twenty garages and cars and the risk of collapse of top hill residential houses. In the outer urban areas, similar movements occurred, producing damages to buildings, or force mitigation actions. Road were affected by more than one hundred movements (falls, slides and flows) having some of them caused total erasure of the pavement and the destruction of the infrastructures.

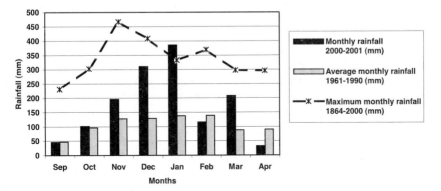

Figure 1. Monthly rainfall data in the 2000-2001 winter.

This evidence made clear to the local government that physical characterization and hazard identification must be handled and translated into the planning frameworks.

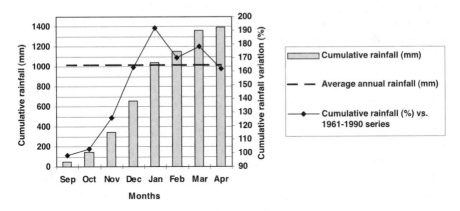

Figure 2. Cumulative rainfall in the 2000-2001 winter.

Instability methodological evaluation

A methodological evaluation of instability associated with geomorphic processes was developed, using the GIS technology, which made possible a composite identification of areas with growing levels of susceptibility (Brabb, Pampeyan & Bonilla, 1972; Tavares, 1999 acceptation) with repercussions on land use and comprehensive plan.

The database, with two levels of attributes, included: a lithological map (outcropping units, structural maps, geotechnical soils and rocks characterisation); a digital terrain model (slopes steepness and aspects); the hydrological network (drainage density); a multi-spectral Landsat satellite imagery (surface moisture evaluation, artificial and natural areas classification, bare or sparse vegetation cover projection); the collection of mass movement occurrences (falls, slides, lateral spreads, flows, gully erosion and karst subsidences and sinkholes); the land use plan (anthrophic occupation percentage, road infrastructures density).

For this purpose, twenty-seven physical determinant factors in instability associated with geomorphic processes were selected, analysed by means of qualitative scales and weighed according to field observations. The data retrieval and analyses displayed a composite surface map with a given index value. The arrangement in hierarchical ranks defined three levels (low, moderate and high) of relative instability or susceptibility, whose representation is shown in Figure 3.

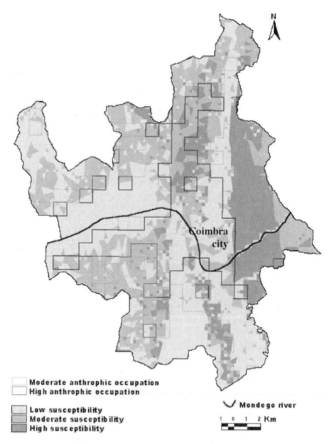

Figure 3. Susceptibility classification vs. anthrophic occupation in Coimbra Municipality

The composite susceptibility outcome vs. the disturbed land with anthrophic occupation, points out: (1) the small representation of areas with high susceptibility (less then 20% of the studied area); (2) the high susceptibility level location in outer urban areas to the N, SW and E of Coimbra city; (3) the importance of moderate susceptibility in areas with increasing urbanization and an high density of road infrastructures; (4) the progressive anthrophic translation, into areas with growing levels of susceptibility, harming others characterized by minor levels; (5) the forsake of the historical areas of occupation with known physical characteristics.

CONCLUSION AND LAND PLANNING APPLICATION

This evidence made clear to the local government that physical characterization and hazard identification and assessment must be handled and translated into the planning frameworks. Last winter instability occurrences developed a major sensibility in technical agents and in public opinion towards the appropriate management of unstable lands as well as and the understanding of the disturbance. The accuracy of good practice measure and the evolution to a postdisturbance period (Toy & Hadley, 1987) became goad to be reached.

The physical characterization and the susceptibility classification of the studied area (with a checked methodology) allow a better approach to the management criteria (efficiency, equity and biodiversity). In Figure 4 several items are represented whose answer will improve the criteria articulation and the sustainable development of the studied area.

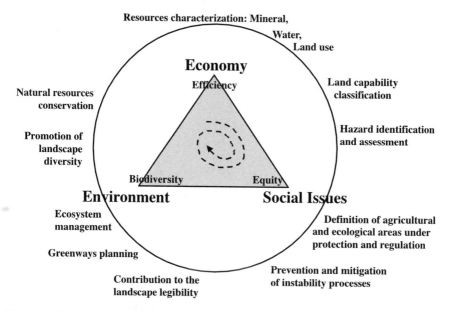

Figure 4. Improvement on management criteria articulation.

REFERENCES

Brabb, E.E., Pampeyan, Bonilla, 1972. Landslides susceptibility in San Matero Country, California. U.S. Geolog. Survey Misc. Field Studies Map, MF-360.

Cruden, D.M., Varnes, D.J., 1996. Landslides types and processes. Turner, A.K., Schuster, R.L. Eds., Landslides: Investigation and Mitigation. Transportation Research Board, National Research Council, Washington DC, pp. 36-75, Special Report 247.

Fannin, R.J., Rollerson, T.P., 1993. Debris flows: some physical characteristics and behaviour. Canadian Geotechinal Journal, 30, pp. 71-81.

ISRM, 1981. Basic Geotechnical description of Rock Masses (BGD). International Journal Rock Mechanics Mining, Science & Geomechanics Abstracts, vol. 18, pp. 55-110.

Tavares, A.O., 1999. Condicionantes físicas ao planeamento. Análise da susceptibilidade no espaço do concelho de Coimbra. Ph.D. Thesis University of Coimbra, Portugal, 346p.+26maps.

Toy, T.J., Hadley R.F., 1987. Geomorphology and reclamation of disturbed lands. Ed. Academic Press Inc, Orlando, 480p.

Varnes, D.J., 1978. Slope movements : type and processes. Schuster, R.L. Krizek, R.J. Eds., Landslides Analysis and Control. Transportation Research Board, National Research Council, Washington DC, pp. 11-35, Special Report 176.

WP/WLI UNESCO, 1993a. A suggested method for describing the activity of a landslide. Bulletin of the International Association of Engineering Geology, 47, pp. 53-57.

WP/WLI UNESCO, 1993b. Multilingual landslide glossary. International Geotechnical Societies, Canadian Geotechnical Society, Richmond.

Land instability and safety:
The role of the Geotechnical Specialist

DR ALAN THOMPSON, Symonds Group Ltd, East Grinstead, England

SUMMARY
In 1993, the Institution of Civil Engineers promoted the need for professional standards in the assessment of land instability and formally defined a Geotechnical Specialist in terms of qualifications, Chartered status and relevant experience. Since then, a number of procedures relating to the assessment of geohazards by Geotechnical Specialists have begun to emerge. One of the first examples of this was the policy for development on unstable land within the Harrogate District Local Plan, 1996. This included a requirement for developers to submit a Ground Stability Declaration Form, completed and signed by a Geotechnical Specialist. The form serves the dual purpose of certifying to the Local Planning Authority that the issue of land instability has been assessed by a competent person, whilst giving assurance to the developer that the site is worth investing in. Similarly, Part VI of the Quarries Regulations 1999 sets out formal requirements for Geotechnical Assessments to be undertaken, and signed off, by Geotechnical Specialists. The assessments are required to identify 'significant hazards' and to make recommendations regarding how these should be addressed in order to minimise the risk to persons within the quarry. This paper examines the role of the Geotechnical Specialist in the effective management of land instability problems, and asks whether procedures similar to those outlined above should be extended to other areas.

INTRODUCTION
In 1993, the Institution of Civil Engineers' Site Investigation Steering Group publication '*Without Site Investigation, Ground is a Hazard*' promoted the need for professional standards in the assessment of land instability. That work formally defined a Geotechnical Specialist[1] in terms of qualifications, Chartered status and relevant experience. Since then, a number of procedures relating to the assessment of geohazards by Geotechnical Specialists have begun to emerge in Development Plans and, most recently, in the Quarries Regulations 1999.

---

[1] defined by the I.C.E. as "*A Chartered Engineer or a Chartered Geologist, with a postgraduate qualification in geotechnical engineering or engineering geology, equivalent to at least an M.Sc., and with three years' post-Charter practice in geotechnics; or a Chartered Engineer or Chartered Geologist with five years' post-Charter practice in geotechnics*".

LAND INSTABILITY IN RIPON, NORTH YORKSHIRE

One of the first applications was the policy for development on unstable land within the Harrogate District Local Plan, 1996. This policy, which makes specific reference to the problem of subsidence caused by the progressive dissolution and collapse of natural gypsum deposits beneath the city of Ripon, North Yorkshire, was introduced in line with the recommendations of a two-year investigation funded by the former Department of the Environment and Harrogate Borough Council, and carried out by Symonds Travers Morgan (Thompson *et al.*, 1996).

That study confirmed the highly localised extent and infrequent (but occasionally dramatic) occurrence of subsidence within the Ripon area, and identified the need for a pro-active planning response in order to limit the adverse impact this could have on both existing and future development.

PPG14 "*Development on Unstable Land*" makes it clear that responsibility for the safe development and secure occupancy of a site rests, not with the Local Planning Authority, but with the developer and land owner. PPG 14 also draws attention, however, to the fact that Local Planning Authorities can help to ensure that information on land instability is taken properly into account in the planning of new development, through appropriate forward planning and development control procedures.

It was already clear, from earlier work by the British Geological Survey (Cooper, 1986, 1989), that the problem in Ripon was geologically complex. The Local Planning Authority realised that it did not have the necessary in-house expertise to judge the level of risk involved, or the nature and extent of any remedial action that might be required to allow the safe development of individual plots of land. Neither could it afford to seek external expert advice on each individual planning application.

What was needed, therefore, was a means of ensuring that the spatial pattern of future development was guided by the best available advice; and that decisions on individual development proposals were guided by reliable, independent, site-specific advice on the level of risk involved and the best way of dealing with this.

The first of these requirements was dealt with directly by the Symonds' study. This involved remapping the solid geology of the area, in conjunction with the British Geological Survey; undertaking a detailed hydrological, geomorphological and hydrochemical study to establish the probable mechanisms and spatial patterns of gypsum dissolution within the area; and the preparation of a 1:10,000 scale development guidance map to support the new Local Plan. The map identified broad areas in which specific development control policies would need to apply (see below), and also indicated (by means of gradational shading) spatial variations in the likely rates of gypsum dissolution within the area most at risk. A full account of method of assessment and the findings of the study is given in the Symonds Travers Morgan report (*ibid.*), and a more concise review is given in the paper by Thompson *et al.*, 1998a.

The second requirement was fulfilled by the introduction of systematic development control procedures, developed by Symonds.   These included requirements for desk studies, walk-over surveys and site investigations, designed to assess the specific risk of subsidence for each proposed development, and to assess the most appropriate type of foundation design required to minimise such risk.   It was recommended that these studies be undertaken by a Geotechnical Specialist, as defined by I.C.E., with specific knowledge of the geological conditions and processes operating within the Ripon area.

A crucial difference between these requirements, and similar procedures adopted previously in other locations - Ventnor, on the Isle of Wight, for example - is that the report by the Geotechnical Specialist has to include a signed declaration form (Table 1). This requires a named Geotechnical Specialist to state clearly whether or not ground instability can reasonably be foreseen within the lifetime of the proposed development and, if so, whether the mitigation measures proposed can reduce this risk to a reasonable level. The specialist is also required to state whether or not the information available to him is adequate for the purposes of the assessment.

The Local Planning Authority has a checklist relating to the answers given on the declaration form, and uses this to determine what action to take (e.g. to request further information if the Geotechnical Specialist is unable to declare that he has adequate site investigation data, or to recommend that consent be granted, if everything is in order).

These procedures enable the Local Planning Authority to exert the necessary control, whilst ensuring that decisions regarding the assessment of potential instability on individual sites are taken by those who are best qualified to do so. The declaration form serves the dual purpose of certifying to the Local Planning Authority that the issue of land instability has been assessed by a competent person, and of giving assurance to the developer that the site is worth investing in.

To date, the procedures appear to be working well, although concerns have been raised that the declaration forms are sometimes signed by experienced structural engineers who are not properly qualified as Geotechnical Specialists. This, clearly, is a matter for individual developers and their consultants (since it is they who are ultimately responsible), but it may be appropriate for the Planning Authority to request proof of qualifications.

Although relating specifically, in this instance, to subsidence associated with gypsum dissolution, the fundamentals of this approach have much wider application, as noted in the DETR Guide to Good Practice on Environmental Geology in Land Use Planning (Thompson *et al.*, 1998b). Whilst individual local authorities would not usually have access to the level of funding that was needed in Ripon to identify the precise nature and extent of the hazard, this need not deter the basic principle of using a policy that requires expert input to land use planning decisions.

GEOTECHNICAL ASSESSMENTS, QUARRY REGULATIONS 1999

The Quarries Regulations 1999 introduce the concept of pro-active risk management as a means of reducing the likelihood of injuries within excavations and tips. Part VI of the Regulations deals specifically with the risks associated with instability or movement of proposed and existing slopes within quarries.

In order to facilitate the control of such risks, Regulation 32 requires the quarry operator to ensure that a suitable and sufficient *Geotechnical Appraisal* is carried out to determine whether any part of the excavation or tip is a 'significant hazard'. 'Hazard', as defined in Regulation 2 (1), means "*having the potential to cause harm to the health and safety of any person*". Any unstable or potentially unstable slope has the potential to cause harm.

'Significant hazards' are more specifically defined in Paragraphs 293 to 301 of the Approved Code of Practice (ACoP) published by the Health & Safety Executive (HSE, 2000). This automatically classifies certain features as significant hazards purely on the basis of geometric criteria, such as the overall height and angle of a quarry face, or the total volume of a tip (which may include overburden mounds, stockpiles and silt lagoons). The ACoP also provides guidance on the kinds of evidence that may be indicative of actual or potential instability, including tension cracks behind the crest of a slope, bulging at the toe, groundwater seepage etc.

The person undertaking a geotechnical appraisal is required, by Regulation 2 (1), to have "*sufficient training, experience, knowledge and other qualities to enable him properly to undertake the duties assigned to him*". In this case, relevant knowledge and experience should include an understanding of the principles of geotechnics, particularly in relation to the geological conditions likely to be encountered within the quarry concerned, and familiarity with the operational aspects of that quarry. The latter is important, particularly with regard to distinguishing between significant and non-significant hazards. For these reasons, the initial appraisals are often carried out by the operator's own personnel.

Where an appraisal concludes that part(s) of an excavation or tip represent a significant hazard, Regulation 32 (4) requires the operator to ensure that a more rigorous '*Geotechnical Assessment*' is carried out as soon as is reasonably practicable. A Geotechnical Assessment is defined, in Regulation 33, as "*an assessment carried out by a Geotechnical Specialist, identifying and assessing all factors liable to affect the stability and safety of a proposed or existing excavation or tip".*

The Geotechnical Specialist is as defined by the Institution of Civil Engineers (anon, 1993), but is specifically required, by the Regulations, to have experience in soil mechanics, rock mechanics or excavation engineering, and to be competent to perform a geotechnical analysis to determine the hazard and risk arising from the excavation being assessed. Regulation 33 (1), (a) to (e) identifies the specific duties of the Geotechnical Specialist, in relation to a geotechnical assessment.   Briefly summarised, these comprise:

- preparing the documents and particulars specified at Schedule 1 of the Regulations or supervising their preparation by others, as appropriate;
- forming a conclusion as to the safety and stability of the proposed or existing excavation or tip being assessed, and as to whether this represents a significant hazard;
- where appropriate, forming a conclusion as to whether any remedial work is required and the date by which such work should be completed;
- where appropriate, forming a conclusion as to the date by which the next geotechnical assessment should take place; and
- consideration of the excavations and tips rules.

As with any geotechnical investigation, the accuracy of the assessments will vary according to the quality and quantity of data that goes into them, as well as the skill of those carrying out the work. The quarry operator thus has a duty to provide the Geotechnical Specialist with any information available to him that may be relevant for the purposes of the geotechnical assessment [Regulation 33 (3)].

The Geotechnical Specialist needs, however, to adopt a pragmatic approach that recognises the practical limitations on data collection within a working quarry, and the temporary nature of many of the slopes, but without compromising the fundamental health & safety objectives that underpin the Regulations.

In doing so, he or she may need to exercise some degree of judgement as to what constitutes a significant hazard and as to how such hazards should be dealt with.   In many cases (for example, working faces within a shallow sand & gravel quarry), it may be reasonable for the hazard to be managed by adherence to sensible working practices set out within the excavations and tips rules for that site. An unstable slope may thus be defined as a hazard, but it need not be regarded as a 'significant hazard' if active measures are taken to prevent unsafe access to the affected area, thus reducing the 'risk'.   For this reason, an important part of the Geotechnical Specialist's role is to inspect the excavation and tips rules for each site and, where necessary, recommend improvements or changes to these. It is then for the operator to ensure that these are properly implemented.

In other cases, a feature that has been identified as a significant hazard, by virtue of the geometric criteria defined in the ACoP, may need to be 'down-graded' to a non-significant hazard during a Geotechnical Assessment. This may occur where, despite the physical size or shape of a particular feature, there is no foreseeable failure mechanism and/or where a detailed slope stability analysis has demonstrated an adequate factor of safety.

It is perhaps too early to tell whether Geotechnical Assessments are providing any real benefit to the quarrying industry. They certainly ought to, by requiring consultants to give clear and justified opinions on slope stability issues, and by formalising best working practices among quarry operators. Ambitions are often thwarted, however, or at least modified, by commercial realities: the desire of quarry operators to minimise the financial burden placed on them by new legislation, and the commercial imperative among consultants to undercut the price of their competitors, thus potentially undermining the quality of the work produced.

THE FUTURE ROLE OF GEOTECHNICAL SPECIALISTS IN THE MANAGEMENT OF LAND INSTABILITY

The examples outlined above demonstrate the beneficial role that Geotechnical Specialists can play in helping to implement both land use planning policies relating to land instability, and formal legislation relating to the control of associated risks in quarries. The question therefore arises: might there be benefits in adopting similar procedures more widely, either as a planning tool, or as a more formal health and safety obligation? Should coastal cliffs, for example, be subject to regular inspections by Geotechnical Specialists, in the interests of public safety?

The answer, as with most things in the sphere of engineering geology, is probably "it depends..."

It depends, for one thing, on whether Geotechnical Specialists, as defined by I.C.E., are always the best suited to give advice on land instability problems. In the UK there are acknowledged experts on coastal instability who would fail to qualify as Geotechnical Specialists, because they have not achieved either Chartered Engineer or Chartered Geologist status. It should be noted, however, that suitably experienced geomorphologists, with a first degree in either geology or physical geography, can qualify as Chartered Geologists and thereby eventually become recognised as Geotechnical Specialists, if they have the right experience. The onus is on those with an interest in land instability to apply for the necessary professional qualifications.

It also depends on cost. Would it be reasonable to impose the costs of regular geotechnical assessments on coastal authorities and/or landowners? Would it be justified in terms of health and safety statistics? The justification would seem to be much less clear than for quarrying, where slope instability is directly attributable to the actions of the mineral operator, and where a health and safety obligation clearly exists. Coastal planning authorities, by comparison, have no obligation to prevent coastal erosion and instability, except where they are also the land owner. The recent case of <u>Holbeck Hall Hotels and Another-v-Scarborough Borough Council</u> (February 2000) clarified that the local Council (as landowner) may owe a duty of care to its neighbour, but that this duty depends on the extent to which instability can be foreseen.

As we move forward into a more enlightened age, will it be reasonable, in future, to expect such risks to be foreseen? If so, this might signal a requirement for the more widespread use of the approaches outlined in this paper, and an increasingly important role for the Geotechnical Specialist.

ACKNOWLEDGEMENTS
The author is grateful to the Department of Transport, Local Government and the Regions (formerly the Department of the Environment) and Harrogate Borough Council for permission to publish this paper, and to Lindsay Frost (formerly of Harrogate BC, now at Lewes District Council), for helpful comments on an earlier draft of this paper.

REFERENCES
Anon (1993): **Without Site Investigation, Ground is a Hazard**. Institution of Civil Engineers' Site Investigation Steering Group. Thomas Telford, London (45pp).

Cooper, A.H. (1986): Subsidence and foundering of strata caused by the dissolution of Permian gypsum in the Ripon and Bedale areas, North Yorkshire. in **The English Zechstein and related topics**. Harwood, G.M., and Smith, D.B. (editors) Special Publication of the Geological Society of London, No. 22, 127-139.

Cooper A.H. (1989): Airborne multispectral scanning of subsidence caused by Permian gypsum dissolution at Ripon. North Yorkshire. **Quarterly Journal of Engineering Geology, London**. Vol 22, 219-229.

Department of the Environment (1990): **Planning Policy Guidance: Development on Unstable Land (PPG 14)**. HMSO (25pp).

Thompson, A., Hine, P.D., Greig, J.R., and Peach, D.W. (1996): **Assessment of Subsidence Arising from Gypsum Dissolution (with particular reference to Ripon, North Yorkshire)**. Technical Report to the Department of the Environment by Symonds Travers Morgan (288pp).

Thompson, A., Hine, P.D., Peach, D.W., Frost, L. and Brook, D. (1998a): Subsidence Hazard Assessment as a Basis for Planning Guidance in Ripon. In Maund, J.G. and Eddleston, M. (eds): **Geohazards in Engineering Geology.** Geological Society of London, Engineering Special Publications 15, 415-426

Thompson, A., Hine, P.D., Poole, J.S. and Greig, J.R. (1998b): **Environmental Geology in Land Use Planning: A Guide to Good Practice**. Report to the

Department of the Environment, Transport and the Regions, by Symonds Travers Morgan, East Grinstead (80pp).

Table 1
Ground stability declaration form for use in Ripon

Site Name	Site Address	Development control area

CATEGORY		QUESTION	YES/NO/ ?/N/A
A) Competent Person	(i)	Has the report been prepared by a Geotechnical Specialist, as defined by the ICE Site Investigation Steering Group, (Anon, 1993) (*see overleaf*)	
B) Site History	(i)	Has the site been affected by known historical ground instability problems ?	
	(ii)	Is the site located within or adjacent to, or does it contain any subsidence features as identified on the development Guidance Map, or on updated information held by Harrogate Borough Council ?	
C) Site Inspection	(i)	Has a detailed site inspection been carried out?	
	(ii)	Does the land within or adjacent to the site bear any geomorphological evidence of former, on-going or incipient ground instability ?	
	(iii)	Does the site or neighboring property bear any evidence of structural damage or repairs which might be associated with ground instability ?	
D) Geotechnical Desk Study	(i)	Have any previous ground investigation reports and/or borehole records from this site been consulted ?	
	(ii)	If yes, is this information adequate, reasonably to confirm the presence or absence of gypsum* and the presence or absence of cavities or foundered strata which could affect the stability of the site ?	
	(iii)	If yes, have any cavities or foundered strata been identified beneath the site ?	
	(iv)	Have any massive gypsum beds (greater than 1m thick) been identified in contact with underlying limestone beneath the site ?*	
E) Ground Investigation	(i)	Has a ground investigation been carried out in support of this application ?	
	(ii)	If yes, is the information obtained adequate, reasonably to confirm the presence or absence of gypsum* and of cavities or foundered strata which could affect the stability of the site ?	
	(iii)	If yes, have any cavities or foundered strata been identified beneath the site ?	
	(iv)	If yes, have their locations and dimensions been properly identified ?	
	(v)	Have any massive gypsum beds (greater than 1m thick) been identified in contact with underlying limestone beneath the site ?*	
F) Evaluation of Stability	(i)	Is the information, available under B, C, D and (where applicable) E above adequate, reasonably to assess the stability of the site ?	
	(ii)	If yes, can ground instability reasonably be foreseen within or adjacent to the site within the design life of the proposed development, allowing for any deterioration of ground conditions caused by the development itself ?	
	(iii)	If yes, can such instability be reduced to a reasonable level by the use of appropriate and cost effective mitigation measures ?	
G) Mitigation Measures	(i)	Have mitigation measures been proposed with respect to ground instability issues ?	
	(ii)	If yes, are these designed to reduce the effects of any actual or potential instability to a reasonable level ?	
	(iii)	If yes, are these likely to have any adverse effects on the stability of other, adjacent sites (for example by affecting the existing groundwater regime within the area) ?	
H) Name, qualifications and signature of person responsible for this report		Full Name	
		Qualifications	
		Year of becoming a Geotechnical Specialist as defined by the ICE Site Investigation Steering Group (*see overleaf*)	
		Signature	
		Company Represented (*if applicable*)	

Session 3:

Hazard identification and risk assessment

Systems for hazard identification and assessment are continually being improved in terms of accuracy, reliability, efficiency and affordability. Improving risk assessment is essential to assist the protection of life, property and economies, particularly in view of the predicted impacts of climate change.

Modelling of two complex gravitational phenomena in Marchean coastal areas (Central Italy)

D. ARINGOLI, Department of Earth Science, Camerino University, Italy
M. CALISTA, Department of Earth Science, Chieti University, Italy
U. CRESCENTI, Department of Earth Science, Chieti University, Italy
B. GENTILI, Department of Earth Science, Camerino University, Italy
G. PAMBIANCHI, Department of Earth Science, Camerino University, Italy
N. SCIARRA, Department of Earth Science, Chieti University, Italy

INTRODUCTION

Along the central portion of the Adriatic coast, where the highest sea cliffs are found, numerous landforms have been identified, the genesis of which cannot be entirely fitted into the tectonic-structural evolutional framework of the area. An evolutional model has been proposed for them, in the complex kinematics of which they are credited with a role of by no means secondary importance with respect to the tectonic and gravitational factors, to the strong Late Pleistocene sea level oscillations, and to the significant seismicity of the area. The proposed theme has been already analyzed in specialist literature (Harrison & Falcon, 1934; Jahn, 1964; Guerricchio & Melidoro, 1981; Sorriso-Valvo Ed., 1989; Crescenti et al., 1994; Dramis et al., 1995). The present work deals with two areas that are considered representative samples, namely the "North Area" (Fig. 1) and the "South Area" (Fig.2), for which the results of the study are reported; these derive from initial geomorphologic considerations and interpretations whose evolutional models have been verified by applying a finite difference numerical analysis code, with a view to reconstructing the precise slope dynamics. These numerical simulations have proved particularly useful for establishing the future evolution of such dynamic systems and identifying their intensity with regard to danger and the hazard connected with it.

GEOLOGICAL SETTING

North area. This area is situated in the neighborhood of the city of Pesaro (Fig.1), on the Adriatic margin of a minor chain where marine sedimentary units belonging to the Miocene terrigenous cycle can be observed cropping out, formed by marls with clayey and chalky intercalations (Tortonian-Messinian) and turbiditic weakly cemented with clayey-marly intercalations (Messinian). There follow continental deposits made up of: sandy and partially gravelly Pleistocene alluvial terraces, Holocene silty-sandy colluvial and sheets of detrital material (Centamore & Deiana, 1986). The dominant structural elements are eastern verging faults and folds, approximately parallel with the coastline, that on the whole point to an anticline structure bordered by direct faults and connected with the extensional tectonic phase which developed from Upper Pliocene and with maximum intensity to the end of Lower Pleistocene. The lithotypes are highly tectonized with faults parallel to the anticline axis. In depth, these elements are presumably listricated on the underlying thrust plane, which is characterized by a slight western dip direction. Those folding structures derive from an intense compressional tectonics, which persistent activity along the Adriatic coast is probably responsible of actual and recent deformational phenomena (Riguzzi et al., 1989).

South area. This area is located in the southeastern sector of the external Marchean basin, characterized by the terrigenous sedimentation that closed the Plio-Pleistocene marine cycle, and is made up of: prevalently sandy-conglomeratic Sicilian-Crotonian deposits, overlying prevalently clayey Emilian sediments.

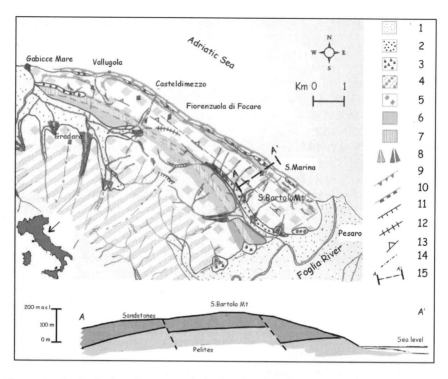

Figure 1. Geological and geomorphologic simplified map of the *North Area:* 1) recent alluvial deposits; 2) terraced alluvial deposits; 3) landslides and detritic deposits; 4) pelites (Lower Pliocene); 5) sandstones (Middle Messinian); 6) pelites and sands (Upper Messinian); 7) Schlier (Tortonian-Langhian); 8) alluvial and debris fan; 9) erosional coast line; 10) scarp > 10 m; 11) scarp < 10 m; 12) trenches; 13) counter-slope; 14) fracture or faults; 15) cross section for numerical modelling.

Recent and real alluvial deposits are present along the floor of the San Biagio Stream valley interbedded with recent sand deposits. During the Upper Pliocene, the compressional tectonic has strong influenced the sedimentation on the area and that produced folds, thrusts and transform faults (Fig.2). The compression led to the uplifting of the area and to the resultant erosion of the summit portion of the Upper Pliocene sediments, contemporaneously, Apennine-trending direct faults were activated above all on the Adriatic slope of the folds; moreover, beginning from Pleistocene, a reactivation following normal kinematics of preexistent compressional transversal discontinuity can be observed (Centamore & Deiana, 1986; Bigi et al., 1995).

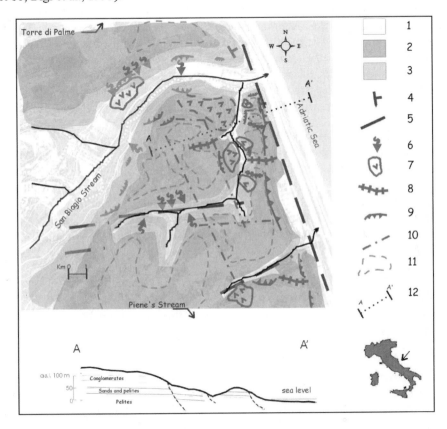

Figure 2. Geological and geomorphologic simplified map of the *South Area:* 1) recent alluvial, colluvial and beach deposits; 2) conglomerates and gravel: coastal and fluvial deposits (Sicilian); 3) sands with gravell levels: beach deposits (Sicilian); 4) strata dipping; 5) fracture or faults; 6) surficial deformation; 7) landslide; 8) trenches; 9) degradational scarp; 10) gravitational discontinuity or fracture; 11) low relief energy surface; 12) cross section for numerical modelling

GEOMORPHOLOGIC ANALYSES

The most important morphological element in the *North area* is a high coast (cliff) corresponding to the above-mentioned anticlinal fold with a gravel beach of small thickness present at its foot. At the top, significant geomorphologic elements, representing past and recent slope dynamics, can be seen. Steps, trenches and limbs of an ancient sub-horizontal erosion surface ("paleosurface"), intercalated with evident small valleys both parallel and perpendicular to the coastline (Fig. 1), can be recognized. Another diagnostic elements are represented by anomalously shaped small valleys, mostly following the orientation of the tectonic fracturing systems, valleys draining inland (with "box-shaped" valley heads) and anomalous courses of streams. The slope dynamics is particularly intense on the whole eastern slope of the morphostructure, also due to coastal dynamics interference; in fact, there has been a progressive withdrawal of the cliff because of the motion of the sea waves that has removed material at the foot. Associated with this are numerous landslides, determining serious situations of danger for the built-up areas on the promontory. In particular, the ongoing landslides prevalently affect the portions: Pesaro-Santa Marina, where translational slides predominate; these are prearranged and guided by deposits dipping out of the slope and having a dip angle of between 5° and 20°; Santa Marina-Vallugola (farther north) where a substratum dipping into the slope crops out (the slope of the layers being from 20° to 40° on subvertical walls) and where rotational slides occur, giving rise to powerful accumulations that are at times comparable with marine terraces (e.g., Casteldimezzo). Still farther north, between Vallugola and Gabicce, local falls and translational slides of detrital material and clayey lithotypes can be found. The slopes down to the west are less steep, not subject to the action of the sea, and locally display only slow deformation of the eluvial-colluvial cover.

The overall morphological setting of the *South area* is marked by an inactive trending NNW-SSE cliff, which modeling was probably driven by normal faults subsequent the up-rise. To NNW, it is cut by the San Biagio Stream and then continues until the Torre di Palme relief; to the south, it is dismembered into more regularly shaped reliefs until the Piene's Stream morphological interruption (Fig. 2). The reliefs are distinguished by sub-planar summits and steep slopes (up to and over 30°), above all to the east, where they run parallel to the coast. From the hydrographic standpoint, some deviations are evident in the minor hydrographic network; with respect to gravity, phenomena of a different typology and extent are present: minor landslides attributable to a deepening of the minor hydrographic network, and more extensive slope deformation revealed by trenches that can be noted in the upper portions of the reliefs. These elements have been conditioned by geological joints, but the major correspondence can be found in the fracturing and degrading scarps. The above mentioned geomorphologic elements supplied useful data in order to hypothesize a first probable slope dynamics for the two areas, due to deep deformational mechanism. In this aim two sample sections had been numerically modelled, to verify as the evolution models as the exact understanding of surveyed geomorphologic elements.

NUMERICAL MODELLING

The code used (FLAC, 2000) is a two-dimensional method of numerical analysis of finite differences for calculations of mechanics of continua. FLAC is based on a *"lagrangian"* calculation scheme which is very adaptable in the modelling of large-scale deformations and collapse of materials (Sciarra & Calista, 1999). Using this code in seismic situations two particular behaviour patterns of the model in differing conditions of sea level were analysed. The positions analysed are one relative to a level of water table coincident with the real level and the other without the water table (simulating the Pleistocene age). The properties of the formations are summarised in table 1; the data are relative to laboratory and in situ tests.

Table1 – Medium values of utilized geo-mechanical parameters

Parameters	γ unit weight (kNm^{-3})	c' cohesion (kPa)	φ' friction angle	E Young modulus	$\sigma^t_{max} = c'/tg\varphi'$
sandstones, conglomerates	20.1	20	30	7.5 (GPa)	35 (kPa)
pelites	19.5	10	23	600 (kPa)	23 (kPa)

The general analysis carried out, as imposed by the code, consists in an overall re-equilibrium of the system and therefore in the study of failure conditions. The analysis of the process of global re-equilibrium is divided into three phases. The first phase analyses the filtration within the model to identify, once the hydraulic equilibrium has been reached, the negative pressure distribution irrespective of any mechanical effect. In the second phase the model operates a mechanical adjustment using the values of neutral pressure obtained from the first phase and imposing zero the compressibility module of the water (any variations of the negative pressures is avoided). In the third phase there is contemporary analysis of the process of filtration and the relative mechanical adjustment. At the end of the process of global equilibrium it is possible to represent the vectors of movement. In this way it is possible to observe the behaviour of the system being studied. Following the phase of global equilibrium and wishing to analyse the conditions of limit equilibrium, it is possible to move onto the analysis of the stress-strain behaviour of the system.

DISCUSSION

In figures 3 and 4, which diagrams show the final state deformation and the maximum shear strength subsequent an ideal local seismic input, results are clearly visible. In figures can be observed the different responses connected to models checking the sea level influence on global system behaviour. In both studied cases can be find how the actual sea level standing involve an important increase of deformational condition and then a major genesis of shear stress concentrated zone.

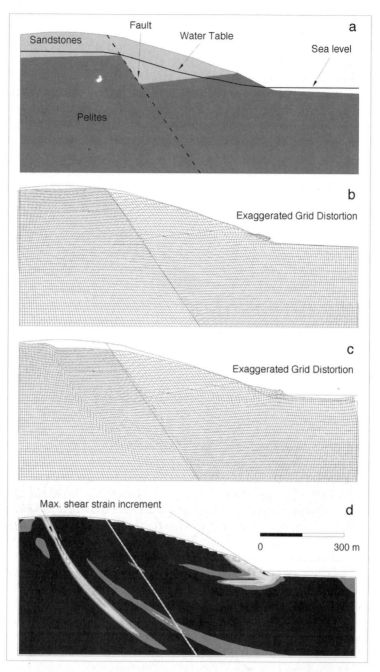

Figure 3. Modelling of a cross section along North area (see fig.1): a) geological section; b) meshes deformation in dry conditions; c) meshes deformation utilising a table water; d) representation of the shear strain increments.

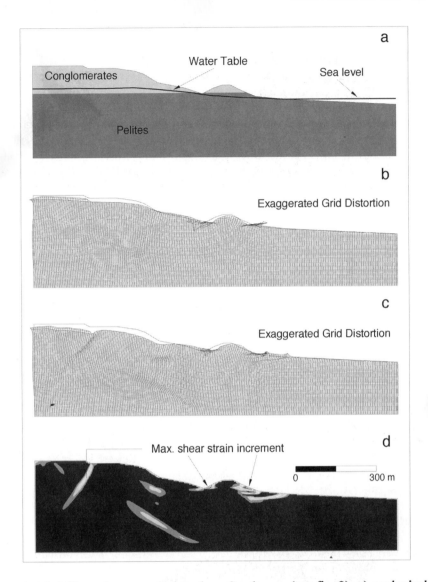

Figure 4. Modelling of a cross section along South area (see fig. 2): a) geological section; b) meshes deformation in dry conditions; c) meshes deformation utilising a table water; d) representation of the shear strain increments.

This according what already discussed in the geomorphologic analysis, supposing deep-seated kinematics. Along the *North area* cross section (Fig. 3) can be observed how the late Pleistocene sea level up-rise produced a general bedrock mechanical properties decline, related to decrease of pore water pressure. This allowed an easier mobilization of the system that shows points of weakness also in portion not corresponding preexistent tectonic dislocations. Anyhow, on the foot of cliff, good evidences of failure phenomena are present, related to surficial mass movements. On the *South area* cross section (Fig. 4) analogous behaviors, even if less intense, can be observed probably related to the minor relief energy.The presence of a table water confirms a general weakening of the model, with settlements on the left zone of the model and a general translation toward right. The mechanism is typical of a lateral spread. It is also observable the generation of failure planes producing trenches on the topographic surface. We can conclude that the geomorphologic peculiarities of the surveyed areas, interpreted like a natural geological evolution due to the gravity, are confirmed by the numerical modelling carried out. Therefore the pleistocenic sea level variation, associated to seismic conditions, have had a fundamental role on the real morphological settings of the analysed zones.

BIBLIOGRAPHY

Bigi S., Cantalamessa G., Centamore E., Didaskalou P., Dramis F., Farabollini P., Gentili B., Invernizzi C., Micarelli A., Nisio S., Pambianchi G. & Potetti M. (1995). *La fascia periadriatica marchigiano-abruzzese dal Pliocene medio ai tempi attuali: evoluzione tettonico-sedimentaria e geomorfologia.* Studi Geologici Camerti, Vol. spec. 1995/1, 37-49.

Centamore E. & Deiana G. (1986). *La geologia delle Marche.* Studi Geologici Camerti, Vol. spec. 145 pp..

Crescenti U., Dramis F., Prestininzi A. & Sorriso-Valvo M. (1994). *Deep-seated gravitational slope deformations and large-scale landslides in Italy.* Chieti University, Spec. Vol. 7th Int. Congr. of IAEG., Lisboa (Portugal), 36-41, 1994.

Dramis F., Farabollini P., Gentili B. & Pambianchi G. (1995). *Neotectonics and large-scale gravitational phenomena in the Umbria-Marche Apennines, Italy.* In: Slaymaker O. (Ed.), Steepland geomorphology. J. Wiley & Sons Ltd, 199-217.

FLAC (2000). *Fast Lagrangian Analysis of Continua.* Release 4.0, Itasca Consulting Group, Minneapolis, Minnesota USA.

Guerricchio A. & Melidoro G. (1981). *Movimenti di massa pseudotettonici nell'Appennino dell'Italia meridionale.* Geol. Appl. Idrogeol., 16, 251-294.

Harrison J.V. & Falcon N.L. (1934). *Collaps structures.* Geological Magazine, LXXI, 529-539.

Jahn A. (1964). *Slope morphological features resulting from gravitation.* Zeit. Geomorph., suppl. Bd., 5, 59-72.

Riguzzi F., Tertulliani A.Z. & Gasparini C. (1989). *Study of seismic sequence of Porto San Giorgio (Marche) - 3 July 1987.* Il Nuovo Cimento della Soc. It. di Fisica, 12 (4), 452-466.

Sciarra N. & Calista M. (1999). *Modellazione del comportamento di formazioni rigide su di un substrato deformabile: il caso di Caramanico Terme (PE).* Mem. Soc. Geol. It., vol.54, 139-149, Roma.

Sorriso-Valvo M. (1989). Atti del III *Seminario del Gruppo Informale del C.N.R. "D.G.P.V.".* Boll. Soc. Geol. It., 108, 369-451, Roma.

Can we predict coastal cliff failure with remote, indirect measurements?

DR JON BUSBY, British Geological Survey, Keyworth, Nottingham, NG12 5GG, UK
MR JEAN CHRISTOPHE GOURRY, Bureau de Recherche Géologiques et Minières, Aménagement et Risques Naturels, 3, Avenue Claude Guillemin, B.P.6009 – 45060 Orléans cedex 2, France
DR GLORIA SENFAUTE, Institut National de l'Environnement Industriel et des Risques, Ecole des Mines de Nancy, Parc de Saurupt, 54042 Nancy cedex, France
DR STIG PEDERSEN, Geological Survey of Denmark and Greenland, Thoravej 8, DK-2400 Copenhagen NV, Denmark and
PROFESSOR RORY MORTIMORE, School of the Environment Research Centre, Cockroft Building, University of Brighton, Moulsecombe, Lewes Road, Brighton, BN2 4GL, UK

ABSTRACT

Over the years a considerable body of data has been collected on the nature and lithology of coastal cliffs. This has enabled geologists and engineers to develop a better understanding of cliff collapse mechanisms and the factors that determine rates of erosion. However there has been very little research into physical property changes within the rock mass behind a cliff face prior to collapse. If any such changes can be quantified they could be used as pre-cursors to impending collapse.

A European Union 5th Framework Research and Development project 'PROTECT' has been initiated to research into the applicability of two geophysical techniques for monitoring coastal cliffs. The first is based on the assumption that in the upper part of a cliff sub-vertical fractures within the rock will gradually dilate with time until a collapse is initiated. Since fractures often occur in sets with a preferred orientation they impose anisotropic physical properties to the rock mass. It has been shown that the apparent resistivity of the rock will vary with azimuth reflecting the dominant fracture orientation. A factor of anisotropy can be calculated from the measurements and this would be expected to vary with time if the fractures are dilating. The second technique is based on measuring acoustic crack emissions as the rock cracks and also on monitoring actual movements within the cliff. A series of high frequency transducers are placed in vertical boreholes behind the cliff face. Any acoustic crack emissions recorded should enable the location of the crack to be determined. In addition an iron bar, drilled into the cliff face and instrumented with transducers will record the movements of any cracks that are intersected by the bar. Verification of cliff movements will be attempted with accurate, repeat surveying grids at the cliff edge.

Three research sites on chalk cliffs have been established. These are located at Birling Gap and Beachy Head on the south coast of the UK; at Mesnil-Val on the Normandy coast of France and at Møns Klint on the Baltic coast of Denmark.

INTRODUCTION

The erosion of hard rock cliffs is inevitable and to-date it has been considered to be relatively unpredictable. A large proportion of the European coastline is subject to erosion and cliff recession. This dynamic process continually changes the hydrogeological and stress regimes and exposes fresh geological features and materials to changes in environment and stress. The assessment of cliff recession is an important factor in hazard assessment, conservation, amenity, and land-use planning. Hard rock cliffs erode through catastrophic collapse along pre-existing discontinuities in the rock mass. These may be ancient faults or fractures, orientated at a variety of angles to the cliff face, or relatively new tension fractures formed during cycles of cliff recession, sub-parallel to the cliff face. Glacial, periglacial and weathering processes have frequently deposited a layer of reworked rock and soil onto the bedrock that makes direct observations of discontinuities in the bedrock difficult.

Research projects to-date have concentrated on assessing the factors that will indicate the relative hazard due to erosion of cliff sections. These factors are geological (lithology and fracturing of the rock mass), terrestrial (climatic variation, water table level and underground flow) and marine (tidal and wave action, amount of shingle present). This paper reports on a new research project, PRediction Of The Erosion of Cliffed Terrains (PROTECT), supported by the European Union 5th Framework Programme. The principal objective is to provide long and short term alerts of impending coastal cliff instability. It aims to measure alterations within the rock mass prior to a collapse. Two geophysical techniques are being deployed. The first is Azimuthal Apparent Resistivity (AZR) that measures the variation of apparent resistivity with orientation of measurement due to the anisotropy imposed by parallel fracturing. The second is high frequency microseismics that measure the acoustic crack emissions of propagating cracks.

AZIMUTHAL APPARENT RESISTIVITY

It is known that fractures in hard rocks such as chalk and limestone occur in parallel sets, which impose anisotropic physical properties to the rock mass. The catastrophic failure of the cliff will be along one of these sets of fractures. Increased tension (dilatency) within the fracture network will increase the anisotropy of the rock mass. Hence a relative measure of the increased anisotropy will indicate sections of cliff where the fracture tension is increasing. It is highly likely that increased tension within the fracture network will eventually lead to rock failure, although the timing (months or years) is not currently known. However knowledge of fracture tension will help to identify vulnerable sections of coastline. An apparent resistivity measurement is made by imposing a low wattage direct current flow between two electrodes implanted into the ground surface and measuring the resultant potential difference between an additional pair of potential electrodes (see Figure 1). When the fractures posses resistivity contrasts with the host rock the measured apparent resistivity will vary with the orientation of the electrode array. Measurements are made by rotating the electrode array through 180° or 360° and taking measurements along a sufficient number of azimuths to define any variation of apparent resistivity with orientation (Taylor & Fleming 1988). From this the azimuth of the principal fracture set can be defined and a measure of the anisotropy can be obtained from the ratio of the maximum and minimum apparent resistivities. If the measurements are repeated over a period of time, any variations indicate an alteration of the physical properties of the rock mass, one of which would be changes in dilatency within the fracture network.

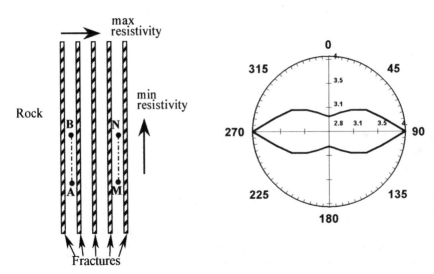

Figure 1. A set of parallel fractures will impose anisotropy to an apparent resistivity measurement, conducted through current electrodes A, B and potential electrodes M, N. When the electrodes are rotated about the centre of the square A, B, M, N and the results plotted against azimuth in a polar diagram, an ellipse results. The major axis of the ellipse is perpendicular to the fracture strike.

The Square Array

The majority of AZR measurements have been taken with the electrodes arranged colinearly (eg. Nunn *et al.* 1983, Busby 2000). However on a cliff top, space is restricted and Habberjam & Watkins (1967) have shown that with the electrodes arranged in a square they are more sensitive to anisotropy and require about 65% less surface area than an equivalent colinear array. Lane *et al.* (1995) have applied the azimuthal square array to the mapping of bedrock fractures and have extended the interpretational analysis. By using a switch box there are three electrode configurations for each square (see Figure 2). The α and β configurations are perpendicular measurements, whilst the diagonal γ configuration acts as a check on α and β. If the square is rotated about its centre point in increments of 15° it only requires six rotations to define all orientations.

The volume of rock involved in the measurement (and hence the number of fractures crossed) as well as the depth of penetration are determined by the array spacing, that is the length of a side of the square. In practice all array spacings at a particular orientation would be completed before rotating the array.

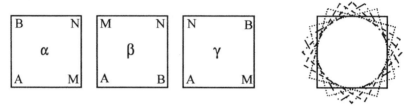

Figure 2. The α, β and γ configurations of the square array where A, B are current electrodes and M, N are potential electrodes. By rotating the array in increments of 15° all orientations are defined with six configurations.

Measurement Procedure

Figure 3. A series of square array azimuthal apparent resistivity measurements carried out perpendicular to the cliff edge.

Data collection consists of a series of rotated square arrays orientated perpendicular to the cliff edge, as illustrated in Figure 3. Those near the cliff edge should sample ground that is likely to be affected by fracture dilatency. In addition a square array measurement conducted sufficiently far inland from the cliff edge will act as a control. Some temporal variations are likely to be observed due to influences such as seasonal changes in saturation levels, but these should be observed in all the measurements. A time series will be built up by repeating the measurements every two months for a period of two years. Three research sites have been established at Birling Gap and Beachy Head on the Sussex coast of the UK; at Mesnil-Val on the Normandy coast of France and at Møns Klint on the Baltic coast of Denmark.

MICROSEISMICS
The purpose of the microseismics are to detect cliff collapse precursor seismic signals in an area of a few hectares near the cliff edge, a few minutes or hours before a rock collapse occurs. Two complimentary techniques are to be attempted comprising a high frequency microseismic monitoring network and a waveguide inserted into the cliff face.

Microseismic monitoring network
The network is established by installing tri-directional transducers in vertical boreholes drilled from the cliff top to three-quarters of the cliff height. As the low frequency cut-off is less than for the waveguide tool, the range of the microseismic network is greater. It is possible to install up to 16 transducers to produce a regular acquisition over a large area (more than 1 ha). The recording is triggered on the threshold of the amplitude of the seismic signal. The geometry of the transducers in the ground, the threshold of amplitude, the frequency of the accelerometers and the signal processing required to enhance the signal to noise ratio and to extract the signal will be determined at an early stage of the project. Continuous recording of the acoustic crack emissions in the rock mass will be conducted for a period of a year at the Mesnil-Val site on the Normandy coast.

Modelling of the acoustic crack emissions in the cliff rock mass will be carried out in order to evaluate the emission mechanism, the surface and the location of the fracture. This research will be based on a seismological approach, using knowledge obtained in mine research. The different rock fracturing mechanisms will be examined and their influence on the microseismic signals will be explained.

Waveguide
The waveguide consists of a metallic bar equipped with three accelerometer transducers. These are connected on each end and in the middle. The bar is fixed in a horizontal borehole drilled into the cliff face near a fracture that is suspected to be unstable. The cracks and displacements will be transmitted to the bar and then recorded by the accelerometers. As with the microseismic monitoring network continuous recording of the measurements will be undertaken for a year followed by modelling of the results.

DISCUSSION

The two techniques that are being applied in the PROTECT project are well established in other fields. Azimuthal apparent resistivity is used in hydrogeological studies and microseismic networks are used to monitor acoustic crack emissions in underground mines. Their application for monitoring the stability of cliffs is novel and many factors have yet to be determined. Accurate surveying grids have been established at all the research sites in order to directly detect any small, surface movements. These grids will be surveyed three times per year.

Rockmass and external parameters may influence the measurements taken and hence additional data will also be collected. These will consist of hydrogeological, meteorological and geological data. Laboratory tests will be carried out on samples with different characteristics (water saturation, porosity, calcimetry of chalk etc.). These tests will be useful to identify the rupture mechanisms of chalk and limestone and help understand the evolution of acoustic crack emissions from the deformation to the rupture. Data on the structure of the cliffs will be compared with the geophysical measurements to determine the relationship between structure and vulnerability to collapse. The distribution and orientation of fractures, is likely to be a key parameter in the structural assessment. Fracture characteristics will be logged, including orientation, aperture, fill and weathering. Lithological classification will record the variation of lithology within and between cliff sections. The geophysical signature will be affected by both lithology and the vulnerability to collapse.

ACKNOWLEDGEMENTS

This research is partly funded by the Commission of the European Communities (Contract EVK3-CT-2000-00029). This paper is published by permission of the Director of the British Geological Survey (NERC).

REFERENCES

Busby J. P., 2000. The effectiveness of azimuthal apparent resistivity measurements as a method for determining fracture strike orientations. *Geophysical Prospecting* **48**, 677-695.

Habberjam G. M. and Watkins G. E., 1967. The use of a square configuration in resistivity prospecting. *Geophysical Prospecting* **15**, 445-467.

Lane J. W. Jr., Haeni F. P. and Watson W. M., 1995. Use of a square-array direct current resistivity method to detect fractures in crystalline bedrock in New Hampshire. *Ground Water*, **33**, 476-445.

Nunn K. R., Barker R. D. and Bamford D., 1983. In situ seismic and electrical measurements of fracture anisotropy in the Lincolnshire Chalk. *Quarterly Journal of Engineering Geology*, London **16**, 187-195.

Taylor R. W. & Fleming A. H., 1988. Characterizing jointed systems by azimuthal resistivity surveys. *Ground Water* **26**, 464-474.

Probabilistic and fuzzy reliability analysis in stability assessment

C.CHERUBINI, Department of Civil and Environmental Engineering, Technical University of Bari, Italy, e-mail: c.cherubini@poliba.it
P.MASI, Geotechnical Engineer, Italy, e-mail: paolo_masi@libero.it

INTRODUCTION

Stability assessment is a difficult geotechnical problem thanks to many uncertainties. Some of them are connected to variability and uncertainties in evaluation of soil parameters.

In literature there are a collection of probabilistic methods - some of which are reported below - for uncertainty modelling, but few practice–oriented studies. So it seems difficult to apply the proposed solutions in term of Reliability Index and probability of failure to other real cases, owing to the difference, for example, between soil parameters correlation values in sample studies and real cases. Literature, in fact, seems to take no interest in investigating the stability of the different reliability approach solutions, with correlation coefficient varying. Such a study would also give interesting indications in soil cross correlation influences on slope stability. The purpose of the present study is, first, to supply such a parametric study, also in order to allow a comparison between the solutions coming from different reliability approaches, and secondly, to propose a simple but useful algorithm which allows the uncertainty propagation in the solution.

SOIL VARIABILITY

The first step in a stability assessment is to model soil correctly. To this purpose three primary sources of geotechnical uncertainty have to be distinguished: inherent variability, measurement error and transformation uncertainty (Phoon and Kulhawy, 1999). The first of these sources results from the natural geologic processes that produce and continually modify the soil mass in situ, the second is caused by equipment and random testing effects, the third is connected to hypothesis, idealizations and simplifications introduced by the solution models.

The spatial variation connected to inherent variability and measurement errors can be conveniently decomposed into a smoothly varying trend function and a fluctuating component as follows:

$$\xi (z) = t(z) + w(z);$$

in which ξ is the soil property, and $w(z)$ is a homogeneous random field (i.e. a random field with mean and variance constant with depth, and for which the correlation between the deviations at two different depths is a function only of their separation distance). An important consequence of this representation is the possibility of determining the soil property scale of fluctuation. This parameter supplies useful indications about the correlation distance of the property in study and can be used in order to apply a variance reduction algorithm. A simple but approximate method to calculate the scale of fluctuation is given by Vanmarcke (1977) as:

$$\delta_v = 0.8 \ d;$$

in which δ_v is the vertical scale of fluctuation, and d is the average distance between intersections of the fluctuating property and its trend function.

Once the scale of fluctuation has been calculated, the successive step is to reduce the variances of soil parameters using the δ_v values. An effective and well documented formula to the purpose is:

$$Var(X_i) = \frac{\text{var}}{n} + \text{var} \, \Gamma^2(L) \; ;$$

in which
Var is the reduced variance for the i - th parameter
X_i represents the examined soil parameter;
var is the X_i variance;
n is the number of considered samples for the $i - th$ layer;

Γ^2 is the function : $\Gamma^2(C_i) = \dfrac{\delta_x}{L_x} * \dfrac{\delta_y}{L_y} \; ;$

δ_x e δ_y are the horizontal and vertical scale of fluctuations.

A correct soil variability representation supplies good indications about trend and random field, which allows correct reduction of the variance using the above reported formula.

RELIABILITY ANALYSIS
In order to perform a reliability analysis a lot of indications have to be considered. Some of them are contained in studies like Hassan and Wolff (1999). In these studies the importance of β calculation chosen algorithm is highlighted. In fact there is the possibility of obtaining substantially different values of Reliability Index, in accordance with the chosen deterministic or probabilistic surface (or band) on which a reliability analysis has to be performed. Obviously the aim of such an analysis is to find the minimum β value, which indicates the most critical surface. In particular Hassan and Wolff suggest a simple but effective algorithm in order to calculate β_{min} value and to find the most critical slip surface. He also found an indicator of β_{min}, which is β_f. This is a Reliability Index that *"can be associated with a slope – but*

not any specific surface ". In the present study, in order to determine the most critical surface, the Hassan and Wolff algorithm will be used, furthermore the Reliability Indexes, coming from different procedures, will be examined in order to demonstrate the stability of FOSM (both with normal and lognormal distributions, using the Taylor series approximate method) and MC solutions, with the correlation coefficient between c and ϕ varying. This type of study will also allow investigation of the variation in Reliability Indexes and P_f values with correlation coefficient varying.

PARAMETRIC STUDY
To reach the mentioned results a sample slope is used, on which all the analyses in question are performed. The sample slope is near Aliano (Cherubini et al.,1994), a small town in Basilicata Region. This is a two-layer slope, whose soil characteristics are summarized in figs.1 and 2.

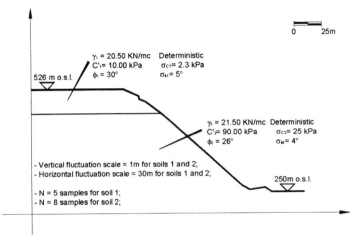

Fig. 1. A View of Aliano slope and a cross section with soil characteristics.

In the study the only random modelled parameters will be effective cohesion c and friction angle ϕ, unit weight being considered deterministic. The above mentioned procedure for variance reduction, is therefore applied only to c and ϕ.

The next step consists of running the search algorithm in order to find the most critical surface. To the purpose, with the strength parameters set to a specific combination, the critical surface was searched to find a minimum factor of safety FS_i for each combination. These combinations are obtained by sequentially setting each parameter at weak value $(E(X_i) - m\sigma(X_i))$ with the remaining parameters set at their mean values. Each parameter combination gives one critical surface with factor of safety FS_i, for a total of n surfaces and factors of safety. According to the indications found in literature the m value, in the present study, is set to 1. Each so found surface is fixed and a full evaluation of β_i is performed using the above mentioned Taylor series method.

Among these surfaces the most critical is chosen to be that of lowest β_i.

The calculation of Reliability Index can't supply a complete idea of the problem in a slope stability assessment, more complete indications about the subject are given by the Probability of failure. In order to determine Probability of failure it is necessary to make some hypothesis about the β distribution. Such an assertion has noticed that Reliability Index is something like a normalized random variable whose distribution is unknown. In literature the FS or β probability density function is commonly considered as lognormal (U.S. Army (1993)), even though a common approach is to assume it is a normal probability density function.

In any case P_f value can be supplied simply using the well known diagram (Harr, 1987).

In the case of lognormal probability density function it has to be remembered that Reliability Index is given by:

$$\beta = \frac{E[\ln(FS)]}{\sigma_{\ln FS}};$$

where:

$$E[\ln(FS)] = \ln(E[FS]) - \frac{\sigma_{\ln FS}^2}{2};$$

and

$$\sigma_{\ln FS} = \sqrt{\ln\left(1 + \left(\frac{\sigma_{FS}}{E[FS]}\right)^2\right)}.$$

In the present paper, in order to verify the stability of the solution coming from a reliability analysis, we calculated the variation of Reliability Index and P_f vs correlation coefficient between the soil properties. The only considered correlation is the cross correlation between within-a-layer parameters, ignoring the correlation between different layers soil parameters.

In order to determine the Reliability Index by means of Monte Carlo approach, a SLOPE/W special routine has been used. This routine allows, for each parameter, to extract a sample from a parameter population with specified mean and standard deviation, and to calculate the FS for each fixed set of parameter values.

In order to determine the most critical surface the present routine can't be totally used. In fact in SLOPE/W the critical slip surface is first determined based on the mean value of the input parameters. Probabilistic analysis is then performed on the found critical slip surface, taking into consideration input parameter variability. Not according this procedure to our search algorithm, we decided to make SLOPE/W perform an MC simulation on each critical surface before determined using specified parameter sets. Each of these sets has been determined by alternatively setting all soil parameters to mean values except one which is set to $E(X_i) - m\sigma$ (where m is set to -1).

By default SLOPE/W considers a FS normal distribution. Having verified the small shift between normally and lognormally distributed FS, we accepted the present approach. The determination of number of Monte Carlo trials is an important preliminary step which has to be highlighted: this number has to be so great to allow a small system error in estimating the FS distribution. About the subject in the present study we used a well documented formula which is:

$$N = \frac{h_{\alpha/2}^2}{4\varepsilon^2};$$

where:

$x_{\alpha/2}$ is the number of success in N trials such that the probability they will be above or below that value is no greater than $\alpha/2$;

φ is the normal cumulative distribution function;

$$h_{\alpha/2} = \varphi^{-1}\left(\frac{1}{2} - \frac{\alpha}{2}\right);$$

$$\varepsilon = R - \left(\frac{x_{\alpha/2}}{N}\right).$$

In order to reach a good approach level we decided to realize a Monte Carlo simulation of which the result doesn't differ by more than 1% from the estimated value ($\varepsilon=0.01$) with 99% confidence ($1 - \alpha = 0.99$). So the MC procedure required $h_{\alpha/2}=2.58$ and $N = 16641$ trials. So it has be assumed to realize 20000 MC trials for each MC simulation. In order to confirm this assumption a preliminary study on the stabilization of MC solution has been carried on making the number of MC trials grow from 10 to 20000 MC trials. The simple diagram in fig. 4 highlights the agreement between the results and theory.

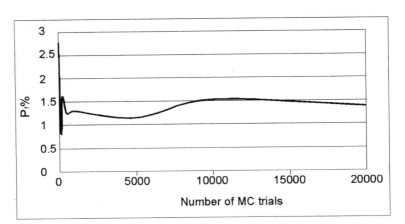

Fig. 2. Failure Probability vs Number of Monte Carlo trials

A comparison between the results of the stability analysis are shown in figures 3 and 4.

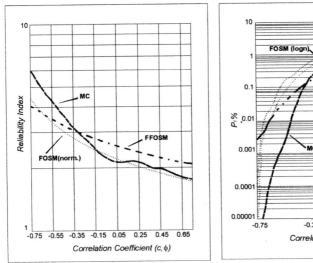

Fig. 3. Reliability Index vs Correlation Coefficient

Fig. 4. P_f% vs Correlation Coefficient

FUZZY ALGORITHM FOR SOIL PARAMETER TREATMENT

A couple of different considerations induced us to apply this type of uncertainty modelling. The first of them is that soil parameters are mostly manually measured using procedures which allow a great part of subjectivity. The usual probabilistic reliability approach stops the propagation of the subjectivity-connected uncertainty in first steps of the analysis, this fact being strictly connected with the nature of the usual mathematical logic. The second consideration is that some of the uncertainties connected to measured geotechnical parameters may be nonstochastic in nature. A reliability analysis which allows nonstochastic uncertainties to propagate in a reliability solution can supply the great advantage of using a well known algorithm (FOSM) for stability assessment together without the limits connected to an exclusive logic data treatment.

The proposed original algorithm consists in defining soil parameters as fuzzy numbers. To the purpose a useful membership shape is the trapezoidal as shown in fig. 7. In this figure a,b,c,d are calculated as:

$$a = E(X_i) - m_1 \sigma(X_1);$$
$$b = E(X_i) - m_2 \sigma(X_1);$$
$$c = E(X_i) + m_2 \sigma(X_1);$$
$$d = E(X_i) + m_1 \sigma(X_1);$$

m_i being the number of standard deviations which, in the present study, are assumed to be:

$$m_1 = 0.5;$$
$$m_2 = 1.5.$$

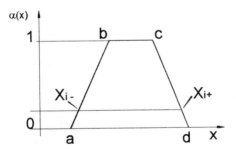

Fig. 5. Trapezoidal fuzzy number

In order to find the most critical surface the analysis will follow the same above mentioned algorithm which allows the calculation of β_{min} . In this case there is the problem of each β value calculation, because Taylor's method for approximate derivative determination can't be applied.

To the purpose a useful, but simple way to proceed is to realize some fuzzy number α - *cuts* and to calculate the FS corresponding to the so found parameter values. Each α - *cut*, in fact, supplies two determinations of each X_i parameter, named X_i^+ and X_i^-. In the present study we assumed to do 10 α - *cut* (from 0.1 to 0.95 possibility level), so we obtain $2*4 = 8$ parameter determinations for each investigated possibility level.

So each possibility level allows the calculation of two FS named FS^+ and FS^-. Now the derivatives calculation for each α - *cut* is possible simply applying the approximate method to each α - *cut* couple of values . The final step consists in assuming for each derivative value the fuzzy mean value (which is also the most possible value) calculated with the formula:

$$E\left(\frac{\partial F}{\partial X_j}\right) = \frac{\sum_{i=1}^{N} \alpha_i \cdot w_{\alpha i}}{N} \; ;$$

where

α_i is the investigated possibility level

$$w_{\alpha i} = \frac{FS_{\alpha i}^+ + FS_{\alpha i}^-}{2 m_i \sigma(X_i)} \; .$$

m_i is the number of standard deviation of which X_i^+ and X_i^- are greater or smaller than mean of X_i.

N is the number of α - *cut*.

This simple algorithm allows the application of the FOSM model without making assumption about nonstochastic parameter uncertainties.

In order to verify also the stability of this solution, a parametric study has been made making the correlation coefficient between c and ϕ vary from –0.75 to 0.75.

The obtained curves, represented in figs. 5 and 6, seem also to confirm a substantial agreement between different methods, being of few percentage points the difference in P_f % obtained for a 0.75 correlation coefficient value. The real correlation coefficient values, included between –0.3 and –0.75 (Cherubini, 1997), make the results band - between the lowest to the greatest value - of 0.3 P_f percentage points.

CONCLUSIONS

The present study guidelines have been to supply and compare useful model application results to a stability reliability analysis. Although these models are greatly used and diffused, in literature there aren't ever made clear a lot of aspects connected to their application. Despite to this fact, such aspects (the searching for most critical probabilistic surface, β_{min} calculation, data uncertainty propagation and modelling) seem to play an essential role in a good quality solution. Another frequent assumption in literature reliability analysis is the absence of correlation between soil parameters. Such an assumption, being very far from the reality, didn't allow verification of the stability of reliability solutions in cases of a not zero correlation coefficient. In the present study, in order to supply practice – oriented indications we investigated reliability solution behaviour within a large enough band of correlation coefficients.

The last step we decided to investigate was the uncertainty propagation in reliability solutions. The proposed algorithm seems to supply all the advantages of a fuzzy treatment of nonstochastic manually measured soils parameters, together with the possibility of a well known model application, the stability of which has just been demonstrated.

REFERENCES

Cherubini C., Cotecchia V. Guerricchio A., Mastromattei R., (1994) *"The stability conditions of the town of Aliano (Southern Italy)"* 7th International IAEG Congress pp. 2145 - 2153, Balkema, Rotterdam.

Cherubini C. (1997) *"Data and considerations on the variability of geotechnical properties of soils"*. Advances in Safety & Reliability. ESREL vol.2, pp.1583 – 1591. Lisbon

Harr (1987). *"Reliability – Based design in Civil Engineering"*. Mc Graw Hill Book Co. N.Y.

Hassan A. and Wolff T.F.(1999), "Search Algorithm for Minimum Reliability Index of Earth slopes". Journal of Geotechnical and Geoenvironmental Engineering, April 1999, pp. 301-308.

Phoon K.K. and Kulhawy F.H.(1999) *"Characterization of geotechnical variability"* Canadian Geotechnical Journal 36 (1999), pp. 612 – 624

Vanmarcke E.H., (1977). *"Probabilistic modeling of soil profiles"*. Journal of the Geotechnical Engineering Division, ASCE, 103 (GT11), pp. 1227 – 1246.

U.S. Army (1993). *"Reliability assessment of navigation structures: Stability assessment of existing gravity structures"* Eng. Tech. Letter No. 1110 – 2 – 310, U.S. Army Corps of Engineers, CECW – ED.

Strategies for quantitative landslide hazard assessment

PROFESSOR R.N. CHOWDHURY, University of Wollongong, Wollongong NSW Australia 2522
DR P. FLENTJE, University of Wollongong, Wollongong NSW Australia 2522
DR M. HAYNE, Geoscience, Australia
D. GORDON, Geoscience, Australia

INTRODUCTION

In recent years there has been an increasing trend towards developing systematic approaches for assessing landslide hazard and risk. Probability and hazard assessment requires an understanding of the factors which increase the failure susceptibility of slopes as well as the frequency of triggering factors such as rainstorms and earthquakes. This paper is concerned primarily with rainfall triggered landslides and attention is focussed on the transition from qualitative to quantitative assessments.

A multi-level strategy comprising of methods of different complexity with different requirements of investigation and data, has already been emphasised in previous published work of the authors. A hazard-consequence matrix approach has also been advocated by the authors and by other researchers (Flentje et al 2000, AGS 2000).

In this paper attention is given to quantitative hazard and risk assessments concerning existing landslide sites and potential areas of landsliding with emphasis on the latter. For the first type of assessment, historical frequency of landsliding combined with observational subsurface monitoring provide the main elements of a direct approach which has been successfully implemented but for which further refinements are in progress.

Identifying potential areas of landsliding with associated levels of hazard and risk is far more difficult and requires careful modeling and interpretation. Attributes of potential areas of landsliding may be compared to corresponding attributes of existing landslide areas. Most important attributes are first selected and the critical ranges or values of an attribute can be identified using GIS-based approaches for querying and analysis. For example, different geological units can be ranked in order of failure susceptibility and the same can be done with slope inclination, geomorphology and other factors.

Instability – Planning and Management, Thomas Telford, London, 2002, 219–228

This paper refers to research work carried out and in progress at the University of Wollongong with particular reference to an urban area, the Wollongong Local Government Area, within the State of New South Wales (Australia). The research work is carried out with the support of external partners which include the Wollongong City Council (WCC), the Rail Infrastructure Corporation (RIC) and Geoscience Australia (GA) previously known as the Australian Geological Survey Organisation (AGSO). Strategic partnership with GA is of particular value for modeling and analysis within the framework of Geographical Information Systems (GIS).

The Wollongong Local Government Area is under the control of WCC and their management framework for land-use in the hilly suburbs has been discussed in a companion paper (Flentje et al 2002). A brief description of the study area has been presented by Flentje and Chowdhury (1999) and includes topography, geology and the distribution of main landslide types. Uncertainties related to the assessment of landslide hazard and risk have been discussed by Chowdhury and Flentje (2002).

BASIS FOR QUANTITATIVE ASSESSMENTS
The assessment of landslide susceptibility and, therefore, of hazard requires the assessment of surface and subsurface conditions, geology and existing landslides. The occurrence of landslides in the study area is primarily associated with increasing pore water pressures and, therefore, rainfall is the dominating trigger factor. Many existing landslides are reactivated during periods of intense and/or prolonged rainfall. Thus the frequency of landsliding for an existing landslide site is often controlled by rainfall. Considering a large area or region, frequency of widespread landsliding is also controlled by the frequency of rainstorms. The locations of individual landslides are often difficult to predict. There are a number of influencing factors which contribute to landslide susceptibility and the interactions between these different factors are often complex.

Geotechnical analyses based on careful geological and hydrologic models can be useful for assessing the landslide susceptibility of individual sites provided reliable and accurate subsurface data are available. However, such an approach is unduly cumbersome for hazard zoning of large areas or regions and also prohibitive in terms of cost and time.

Thus it is necessary to assess hazard and risk in urban sloping areas with very limited subsurface information. Therefore, it is necessary to compile the historical data on the occurrence and distribution of landslides and combine it with information on topography, geology, rainfall and other important factors. At known landslide sites, the frequency of recurrence of landsliding can also be ascertained from an analysis of historical data. In this regard, three important aspects of the University of Wollongong research effort are

(a) GIS-based landslide inventory (Database)
(b) observational approach based primarily on subsurface monitoring, of inclinometers and piezometers and
(c) rainfall analyses ~ Records.   Long term monitoring

LANDSLIDE HAZARD AT AN EXISTING SITES OF SLOPE INSTABILITY

The number of recurrences of landsliding at each site can be used as a basis for assessing the frequency and, therefore, the hazard of landsliding. A more appropriate parameter is the historical landslide frequency, the ratio of number of recurrences to the time period in which these occurred. For a slow moving landslide which reactivates intermittently, the number of recurrences should include significant shear movements detected by inclinometers even if surface observation did not indicate a full-scale failure.

Flentje (1998) ranked all the landslides recorded in his database according to frequency of landslide recurrence. He found that the highest ranked 3% of landslides were associated with 62% of the houses destroyed and 52% of the houses damaged.

PREDICTION OF LANDSLIDING BASED ON RAINFALL THRESHOLDS

For a particular site which has been monitored for a significant period of time including major rainstorm events, the dates of landslide recurrence are known. Rainfall data from nearby rainfall stations are then analysed in such a way that magnitudes of cumulative rainfall can be calculated, for different antecedent periods (eg 3 days, 7 days, 30 days etc). Consequently the most appropriate antecedent period and the associated threshold rainfall magnitude can be identified. A similar exercise has been carried out for a number of individual landslides. Considering all of them together within the study area, threshold rainfall magnitudes have been proposed for

(a) reactivation of shear movements, and
(b) reactivation of disruptive slope instability.

As a first approximation, these predictive rainfall magnitudes may also be considered as thresholds for the occurrence of first-time landsliding during major rainstorm events. This approach is particularly useful for real-time or near-real-time predictions during rainstorms. This approach can facilitate landslide management and risk mitigation through warnings and evacuation procedures where necessary. However, this approach does not facilitate identification of the severity of the hazard based on location. In other words, the spatial variability of the hazard remains largely unknown outside the existing landslide areas which may have been mapped.

LANDSLIDE HAZARD OUTSIDE KNOWN SITES OF SLOPE INSTABILITY

Introduction

It is very important to identify the severity of landslide hazard in different parts of a large urban area or region or along a transportation route. Qualitative approaches for zoning of land into different hazard categories rely primarily on local experience and judgement of geologists and geotechnical engineers. Progress towards a more systematic and quantitative methodology requires a knowledge-based approach. Attributes of known landslide areas must be identified and a search for similar attributes can then be made outside the landslide areas. Such an approach was tried with some success by the University of Wollongong team for a small part of the study area in 1997 (Stevenson 1997). It was then decided to carry out a more comprehensive study which would require, as a first step, the GIS-based identification of attributes of existing landslides related to important influencing factors or GIS-based themes such as, geology, geomorphology, slope inclination, curvature and vegetation. The approach was proposed by the UOW team to the GA team and has been tested in another study area, Southeast Queensland, as reported in Hayne and Gordon (2001).

Methodology

Maps showing the spatial distribution of these attributes are overlayed with one another within a GIS framework, generating a complex raster based map containing, for the Wollongong study area, approximately 2.5 million 2m × 2m pixels. Each of these pixels has a specific character string comprising the input themes. Analysis of the relationships between these character strings and existing landslides is carried out by a statistical computer package called "See5", as described briefly in a following section.

The nature of a regional assessment of potential landslide occurrence necessitates data input, storage and manipulation in a GIS environment and the application of sophisticated analysis tools. The application of GIS is critical in this instance as the susceptibility to landslides is due to various combinations of a number of physical conditions. Relationships, therefore, become very complex and data sets can be very large in order to achieve maximum spatial resolution. Application of analysis tools to the GIS environment allows the evaluation of relationships between physical conditions and landslide occurrences at an appropriate resolution, which is confined by the resolution of the input data. A detailed pixel resolution of 2m × 2m has been selected for the Wollongong study.

With the development of GIS themes that model the physical environment, an awareness of scale is paramount. Digitally generated images are scaleless and can be reproduced at broadly ranging and potentially inappropriate scales with regard to the targeted investigation. Furthermore, the accuracy and reliability of the results reflects the accuracy and reliability of the input data. The best possible assessment will result from the application to the landslide model of the most accurate data sets. In this respect the landslide data gathered by the University of Wollongong are of high quality. The landslides were mapped, initially over a 2 year period, by an experienced local engineering geologist directly onto 1:4000 scale field sheets, which included 2m contours from the Land Information Centre and cadastre managed by the Wollongong City Council. The mapping was also carried out with the aid of 1:4000 orthophoto map sheets (a map with the 2m contours plotted on a black and white aerial photograph), a hand held Global Positioning System, a 30m tape and a compass. The maps were comprehensively validated with extensive field checking and desktop evaluation.

Relevant Influencing Factors And GIS-based Themes

Factors relevant to GIS modeling of slope stability and how these have been applied as GIS themes are discussed briefly below. The relevance of each to the determination of terrain conditions and therefore areas of potential landsliding are discussed below.

Existing Landslides: The accurate and detailed mapping of landslides is paramount as it is the identification of the terrain conditions of each slide that is the basis for the identification of areas of potential landslide activity. Landslide data was mapped and compiled by the University of Wollongong and digitized into the Genamap GIS software available at the Wollongong City Council (Flentje 1998, Chowdhury and Flentje 1998). This data was provided to and attributed by GA (AGSO)and UOW. Related to this is the Landslide database at the University of Wollongong, which has up to 75 fields of data available for each landslide site (Chowdhury and Flentje, 1998, Flentje and Chowdhury, 1999). This data includes; site reference code, description, volume, classification, occurrence and recurrence. Individual landslide classification has been provided and added to the attribute list for each landslide site by GA (AGSO).

Digital Elevation Model (DEM): The DEM acts as the source theme for the development of a number of the terrain conditions, which include, slope, geomorphology and curvature. Developed from digital topographic data (2m contour interval) devised from digital photogrammetry and scanning of existing mapping, the DEM has been generated using ArcInfo TOPOGRID algorithms from the digital contours. WCC supplied this data under a digital data license agreement.

Slope Angle: Slope angle is calculated either on a floating point basis or in increments (i.e. 0°-3°), and is based on a pixel by pixel evaluation (3 x 3 pixel matrix). The slope angle for each landslide is measured from a series of distributed points within the landslide. Slope angle was derived directly from the DEM at 5m pixel size within ArcInfo GIS.

Geology: The University of Wollongong (UOW) has quantitatively demonstrated that specific geological formations have a significant relationship to landsliding in comparison to other formations. They carried out detailed and comprehensive geological mapping that was drafted on to the 1:4000 base maps. This mapping was based on field inspections, borehole locations and logs, and reference to various existing maps ranging in scale from 1:6336 to 1:100000. The University of Wollongong (UOW) digitized this data into the Genamap GIS software of the Wollongong City Council. This attributed data was passed on to GA (AGSO). Geology, as it relates to individual landslides, was represented with respect to a series of distributed points within the landslide.

Macro Scale Curvature: This is essentially an analysis of the shape of catchments or stream valleys within the study area. In the context of the present study, it is a semi-quantitative definition of major ridges, valleys and planar landforms. Ordinarily, valleys that contain the headwaters of streams define areas identified as concave; ridge lines and planar slopes separate these. Concave areas accumulate water and colluvial material and often are more prone to landslide because of concentration of runoff, undercutting by streams and higher subsurface pore pressures than convex slopes. Ridgelines generally define areas having a relatively shallow depth to bedrock and which shed water. Ridge forms were developed by inverting the DEM and developing a drainage pattern (thalweg) based on a flow accumulation of 200 pixels with a 10m buffer applied either side of this line. Valley forms were created by developing a drainage pattern (thalweg) from the DEM based on a flow accumulation of 150 pixels and a 30m buffer either side of this line. The remaining areas were tagged as planar. The geomorphology of each landslide was calculated from a series of distributed points within the landslide.

Micro scale Curvature: Curvature is a measure of concavity and convexity and has been broadly considered by the analysis of above. However, a micro-specific analysis was also considered to be desirable. The curvature of each cell (2m x 2m) in the DEM has been developed by fitting a parametric surface to the neighbourhood of that cell (the 8 surrounding pixels). It is calculated as a floating point that is based on a pixel-by-pixel evaluation. Curvature was calculated directly from the DEM within ArcInfo GIS. As it relates to individual landslides, curvature was calculated from a series of distributed points within the landslide.

Vegetation: Whilst vegetation cover may increase gross infiltration, through the process of evapo-transpiration it has the effect of lowering net infiltration and hence groundwater tables and pore water pressures. Thus pore water pressures within forested areas are always expected to be lower than those within unforested slopes that are otherwise identical. Consequently soil shear strengths after rainfall will be higher in forested slopes. An increase in the material strength of the upper soil horizons and or near surface material resulting from root systems of the vegetation cover can also be significant.

The description of vegetation cover is based on a vegetation community type categorisation. Cleared vegetation relates to land cover such as pasture, crops, settlements and bare areas, whereas uncleared areas can be defined as variations within woody native vegetation communities. Digital vegetation data, mapped at 1:4000 scale is scheduled for release during 2002 by the New South Wales National Parks and Wildlife Service. The data is being generated with the aid of aerial photography and field analysis carried out during the period 1999 to 2001. For each landslide, the density of vegetation was calculated from a series of distributed points within the landslide.

Machine Learning Methodology

The determination of further potential areas of landslide occurrence has been undertaken using a machine-learning approach. The term 'machine learning' is the science of computer modeling of learning processes. The See5 algorithm (Quinlan, 1993) was selected to carry out this computer modeling. See5 develops a decision tree that, on the basis of answers to questions about the physical attributes associated with landslides as well as the attributes of non-landslide areas, predicts the value of other areas. Questions may take the form of: "is there a landslide at this location?" "what is the slope angle?" "what type of geology is relevant to the site?" etc. It takes the knowledge from existing data (the data that has been compiled in the GIS-based on landslides mapped in the Wollongong area) and uses this as training for the prediction of new areas susceptible to landslide as well as areas with no susceptibility. This was achieved using a decision tree approach as discussed below.

A decision tree is composed of leaves that correspond to classes (either the presence or absence of landslide), decision nodes that correspond to attributes (themes) of the data being classified (ie geomorphology, geology, curvature, slope angle etc), and arcs that correspond to values for these attributes. By selecting the most "informative" attribute at each node the data set can be divided into distinct subsets. *A decision tree is important not because it summarises what we know about the known landslides that are used as training cases, but because it is used to help identify potential landslide areas.* The See5 algorithm makes this prediction by constructing a decision tree. Tests are constructed on each attribute so as to associate a separate branch with each value for which cases are available.

The See5 algorithm selects a number of pixels (the number and distribution of points is a variable) within the landslide (currently it selects 10 points along a central down slope axis of the landslide) for the complete landslide population and statistically analyses the character string for each pixel to determine the importance of the relationships between the contributing factors. With a population of, say 200 landslides, then 2000 landslide-related points will be examined.

The See5 algorithm then selects the same number of randomly distributed points within the trial study area but outside of known landslide boundaries and statistically analyses the character string for each pixel to determine the percentage importance of the relationships between factors. The program then compares the rules generated by each set of data and compares the result between datasets. This comparison provides a level of confidence. The rules developed provide a set of prior probabilities generated on a set of sample points.

Within the GIS environment individual rules are located spatially for every pixel in the study area. Rule sets are ordered and spatially located from lowest to highest confidence. Ranges of potential hazard are specified and plotted based on the confidence limits. Lower confidence limits may over-estimate the spatial extent of potential landslide areas and limits are based on field assessment.

Further details of the procedure and examples will be provided elsewhere. The preliminary results of modeling potential hazard are not included here due to space limitations. However, some of the important findings are discussed below.

Preliminary Trial Application
A preliminary trial of this methodology covering 43km^2 area with 200 known landslides between Thirroul and Woonona within the City of Wollongong Local Government Area was undertaken. A total of five of the above themes were incorporated in this initial trial; DEM, geomorphology, slope angle, geology and landslides. One additional theme, slope length, was also used initially. Slopes were subdivided into those having lengths greater than or equal to 20m and those with lengths less than 20m. This was intended to highlight the higher susceptibility of longer slopes to landsliding relative to the shorter slopes.

Overlaying the five vector based polygon themes within this smaller trial area generated a new map theme with approximately 820,000 polygons. These 820,000 polygons comprised approximately 1500 specific character string categories. This number of polygons confirmed the need for some statistical analyses of the data. An initial analysis of this data using the See5 algorithm produced a distribution of hazard which showed reasonable hazard locations but did not highlight the variability of hazard. This was not considered to be an acceptable outcome by the research team.

The results of this experimental trial revealed the complexity and challenges that exist for the prediction of future landslide areas based on this methodology. Four important lessons became apparent as a result of this preliminary trial.

1. The preliminary trial area and the family of landslides it contained were not representative of the escarpment as a whole.

2. The landslides were assessed as landslides only without specifying whether, for example, a landslide was a shallow or deep-seated landslide or whether it was a slide or debris–flow. Differentiation of landslide types, based on an internationally recognised classification system was, therefore, considered to be essential.
3. Stream catchment style boundaries are appropriate to define the study area as opposed to map sheet boundaries or suburb boundaries.
4. Complex relationships and large datasets generated by the GIS-based methodology necessitate computer based statistical analyses and in particular an expert system or machine learning algorithm.
5. The inclusion of depth of colluvium as a 'theme' or influencing factor is necessary.

These lessons have been considered and each has been addressed in a new trial of a 185km^2 area which contains 446 known landslides is currently in progress. That new trial is currently in progress. This extended trail area generates approximately 46 million pixels.

DISCUSSION
Progress from qualitative to quantitative landslide hazard assessment requires reliable data on surface and subsurface features of an area or region and on the distribution of existing landslides. In particular, for urban areas and transportation routes, the development of a landslide database can be very useful. Approaches for assessing the hazard of reactivation of existing landslides are quite different from those required for assessing the hazard of potential landslides outside the boundaries of existing landslides. For rainfall triggered landsliding, it is very useful to use an observational approach combined with rainfall analyses to estimate threshold rainfall magnitudes for landsliding to occur. However, this approach cannot facilitate the study of how potential landslide hazard varies within the whole study area. To study this spatial variability of landslide hazard in potential areas, a knowledge-based approach may be used. Such an approach considers important influencing factors for landslide occurrence and combines them within a GIS framework. The attributes of existing landslides can be studied and modeling and simulation then carried out for the potential areas as well.

CONCLUSIONS
1. A comprehensive landslide database is essential for quantitative landslide hazard assessment relevant to urban hilly areas and transportation routes.
2. An observational approach based on surface and subsurface monitoring of movements and pore water pressures can be very useful for assessing the reactivation of existing landslide areas.
3. Rainfall analyses combined with historical data on landslide occurrence and recurrence as well as observational data on subsurface movement enable the estimation of threshold rainfall magnitudes for landsliding to occur.

4. Potential landslide hazard outside the boundaries of existing landslides requires modeling and simulation based on a knowledge-based approach. This can best be done within a GIS-based environment using a number of "themes" or attributes representing the key influencing factors for landslide susceptibility.
5. Previous knowledge, research and professional experience within a particular geographical area will facilitate the correct selection of "themes" which are of key importance to landsliding in that area.
6. It is also important to differentiate between different types of landslides before modeling the attributes within existing landslides and within other areas.
7. Research concerning the modeling of hazard in potential areas is still continuing.

REFERENCES

Stevenson, N. (1997), Hazard Assessment of Landslides - Towards a Quantitative Approach, Bachelor of Engineering Thesis, University of Wollongong (unpublished).

Chowdhury, R.N. and Flentje, P. (1998), Effective Urban Landslide Hazard Assessment, Proceedings 8th IAEG Congress, Vancouver, Canada, 2, 871-77.

Flentje, P. (1998), Computer Based Landslide Hazard and Risk Assessment (Northern Illawarra Region of New South Wales, Australia), PhD Thesis, University of Wollongong, Australia (unpublished), 525p.

Flentje, P. and Chowdhury, R.N. (1999), Quantitative Landslide Hazard Assessment in Urban Area, Proceedings 8th Australia-New Zealand Conference on Geomechanics, Hobart, Tasmania, Vol. 1, 115-120.

AGS (2000), Landslide Risk Management Concepts and Guidelines, Australian Geomechanics, Vol. 35, No.1, March 2000, 49-92.

Flentje, P., Chowdhury, R.N. and Chit Ko Ko (2000), A Matrix Approach for Assessing Landslide Risk in the Context of a Comprehensive Strategy, Proceedings of GeoEng 2000 Conference, Melbourne, Australia, Technomic Publishing Co. Inc., CD Rom (Abstracts Volume 52pp).

Hayne, M.C. and Gordon, D. (2001), Regional Landslide Hazard Estimation, A GIS/Decision Tree Analysis: Southeast Queensland Australia, Proceedings 14th Southeast Asian Geotechnical Conference, Hong Kong (in press).

Chowdhury, R.N. and Flentje, P. (2002), Uncertainties in Rainfall-Induced Landslide Hazard, Symposium -in-Print, Quarterly Journal of Engineering Geology and Hydrogeology (in press).

Flentje, P., Chowdhury, R.N. and Tobin, P. (2002), GIS Framework for Landslide Risk Management, Proceedings of the Conference on Instability - Planning and Management, Ventnor, Isle of Wight, UK, (contribution to this volume)

Analysis of land sensitivity through the use of remote sensing derived thematic maps: landslide hazard

L.FERRIGNO and G. SPILOTRO, Dept. Structural and Geotechnical Engineering and of Engineering Geology, Basilicata University, Potenza, Italy

INTRODUCTION

The concepts of forecast and prevention of risk events are very important in the perspective of territorial planning and civil protection. By the term forecast the Italian law (law 24/2/92 n°225) means also "the activity directed to the study and the determination of the causes of disastrous phenomena, to the identification of risks and to the individuation of those areas subject to risks".

In this study the main attention was given to the evaluation of hazard assessment, meant as spatial prediction from which, in a particular area, a phenomenon of soil instability can happen, without references to the occurrence time.

On a regional scale, as the deterministic models cannot be proposed for obvious reasons, morphometrical analyses and statistic, empiric or crossed models can be used.

In general terms two main approaches can be distinguished: direct and indirect methods. The former are exclusively based on the geomorphological analyses of the site of interest, leaving to the interpreter the temporal and spatial prediction of the areas potentially unstable. The indirect methods, on the other hand, are based on the analysis of thematic maps concerning the spatial distribution of the values of the different factors related to the landslides: lithology; slope; tectonic history and morphological evolution of the area; climatic and hydrogeological factors; type of soil and land use, human interactions. According to the used methodology to assign the weight of each factor and for the relationships applied between them for the spatial forecast of unstable areas, euristic or statistic models can be applied.

The main difference between the two approaches lies in the fact that in euristic models the focus is, a priori, on the role of each factor in the instability, not considering the landslides already happened; the risk is expressed by a mathematical relationship among the factors related to the instability. In statistic models, on the other hand, the weight to be associated to each factor is deduced from the spatial distribution of the instability phenomena already happened; univariate or multivariate statistic procedures can then be used.

The above mentioned methodologies use layers of information, the so-called thematic maps and nowadays they are able to use the wide range of territorial informative systems and related methods connected to them.

REMOTE SENSING

The remote sensing techniques can be very useful for the redaction of thematic maps and thus for the elaboration of risk maps. In fact, the images of wide areas can be acquired and elaborated rapidly, allowing a continuous updating of data, offering an homogeneity which is difficult to keep in soil investigations and eliminating the problems deriving from impossibility or difficulty in the accessibility.

It is necessary to keep in mind that the results obtained from the analysis of remote sensed data, obtained by images or photo at high or low altitude, need soil based validation; moreover, the choice of images and of methods of interpretation has to be done according to the specific characteristics of the investigated area and of the informative plan required.

Literature on the applications of remote sensing is very wide and increases continuously; each application has its own specific requirements as to the spatial, spectral and geometric resolution. In general terms, the sensors, useful for the observation of environmental parameters, can be divided in two main categories: those which operate in the microwave frequencies or those which operate in the visible or infrared frequencies (in the latter case the instruments are called "optical").

With reference to the basic thematic layers in the analysis of land sensibility, the possibilities are, at present, many and implemented by the birth of new satellites at a high geometric resolution.

In particular for the detection of structural features and of geomorphology it is possible to obtain interesting information either by the use of optical data (Landsat TM, Ikonos, Orb-View 3), through the contrast of reflectivity which highlights rightly faults or fractures, or by the Spot images in pancromatic, performing photogrammetrical stereo operations.

Further information derives also by thermal images, particularly from the monitoring of thermal inertia, which allows the recognition of morphostructural features and of the drainage net.

Also radar technology gives indication of   linear earth structures; it allows, moreover, the realisation of a DEM (Digital Elevation Model), from which information on slope and on the wideness of drainage channels are obtained. A particular application of SAR data regards, moreover, the Interferometry which gives data on topography and on deformations of land surface (subsidence). The topography deduced from an interferogram can lead to a DEM, which, together with those data coming from SPOT images, offers an acceptable degree of approximation (currently less than 10 m).

The main applications of Interferometry are related to the topographic field, but, with better and more sophisticated techniques, improving the geometrical resolution, they can be applied to monitoring of displacement (subsidence, landslides) and to the characterisation of the nature of the soil.

The condition of soil humidity can be deduced from information SAR; in fact the coefficient of backscatter is influenced by the soil humidity, even though the capacity of penetration of the signal in the soil is a function of its composition. At present it is possible to measure the humidity only at the first 5 cm of depth (Schmugge, 1990; Le Toan et al., 1998).

Analyses on the soil humidity can be taken even by using passive sensors, such as SSM/I (Special Sensor Microwave Imager). This microwave radiometer is formed by seven radiometers which measure, simultaneously, the emission of land surface and of atmosphere to the frequencies of microwaves; the spatial resolution changes with frequency and it changes from a minimum of about 13 km to a maximum of 70 km; it is thus useful for studies on a wide scale.

Interesting applications have been carried out on the soil humidity evaluation by the correlation of brillance temperature (Tb) with API index (antecedent pluviometrical index (Gabriele S. et al., 1997)). Another example is given by the extension of inundated areas by means of radiometric measurements (Ricciardi, 1998).

Geolithological information and information on the conditions of soil humidity can be obtained through interpretation of optical data, but it may be strongly affected by errors in areas modified by human activity, also for only agricultural purposes.
Information on the land use can be entirely deduced from the interpretation of optical data at a high resolution, which allows the survaillance and the monitoring of the area in different spectral regions, in different times and with different geometrical resolutions.

The optical data used in this study have been acquired from the satellite Landsat TM which, even though it provides a low geometrical resolution, gives different applications, thanks to information distinct in seven bands of the electromagnetic spectrum, which allows the interpretation of images with a different composition of colors, or with the results coming from the application of operations on different bands, such as the derivative, the division, the sum , the product, the difference.

STEVENSON'S MOD'FIED METHOD
In a test area of Basilicata Region, South of Italy, Stevenson's model (1977) of landslide hazard has been applied; it uses the following factors: index of plasticity (P), annual max piezometric condition (W), slope (S), complexity of surface (C) and land use (U).

The index of landslide hazard according to Stevenson is:

R= ((P + 2 x W) x ((S + 2 x C)) x U

Areas for which R > 50 belong to the maximum level of risk. Stevenson validated his model on an area in Tasmania, on the basis of several known landslides.

For the present study the weights associated to each factor have been determined according to the Van Westen's method (1993) on the basis of the historic series of landslides applied on the test area. (Fig. 1, Guida D. et al., 1995).

The test site belongs to the Municipality area of the city of Potenza , which is about 200 km2 wide. In this area there are outcropping soils which ranges between the Cretaceous and Middle Pliocene age, belonging to different formations and tectonic-stratigraphical units.

BASIC THEMATIC LAYERS
In order to use and validate Stevenson's model for the determination of landslide hazard assessment the following basic factors have been evaluated and mapped: slope, lythological and geotechnical characters, underground water condition, complexity of topographic profile, land use.

The plasticity index, P, which symbolizes the potential strength of a soil, varies in three ranges, corresponding to high, medium and low values.

Water factor (W) is referred by the authors to the max height of the water table towards a possible local slip surface. Score 3 has been attributed for a depth of water table less then 5 m; W = 2 for depth ranging from 5 and 10 meters and W = 1 for depth greater than 10 meters.

Slope factor (S) is representative of the driving forces; it has been determined starting from a DEM obtained by cartographic data. Slope, ranging from 0 to 90°degrees, has been calculated in each nodal point of the DEM grid.

Figure 1: Landslides areas in on the Municipality area of the city of Potenza (Italy), (from Guida et al., 1995).

In the model three classes have been defined: for slopes ranging from 0 to 6° risk factor S = 1 has been given, for slopes ranging from 6° to 9° S = 2 and for slopes greater than 9°, S = 3.

The index of complexity, characteristic of the geomorphological history of the area, has been calculated by means of the surface interpolating each point and the six nearest points around it, having the equation

$$z = a + bx + cy + dx^2 + ey^2 + fxy$$

The measure of the dispersion around the regression surface, the standard error of z, calculated for every point of the grid, has been considered as the complexity of the slope. The calculated values range from 0.13 to 10. Also for complexity three intervals have been chosen: C = 1 for those areas whose complexity index ranges from 0 to 1, C = 2 for the interval 1- 4 and C = 3 for values greater than 4.

The soil use chart has been derived from multispectral images Landsat 5 TM taken in January, June and July 1992.

On the above mentioned images all the necessary techniques for a right photointerpretation have been used: georeferencing (UTM), Principal Components analysis, classification.

The images were acquired in three different periods of the year, to allow a distinction to be made between those areas of bare soil and those used for agriculture; in fact the latter shows, in some periods of the year, a lack of vegetal cover and could be confused with the former.

Five different classes have been defined useful for the application of the risk model: urban (U=1.25); bare soil (U=1.5); sown land , pasture and uncultivated bushy land (U=1.5); wood (U=1). In Fig. 2 the result is reported.

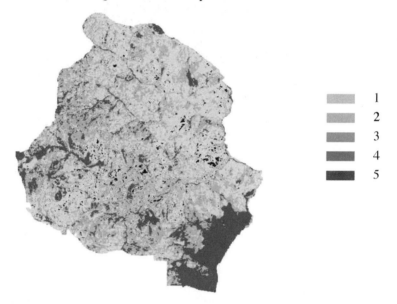

Figure 2: Land use map: 1) Urban; 2) Bare soil; 3) Sown land; 4) Pasture and uncultivated bushy land; 5) Wood.

VALIDATION OF THE METHOD

For the application of the model a methodology of matricial analysis associated to GIS has been used. (Spilotro et Al, 2001). The study area has been overlapped by a grid of nodal points with a regular mesh of 250 m wide. The matrix method allows us to sum in the high gradient context nodes or partial grids with reduced nodal distance. In our case and in a more restricted dominion, a regular grid with side of 50 m has been added. The degree of risk has been evaluated in each nodal point. The validation on the benchmark of the test area leads to the following modified Stevenson's equation:

$$R = ((0{,}9 \times P + 0{,}99 \times W) \times (0{,}6 \times S + 0{,}77 \times C)) \times 0{,}32 \times U$$

Three classes of risk have been introduced: the first, for R < 4.4, which corresponds to low risk. The second class of moderate risk, is defined for 4.4 < R < 5.9 . For R > 5.9 there is the class of high risk.

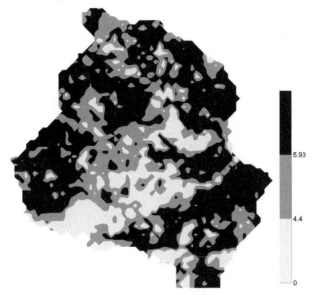

Figure 3: Risk map. Low risk for R < 4.4, Moderate risk for 4.4 < R < 5.9; High risk for R > 5.9.

CONCLUSION

The map of the hazard assessment of instability resulting from the use of the Stevenson's modified method on five basic informative layers is shown in fig. 3 with the risk index associated to a shade of colour.

The relative compliance of the model is very good, with 80 % of the landslide area belonging to high and middle classes of risk and with an error (landslides in low risk area) of only 4%.

On the basis of the present technological possibilities, three out of the five informative layers used in the method can be directly deduced from remote sensed images, and a fourth, the lithology, has very good perspectives of being deduced from remote sensing in non anthropized areas.

Like in every type of model, the evaluation of instability conditions of a specified area, has a reliability directly correlated to the degree of definition and precision of the information in every single thematic layer. Remote sensed information are fully comparable to the scale of the current risk map with those obtained by means of earth based surveying.

REFERENCES

CANORA F., LEANDRO G., SPILOTRO G., 2001. *Metodologia matriciale associata ai GIS per il processo di carte di vulnerabilità e di rischio.* FIST-GEOITALIA, Chieti, sept. 2001.

CARRARA A.,CARDINALI M., GUZZETTI F., REICHENBACH P, 1995. *Gisbased tecniques for mapping landslide hazard* .Accademic Pub., Dordvecht, Netherlands, pp.40.

CROSETTO M., ARAGUES F. P. (1999) *Radargrammetry and SAR interferometry for DEM generation: validation and data fusion.* Whorking group on Calibration and validation. SAR Whorkshop, oct 1999, Tuolouse, France.

ENGMAN E.T., MATTIKALLI N.M. (1997) *Microwave remote sensing of soil moisture for estimation of soil properties* . IEEE Transaction on Geoscience and remote sensing.

EUROPEAN SPACE AGENCY (1998) *ERS and its Applications: Land, subset :Flood mapping*, ESA BR-128/II.

GABRIELE S., NATIVI S., MAZZETTI P., PERGOLA N., ROMANO F., TRAMUTOLI V. *Satellite and ground based observations for extreme events interpretation:one application on precipitation anoomalies and floods in the south of Italy.*

GUIDA D., IACCARINO G. , LAZZARI S. (1995) Carta inventario delle frane dell'alta valle del fiume Basento (Potenza). Atti del convegno Ambiente fisico, uso e tutela del territorio, Potenza, 1995.

LE TOAN T., DAVIDSON M., BORDERIES P., CHENERIE I., MATTIA F., MANNINEN T., BORGEAUD M.,(1998) *Improved observation and Modelling of bare soil moisture retrieval.* Proc. 2nd Whorkshop on Retr.of Bio and Geo-physical Parameters from SAR data for Land application, oct 1998.

RICCIARDI M. (1998) *Evaluation of 19 and 37 GHz observations to study Seasonal and Interannual Flooding in the Pantanal.* Tesi di dottorato.

TENG W.L., WANG J.R., DORAISWAMY P.C., 1993, *Relationship between satellite microwave radiometric data, antecedent precipitation index, and regional soil moisture*, Int J. Remote sensing, 14,13, pp.2483-2500

SCHMUGGE (1985) *Remote sensing of soil moisture* .Hydrological forecasting, Jhon Wiley & sons.

STEVENSON (1977) *An empirical method for the evalutation of relative landslip risk* Bulletin of the International Association of Engineering Geology, n°16 Krefeld.

VAN WESTEN C.J. (1993) *Application of Geographic Information System to landslide hazard analysis*, ITC- Publication n.15, ITC, Enschede, pp 245.

VARNES D. J. , IAEG Commission on Landslides (1984) *Landlides Hazard Zonation, a review of principles and pratice*, UNESCO Paris.

GIS framework for landslide risk management

DR P. FLENTJE, University of Wollongong, Wollongong, NSW, Australia 2522
PROFESSOR R. CHOWDHURY, University of Wollongong, Wollongong, NSW,
Australia 2522
P. TOBIN, Wollongong City Council, Wollongong, NSW, Australia 2500

INTRODUCTION
The management of sloping land must be considered in the context of slope stability,
public safety and sustainable development. The social, economic and environmental
consequences of landslides can be enormous besides the threat to human safety.
Therefore, it is important to develop systematic and effective approaches within a
rational framework for risk management of landslides.   Sloping land can be
subjected to different types of use. For example, it may be used for agricultural
purposes, urban development, opening up or extending transportation routes or for
the construction of hydro-electric projects. Each type of use may have a different
type and extent of impact on the developed area and on areas adjacent to it.
Moreover, the consequences to human life, property and the environment are likely
to be very different in each case. Therefore, policies and procedures for hazard and
risk management including issues of acceptable/tolerable risk levels will depend on
the type of land-use.

This paper is concerned primarily with landslide risk assessment and management of
urbanised hilly areas. In particular, reference is made to the Wollongong Local
Government area in the state of New South Wales, Australia. While the main
strategies, policies and procedures are outlined briefly here, recent developments
concerning quantitative methods of hazard and risk assessment are outside the scope
of this paper. Geographical Information Systems (GIS) are an integral component of
landslide risk assessment and management in Wollongong.

KEY QUESTIONS FOR LANDSLIDE RISK MANAGEMENT
The main elements of landslide risk are hazard, elements at risk, vulnerability of
those elements, the consequences of landsliding and the levels of hazard and risk
which will be acceptable or tolerable to individuals and the community (Fig. 1).
Understanding, identification and assessment of hazard can not be limited to sites
and areas which are currently showing evidence of slope instability. It is extremely
important to recognise areas and individual sites which have suffered instability or
landsliding in the past even if there are no signs of continuing instability. More

importantly, landsliding may occur in new areas as a consequence of natural geological processes and triggering factors such as rainfall, earthquakes and stream erosion. Thus identification and assessment of landslide hazard in a region is a complex task which includes careful observation, analysis, modelling and judgement. Identifying elements at risk and their vulnerability requires careful consideration of landslide processes, mechanisms and types. In particular, the velocities and travel distances of landslide masses are important in this regard. Information can be gathered from the performance of existing landslide sites. However, such information is of limited value when considering potential landslides in areas which have not been affected in the past and are currently stable. Therefore, careful analysis and modelling may be required to identity source areas of landsliding and the areas in the vicinity which may be affected if and when landslides at new sites do occur. In the absence of reliable models, reliance is placed on past experience and judgement.

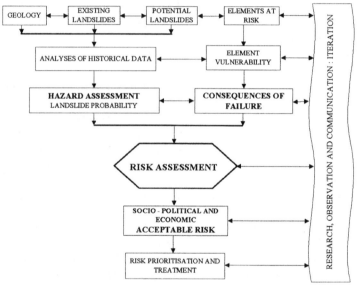

Figure 1. General Risk Assessment Methodology

Consideration of consequences of landsliding similarly requires engineering judgement based on understanding of hazard, elements of risk and previous experience. Issues concerned with acceptable and tolerable risk require an understanding of local communities and of the manner in which previous landsliding and individual disasters have affected public perception of risk.

Decision-making is influenced by current state of knowledge in the national and international context. It is also influenced by available information about a range of influencing and triggering factors, lessons learnt from previous landslide events and community expectations.

From an overall management perspective, the following key questions must be addressed on a continuing basis:

- How should landslide hazard and risk be assessed in cities and towns and along transportation routes?
- What is the most effective way in which geological and engineering assessments will be included by decision-makers, while considering critical planning issues concerning the physical development of cities?
- How can landslide management be carried out to protect areas of ecological and heritage significance and other sensitive areas?

MANAGING URBANISATION IN HILLY AREAS

With regard to the management of land, often the choice is not between complete conservation or unrestricted development but between reasonable development planned on the basis of a careful assessment on the one hand and unsustainable extent or intensity of development on the other.

A striking example of intensive development of sloping areas is Hong Kong. Urbanisation has led to extensive slope stability problems in different areas which include natural slopes of tropical residual soil as well as constructed fill slopes. Enormous economic losses due to slope instability and landsliding have occurred and some catastrophic failures have resulted in loss of life. This situation has necessitated major expenditure on engineering control measures. Considerable investment in research and planning has been made and detailed guidelines have been developed for the management of hilly areas. However, continued research is considered necessary for further increase in knowledge and continued improvement in geotechnical practice so that hazards and risks can be managed better.

A recent research study of the University of Wollongong has included the development of a GIS based geotechnical-landslide database concerning the occurrence of slope instability within the Wollongong Local Government area managed by the Wollongong City Council (WCC). The number of reported landslides per year from 1880 to 2000 is shown in Fig. 2 (Chowdhury and Flentje, 1998). This figure shows the dramatic rise in the occurrence of landslides since about 1950 with increase in urbanisation of Wollongong and especially of the hillslopes above the narrow coastal plain which was urbanised in the early decades of settlement. The various peaks in this graph are, of course, associated with very high rainstorm events in particular years.

There are, of course many countries where significant and catastrophic landslide have occurred and these include USA, Canada, China, Italy, Switzerland and India.

However, catastrophic slope failures may also occur in countries like Australia which are generally considered to have relatively low to moderate landslide hazards from an international perspective. The Thredbo landslide disaster of August 30, 1997 is a striking example. It occurred at the Thredbo ski resort located 3 hours drive south of Canberra. Eighteen lives were lost in this disastrous event. It is interesting to note that the landslide site is located on a hill slope which has been intensively developed to construct holiday lodges. This may be seen from the photographs (Figs. 3 and 4). It is also interesting to note that an external natural agent such as high rainfall or an earthquake did not trigger the landslide.

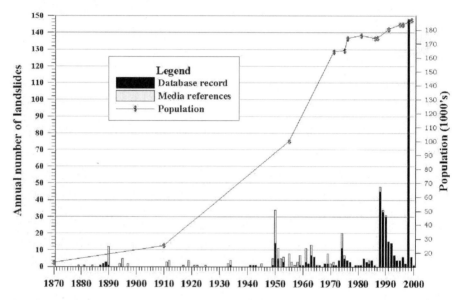

Figure 2. Wollongong population 1870 – 2000 and annual landslide occurrence

A comprehensive coronial enquiry assisted by specialist engineering geologists and geotechnical engineers highlighted the failures of landslide assessment and management along the Alpine way, the road linking Thredbo village to adjacent towns (Hand, 2000). These failures of management created the conditions under which a catastrophic landslide such as the Thredbo landslide of August 30, 1997 could occur. Inadequate landslide assessment and management occurred over several decades of roadway construction and upgrade and while the village of Thredbo was also being developed.

Thus the disaster was a consequence of long-term natural processes as well as human intervention with inadequate management. The main triggering factor leading up to the disaster was a regime of high pore water pressures created by a leaking water main, reflecting lack of monitoring and maintenance relevant to potential landsliding.

A MODERN GIS-BASED FRAMEWORK

The development and use of any methods for the assessment and management of hazard and risk is an interdisciplinary task requiring an understanding of diverse concepts, information about many spatial and temporal features and processes as well as diverse skills of analysis and synthesis. For example, surface and subsurface details have a great deal of variability within a region. These include topography, elevation, geology, depths of soil layers, groundwater patterns, pore water pressures, shapes of valleys etc.

Figure 3. The Thredbo landslide, August 30, 1997, (centre of photograph) occurred on a densely developed slope in the village of the Thredbo, New South Wales, Australia.

The collection, updating, manipulation, processing and analysis of large quantities of spatial and temporal information can be handled very well within a GIS. Consequently, the use of GIS systems by geologists and geotechnical consultants is growing rapidly. Many researchers have adopted GIS-based procedures for modelling of hazard and risk related to natural disasters such as landslides, floods and earthquakes. Organisations at the level of state government and local government with responsibility for the use of land and related infrastructure have started using GIS for storing, updating, manipulation and use of different types of data. The pace of change, technological developments and progress in research are encouraging such organisations to make more effective use of GIS-based approaches for planning of land use and for management of landslides.

Identification and assessment of hazard can be facilitated by superimposing different types or layers of data. Similarly, elements at risk and their vulnerability can be identified and assessed more efficiently with overlapping layers of information. In turn, systematic and efficient methods and procedures will facilitate the development of policies and priorities which reflect current knowledge and take account of all available information. The use of modern technology facilitates interaction between researchers on the one hand and decision-makers at local and state levels on the other. The use of GIS, in particular, provides a flexible and versatile framework for incorporating the latest data and most advanced analytical techniques. Thus technology transfer can take place rapidly. Moreover, a GIS framework would improve the development of more efficient warning systems including real-time or near real-time applications.

Figure 4. The Thredbo landslide, view upslope from Bobuck Lane midway along the path of the debris flow. Carinya Lodge was located approximately in the centre of the image before it was destroyed and the head of the slide is draped in plastic in the upper left side of the image.

HAZARD AND RISK ASSESSMENT FRAMEWORK
The important elements of a landslide hazard and risk assessment framework include (a) reliable and accurate maps of topography, geology and existing landslide areas (b) geological, geotechnical and landslide databases (c) rainfall data (d) observational data including those from piezometers and inclinometers (e) methods of modelling, analysis, synthesis and interpretation based on key definitions of hazard and risk (f) data concerning elements at risk and their vulnerability. The context here is primarily that of rainfall-triggered landslides.

The use of modern computer-based methods and, in particular, the role of a Geographical Information System (GIS) has been discussed by Chowdhury and Flentje (1996) and Flentje and Chowdhury (1996).

The requirements of a landslide database, essential and desirable, have been discussed by Chowdhury and Flentje (1998) and the benefits of a comprehensive database have been demonstrated in relation to the Wollongong Local Government area in New South Wales, Australia.

A comprehensive GIS-based approach has facilitated many important and useful outcomes including the following:-

- Ranking of landslide sites on the basis of probability of landslide reactivation using historical data on the number of recurrences of each landslide
- Ranking of landslide sites on the basis of probability of landslide reactivation using observational data from inclinometers along with recorded rainfall data over a significant period of time
- Probability of occurrence of landsliding in different areas on a spatial rather than time basis

RISK MANAGEMENT INCLUDING ENGINEERING SOLUTIONS
Informed decisions on the management of sloping land can be made on the basis of a comprehensive hazard and risk assessment procedure. It is, of course, very useful to study individual sites in detail with regard to their performance over time. Such studies should include the assessment of remedial or preventive measures which have already been implemented. Decisions can then be made on any further measures that may be required. Several case studies within the Northern Illawarra have been discussed by Flentje (1998) and a number of remedial measures have been implemented at some of these sites in the past. These include both surface and subsurface drainage measures which are usually the most effective for rainfall-triggered landsliding. These case studies have validated the GIS-based framework adopted for the whole study area.

Landslide risk management poses difficult decisions for various authorities. For example, in the case of the Wollongong Local Government area, the Roads and Traffic Authority (RTA), the Rail Infrastructure Corporation Authority (RIC) and the WCC have responsibility for management of various parcels of land and the related infrastructure. For example, the RTA has responsibility for the coast road (Lawrence Hargrave Drive) which traverses the escarpment (Fig. 5). The railway line, for which the RIC is responsible is located immediately above this road. Each organisation must make difficult choices concerning land use planning, upgrading of infrastructure, new residential subdivisions, development applications in existing suburbs and, most importantly, the use of scarce resources for hazard and risk management.

Both short-term needs and long-term benefits must be considered and this requires a thorough understanding of the basic principles of slope performance and the role of influencing factors. Secondly, there must be a way to identify hazardous sites or areas in such a way that priorities can be established. Thirdly, the approaches must be systematic, reliable and simple so that the authorities concerned can disseminate information and, on that basis, have community support for the decisions, priorities and proposed solutions.

It is obvious from what has been stated before that, a modern computer-based framework for hazard and risk assessment can be a very powerful tool for decisions related to the management of sloping land. Whether the land is urbanised or whether it is subject to other functions or impacts, a systematic GIS-based approach, which recognises the interdisciplinary nature of slope stability, is very valuable for land management in hilly areas. However, it is extremely important to recognise that these methods, techniques or approaches must be firmly based on modern geo-engineering principles which govern the stability and performance of slopes. A basic risk management framework is shown in Fig. 1.

Figure 5. The Illawarra Escarpment above Clifton and Scarborough looking south. The Wollongong Central Business District is in the top left of the image.

CURRENT AND EMERGING PRACTICE, WOLLONGONG LOCAL GOVERNMENT AREA IN NEW SOUTH WALES, AUSTRALIA

Over the last three decades, the Wollongong City Council (WCC) has continually improved the assessment and management of land instability. In the early 1970's a comprehensive study of geology and slope stability was commissioned.   Maps showing six zones with varying susceptibility to landsliding were an important end product of that study.   In-house information was updated from observations of landslides following significant rainstorm events and the limitations and errors in the original landslide maps have also progressively been identified.   Geotechnical specialist firms have been commissioned from time to time to carry out specific slope stability studies in suburbs with significant landslide problems and potential.

Most importantly, Wollongong City Council has supported research on basic and applied aspects of slope instability at the University of Wollongong (UOW).   During the years 1993-1997, WCC was the industry partner of a project supported by the Australian Research Council (ARC) which led to a PhD thesis incorporating GIS-based maps of geology and existing landslide and a comprehensive and unique landslide database.   Moreover, WCC has been one of the industry partners of a comprehensive research project undertaken at UOW under the Strategic Partnership Industry Research and Training (SPIRT) scheme over the three years 1998-2000 as well as of the current project over the period 2001-2003.

In 1985, the Australian Geomechanics Society (AGS) published a paper entitled "Geotechnical Risk Associated wit Hillside Development" which suggested different categories of risk, primarily for urban areas.   WCC was one of several local government bodies which incorporated these guidelines in its management practice for dealing with development applications.   This 1985 paper had significant limitations primarily because hazard and consequence were not considered separately.   A comprehensive paper has now been issued by the Australian Geomechanics Society which makes the 1985 paper obsolete (AGS, 2000)

By 1995 a generic Australia and New Zealand standard for risk management was available which was subsequently updated in 1999 (AS/NZS, 1999).   Although this standard does not refer to or deal with landslide risk or even with geotechnical risk, it suggested an appropriate Hazard-Consequence matrix framework for risk assessment and management.   For example, in the aftermath of the 1997 Thredbo tragedy, this framework was used to assess the post-disaster risk to the remaining residential properties in Thredbo village.

In August 1998, there was a major rainstorm event in Wollongong which led to a number of landslides including debris flows and deep-seated movements, reactivations of existing landslide sites as well as development of new landslides. Assessment of future hazard and risk associated with these problem sites was carried out by a three member team which included representation from WCC and UOW (GTR, 1999).   The assessment was based on the hazard-consequence matrix approach.   Decisions or management options for each site were based on this systematic assessment approach.

It is also important to note that during the 1998 storm event, real-time warning concerning the occurrence of widespread landsliding was issued by UOW research team. This was a vindication of the systematic methods developed by UOW including the use of a comprehensive database, GIS-based mapping and analysis as well as consideration of historical rainfall events in relation to monitored subsurface movements leading to the concept of Antecedent Rainfall Percentage Exceedance Time (ARPET). A discussion of uncertainties in the hazard assessment of rainfall-induced landslides has been covered in another paper (Chowdhury & Flentje 2001). Recent developments concerning assessment of rainfall induced landsliding are discussed in Flentje and Chowdhury (2001).

GIS-BASED RISK MANAGEMENT SYSTEM AT THE WOLLONGONG CITY COUNCIL

The Wollongong City Council (WCC) has identified the following 5 hazards as important development constraints.

- Landslides
- Bush Fire
- Contaminated Land
- Filled land
- Flood

Each of these hazards has been mapped by specialists in their respective disciplines. Maps of each of these hazard areas have been incorporated into the WCC's Geographic Information System (GIS) as appropriate map 'layers'.

Landslides are represented in the GIS by two layers. The first layer contains reported landslides only, which are classified as Very High hazard. The second layer contains suspected areas of slope instability, which are classified as either Low, Medium or High hazard of instability.

The University of Wollongong research team is conducting a quantitative landslide risk assessment of the whole city. This is being carried out with data coverage generally accurate to a scale of 1:4000. Over 500 landslides have been mapped and the historical probability of landsliding has been determined for each of these sites. Potential landslide areas are being identified with the aid of the GIS and several map layers including a) existing landslides, b) geology, c) macro and micro slope shape, d) slope angle, e) vegetation and f) depth to bedrock. The elements at risk have been identified on the basis of a combination of WCC land zoning and a database of rubbish bin allocations across the city. Consequence assessment is still to be carried out, but this will include an assessment on the basis of recorded damage/destruction of houses and the estimated run-out distance of each landslide or potential landslide. Obviously, the scope of the GIS-based system of the WCC will develop further as they adopt a more comprehensive hazard and risk assessment and management system following on from the results of funded research such as that at the University of Wollongong.

CONCLUSIONS
The assessment and management of landslide risk requires a systematic approach which incorporates assessment of hazard, elements at risk and the consequences of landsliding. Well defined principles of generic risk management exist in national standards of several countries including Australia. Such principles are now being incorporated in the geotechnical field including their application to landslides. Geographic Information Systems are perfectly suited for use as integral tools for hazard and risk assessment and management, and the Wollongong City Council (WCC) uses such a system for a variety of hazard categories. With support from the WCC and other agencies, the University of Wollongong is developing and applying an innovative and comprehensive GIS based landslide risk management system in Wollongong. While still a work in progress, this University of Wollongong landslide risk management system will provide a considerable improvement to the landslide risk management capability of the WCC.

REFERENCES
AGS, 2000. "Landslide Risk Management Concepts and Guidelines". Journal and News of the Australian Geomechanics Society (AGS) Volume 35, Number 1, March 2000, pp 49-92

AS/NZS, (1999). "Risk Management Standard", Standards Association of Australia/New Zealand, Standard 4360: 1999.

Chowdhury, R. N. and Flentje, P. 1996. Geological and Land Instability mapping using a GIS package as a building block for the development of a risk assessment procedure. Proceedings of the Seventh International Symposium on Landslides, Trondheim, Norway. Editor: Senneset, K. June, Volume 1, 177 - 182, Balkema, Rotterdam ISBN 90 54108185.

Chowdhury, R. N. and Flentje, P. 1998. A landslide database for landslide hazard assessment. Proceedings of the Second International Conference on Environmental Management. February 10 -13. Wollongong Australia. Editor: Sivakumar, M. and Chowdhury, R. N. Pages 1229 - 1239. Elsevier London.

Chit Ko Ko, Flentje, P and Chowdhury, R., 2002. Uncertainties in Rainfall-Induced Landslide Hazard. Quarterly Journal of Engineering Geology and Hydrogeology. Symposium in Print on Landslides. Volume 35/1. Paper accepted.

Commission of Inquiry (1999). The Long - Term Planning and Management of the Illawarra Escarpment, Wollongong Local Government Area, Report by the Commission of Inquiry for Environment and Planning, NSW Government, May 1999, 145 pages (plus maps and appendices).

Flentje, P. (1998). Computer Based Landslides Hazard and Risk Assessment (Northern Illawarra Region of New South Wales, Australia). Doctor of Philosophy Thesis, University of Wollongong, New South Wales, Australia. Unpublished, 525p.

Flentje, P. and Chowdhury, R. N. 1996. Preparation and validation of digital maps of geology and slope instability. Proceedings of the Eighth International Conference and Field Trip on Landslides, Madrid, Spain, December, 15 - 28, Editors: Chacon, J, Irigaray, C and Fernandez, T. Published by Balkema, Rotterdam ISBN 90 54108320.

Flentje, P. and Chowdhury, R. N. 1999. Quantitative Landslide Hazard Assessment in an Urban Area. Proceedings of the Eighth Australia New Zealand Conference on Geomechanics. Editor: Dr. Nihal Vitharana. February 15 - 17, Hobart, Tasmania. Institution of Engineers, Australia.

Flentje, P. and Chowdhury, R. N, 2001. Aspects of Risk Management for Rainfall - Triggered Landsliding. Proceedings of the Engineering and Development in Hazardous Terrain Symposium, New Zealand Geotechnical Society Inc. University of Canterbury, Christchurch, New Zealand. The Institution of Professional Engineers New Zealand. August 24-25, pp 143-150.

Flentje, P, Chowdhury, R. N. and Chit Ko Ko, 2000. A Matrix Approach for Landslide Risk Assessment. Proceedings of GeoEng2000, an International Conference on Geotechnical and Geological Engineering. Melbourne, November 19 – 24. Technomic, Pennsylvania, on CD ROM paper number SNES0252.Pdf.

Flentje, P, Chowdhury, R and Tobin, P., 2000. Management of Landslides Triggered by a Major Storm Event in Wollongong, Australia. Proceedings of the Second International Conference on Debris-Flow hazards Mitigation: Mechanics, Prediction, and Assessment. Taipei, Taiwan. Balkema, Rotterdam, p 479 - 487.

GTR, 1998. Report concerning Hazard and Risk associated with sites identified after the August 1998 rainstorm in Wollongong, NSW, Australia. Report commissioned by Wollongong City Council Emergency Services Department. Confidential report prepared by a Geotechnical Team including one of the authors. Unpublished report.

Hand, D., 2000. Report of the inquest into the deaths arising from the Thredbo Landslide. State of New South Wales, Australia, Attorney Generals Department - Office of the NSW Coroner. Web access to coroners report: http://www.lawlink.nsw.gov.au/lc.nsf/pages/thredbo_cor.

The application of landslide modelling techniques for the prediction of soft coastal cliff recession

ROBERT FLORY, Department of Civil Engineering, University of Bristol, UK
DAVID NASH, Department of Civil Engineering, University of Bristol, UK
MARK LEE, Department of Marine Science and Coastal Management, University of Newcastle, UK
DR JIM HALL, Department of Civil Engineering, University of Bristol, UK
DR MIKE WALKDEN, Department of Civil Engineering, University of Bristol, UK
MARCUS HRACHOWITZ, Universität für Bodenkultur, Institut für Angewandte Geologie, Vienna, Austria

INTRODUCTION

Quantitative landslide modelling can provide valuable insights for use in hazard identification and risk assessment. Because of the site-specific nature of the landsliding process, model application begins with a detailed field investigation of the local conditions and development of conceptual models, before these are implemented quantitatively. This paper describes the development and implementation of a landsliding model that forms part of a probabilistic tool for predicting soft coastal cliff recession.

Intensive field monitoring and data analyses have been applied to the soft rock coastal cliffs situated at Walton-on-the-Naze in Essex, UK (Figure 1). As with much of the coastline of East Anglia the cliffs are extremely vulnerable to intense and varied marine action. Recession rates are therefore high and vary from 2m per year at the north end of the site to an average of 1.2m at the south. The cliffs are about 1 km in length and vary in height from 22m at the south end, sloping gently to beach level at the north.

Figure 1. Location of Walton-on-the-Naze

SITE DESCRIPTION
Geology

The cliffs at the Naze are made of a sequence of Eocene to recent sediments. Red Crag sediments overly the London Clay in a bed that is about 7m thick at the south end of the site and gradually reduces to nothing at the North end.

The pocket of Red Crag found on the Naze is the southernmost trace of the sediment covering most of Norfolk. The uppermost layer of the Red Crag contains periglacial sands and gravels and ice wedges. These overlie crio-terbinated clays and Hoxnian deposits. Below these layers, the Red crag contains shelly, quartz-rich sands with iron oxides from the Pleistocene Epoch. The basement bed contains phosphatic nodules and is equivalent in age to the Pliocene Coralline Crag. These younger sands and gravels are separated from the London Clay by an undulating erosional contact (Daley & Balsan 1999).

The London Clay at Walton-on-the-Naze is part of the extensive outcrop in East Essex. Most of the town is built on London Clay, which in this area is a very stiff, brownish, bluish, fissured dark grey clay of extremely high plasticity. At the Naze, the clay is also is very jointed in an approximately northwestern direction. The bedding planes dip slightly to the south at an angle of less than 5°.

The London Clay is heavily overconsolidated, having had huge overburden loads during glacial periods, where the material was covered by an ice sheet several hundred metres thick. At the end of the ice age this ice sheet receded, resulting in unloading and expansion of the clay. Clay particles that were not beyond their elastic limit regained their original shape producing large local strain forces. One of the consequences of this behaviour is "progressive failure" where average shear strength can be well below peak at failure. This phenomenon is of particular relevance to the landslides at Walton-on-the-Naze and will be discussed later in this paper.

SITE INVESTIGATION
Six boreholes were sunk in to the cliff and foreshore to gain information on material properties. The boreholes were equipped with piezometers and an inclinometer to establish ground water conditions and investigate mechanism. Ongoing surveys and site experiments have been used to establish the main geomorphological processes at the site.

The results have shown that landsliding mechanisms can be different on adjacent landslides as well as varying at different times of the year, which highlights the difficulty of modelling such a complex and site-specific problem. The inclinometer results and peg surveys have been particularly useful at highlighting this issue. The inclinometer installed showed that the landslide was much shallower than first thought with the shear surface at 1.8m depth. It also showed that the landslide was moving as a single block. This observation was confirmed by surveying an array of pegs driven into the surface of the landslide. Identical displacements were recorded on all the pegs. Meanwhile, on the adjacent landslide, displacements were found to vary. Greater displacements were recorded at the centre of the landslide and smaller ones at the edges showing the material to be flowing like a mudslide.

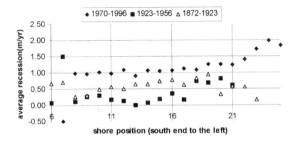

Figure 2. Recession rates in three different periods

Other site data

Historical data from aerial photographs, digital terrain models and Lidar data have also been used to provide accurate top and toe positions of the cliff for over 30 years. These have been combined with maps dating back to 1872 in order to calculate recession rates.

Figure 2 shows the average toe recession rate per year during three different periods at sections 50m apart along the shore. Between 1872 and 1923 there were wooden groynes. Between 1923 and 1956 there were rock groynes at, or close to points 7, 12 and 17 along the frontage. These groynes were destroyed in the 1953 flood and between 1970 and 1996 there is no protection. The cliff shows rapid readjustment to the removal of the groynes, with recession rates increasing from 0.4-0.6m/yr to 1.2m/yr.

Geomorphology

The geological structure has enabled a series of mutually competing complex landslides to form. The cliff is characterised by a series of embayments, which vary in width from 65m at the southern end to 30m at the north and in length from 55m to 20m, although the average slope is fairly constant at about 23°. Rotational slides at the back of the cliff (Figure 3) are deposited as debris on the shore via a complex translational landslide (Figure 4). These landslides have been measured to have moved at up to 200mm per day during periods of heavy rainfall.

Figure 3. Typical rotational failure at the Naze

Deep seated landslides.

The formation of deep-seated landslides has also been observed at the site. Deep-seated landslides occur as a result of a slip surface forming through the entire length

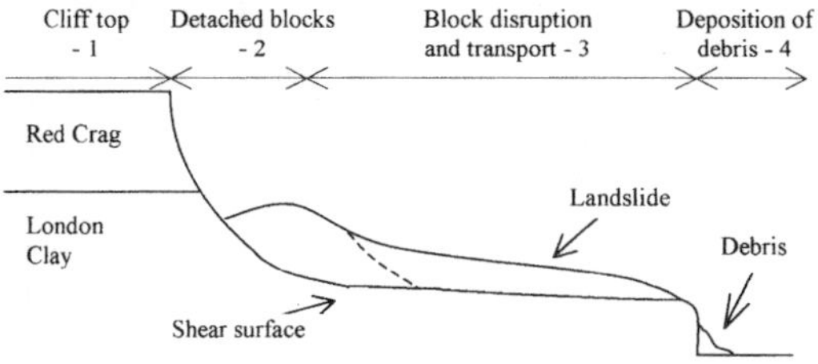

Cliff top   Detached blocks   Block disruption   Deposition of
- 1   - 2   and transport - 3   debris - 4

Red Crag

Landslide

London
Clay

Debris

Shear surface

Figure 4. Cliff recession processes at Walton-on-the-Naze

of the embayment rather than just at the back of the cliff. Such movement represents failure, presumably progressive, through the London Clay and are therefore probably a result of progressive failure. The residual strength of London Clay is very low and therefore the run-out of the failure can be very long.

Two deep-seated landslide events have been identified during the past decade. The first of these events seems, from evidence in aerial photographs, to have occurred in 1993 when about 20m of land was lost from the top of the cliff. The second occurred in 1999 with aerial photographs before and after showing that the embayment had significantly lengthened as a result of the failure. Figure 5, which shows cliff profiles for 1970,1992,1996, and 1999, shows how following a deep-seated landslide the cliff toe is progressively eroded. As well as raising the level of the toe this steepens the overall angle of the cliff, preparing it for a subsequent landslide event.

Diagrammatic drawings of the landsliding sequence are shown in Figure 6. Stages 1 and 2 show the process of the increasing height of the toe up until failure. Reports from local residents of a deep-seated landslide in 1999 suggest that the landslide might have run out by up to 20m in advance of the usual cliff toe line. Such material is very disturbed and soft so is removed rapidly by marine erosion (stage 3). Records of the 1999 slide indicate that it was cut back by between 1 and 1.5m per month in August and September 2000. The landslide was continuously on the move with height of the toe of the landslide halved between the months of September and November as well as having been cut back further past the normal line of the toe (stage 4). As the landslide was continuously moving forward, the back face of the cliff became more exposed and much steeper making it vulnerable to more falls (stage 5). Between the months of November 2000 and March 2001 there were several falls from the Crag and one large slip of particular interest. The shear surface of

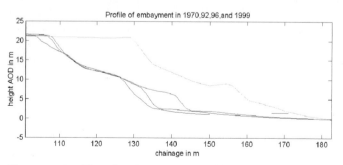

Figure 5 – Profiles of embayment preceding a deep-seated failure

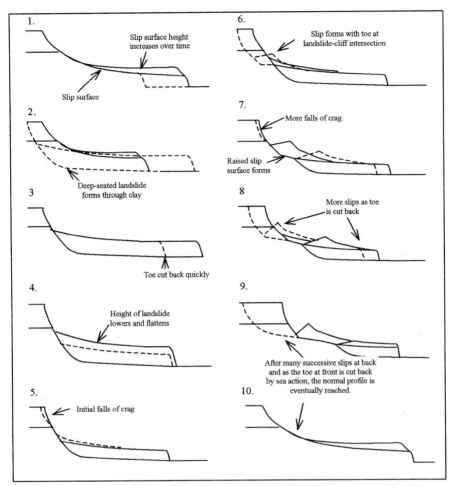

Figure 6. Process of landsilding at the Naze – dashed line indicates profile of succeeding stage

the slip clearly went through the London Clay, though the top of the failure was above the toe of the cliff (stage 6 and 7). At this stage, deeper-seated slides are prevented by the large amount of material at the toe. The next stages leading to normal recovery are postulated in stages 8 to10 of Figure 6. The landslide debris at the toe is slowly cut back as well as successive falls on the raised shear surface at the back until the original profile is recovered.

Migration of the embayments

Analysis of aerial photographs and Lidar data from 1970 to 1999 suggests that in the medium to long-term the embayments do not recede perpendicular to the shoreline but migrate southwards. This behaviour is not fully understood although the most likely reason is that slips are forming along joints in the London Clay; although the

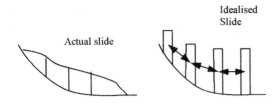

Figure 7. Shape of slice

direction of ground water seepage may also be a factor. Aerial photographs of the show that the headlands are not perpendicular to the line of the toe but at an angle as a result of the migration or the jointing of the clay. This explanation suggests a different mechanism to that observed by Pethick (1996) on the Holderness coast, which is related to direction of sediment drift.

Back-analysis
Back-analysis to estimate the ground water conditions at failure were done on several of the landslides. The OASYS slope stability package was used to analyse cliff geometries before and after failure. Soil properties typical of those found in other slopes of London Clay that had failed due to progressive failure were assumed. On one embayment, the predicted slip surface matched the profile shown after the event, providing some evidence of the ground water condition at failure. For other events slip surfaces had to be assumed using survey data of the cliff top. For the profile produced by the large 1999 event, the only realistic scenario producing a deep-seated failure was that large suction forces existed within the clay.

NUMERICAL MODEL OF THE RECESSION PROCESS
The geomorphological analysis outlined above identified two main mechanisms for more detailed numerical analysis: first the conveyance of debris towards the sea on pre-existing slides and secondly the initiation of deep-seated slides. Progress has been made on modelling the first of these processes. To do so involves representing the conditions under which movement will commence and quantifying the amount of deformation until equilibrium is re-established. Bishop's method of slices has been used to estimate the initiation of movement and combined with a force equilibrium method where the slices are assumed to be connected with stiff rods, to simulate the inter-slice forces. Resolving the forces (Figure 8) produces a non-linear differential equation of second order, in the form of a free vibration with viscous damping. Butterfield (2000) developed a similar

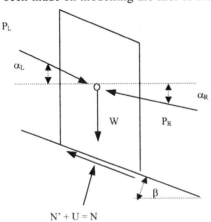

Figure 8. Force diagram for a slice

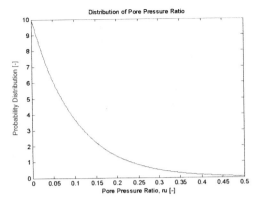

Figure 9. Encounter Probability for pore water pressure ratio

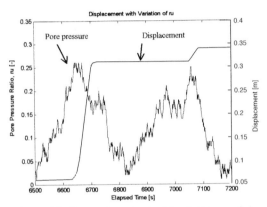

Figure 10. Example output of landslide model

equation describing a damped vibration using similar assumptions:

$$X'' + G(X')^n = H(t)$$

where: X'' is acceleration, $G(X')^n$ is a velocity dependent damping term, and $H(t)$ is a time-dependent force.

The model attempts to take into account two particular properties of the London Clay related to shear strength. Firstly, the shear strength of the soil will vary with displacement up to a peak strength and eventually down to the residual strength (Skempton 1964) and secondly, because of the cohesive properties of the soil, the shear strength will increase logarithmically with increasing rate of shear (Vaughn and Tika 1996).

Simulation of groundwater level variation

Groundwater conditions are simulated using an exponential probability distribution to estimate the pore pressure ratio:

$$f(x \mid \lambda) = \lambda e^{-\lambda x}$$

where x in the pore pressure ratio and λ the distribution parameter.

Based on evidence from borehole data, the groundwater level was set low most of the time. The encounter probability distribution for the pore pressure ratio is displayed in Figure 9. For each time-step the pore pressure, slip surface geometry, shear strength and acceleration for each slice are re-calculated to estimate the total displacement during each landslide movement. Figure 10 shows example output where two particular episodic movements of the landslide are predicted as the pore pressure ratio varies. The model is currently able to predict movement up to 70 days in advance, although optimisation of the code should improve this.

CONCLUSIONS

Intensive field measurements, laboratory test data and stability analysis of the cliffs at Walton-on-the-Naze have been used to establish a conceptual model of the processes of soft coastal cliff erosion. Two main mechanisms of landsliding have been observed. Firstly, complex shallow landslides, which convey material from the cliff top to the shore and secondly, deep-seated landslides, where failure occurs through the entire length of the cliff. Progress has been made in the development of a numerical model of these processes, which combines Bishop's method of slices with a force equilibrium method. Groundwater conditions have been represented by a synthetic time series, which was generated using an appropriate probability distribution.

FURTHER WORK

Progressive failure models of smaller slumps at the rear of the embayments and the formation of less-common deep-seated slips are being developed. This is being done by combining the force equilibrium method with a model introduced by Verani and Watson (1999) in which slices are assumed to be interconnected by springs, which simulate the strain energy stored in the soil. The new landsliding model will ultimately be integrated with a model of shoreline erosion (Walkden *et al*, 2002).

Further site and laboratory work will explore the ground water conditions in the cliff to reduce the number of necessary modelling assumptions.

REFERENCES

Butterfield, R. (2000), "A Dynamic Model of Shallow Slope Motion Driven by Fluctuating Ground Water Levels", Landslides in Research, Theory and Practice, Thomas Telford, London.

Daley, B. and Balsan, P. (1999), "British Tertiary Stratigraphy", 1st ed., pp.291-295, Joint Nature Conservation Committee, Peterborough.

Pethick J. (1996), "Coastal Slope Development: Temporal and Spatial Periodicity in the Holderness Cliff Recession", Advances in Hillslope Processes, Vol. 2. pp.897-917. John Wiley & Sons Ltd.

Skempton A.W. (1964), "Long-term Stability of Clay Slopes", Geotechnique, Vol.14, pp. 77-101, The Institution of Civil Engineers, London.

Tika, T.E., Vaughn, P.R. (1996), "Fast Shearing of Pre-existing Shear Zones in Soil", Geotechnique, Vol.46, The Institution of Civil Engineers, London,

Verani, C., and Watson, D (1999), "Simulation and Analysis of Progressive Failures of Slopes", 3rd Year Major Research Project, University of Bristol, Department of Civil Engineering, Bristol.

Walkden, M.J., Hall J.W., and Lee, E. M. (2002), "A modelling tool for predicting coastal cliff recession and analysing cliff management options." Proc. Conf. Instability-planning & management. Ventnor, U.K.

Modelling the impact of forest management changes on landslide occurrence

P. FRATTINI, Dip. Scienze Geologiche e Geotecnologie, Univ. Milano Bicocca, Italy, and
G.B. CROSTA, Associate Professor, Dip. Scienze Geologiche e Geotecnologie, Univ. Milano Bicocca, Italy

ABSTRACT
Simulations of the probability of slope failures related with timber harvesting (complete clear-cutting and partial cuttings) were carried out within a pilot area, nearly 25 km^2 in size, in the Lecco province (Lombardy, Central Alps, Italy). Information on both forest management changes and landslide occurrence was acquired through the interpretation of a historical series of aerial photographs and field surveys.

In order to analyse the effects of forest changes on the triggering of landslides by intense rainfalls, a simple distributed model, which simulates the response of root cohesion to different timber management systems, was integrated with an infinite slope stability model. Subsequently, a distributed physically-based transient hydrological model was coupled with the stability model. This hydrological model simulates the diffusion of pore pressure head within the soil profile in condition of quasi-saturated soils.

By using such simulation tools, root cohesion response to changes in timber harvesting was investigated for the whole period 1950-1999, and the modelled values were used for the stability of rainstorms occurred in that period. Lastly, in order to provide land managers with a support for planning suitable strategies of vegetation management in landslide prone catchments, different scenarios of future land use were developed.

INTRODUCTION
The analysis and forecast of temporal and spatial distribution of shallow landslides (soil slips) is an important aspect in land management of mountainous areas. The aim of hazard assessment is the evaluation of both the susceptibility of the slope to fail and the probability of occurrence of landslides in time.

Instability – Planning and Management, Thomas Telford, London, 2002, 257–264

Many methods for landslide susceptibility assessment have been proposed in literature (Brabb et al., 1984; Carrara, 1983; Bonham-Carter, 1990; Hammond et al., 1992), but few attempts have been made for the evaluation of time probability of occurrence of phenomena. This probability can be assessed through physically based mathematical models that explicitly incorporate the dynamic variables such as the degree of soil saturation and the cohesion due to the presence of the roots. Regarding the degree of saturation, different models have been developed (Montgomery and Dietrich, 1994; Wu and Sidle, 1995; Borga et al., 1998; Pack et al., 1999), generally based on the coupling of the infinite slope stability analysis with hydrological models able to modulate water table heights in steady or quasi-steady conditions with groundwater flows parallel to the slope. The assumptions of these models are usually too restrictive under certain conditions (i.e., rapid response of pore water pressure to transient rainfall). Iverson (2000) has recently developed a flexible modelling framework, with different approximations of Richards' (1931) equation valid for varying periods of time and for different hydrological conditions. In order to analyse the role of root cohesion changes on the triggering of landslides by intense rainfall, a model which simulates the response of root cohesion to different timber management systems was proposed by Sidle (1991, 1992).

In this study we apply and compares an integrated modelling strategy in an area of the Lecco province, Lombardy, in order to investigate the possibility to use physically based models to assess the rainfall and land use control on landslide occurrence. Eventually, we illustrate and discuss different rainfall scenarios (with different recurrence time) under different land use strategies for the study area.

STUDY AREA
The study area is located in the Lecco Province (Lombardy Region, Northern Italy) and includes the Pioverna Occidentale basin (figure 1). The whole study area has an extension of 25 km^2 and is situated between 620 m and 2160 m a.s.l.. The mean annual precipitation ranges from 1500 mm in the lower part of the basin, to 1750 mm in the upper part. Statistical analysis of maximum intensity values for Barzio rain gauge allowed to determine the recurrence time of intense rainfall events. By applying Gumbel's statistics, it has been shown that daily intensities with recurrence time of 50 years and 100 years are 168 mm and 195 mm, respectively.

The bedrock prevalently consists of massive limestones and dolostones (Esino Fm. and Dolomia Principale Fm., respectively) and, subordinately, of well bedded dark limestones intercalated with black shales (Zu Fm). Sandstones and conglomerates outcrop locally. Significant karstic processes exist in the upper part of the basin. A large part of the lower basin is occupied by thick glacial deposits (figure 2a). The middle basin is characterised by colluvial soils and scree slope deposits and the upper part consists of outcropping rocks and very shallow colluvial deposits. Field survey, in situ and laboratory tests (direct shear, grain size analyses, etc.) have been performed within the area.

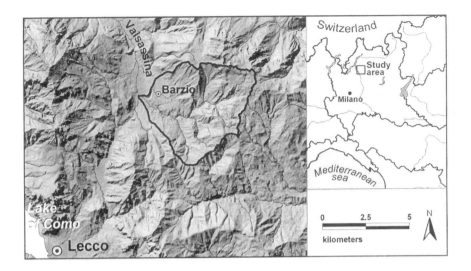

Figure 1. 20 x 20 m cell size DTM for the study area.

Land-use is represented by meadows, alpine grasslands, mixed beech forests and pastures, with urban settlements in the lower part of the basin (figure 2b). The analysis of historical information (photographs and reports) showed that the land use was considerably changed during the first half of XX century. Especially between the two World Wars, an intense exploitation of forest resources led to an almost complete rejuvenation of beech forests. Since then, very few changes occurred in land use. A progressive ageing of beech forest could induce an extensive forest harvesting in the next years. A purpose of this paper is to provide land managers with a support for planning suitable strategies of vegetation management for the future.

MODELLING RAINFALL CONTROL ON LANDSLIDE TRIGGERING
A grid-based distributed transient hydrological model has been implemented, namely a pore-pressure diffusive model (Iverson, 2000). The model has been coupled with an infinite slope stability analysis (Skempton and Delory, 1957). A 20*20 m cell size DTM (figure 1) was prepared and adopted for the analyses. Terrain units for parameter calibration have been generated starting from field and laboratory data and from land use and geological maps. The model (Iverson, 2000) simulates the transmission and distribution of pore pressure under wet conditions within the soil profile during and after the rainstorm. The model result is the groundwater pressure head value at a specific depth for a certain time. The calibration was essentially performed on the basis of prior information about soil and vegetation, with adjustments made to improve the distribution of computed Safety Factor with respect to the actual distribution of observed landslides (table 1).

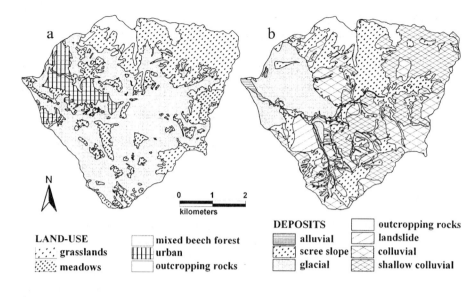

Figure 2. a) Land-use map; b) map of superficial deposits and areas of outcropping rocks.

Table 1. Soil parameters adopted for the simulations

Soil typology	d_s	γ_s	φ	c_s	K_s	D_0
shallow colluvial soils	0.5	18	30	3.0	$5.5 \cdot 10^{-4}$	$2 \cdot 10^{-5}$
colluvial soil	0.7	18	25	3.0	$2.7 \cdot 10^{-4}$	$2 \cdot 10^{-5}$
glacial deposits	1.0	18	30	3.0	$5.5 \cdot 10^{-4}$	$2 \cdot 10^{-5}$
scree slope deposits	1.0	18	35	3.0	$2.7 \cdot 10^{-4}$	$2 \cdot 10^{-5}$

d_s (m): soil depth
γ_s (kN/m³): soil density derived from laboratory test
φ (°): angle of internal friction derived from direct shear tests with material finer than 2mm
c_s (kN/m²): soil cohesion derived from direct shear tests
K_s (m/s): saturated hydraulic conductivity estimated from grain size distribution
D_0 (m²/s): maximum hydraulic diffusivity obtained from calibration

MODELLING LAND-USE CONTROL

A simple grid-based distributed model of changes in root cohesion in response to vegetation management (Sidle, 1991) has been implemented within a GIS environment. An exponential decay function describes the rate of root strength deterioration following vegetation removal (Ziemer and Swantson, 1977; O'Loughlin and Watson, 1979), whereas a sigmoid relationship describe the root strength regrowth due to newly planted or invading vegetation (Ziemer, 1981; Sidle, 1992). The coefficients for root cohesion model have been derived from literature (O'Loughlin, 1981; Sidle, 1991) and adapted to European beech forest. A distinction has been made between fertile sites, with deep soils ensuring water and nutrient supply, and poor sites.

MODEL APPLICATION

Root cohesion changes in response to different vegetation management practices are modelled for European beech forest. An initial value of 3.3 kPa is considered (O'Loughlin, 1981). This value corresponds to the root cohesion of an old-grown forest. A complete clear-cutting is compared with partial cuttings with rotation time interval of 100 years. The clear-cutting causes a strong drop of root cohesion within few years, with minimum value of 1.23 kPa after 13 years (figure 3a). The root strength regrowth is faster in fertile sites, with a complete recovery of initial cohesion in 40 years. Poor sites need almost 100 years to recover the initial cohesion. Partial cuttings with removal of 50% of overstory vegetation every 50 years (figure 3b) show a smaller drop of root cohesion (minimum value of 2.2 kPa after the second cut in poor sites). The final recovery is complete for fertile sites, but incomplete for poor sites (3 kPa). In this last case, a 100-years rotation will lead to a progressive decline of root cohesion in time. This effect is more evident with shorter time length between partial cuts. In the case of removal of 25% of overstory vegetation every 25 years (figure 3d), the final root cohesion amounts to 2.6 kPa for poor sites and 3.1 kPa for fertile sites (with minimum root cohesion after 80 years of 2.33 kPa). The faster regrowth of plants on fertile sites permits to obtain a complete recovery of root cohesion both with 50 year-interval cuts (figure 3b) and 33 years-interval (figure 3c), with minimum root cohesion amounting to 2.37 and 2.65 kPa.

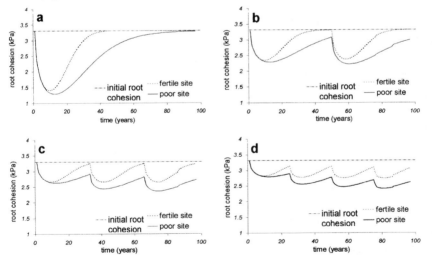

Figure 3. Changes in root cohesion of beech forest through different vegetation management: a) clear-cutting; b) partial cutting (50% of overstory vegetation every 50 years); partial cutting (33% of overstory vegetation every 33 years); partial cutting (25% of overstory vegetation every 25 years).

In general, a reduction in the time interval between partial cuttings minimises the lost of root cohesion. After a certain threshold value of cutting interval, a complete recovery is not possible, and a progressive long-term decline of root cohesion occurs. The optimal vegetation management strategy should consider both these effects. The minimum values of root cohesion for each management system are summed to the soil cohesion of forested areas and introduced as input for the stability models. Figure 4 reports the results of the slope stability simulations for present situation (with root cohesion set to the maximum value in forested areas), for clear-cutting and for partial cuttings with removal of 25% of overstory vegetation every 25 years. The percentage of area modelled as unstable (Safety Factor, SF <1) or quasi-unstable (1<FS<1.3) are reported in table 2.

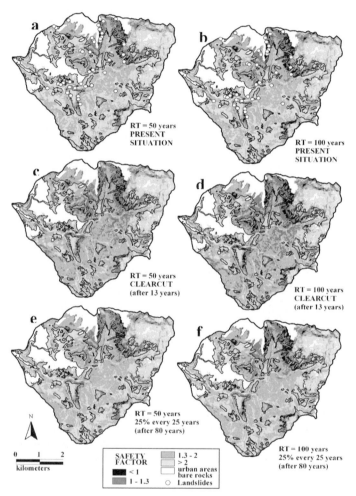

Figure 4. Results of distributed slope stability analysis with the diffusive hydrological model using daily rainfall values with 50 years (left maps) and 100 (right maps) years recurrence time. Forested areas are contoured (see figure 2a). c) and d) correspond to case a) in figure 3; e) and f) correspond to case d) in figure 3. Observed landslides are reported in a) and b).

Table 2. Percentage of the total area modelled as unstable under different rainfall conditions and different vegetation management practices (CC = clear-cutting; PC-25 = partial cuttings with removal of 25% of overstory vegetation every 25 years)

Recurrence time (years)		100			50		
Daily rainfall (mm)		195			168		
Management system		present	CC	PC-25	present	CC	PC-25
Modelled unstable areas (%) with	SF < 1	2.8	3.5	2.8	1.6	1.8	1.6
	1 < SF < 1.3	9.6	18.9	10.7	8.4	15.7	9.3

A significant increment in slope instability is observable after clear-cutting, with a consistent part of the basin modelled as unstable after 13 years, when the minimum value of root cohesion is reached

Partial cuttings do not cause a strong increase in slope instability but they determine a progressive long-term deterioration of root cohesion with cumulative effects in the successive rotations.

CONCLUSION
The description of shallow landslide hazard in terms of spatial (distributed model) and temporal (transient model) probability is a very important issue. This can be done in order to evaluate future scenarios of land-use and to couple these scenarios with rainfall events characterised by a known probability of occurrence.

This issue is pursued integrating a transient hydrological model, a root cohesion model and a stability model within a GIS. The comparison of different vegetation management practices permits the identification of the level of potential hazard associated with each of them and the identification of the optimal solutions in terms of slope stability and risks.

A substantial lack of accurate data on root decay and root regrowth for many kind of forest typology, together with the difficulty in hydrological model calibration, constitute the major limitations in the application of such modelling tool. Further investigation on root response to harvesting systems is therefore necessary in order to obtain a reliable and usable support for planning suitable strategies of vegetation management in landslide prone catchments. This is especially important for alpine and prealpine areas where a progressive land abandonment causes a general carelessness of vegetation management.

ACKNOWLEDGMENTS
The authors wish to thank dr. Giambattista Bischetti and dr. Serena Zanella for useful suggestions. The study has been partly funded by the DAMOCLES Project, EVG1-1999-00027.

REFERENCES

Bonham-Carter, G.F., Agterberg, F.P., and Wright, D.F. (1989) - Weights of evidence modeling: a new approach to mapping mineral potential, in Agterberg, F.P., and Bonham-Carter, G.F., (eds), Statistical applications in the earth science, Geological Survey of Canada Paper 89-9, Ottawa, Canada.

Borga, M., Dalla Fontara, G., De Ros, D., and Marchi, L. (1998) - Shallow landslide hazard assessment using a physically based model and digital elevation data, Environmental Geology, 35, 81-88.

Brabb, E.E. (1984) - Innovative approaches to landslide hazard mapping, Proc. IV Int. Symp. Landslides, Toronto, 1, 307-324.

Carrara, A. (1983) - Multivariate models for landslide hazard evaluation, Math. Geol., 15, 402-426.

Hammond, C., Hall, D., Miller, S., and Swetik, P. (1992) - Level I Stability Analysis (LISA) documentation for version 2.0. Gen. Tech. Rep. INT-285. For. Serv. U.S. Dep of Agric., Ogden, Utah.

Iverson, R.M. (2000) - Landslide triggering by rain infiltration, Water Resources Research, 36, 1897-1910.

Montgomery, D.R., and Dietrich, W.E. (1994) - A Physically based model for the topographic control on shallow landsliding. Water Resources Research, 30, 83-92.

O'Loughlin, C.L. (1981) – Tree roots and slope stability. What's new in forest research, Forest Res. Inst. Publ., 104.

O'Loughlin, C.L., and Watson, A. (1979) – Root wood strength deterioration in radiata pine after clearfelling, N.Z. J. For. Sci., 9, 284-293.

Pack, R.T., Tarboton, D.G., and Goodwin, C.N. (1998) - The Sinmap Approach to Terrain Stability Mapping. Proceedings 8th Congress of the International Association of Engineering Geology, Vancouver, British Columbia.

Richards, L.A. (1931) - Capillary conduction of liquids through porous mediums, Physics, 1, 318-333.

Sidle, R.C. (1991) - A conceptual model of changes in root cohesion in response to vegetation management, J. Environ. Qual., 20, 43-52.

Sidle, R.C. (1992) - A theoretical model of the effects of timber harvesting on slope stability, Water Resources Research, 28, 1897-1910.

Skempton, A.W., Delory, F.A. (1957) - Stability of natural slopes in London Clay, Proc. 4th Int. Conf. SMFE, London, 2 , 378-381.

Soil Conservation Service (1972) - National Engineering Handbook, section 4, Hydrology. U.S. Dept. of Agriculture, Washington D.C..

Wu, W., and Sidle, R.C. (1977) - A distributed slope stability model for steep forested basins, Water Resource Research, 31, 2097-2110.

Ziemer, R.R. (1981) – Roots and the stability of forested slopes, IAHS publ., 132, 343-361.

Ziemer, R.R., and Swantson, D.N. (1977) – Root strength changes after logging in southeast Alaska, USDA-For. Serv. Res. Note, 306.

Slope stability management using GIS

ANDREAS GUENTHER, ANTJE CARSTENSEN and WALTER POHL,
Department of Geosciences, Technische Universität Braunschweig, Germany

ABSTRACT
We present here GIS-based methods for a regional risk-evaluation of large slope areas, e.g. water reservoir slopes in mountainous terrain. The tools presented here allow a rapid regional evaluation of geometrical slope-properties and are therefore suitable for managing qualitative assessments on their stability. Further attempt will be made to integrate rock mechanical and hydrological data within the GIS to quantify the regional sliding-susceptibility of critical areas.

INTRODUCTION
Through the last decade, GIS (Geographic Information Systems) achieved increasing importance concerning the management of space-related geodata. Our project deals with the development of a GIS-tool that will be suitable to evaluate the long-term stability of water reservoir slopes in hard-rock terrain. Of fundamental interest for stability-management of those slopes are regional evaluations of their internal geometry, becoming particularly evident when the rocks include planar geological structures which might, under certain geometric conditions, act as sliding planes and cause enormous hazards (HOEK & BRAY 1981). In this paper, we will first present methods for rapid regional modelling of complex tectonic structures based on geological field-data and the derivation of geometrical relationships between the slopes and potential sliding-structures. Second, we present an application of the described methodology at a water reservoir in Northern Germany.

METHODOLOGY
Geometric informations of the orientation of geologic fabrics cannot be obtained at any location of an area of interest, therefore methods for interpolating geologic structural data over extensive regions must be applied. After generating Digital Structural Models (DSM) for sliding-relevant geological fabrics, these can be combined with slope orientations derived from Digital Elevation Model (DEM)-data to obtain spatial continuous information about the internal geometry and the sliding potential of the slope-areas.

Instability – Planning and Management, Thomas Telford, London, 2002, 265–272

Structural modelling

Tectonic field-data consists of the dip-direction and the dip of a specific geological structure at a certain point. To interpolate these kind of data, any fabric-measurement must be regarded as a unit-vector that could be subdivided into three independent direction-cosines (cosα, cosβ, cosχ) within the cartesian system (Fig. 1, DE KEMP 1998). Each of these cosines must then be separately interpolated over the whole area and a continuous DSM can be obtained by recalculating dip-direction and dip from the cosines at any point of the area.

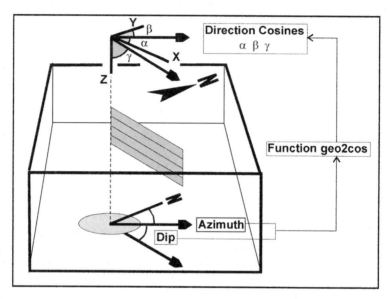

Figure 1: Decomposition of azimuth and dip (or trend and plunge) angles into component direction cosine α, β and γ (after DEKEMP 1998).

Geometrical modelling

To obtain spatial information of the internal geometry of the slope-areas, interpolated DSMs must be crosscut with DEM-data. The angle between potential sliding-planes (e.g. sedimentary bedding) and topography, varying between 0-90°, can be measured in a plane perpendicular to the cutting-edge of DEM- and DSM planes (WALLBRECHER 1986, Fig. 2). An algorithm was employed which allow regional continuous calculations of cutting edge- and angular relationships between topography and potential sliding- and cutoff-planes.

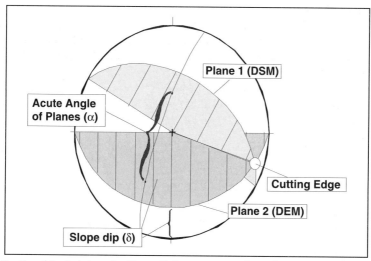

Figure 2: Graphical determination of the acute angle α between a slope surface (derived from DSM) and a geological fabric plane (from DEM). After WALLBRECHER (1986), equal-area projection, lower hemisphere

APPLICATION

The above described methodology for mapping geometrically critical slope-sections was established and tested at the Oker water reservoir (OWR) in the Harz-Mountains, Germany (Fig. 3). The slopes of the OWR consists of strongly folded and faulted layered sandstones and slates of Carboniferous age, overlain by a thin sheet of soil and forest-vegetation (HINZE 1971). The sliding-potential of these slopes is directly controlled by the orientation of geological fabrics with respect to morphology, where major sliding-planes are bedding-planes and cutoffs are tectonic fractures.

Due to their smooth morphology and the overall relatively shallow topographic gradient, the slopes of the OWR do not inhibit a high sliding-potential. In the northern part of the test-area north of the dam-building, however, steeper gradients do exist and shallow rockslides did occur recently (Fig. 3).

On June 1996, a shallow rockslide took place on the W-side of the Oker-valley (Fig. 3). Some 2500 tons of rock material was translated from an 80x30 m area towards the valley. This rockslide allows examination of the mechanical behaviour of anisotropy-planes within the rock-pile (sedimentary bedding-planes, tectonic fractures) during sliding. Sliding occurred on the 35-45° to the SE dipping backlimb of an anticline, having an amplitude of some 100 m with its axis being oriented

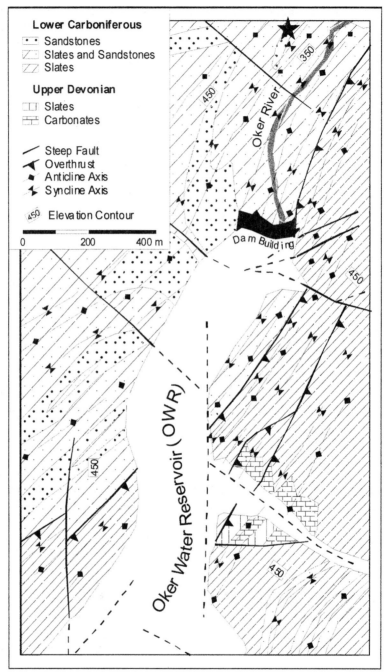

Figure 3: Tectonic map of the northern part of the Oker water reservoir (based on the geological map of Lower Saxony 1:25000, sheet 4128 (Hinze 1971). Also shown: elevation contours (10 m-spacing). Star indicates location of 1996 rockslide.

subhorizontal and subparallel to the slope (Fig. 4). Sliding-planes were sedimentary bedding-planes, whereas the hanging-wall cutoffs are located near the hinge of the anticline, at a great angle to the bedding-planes which are nearly horizontal in this position (Fig. 4). Lateral cutoffs of discrete slide-bodies are NW-SE conjugated fractures.

To identify critical slope sections for geometrically possible slope failures along bedding planes, the folded and faulted strata must be modelled for the whole test-area from structural fabric data, and angular relationships with the morphology must be obtained.

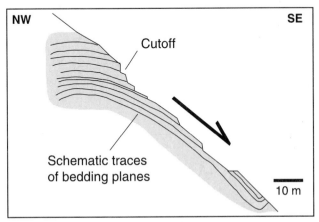

Figure 4: Cross-section of 1996 rockslide (location see Fig. 3). The cutoff is located near the hinge of the NW-facing anticline, bedding-parallel sliding occurred on the backlimb of the structure.

Derivation of DSM
From irregular distributed geological fabric data, a regional azimuth/dip-model for the whole test-area was calculated by transforming field-data into direction cosines, and regional interpolation of these data (Fig. 5). For interpolation, fault-traces from the digitized geological map were incorporated and the hinge-lines of the folds were used as breaklines (Fig. 5). Moreover, the overall strike of the geologic structures was taken into account by specifying a corresponding anisotropy-value during interpolation.

In general, the quantity of field-measurements needed depends on the complexity of the geologic structures and the desired resolution of the models. In our case, a relatively large amount of field-data was needed because of the complex structural setting (Fig. 3). In more simple tectonic settings, fewer data should be sufficient.

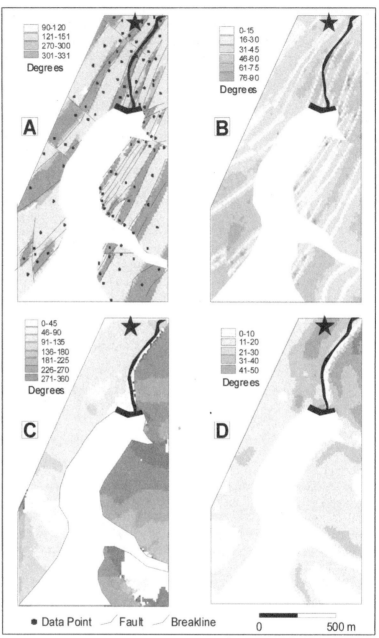

Figure 5: Regional azimuth and dip models from DSM (above) and DEM (below). A, B: Interpolated azimuth and dip models of deformed bedding planes. C, D: Slope azimuth and dip models derived from DGM 50 data of LGN. Pixelsize of all models is 10 m2. Star indicates location of 1996 rockslide.

Derivation of DEM

For our test-area, a DEM was produced from the DGM 50-dataset provided by the LGN with a grid-spacing of 50 m. After creating a topographic surface from these data, regional slope azimuth and dip models can be obtained by gradient calculations on a pixel base (BURROUGH 1986, Fig. 5).

Risk-mapping

After the derivation of azimuth/dip-models of both bedding-planes and morphology, the acute angle α between the two models was obtained with the methodology described above (Fig. 6). After that, potential high-risk areas could easily be mapped by exploring the GIS using Boolian operators (Fig. 6).

From Figure 6 B, it can be seen that extensive high-risk areas (e.g. where $\alpha < 5$ and $\delta > 30$) do only exist on the W-side of the Oker-Valley. This is underlined by the existence of the 1996 rockslide which is positioned in a class 1 area. However, similar geometrical relationships are present at the eastern side of the valley, where the steep SE-dipping forelimbs of the folds include critical angles with morphology.

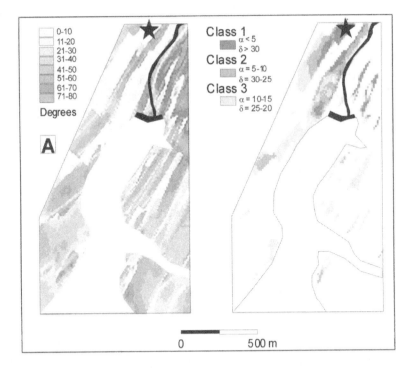

Figure 6: A: Regional angular relationship between bedding planes and topography. B: Classified risk-map based on critical angle between DSM and DEM (α) and critical slope dip (δ). Star: Location of 1996 rockslide.

From this, it can be shown that continuous geometrical modelling of large slope-areas allows the spatial detection of critical slope sections which cannot be directly obtained from field observations alone.

PRELIMINARY CONCLUSIONS AND FURTHER OBJECTIVES

The GIS-methodology described above is useful to obtain regional continuous data on the structural geometry of slopes in hard rock-terrain. It allows a rapid management of spatial risk-assessments for large areas, e.g. for water reservoir slopes. The wide availability of high-resolution DEM-data in many places of the world combined with structural data make our tools an advantage in preventing hazards.

The described spatial modelling of the kinematic slide potential provides the basis for dynamic analyses of reservoir-slopes. This will be done by combining the geometrical models with rock-mass strength classifications (SELBY 1993) and hydrological models to perform predictions on the stability of water reservoir slopes under changing precipitation-scenarios with GIS-based data.

ACKNOWLEDGEMENTS

This study is part of a collaborative project "Developments of methods for monitoring and prognosis of slope stability of water reservoirs", funded by the BMBF (Federal Ministry for Education and Research), grant 02WT0087. Support from the LGN, Hannover, Germany, and the Straßenbauamt Goslar, Goslar, Germany, is gratefully acknowledged. We thank Paulina Avilez for digitizing the geological map.

REFERENCES

BURROUGH, P. A., 1986: Principles of Geographic Information Systems for Land Resources Assessment. – Oxford University Press, 346 p.

DE KEMP, E. A., 1998 : Three-dimensional projection of curvilinear geological features through direction cosine interpolation of structural field observations. – Computers & Geosciences, 24, 269-284.

HINZE, C. (ed.), 1971: Erläuterungen zur Geologischen Karte von Niedersachsen 1:25000 Blatt 4128 Clausthal-Zellerfed. – Nieders. Landesamt f. Bodenforschung, 166p.

HOEK, E. & BRAY, J. W., 1981: Rock Slope Engineering. – IMM London, 358p.

SELBY, M. J. (1993): Hillslope Materials. – Oxford University Press, 451p.

WALLBRECHER, E. (1986): Tektonische und Gefügekundliche Arbeitsweisen. – Enke-Verlag, 244p.

Usoi landslide dam in Tajikistan - the world's highest dam: First stability assessment

DR J. HANISCH, Federal Institute for Geosciences and Natural Resources, Hannover, Germany

ABSTRACT

In 1999 UN/IDNDR Geneva organised a mission to Lake Sarez in the Pamir Mountains of Tajikistan. This 60 km long lake is up to 450 m deep and comprises a volume of about 17 km^3. It was formed from a huge natural dam which came into existence in 1911 when - triggered by an earthquake - a giant landslide of about 2 km^3 of rock and debris blocked the Murgab Valley. The geotechnical hazard assessment from a one-week visit of the dam site and the lower part of the lake resulted in one clear conclusion: The danger of a general dam failure caused by the water pressure or from seepage is low.

INTRODUCTION

In 1911 one of the strong earthquakes, typical of a region of active tectonism, hit the Pamir Mountains of Tajikistan and triggered an enormous landslide (volume: approximately 2 km^3). The landslide blocked the Murgab River valley forming a natural dam, which was named Usoi after a village buried beneath. With up to up to 700 m height it is the highest dam, natural or man-made, in the world. The dam impounded Lake Sarez, named after another village that had to be given up when the water level started to rise. Today, the lake has a surface elevation of 3265 m, is 60 km long and up to 450 m deep with a volume of about 17 km^3, roughly one half the volume of Lake Geneva. Because the dam is not an engineered structure and with a large volume of water it impounded, significant concern has been given to the long-term stability of this natural dam. An outburst could harm up to five million people in four countries downstream.

The data of Usoi landslide dam discussed here were collected during a one-week field visit of the "Disaster Hazard Assessment" sub-team of the UN/IDNDR mission which included two local experts, Dr. A. Ischuk (Dep. Director, Tajik Institute of Earthquake Engineering and Seismology, Dushanbe), Col. Y. Akdodov (Sarez Directorate, Tajik Committee on Emer-gencies, Dushanbe) and Dr. Carl-Olof Söder (SWECO INTER-NATIONAL, Stockholm). The visit covered the extended dam area and the western part of the lake. The geotechnical sub-team had - beside others - the task of asses-sing the likelihood of a failure of the Usoi landslide dam at a reconnaissance level. The duties of the team experts (in that connection) were the following:

- Assessment of the overall stability of the Usoi dam including seismic shaking.
- Assessment of the current state of water filtration through the dam and possible piping.

Fig. 1: Situation map of Tajikistan in Central Asia and of Lake Sarez.

FORMER STUDIES
Extended investigations were carried out on the Lake Sarez problem by Russian and Tajik scientists and engineers from 1915 through 1992. In 1992, however, all installations and investigations were abandoned after the independence of Tajikistan. In August 1998, the Tajik State Committee for Emergencies (SCE) began to install new monitoring systems. The original Russian reports were inaccessible during the UN mission. Therefore, the information and statements from the translated reports obtained during the mission (SCE 1997, 1999) were not fully verifiable. The evaluation of the Russian literature in the libraries and archives of Dushanbe, Tashkent and Moscow should be a primary task for future detailed hazard assessments.

GEOLOGICAL FRAMEWORK
The Pamir Mountains are part of the Himalaya – Hindukush – Karakoram – Pamir mountain belt and belong to one of the tectonically most active zone in the world (e.g. Lukk et al. 1995). It is dominated by strong compressional movements

resulting in a series of active thrust faults and associated wrench faults. As a consequence of this tectonic stress, the area is affected by continual earthquakes, some of which attained Richter magnitude values of 8 or 9 (Fan et al. 1994). The main constituents of the bedrock consist of quartzitic sandstones, quartzites and schists (Carboniferous age), and marbles and shales with some secondary gypsum, anhydrite and dolomite (Permian-Triassic age). The highly active tectonic regime also causes excessive rock fracturing and the formation of abundant shear zones, and intense cleavage.

USOI LANDSLIDE DAM

The Usoi landslide dam developed from a rockslide at the northern slope of Murgab valley (Fig. 2). It separated from the rock mass as a typical wedge failure (Fig. 3). The location of this slope failure was determined by the combination of a series of unfavorable tectonic factors:

(1) The presence of a series of intensive shear zones forming the necessary geometric setting for a wedge failure,
(2) The generally high degree of rock fracturing from the ongoing tectonic movements,
(3) The presence of a SW-NE-trending active wrench fault in the innermost corner of the wedge (Fig. 3).

Physical characteristics of the dam

The landslide mass has been divided into three parts consisting of individual massifs, each of which has its special features in regard to structure, size of blocks, amount of fines, etc. (SCE 1997, 1999):

- *Northern sector.* This is the lowest part of the dam, with a minimum freeboard of approximately 50 m. Large blocks are present at the surface, almost no fines are visible. In the area closest to the source of the old landslide a cone of deposits of recent rock avalanches, debris flows, and mudflows are found.
- *Central sector.* This part rises on the average approximately 100 m above lake level. The surface material in this central part differs from that of the rest of the dam in that there are almost no large blocks, and the surface includes a large amount of fines (silt).
- *Southern sector.* This is the highest part of the dam with a maximum height of about 270 m above lake level. The surface is covered by blocks of various sizes, the largest with a diameter as great as 20 m. Almost no fines are visible on the surface.

Fig. 2. General view of western Lake Sarez showing extensometers for slope monitoring in the foreground, the huge dam (left arrow), and (right arrow) the scarp of the 1911 landslide in the background to the right.

Fig. 3: View from the high central sector of the Usoi landslide dam (camera bag for scale) to the niche from where the huge landslide came from.

Geotechnical stability of the dam
There are two types of stability that need to be addressed: (1) the stability of the entire dam mass, and (2) the local stability of its slopes.

Safety against sliding of the entire mass of the dam
The entire dam mass is subjected to hydrostatic load of the reservoir water. This load could cause the failure of the dam by sliding along a sub-horizontal surface beneath the dam, either at the base of the dam or at greater depth. The hydrostatic load can either act on the upstream face as in a CFRD (concrete-faced rockfill) dam or on the central core, as in a traditional earthfill dam. Because the internal structure of this landslide dam is unknown, the simple CFRD case is used here in a rough stability estimation: The longitudinal cross section of the dam can be simplified as a triangle with a base line   5000 m long and a height of 550 m, impounding a lake with a depth of 450 m (Fig. 4). The specific weight of the dam material is estimated at 22 KN/m^3.

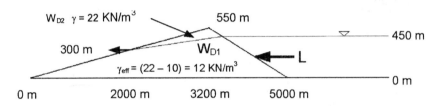

Fig. 4:   Simple scheme for the formulation of the general stability of Usoi landslide dam against the hydrostatic force (water load L) of Lake Sarez. Seepage level at the downstream side of the dam about 150 m below lake level.

The effective weight, W_D, of a 1-m-thick slice of the dam is calculated as follows:
$$W_D = W_{D1} + W_{D2}$$
$$= (F_1 \times \gamma_{eff} + F_2 \times \gamma) \times 1 \text{ m}$$
$$= (1.1 \times 10^6 \text{ m}^3 \times 12 \text{ KN/m}^3 + 0.275 \times 10^6 \text{ m}^2 \times 22 \text{ KN/m}^3) \times 1 \text{ m}$$
$$= 13{,}200 \text{ MN} + 6{,}050 \text{ MN}$$
$$W_D = 19{,}250 \text{ MN}$$
The friction force, F, depends on the weight of the dam and the resisting angle of internal friction, φ, between the dam and its base, which is assumed (conservatively) to be 25°; cohesion is considered to be minor, and is neglected.
$$F = W_D \times \tan \varphi$$
$$F = 19{,}250 \text{ MN} \times 0.466$$
$$F \approx 9{,}000 \text{ MN}$$
The water load, L_W, on a slice 1 m thick averages (as a rough estimate) 2.25 MN/m^2 between the lake surface and a depth of 450 m. This force acts against the vertical component of the slope area of 450 m^2 as follows:
$$L_W = 2.25 \text{ MN/m}^2 \times 450 \text{ m}^2$$
$$L_W \approx 1000 \text{ MN}$$

The factor of safety, FS, is the quotient of the resisting friction force and the horizontal water load:

$$FS = F/L_W$$
$$FS \approx 9$$

Seismic shaking due to a strong earthquake could reduce this safety factor by as much as 50 percent (cf. next paragraph), resulting in what is still a very high safety factor of 4.5, providing enough margins for assessment errors.

The long-term stability of landslide dams can also be assessed from geological case histories: A well-documented example has been reported from the Kali Gandaki Valley in Nepal. There, a natural dam of similar size to the Usoi dam was formed from a huge rockslide in post-glacial times (Hanisch 1998). The lake had a maximum depth of about 600 m and a length of 35 km. It silted up totally; the remains of the lake sediments now exist as widespread horizontal terraces (Fort 1976, Iwata et al. 1982).

Safety against internal sliding processes
The second type of failure of earthfill dams normally occurs as sliding within the body of the dam (Newmark 1964; Fig. 5). The controlling factor in this case is the internal water pressure in the dam that decreases the effective stresses in the material. Internal erosion in a dam can cause the maximum water pressures within the dam body to migrate downstream; if this process continues, the stability will gradually decrease, and finally sliding can occur within the dam. A large slide of this type could also affect the general stability of the dam because the mass of the dam would decrease.

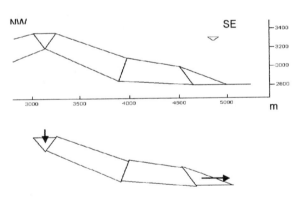

Fig. 5: Kinematics of a potential slide failure of the upstream side of the Usoi landslide dam using the "Method of Kinematic Elements" of Gussmann (1982). Curve of potential failure a first estimate un-til internal structure of the dam becomes available. For explanation, see text.

For this reason, a preliminary calculation of the internal slope stability of the upstream face of the Usoi dam has been performed using the method of "Kinematic Elements" (Gussmann 1982). Differing from classic methods, this technique is able to consider the movements between blocks defined arbitrarily by the analysis of the kinematics of landslide movement (Fig. 5). The couples of friction/cohesion can be defined separately for each interface between elements or at the bases of the elements.

In the present analysis, the following parameters have been applied:

Specific weight of the dam material: $\gamma = 22$ KN/m^3

Property values for interfaces near the bottom of the dam:

Angle of internal friction: $\varphi = 25°$

Cohesion: $c = 10$ KN/m^2

Property values for interfaces between individual elements:

$\varphi = 40°, c = 0$ KN/m^2

The influence of a heavy earthquake has been estimated using the pseudostatic approach, adding to the vertical acceleration, g, a horizontal component the extreme value 0.5 g. In future studies the more accurate dynamic approach by Newmark (1964) and Jibson (1993) should be applied.

Results

The following safety factors have been obtained from these analyses:

FS = 2.48   (water table at elevation 3263 m; no horizontal acceleration)

FS = 1.15   (water table at elevation 3263 m; horizontal acceleration = 0.5 g)

This signifies that the stability of the dam against slope failure toward the lake is sufficient, considering that a horizontal acceleration of 0.5 g is an extreme value in a pseudostatic earthquake analysis (Newmark 1964).

At the downstream slope, along the line of seepage springs, a small canyon has formed, with a depth of approximately 20 m. This canyon has developed to a very small amount in the dam material at its toe but mainly in deposits of a recent flood plain of fluvial and debris-flow that are resting on the downstream slope of the dam. Worries about the influence of this erosion on dam stability (cf. SCE 1997) are not justified.

The local slope stability of the downstream face cannot yet be estimated because the internal structure of the dam is not known.

Seepage

Lake Sarez drains exclusively by seepage flow through the dam. At the top of the dam no indications of a former natural spillway channel have been detected. The annual fluctuations of lake level are reported to be ±6 m. Seepage measurements began on a regular basis in 1943; since then, the discharge has averaged a constant value of 45 m^3/s with annual variation between 28 and 84 m^3/s (SCE 1997). The area in the vicinity of the springs on the downstream slope shows no signs of ongoing erosion, and from a visual inspection no sediment transport can be detected.

No firm conclusion can be drawn as to possible sediment erosion and transport by piping from this brief inspection. However, because the discharge through the dam has remained constant during the period from 1943 until today, the rate of internal erosion must be very low and not apt to create a major problem.

REFERENCES

Fan, G., Ni, J.F., and Wallace, T.C., 1994, Active tectonics of the Pamirs and Karakoram:. Jour.Geophys.Res.,Solid Earth, v. 99, no. B4, p. 7131-7160.

Fort, M., 1976, Quaternary deposits of the middle Kali Gandaki valley (central Nepal): Himalayan Geology, v. 6, p. 499-507.

Gussmann, P., 1982, Kinematical elements for soils and rocks: Proc., 4th Int'l. Conf., Numerical Methods in Geomechanics, Edmonton, p. 47-52.

Hanisch J. (1998): Large-scale river damming of Kali Gandaki (Nepal) by a post-glacial megaslide.- In: Geol.Dynam.Alpine Type Mountain Belt - Ancient and Modern.- Abstr., Proc. 88. Jahrestgg. Geol. Vereingg. Schweiz, Bern, 1998, p. 52.

Jibson, R.W., 1993, Predicting earthquake-induced landslide displacements using Newmark's sliding block analysis: Nat. Res. Council, Transp. Res. Board, Washington, D.C., Transp. Res. Rec., v. 1411, p. 9-17.

Lukk, A.A.; Yunga, S.L.; and Shevchenko, V.I., 1995, Earthquake focal mechanisms, deformation state, and seismotectonics of the Pamir -Tien Shan region, Central Asia: Jour. Geophys. Res., Solid Earth, v. 100, no. B10, p. 20,321-20,343.

Newmark, N. M., 1964, Effects of earthquakes on dams and embankments: 5th Rankine Lecture, Géotechnique, v.15, no. 2, p. 139-160.

SCE 1997: Regional scientific conference on the problems of Lake Sarez and ways of their solution – Final Report. Tajik State Committee for Emergencies, Unpub. Rep., 200 p., Dushanbe.

SCE 1999: Brief report on Sarez Lake problem – summary of studies carried out since 1982. Tajik State Committee for Emergencies, Unplug. Rep., 27 p., Dushanbe.

Landslide hazard assessment for land management and development planning, Scotland District, Barbados

MR IF HODGSON, Scott Wilson, Basingstoke, UK
DR GJ HEARN, Scott Wilson, Basingstoke, UK, and
MR G LUCAS, Soil Conservation Unit, Ministry of Agriculture and Rural
Development, Barbados

ABSTRACT

Scotland District lies on the north eastern side of the island of Barbados. The geology of the District comprises highly contorted sediments thrust upwards during the formation of the Lesser Antilles Island Arc System which have been deformed and intruded by oil bearing clays. The area is largely undeveloped because of the history of slope failures which includes large movements from the peripheral coral-capped escarpment towards the Atlantic together with widespread smaller failures. The paper describes a study commissioned to provide landslide susceptibility and hazard mapping for Scotland District to serve as a guide for agricultural, residential and recreational land management. Remote sensing methods and field mapping were used to collect data on geology, slope angle, groundwater, previous mass movements, current instability and erosion which were analysed in a geographical information system (GIS). Correlations were found between landslide occurrence and geology, slope angle and land use. Correlations between instability and various drainage elements were found not to be significant.

INTRODUCTION

Scotland District occupies some 60km^2 on the north-eastern side of the island of Barbados (Figure 2 and Plate 1). The area has been affected by a history of slope failures that have impacted on land use, building structures and roads for several centuries. The paper describes a study commissioned by the Department of Agriculture to provide a GIS-based landslide susceptibility mapping facility for Scotland District to serve as a guide for agricultural, residential and recreational land management. The emphasis was placed on the mapping of landslide and slope erosion and the definition of the drainage system, as this was considered to be a key factor in landslide initiation in the study area. The study was intended to be primarily desk-study based, making maximum use of remote sensing and existing data sets.

Instability – Planning and Management, Thomas Telford, London, 2002, 281–290

STUDY AREA GEOLOGY AND GEOMORPHOLOGY

Barbados is the only member of the Lesser Antilles island arc group that is not of volcanic origin. The island was formed by Tertiary sediments which were deposited by thrust movements from adjacent rapidly deepening and shallowing trenches. Subsequently a coral limestone cap was deposited before the whole island was raised to its present elevation. Today the coral cap covers most of the island except for Scotland District where the limestone has been removed, principally through landsliding, to expose the underlying Tertiary strata (Figure 1). The stratigraphic succession underlying Scotland District is shown in Table 1 (after Barker, undated).

Age	Unit	Description
Quaternary	Coral Limestone	Massive rock absent (removed) from most of Scotland District but forms the escarpment on its western boundary.
Tertiary	Bissex Hill Formation Oceanic Group	Localised sandy limestones and marls. Grey siliceous mudstones and marls sometimes with volcanic ash.
	Upper Scotland Formation Members: T-Unit, Mount All, Chalky Mount, Murphy's	Predominantly sandstones, sometimes gritty, with shales, marls and conglomerates.
	Lower Scotland Formation Members: Morgan Lewis, Walker's	Dark grey shales and clays with some turbidite sandstones.
Eocene\Miocene	Joe's River Beds	Younger intrusive dark grey to black, oil bearing, structureless, distorted and slickensided clays often containing fragments of the country rocks.

Table 1: Stratigraphy of the Scotland District

Hydrogeology and geomorphology

The escarpment forms a surface watershed to the west of Scotland District but the presence of a number of springs close to the escarpment indicates some underground flow across the surface watershed. The gradual regression of the cliff westwards has meant that many valleys draining westward have been be-headed forming preferred locations of infiltration. Within Scotland District the generally low permeability of the Tertiary strata and surface soils, together with the steepness of slope, encourage runoff rather than infiltration. Surface water appears to be the main provider of stream flow except in close proximity to the escarpment. Springs are generally areal rather than point sources, probably impersistent and often masked by surface water.

The contrasting permeability between strata suggests seepages may be associated with the interfaces, for example, between the Oceanics and the Upper Scotland Formation, and between the Upper Scotland Formation and the Lower Scotland Formation.

Landslides and failed ground occupy a significant proportion of the Scotland District. Tension cracks behind the limestone escarpment are evident on aerial photographs and large detached blocks of limestone can be seen extending some distance down slope, reaching the coastline in some locations. The drainage pattern is controlled by a small number of large valleys draining from the escarpment in the west to the coast in the east. Relative relief is highest in the west, close to the limestone outcrop where valleys are deep and steep-sided. In the east the landscape is more subdued, although knife-edge ridges have been created by progressive erosion and landsliding in some catchments. Devoid of the protective coral cap, erosion is well developed in Scotland District and is most severe in the Joe's River Beds and the softer strata of the Scotland Formation (Mount All and Chalky Mount Members). These strata form steep slopes close to, or at, the limiting angles for soil stability and vegetation support.

DESCRIPTION OF INSTABILITY

The factors outlined above have combined to make large parts of the Scotland District susceptible to a range of slope failure mechanisms, which, according to the OAS report (1971) have affected approximately half of the area. The first recorded landslide in the District, in 1785, was described as large, deep-seated with ground and building displacements of the order of 100 yards. According to accounts at the time the Boscobel landslide of the 1901 was a major deep-seated landslide from the escarpment. Smaller events have been recorded at Sedge Pond in 1934, and at Turner's Hall in 1966.

Most of the larger landslides occur beneath the coral limestone escarpment and appear to involve large masses of Oceanic material, with failure surfaces presumably passing into the underlying Scotland Clays. They appear to be slides, probably mainly translation with local rotation at the back scarp, which are seen as a series of intact rafted blocks. There is some evidence of toppling failure from the cliff and further from the escarpment on flatter slopes, there are examples of debris flows, seen most easily on the aerial photographs. The topography in the 'Panhandle' area is gently undulating, probably as a result of old progressive mass movements. In addition to these large landslides there are numerous smaller failures which have occurred in older colluvium and in-situ strata. These features are nonetheless capable of seriously damaging (Plate 2) or demolishing dwellings and causing major disturbance to road pavements.

STUDY METHODOLOGY

The project was carried out in February to April 2000, and comprised the collection of landslide and mapping datasets from desk study and field survey. The derivation of landslide susceptibility maps required the preparation of a landslide distribution map for the Scotland District against which mapped factors, such as geology, topography, drainage and land use, would be correlated to identify relationships that could then be taken forward in the susceptibility analysis using GIS.

Existing Desk Study Data

The following information was obtained from desk study data sources:

- 1:5,000 digital topographical mapping provided by Lands & Surveys Dept
- Slope category maps taken from Vernon and Carroll (1966)
- 1:20,000 geological mapping (Poole, Barker and Payne 1981)
- Land use and land capability mapping taken from Vernon and Carroll (1966)
- Land use mapping undertaken by Scott Wilson from 1991 aerial photographs during a previous contract for the Ministry of Agriculture
- Records of damages to roads, buildings, land and other facilities obtained from various public authorities
- Details of springs providing potable water supplies from Ministry of Agriculture

Project-Acquired Desk Study Data

Extensive use was made of remote sensing imagery to acquire additional information on geomorphology, landslide distributions and processes, drainage details and vegetation. Landsat 7 ETM+ imagery was acquired on behalf of the Ministry of Agriculture. While this imagery provided some useful information on ground cover, it was partially affected by cloud cover and the resolution (15-30m) was too coarse for detailed interpretation. Aerial photography, however, proved extremely useful. Black and white aerial photographs with a scale of 1:10,000 were taken in 1951 when there was significantly less vegetation in the area. At that time sugar cane had only recently been cleared to make way for legume and fruit agriculture and ground details were recorded extremely well on the aerial photographs. The vast majority of geomorphological interpretation, drainage and landslide mapping was undertaken using this photography. The landslide database recorded from this photography was then updated with colour photography flown in 1991 and fieldwork. Aerial photography flown for a coastal management project in 1997 was not used to any significant degree in this process as it provided only partial coverage of the Scotland District and its scale (1:5000) was considered too large for sensible landslide and geomorphological interpretation. In total 253 landslides and 313 erosion features were mapped. As with all other desk study data, the information recorded from these aerial photographs was digitised for subsequent spatial analysis using GIS.

Fieldwork

Fieldwork was designed on the basis of a sampled ground truthing programme. Quality control required that at least 10% of the aerial photograph interpretation was independently checked and at least 10% of the aerial photograph interpretation required field validation. On the whole, the field validation confirmed the aerial photograph interpretation, but also provided significantly more information on detailed slope, materials and drainage conditions. Access to the site was relatively easy as the Consultant's temporary office was located at the Scotland District Soil Conservation Unit. Accordingly, field studies were increased to include road damage inventories and other geological and slope stability observations. Interviews were held with local landowners to obtain more information on the extent and timing of ground movements affecting their property. This information was entered onto the GIS.

A survey was carried out of observed limiting slope angles on the various lithologies of the Scotland District (See Table 2). These slope values were used to support the susceptibility analysis of the landslide distribution against slope and geology in the factor maps.

Strata Type	Limiting Slope Angle (Degrees)*	
	Inferred Wet Condition	Inferred Dry Condition
Morgan Lewis	11	15
Murphy's Member	22	Variable
Mount All Member	28	Variable
Oceanic Group	15	26
Joe's River Beds	15	18
T Unit Member	11	11

Table 2: Limiting slope angles for various lithologies as measured in the field.

LANDSLIDE FACTOR MAPPING

All desk study and field data were entered into the GIS and the distribution of landslides in the District systematically compared with each of the controlling factors. The results of these analyses are described briefly below.

Landslides on Slope Angle

The slope category map (based on Vernon and Carroll 1966) showed significant correlations with landslide distribution at all levels, but did not show a continuous increase in susceptibility against steepening slope. This apparent anomaly may be due to the role of other factors such as soil/rock properties and structure, or, possibly to the distribution of landslides across slope category boundaries.

Landslides on Geology
Figure 2 shows the distribution of landslide areas on the geological units for which the frequency distribution was found to be highly significant. This susceptibility ranking of the strata is considered appropriate for the smaller, shallower failures that make up the majority of the landslide frequency database. A consideration of the observed to expected ratio (O/E) for the area of failure shows a different pattern which is more relevant for the large, deep-seated failures. It was decided, therefore to combine these two sets by examining the distribution of landslides with increasing slope angle on a stratum by stratum basis, and incorporated into the susceptibility mapping in a similar fashion. Within each stratum type a progressive increase in landslide occurrence was found with increasing slope angle category.

The most susceptible formation was found to be the T-Unit (Upper Scotland Formation) followed by the Joe's River Beds and the Lower Scotland Formation.

Landslides on Superficial Soil Type
A high level of agreement was found between the landslide distribution and superficial soil type. However, the expected correlation between soil type and slope against landslide distribution was not found and the combined soil/slope parameter was not included in the susceptibility analyses.

Landslides on Land Use
The initial relationship between landslide frequency and 'natural' land use appears highly significant with increasing frequency from gullies through grassland, bare ground/soil, terrace to agriculture and woodland. However, when the area of landsliding is examined the O/E values are only marginally greater than unity (ie that expected for a random sample) and this factor was considered unsuitable for use in landslide susceptibility mapping.

Landslides on Drainage
Two main factors were considered, the presence of springs and wet areas, and the proximity to a water course.

Springs and Wet Areas
A total of 395 springs and wet areas were mapped in the study, mostly through aerial photograph interpretation. Concentrations of springs were found in the Oceanics Group, at the base of the Bissex Hill Formation, on the Joe's River Beds and in Head and Scree Deposits and Coral Rock of the 'Panhandle' area. While hydrogeology is known to be an important factor in the control of large, deep seated failures in parts of the study area, this was not reflected in the simple relationships derived from spring and landslide locations. Only 33 of the landslides mapped

coincided with a spring and only 115 of the mapped landslides (ie less than half) have occurred within 100m of a spring. For these reasons, and somewhat reluctantly, it was decided that springs and wet area locations could not be included in the landslide susceptibility mapping.

Distance from a Drainage Channel
This factor was considered because of the possible role of toe erosion in over steepening, and because of the greater propensity of soil saturation. However, no meaningful correlation was found.

EROSION FACTOR MAPPING
Erosion factor maps were produced using similar analyses to those used for landslide susceptibility.

LANDSLIDE SUSCEPTIBILITY AND HAZARD MAPPING
The combination of geology and slope were found to be the most meaningful factors in the preparation of landslide susceptibility maps. A matrix of landslide susceptibility rankings was therefore built up for slope category classes on each strata type for both the larger failures adjacent to the escarpment and the more widespread smaller failures. The GIS was then used to produce a landslide susceptibility map (Figure 2), showing five levels of susceptibility which was then tested against the mapped distribution. This analysis resulted in a highly significant Chi-squared value and a sensible progression of landslide density through the susceptibility classes.

The analysis was further supported by the fact that, where the units coincide between the field mapping and the GIS, a reasonably degree of conformity exists with measured threshold slope angles. This does not hold true in the Oceanics, where partial saturation is believed to play a part in many of the failures.

Outputs from the project included the GIS and its various data layers, the Landsat and aerial photograph interpretation and landslide and erosion distribution maps, and the susceptibility maps prepared for landslide and erosion processes. Hazard maps were produced by combining these maps with known distribution of erosion and landslides. Based on this zonation planning and land management guidelines were developed into a constraint map that summarised levels of hazard, the potential risk they posed to existing land use, and recommendations for further investigation, stabilisation and land development options.

CONCLUSIONS
The study has investigated the means of using relatively easily available desk study and field derived data to provide a zonation map of landslide susceptibility and hazard for the Scotland District of Barbados. The sensitivity of various factors has been examined and the geology and slope angle have been found to be most

influential. Correlations with soil type and land use, which initially appeared useful were, on closer examination, discarded from the final susceptibility mapping. No useful correlation could be found between the mapped landslide and erosion distribution and the drainage pattern, and consequently these factors were excluded from the susceptibility mapping. This outcome was unfortunate, and was probably a result of the extremely complex geology of the area and the relative simplicity of the methods of spatial analysis adopted.

One of the main recommendations made once the project was complete was the need to develop a more filed-based approach to slope hazard assessment, including the importance of geomorphology in the interpretation of historical and current landsliding processes. Nevertheless, the outputs from the study provided an extremely valuable GIS-based procedure for gaining a rapid appreciation of the relative stability of the Scotland District and providing the basis for the development of slope management programmes and land use planning documentation. The use of aerial photograph interpretation and selected field survey, combined with GIS manipulation and interrogation of published data sets as a means of gaining a rapid appreciation of slope hazard for engineering and planning purposes is therefore vindicated by this study which was completed in a total period of two months.

REFERENCES
BARKER LH and POOLE EG (undated): The Geology and Mineral Resource Assessment of the Island of Barbados. Part 1: The Geology of Barbados.
HEARN GJ, HODGSON, IF & WODDY, S (2001): GIS-based landslide hazard mapping in Scotland Disctrict, Barbados. In Griffiths, JS (ed.) Land Surface Evaluation for Engineering Practice. Geological Society, London, Engineering Geology Special Publications, 18, 151-157.
ORGANISATION OF AMERICAN STATES (OAS) (1971): Scotland District of Barbados: Evaluation of the Problems of Erosion and Unstable Ground. Report of the Technical Assistance Mission of the Office of the Regional Development Organisation of American States to the Ministry of Agriculture of Barbados.
POOLE EG, BARKER LH and PAYNE P (1981): Barbados 1:20,000. Geology of the Scotland Area.
VERNON, KC and CARROLL, DM (1966): Soil and Land Use Surveys No 18. Barbados. University of West Indies.

ACKNOWLEDGEMENTS
The authors would like to thank the Ministry of Agriculture and Rural Development for the opportunity to undertake this study. Desk study data sets were obtained from various government ministries and departments through the Ministry of Agriculture, and the authors would like to thank the Department of Lands and Surveys and the Coastal Zone Management Unit (Ministry of Environment) in particular for their assistance. Mr Glenn Marshall and Mr John Warner, both of the Soil Conservation Unit, provided invaluable assistance as project counterparts.

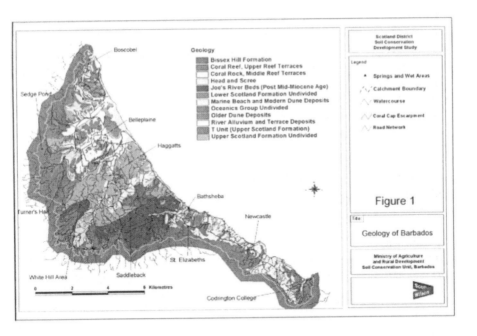

Figure 1

Geology of Barbados

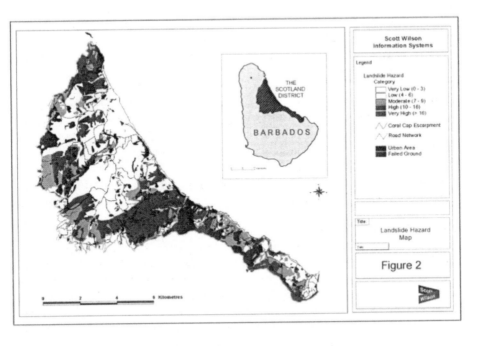

Figure 2

Landslide Hazard Map

Plate 1: View of Scotland District towards the south east.

Plate 2: Landslide damage to house.

Landslide movements affecting the lighthouse at Saint Catherine's Point, Isle of Wight

J. N. HUTCHINSON, Imperial College, London, UK
E. N. BROMHEAD, Kingston University, Surrey, UK, and
M. P. CHANDLER, Former Research Assistant, Imperial College, UK

INTRODUCTION

The landslides of the Isle of Wight Undercliff (Hutchinson, 1991) occupy a 12 km length of the southern coast of the island. They were formed by marine erosion of the Cretaceous outlier of the Southern Downs, with St Catherine's Point close to its western end. A research programme on these landslides was carried out intermittently from 1962 to the early 90s. It was immediately evident that St Catherine's Lighthouse was involved in some slight slide movements and opportunity was taken during our detailed investigations of the St Catherine's Point landslides during 1983 - 85 to bring this to the attention of Trinity House. Their generous funding enabled us to make a detailed surface and sub-surface investigation of the lighthouse site and environs which was reported on in 1985 (Hutchinson *et al.*, 1985).

EARLY HISTORY OF ST CATHERINE'S LIGHTHOUSE

The coast of the Isle of Wight, UK, forms a notorious lee shore, as attested by the numerous wrecks. The first light to meet this danger was built near the top of St Catherine's Hill (c. 238m O.D.) by Walter de Godeton, a local landowner, in about 1323. This appears to have been operated until stopped by sequestration in the Reformation, about 1530. No further light was shown until 1785, when the Trinity Board rekindled the old light and began the erection of a new *pharos* slightly nearer the coast than the earlier one. The latter was never completed, however, as fogs and mists would have rendered it almost useless (Anon, n.d.). Possibly climatic conditions in the 14th century were different from modern ones. The present lighthouse was therefore erected at a much lower elevation, on St Catherine's Point (c. +18m), in 1838 (Fig. 1). It was first lit in March, 1840. As the lantern was frequently obscured by mist, however, the tower was lowered by 13.1 m in 1875. Dunning, (1951) suggested that the decision to lower the tower was also influenced by the fact that it was already showing signs of being affected by landsliding.

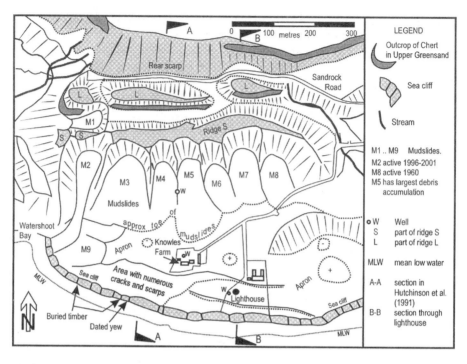

Figure 1. Plan: relationship of the lighthouse to St Catherine's Point landslide.

GEOLOGICAL SETTING & THE ST CATHERINE'S POINT LANDSLIDE

The Cretaceous rocks involved in the Undercliff form part of a gentle syncline, which plunges towards the S.E. and is truncated by the coast. At St Catherine's Point (on the western limb of the syncline) the strata of interest, range from the Ferruginous Sands of the Lower Greensand to the Lower Chalk The seaward component of dip at this site is between 1 and 1.5°.

St Catherine's landslide is a deep-seated compound slide believed to have occurred between 4000 and 1000 years ago (Hutchinson *et al.* 1991). It comprises three main elements: a rearmost subsided and rotated block; a block slide moving along bedding; and finally, a frontal "apron" of debris. The two main slide blocks form prominent ridges sub-parallel to the coast, described by Hutchinson *et al.* (1991) as ridge "L" (landward) and ridge "S" (seaward). Both the main slide and the debris apron have a basal shear seated in a thin clay layer (stratum 2b) towards the base of the Sandrock. The debris apron, up to 36 m thick, predates the formation of the two ridges by deep-seated failure. It is inferred that it was created mainly by rotational failures seated in the Gault, as at Gore Cliff (Bromhead *et al.* 1991). Relics of this system still cap ridge S, and feed mudslides (Fig. 1) that are reactivated by wet weather. A radiocarbon date of 4490 ± 40 years B.P. was determined on a yew log found in the debris at the sea cliff (Figs. 1 & 3). This, with an associated soil horizon and tufa layer, divides the debris of the apron into a lower spread, around 13m thick, and an upper, between 10 and 20m thick. A discussion of the evolution of the landslide is given by Hutchinson *et al.* (1991).

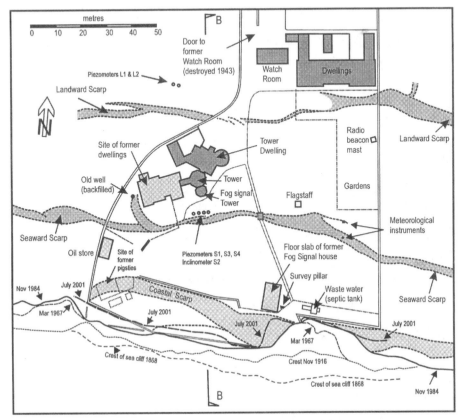

Figure 2. Site plan: geomorphology and boreholes in the lighthouse area.

INVESTIGATIONS OF THE LIGHTHOUSE AREA
Surface Mapping
The relationship of the area of the lighthouse enclosure to the geomorphology of the landslide complex is shown in the map of Fig. 2, mapped in November, 1984 and revised in July, 2001. This also shows positions of the crest of the sea cliffs from 1868. The area of the lighthouse enclosure is traversed by three main landslide scarp systems, termed Coastal, Seaward and Landward Scarps on the plan. The Coastal Scarp is being consumed by more recent, current rotational slipping of the present sea cliff, which is generally 15 to 20 m high. The Seaward Scarp runs 40 to 55 m landward of the crest of the sea cliff and roughly 10 m seaward of the lighthouse. It is generally 0.5 to 1.2 m high and somewhat degraded towards the east. The Landward Scarp runs around 40 to 60 m landward of the Seaward Scarp, *i.e.* about

25 m behind the lighthouse. It is fairly fresh and 0.7 to 0.9 m high in its western part, just behind the lighthouse, but very subdued to the east.

Sub-surface investigations
While the general nature and engineering geology of the landslides and debris apron on the main cross-section had been well established (Section A-A, Hutchinson, *et al.*, 1991), it was clearly desirable to investigate also a Section B-B through the lighthouse. The positions of both sections are shown on Fig 1 and that of Section B - B also on Fig. 2. Subsurface details in the main parts of the slide have been based on Section A-A. Six boreholes were sunk at the positions shown on Fig. 2. Of these, five (LI, L2, S1, S3 and S4) were put down by flight auger, chiefly to permit the installation of Casagrande type piezometers, although some approximate stratigraphical information was also obtained. The sixth borehole, S2, was cored in its important lower part, to investigate the shear zone at the base of the debris apron (Fig. 3), which was confirmed as lying within the stratum 2b clay layer. The inclinometer casing in this borehole was severed by the landslide movements at the level of the basal shear. Groundwater data are summarised on Fig. 3.

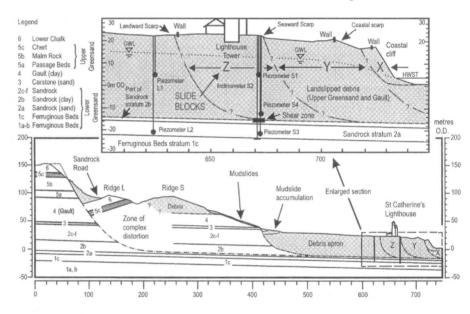

Figure 3. Section B-B through lighthouse, with enlargement of the seaward area.

Coast erosion
The coast of St Catherine's Point has no artificial defences. However, there are numerous large blocks of resistant Upper Greensand, and particularly the Chert Beds, in the debris of the apron which have accumulated on the foreshore to form a natural coast protection. As shown by Fig. 3 the basal slip surface lies around 20 m

beneath the foot of the sea cliff. The distance to seaward of the outermost toe of the St Catherine's landslide complex is not known.

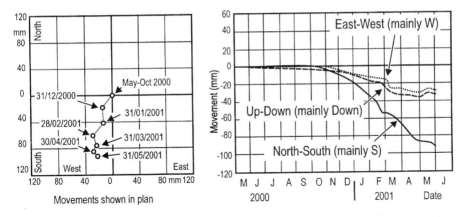

Figure 4. Three-dimensional movements of the lighthouse during 2000-2001.

The original Fog Signal House was built in 1875, near the crest of the sea cliff SSE of the lighthouse (Fig. 2). In 1932, following movement and damage, the fog signal was moved to a new, secondary tower adjacent to the front of the light house (Bowen, 1947), where it functioned until 1987. The original fog signal house foundations and floor are still visible, and tilt inland by approximately 3°. The positions of the crest of the sea cliff over the 180m length of coast fronting the lighthouse installations are shown in Fig. 2 for years between 1868 and 2001. As to be expected in such heterogeneous debris, the amounts of erosion are variable. They range from 9.4m to 23.9m between 1868 and 2001, thus indicating average rates of crest retreat ranging from 70 to 180 mm/year. A small zone of particularly rapid retreat may have been influenced by septic tank discharges.

Past movements
Operation of the lighthouse is insensitive to translation, but hypersensitive to tilt. The light and lenses form a massive unit that floats in a circular, mercury-filled, trough. This can accommodate a small amount of tilt before the lens unit starts to touch a series of rollers, which permit only a further small movement before significant frictions are caused. Hence, Trinity House has been far more concerned with tilt than any lateral component of movement. The first known measurement of the tilt of the lighthouse was made in 1873, two years before the tower was shortened by 13.1 m and indicated a tilt of about 7.6 minutes of arc towards a bearing of 307°. The tilt appears to have fluctuated slightly up to about 1970, to have roughly doubled by about 1975 and again to have fluctuated slightly about the increased value of tilt until the present day.

There is a long history of generally minor cracking in the lighthouse buildings and in the boundary walls. These structures act as useful monitors of differential movement, and are currently being monitored with crack gauges. Observations in 1984 of walls repaired in 1980 showed cracks opening at 3 to 8mm/year., although total crack width since construction of the wall was difficult to assess because of frequent repairs. A summation of crack widths in 1984 indicated that the relative seawards displacement within the enclosure to be 200-250mm, giving an annual average rate of movement of roughly 2 mm/year.

It was clear from geodetic measurements (Bromhead *et al.* 1988) that the lighthouse compound was moving seawards faster (35-50 mm/yr) than the main slide (10-20 mm/yr). That this was mainly basal shear was confirmed when the inclinometer casing was severed in only a few years. In 2000, an Ordnance Survey GPS station was installed in the lighthouse as part of a nationwide system. This station provides accurate three-dimensional positional information. The records of movements over winter 2000-1 are shown in Figure 4. During this time, significant seaward movement (*ca* 90mm) of the lighthouse has taken place at a time of heavy rainfall. Neither the main light nor the secondary steady red light (first exhibited in 1904) that points over Atherfield Ledge towards the Needles is affected by this. Such movements were not detectable by the earlier system of plumb bob monitoring.

It appears that prior to 1939, the field to the NW of the lighthouse complex was far more level than at present. There is anecdotal evidence that the lower windows of Knowles Farm (Fig 1) were then visible from the door of the old Watch House (Fig. 2), which is no longer the case. This evidence points to relative ground level changes in the apron of at least 2m in addition to seaward displacements and tilts.

Origin of the ground movements
Within the debris apron, in-situ lateral stresses have been compressive since at least the occurrence of the main slide. Since then, the state of compression in the debris has been affected by two principal landslide-producing factors, coastal erosion and groundwater fluctuation.

Coastal erosion clearly has a bigger impact on slide activity close to the sea cliff than it does on larger slides in the debris apron, or involving the main slide. It is also responsible for forming the cracks and minor scarps that affect the lighthouse compound. Coast erosion can therefore be considered to result primarily in a reduction in the compressive lateral stress and extension. These effects would tend to intensify with sea level rise and increased storminess, and the local effect of waste water disposal on site close to the cliff edge. Since the lighthouse was turned over to largely remote operation, there are no longer the inputs of waste water from keepers and their families living in the cottages within the lighthouse complex.

Groundwater fluctuations due to rain have a lesser effect on the stability of the coastal margin than on that of the whole of the apron, since a bigger proportion of the former lies under the foreshore with pore pressures controlled largely by sea level. Precipitation falling on ridge S stimulates mudslide activity, and loads the head of the debris apron. Rain falling on this ridge will have little effect on groundwater pressures on the slip surface (in the Sandrock) because of the protective Gault clay cap. However, rain falling on the more permeable ridge L, or on the rear scarp of the main slide and on the Southern Downs adjacent to the main slide, will increase groundwater pressures at its rear and lead to further movements of the main slide system. Streams emerge at each end of ridge L (Fig. 1) and limit the ground water levels behind this ridge. Movements of the main slide, or of the whole of the debris apron, lead to an increase of compressive lateral stress in the apron.

Thus, the oscillations in tilt at the St Catherine's Lighthouse, combined with periodic episodes of seaward movement, are best visualized as the interplay between these two dominant destabilizing processes on the site, the one leading to compression and the other to extension.

CONCLUSIONS

St Catherine's Lighthouse is situated on the debris apron fronting the St Catherine's landslide, near its seaward edge. It can tolerate considerable slow translational movement, but is sensitive to tilt. Retrogressive development of multiple rotational / compound slides in the seaward parts of the apron through coast erosion, has resulted in three main slide blocks, X to Z (Fig 3). Block X is rotated and largely consumed, Block Y has a slight backtilt (*ca.* 3°); Block Z with the lighthouse near its seaward edge, is barely yet backtilted. These blocks are moving very slowly seaward on a deep-seated basal shear surface) at *ca.* −16m OD) with a seaward dip of about 1.5° in Bed 2b of the Sandrock. Movements, none of which have rendered the lighthouse inoperable, comprise:

(i)   A translational component arising from the movements of the main slide (typically 10-20 mm/yr) and apron.

(ii)   Additional movements, also chiefly translational, resulting from the development of slides X to Z.

(iii) Tilts, monitored intermittently from 1873 up to around 15 minutes of arc.

The movements are controlled by both coast erosion and effective rainfall. In (i) above, rainfall predominates; in (ii), coast erosion. Movement (iii) is a function of the interplay of (i) and (ii). The sum of (i) and (ii) is typically 35-90 mm/yr.

Average rates of coast recession over the past 133 years vary greatly, from 70 to 180mm/year, reflecting the heterogeneous nature of the debris. These averages must not be allowed to obscure the essentially episodic nature of the coastal recession. In future, they are likely to increase, particularly through sea level rise and increased storminess. On all these counts, great uncertainty would attach to any extrapolation into the future. The lighthouse will cease to be serviceable as the type (ii) and (iii) movements develop further. This point will be reached when block Z backtilts more, long before the crest of the sea cliff reaches the lighthouse structure.

ACKNOWLEDGEMENTS
We are grateful to Trinity House for permission to publish this paper, and for various data. A grant from NERC supported Dr Chandler's post-doctoral work.

REFERENCES
ANON. n.d. Guide to the Isle of Wight. Ward, Lock & Co., Ltd, London and Melbourne. (secondary ref., find primary).

BOWEN, J. P. 1947. British Lighthouses. Longmans Green and Co., London.

BROMHEAD, E. N., CURTIS, R. D. & SCHOFIELD, W. (1988). Observation and adjustment of a geodetic survey network to monitor movements in a coastal landslide. Proc. 5th International Symposium on Landslides, Lausanne. 383-386. Balkema, Amsterdam.

BROMHEAD, E. N., CHANDLER, M.P. & HUTCHINSON, J. N.. (1991). The recent history & geotechnics of landslides at Gore Cliff, Isle of Wight. International Conference on Slope Stability Engineering - Developments and Applications, 189-196. Thomas Telford, London.

DUNNING, G. C. 1951. The history of Niton, Isle of Wight. Proceedings of the Isle of Wight Natural History and Archaeological Society, 4, 191 - 204.

HUTCHINSON, J. N. 1987. Some coastal landslides of the southern Isle of Wight. In Wessex and the Isle of Wight. Field Guide (ed. K. E. Barber), 123 - 135. Quaternary Research Association, Cambridge.

HUTCHINSON, J. N. 1991. Theme Lecture. The landslides forming the South Wight Undercliff. International Conference on Slope Stability Engineering - Developments and Applications, 197 - 205. Thomas Telford, London.

HUTCHINSON, J. N., CHANDLER, M. P. and BROMHEAD, E. N. 1985. St Catherine's Lighthouse. Report No. 1 on the ground conditions and landsliding at the site of St Catherine's Lighthouse, Isle of Wight (unpublished).

HUTCHINSON, J. N., BROMHEAD, E. N. and CHANDLER, M. P. 1991. Invest igations of the landslides at St Catherine's Point, Isle of Wight. International Conference on Slope Stability Engineering - Developments and Applications, 169 - 179. Thomas Telford, London.

TOOLEY, M. J. 1978. Sea level changes. N-W England during the Flandrian Stage. Clarendon Press, Oxford.

Geophysical exploration of soil stability at open pits – case studies from Yugoslavia

DR. SNEZANA KOMATINA, ZORAN TIMOTIJEVIC, Geophysical Institute, NIS-Naftagas, Belgrade, Yugoslavia

ABSTRACT
In the paper, a short review of applicability of geophysical methods in evaluation of soil stability is presented. Two case studies from Yugoslavia (Majdanpek and Vareš open pits) are discussed.

Keywords
environmental geophysics, soil stability, open pits, case studies, Yugoslavia

INTRODUCTION
Environmental geophysics is known as a group of numerous methods directed to exploration of the first several meters under the terrain surface. As just that part of the terrain is characterized by lithological heterogeneity in horizontal and vertical direction, sophisticated methodology in data acquisition, processing and interpretation is necessary.

Conditions which would be fulfilled for successful application of geophysical methods in landslides and soil instability exploration are the following:
1. Presence of differences in physical characteristics between sliding mass and stable medium;
2. Width of the landslide body would be 5-10 times greater than the body thickness.

Elements important for sanation measures at one landslide are the following:
1. Position of the sliding surface;
2. Distribution of masses with different physical characteristics at the landslide body;
3. Hydrogological characteristics of the landslide.

At the same time, cited elements present the tasks to be solved by geophysical methods application. Most often used methods are: electric soundings, self-potential method (SP), thermometry, seismic refraction method, electric and radioactive well-logging, micro magnetic methods, etc.

On the basis of the results, following data referring to investigated landslide are obtained:

1. Distinguishing parts of the landslide different in lithological and hydrogeological characteristics;
2. Sliding body geometry. Determining lithological content and degree of rock weathering within autochthonous mass;
3. Exploration of hydrogeological characteristics of autochthonous medium;
4. Determining the landslide direction and velocity (total and in parts).

Because of very complex structure of landslides, significant inhomogeniety in horizontal and vertical direction, but also small depth of moving mass and sliding surface relief, adequate density of measuring points and high-precision measurements are necessary

REVIEW OF GEOPHYSICAL METHODS APPLIED IN EVALUATION OF SOIL INSTABILITY

During the last several decades, landslide exploration is more and more based on geophysical methods. General advantage of the methods is possibility to explore rock masses under conditions of original position, in order to acquire continual information in space and time.

Geophysical methods are also applied for exploration of a landslide regime, that is – within regime monitoring. Main task of geophysical monitoring in separated test sites is to define objective relationships between variations of state, characteristics and position of rocks in time, directing to character of landsliding process development, through geophysical measuring of all variants in characteristics of natural and artificial fields, and on the basis of which the landslide regime is to be analysed (Komatina, 1998).

When electric soundings is used, the best results are obtained for landslides in clayey environment with clearly expressed sliding surface. Body of the landslide is characterized by higher moisture and porosity values. Soundings is less efficient for landslides with unclear sliding surface. Finally, at the autochthonous terrain, two-layers sounding curves are obtained.

For evaluation of hydrogeological conditions in the landslide body, SP method is most frequently used, but also Mise-a-la-masse and resistivity method (for determining filtration coefficient).

Landslide dynamics is investigated by electric soundings (so–called regime monitoring), repeated in appropriate time intervals, as also by micromagnetic and SP measurements.

Seismic refraction method is also applied in landslides exploration. Moving mass is characterized by lower velocities of elastic waves propagation in comparison to autochthonous environment. In dependence on velocity values contrast, depth to the sliding surface can be determined.

Beside various well-logging methods, temperature measurements are also carried out. Namely, it was realized that moving mass is characterized by temperature change in dependence on depth (in accordance with local temperature gradient). Irregularities in heat field within the landslide body are explained as a result of different amounts and changing groundwater flow. Higher velocities correspond to better permeability of the rock mass, which is noticeable as a cold area at the temperature profile. In the active landslide, heat is also generated during the friction process along the sliding surface. As the effect is intensive, the surface is defined by local temperature maximum values, obtained by well measurements.

CASE STUDIES FROM YUGOSLAVIA
Majdanpek open pit

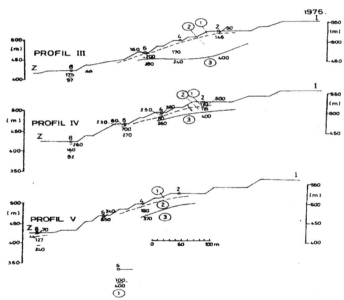

Figure 1. Results of electric measurements along lines carried out athe open pit of Majdanpek copper mine. 1. Electric sonde; 2. Resistivity values (Ohmm); 3. Electric medium.

At Majdanpek copper mine (Eastern Serbia), geophysical methods were applied several times in the last two decades, in order to define thickness of rock mass vulnerable to sliding as well as geometry of the sliding surface. Slope of the pit is made of: phyllite, green shale, quartz-muscovite shale, gneiss, breccia and diabase.

Exploration was performed between 425m and 540m levels. Fifty-two electric sondes were performed, distributed along 14 profiles (Timotijevic, 1976).

In Fig. 1, results of soundings are presented as a cross-section of different electric environments. The first two layers are characterized by resistivity values lower than ones for the third layer (360 Ohmm) - 140 Ohmm and 160 Ohmm, respectively. According to the values, it can be concluded that the first two layers present unstable area, which moves or could move, and the third one is stable. Thickness of the unstable part is 15 to 45m.

Smreka open pit in Vares (Bosnia and Herzegovina)

In order to determine thickness and distribution of the sliding mass, geophysical test measurements were performed. Slope between 695 m and 1010 m levels was investigated by electric soundings and seismic refraction method. Lithologically, slope of the pit is made of limestone and marl layers replacing , deeper – marl and sandstone, while at the deepest level the ore body is located.

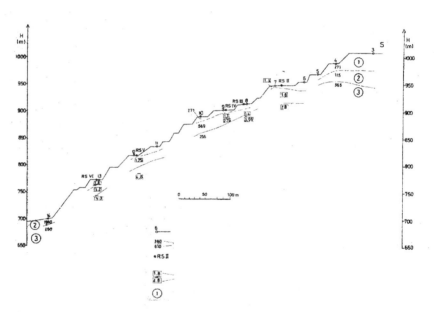

Figure 2. Results of electric and seismic refraction measurements carried out along the line J at the Smreka open pit.

Electric measurements. Along one line, coinciding with J geological one (made on the basis of drilling results) 12 electric sondes (AB/2 =150 m) were measured. Exploration was performed during the rainy period, when the moisture in soil is highest. According to the results of electric soundings, two parts of the rock mass are distinguished: unstable, weathered and moved and another: stable (autochthonous) mass.

In Fig. 2, results of soundings are shown as formations in cross-section. Three formations are distinguished on the basis of their electric values. In engineering-geological sense, layers 1 and 2 present body of the landslide (sliding mass), while the layer 3 is relatively stable part of the rock-mass (autochthonous) medium.

The stable part (the medium 3) is relatively homogenous in electric characteristics. At the northern part of the line, in domain of sondes 3-6, resistivity values are 310-405 Ohmm, while to the south, between sondes 7 and 13, values are: 240-270 Ohmm. Cited value range refer to geological structure of appropriate parts: in domain of sondes 3 and 4 – sandstones; near to sonde 5 – shales, sondes 6 – 9 – replacing limestone and marl layers; sonde 10 – porosity limestones; sondes 11-12 – marls; sondes 13-14 – the ore body (siderite, hematite).

The formation 2 is, electrically, relatively heterogeneous. Within the medium, between sondes 3 and 5, there is homogeneous part (105-125 Ohmm), while at the rest of the line, resistivity values are much higher (141-1034 Ohmm), as a result of geological structure and rock-mass water content.

The formation 1 is a surface part of the rock mass, electrically heterogeneous. Electric values of that part are 92-544 Ohmm.

The highest depth of the medium 3 was determined at the northern part of the line, near to the sonde 3: 60 m, while in domain sondes 4-7, it is 30-40 m. At the rest of the line, depth is 20 – 30 m.

Seismic refraction measurements. The measurements were carried out along 5 lines (each 170 m long), located along the line for electric measurements (Fig.2), making correlation of the results of the two geophysical methods possible.

According to the results of seismics, the following three formations were distinguished:
- Medium 1 – Defined at the lines RS-I and RS-II, according to the propagation velocity of seismic waves: 0.5-1.3 km/s. Thickness of the part: 5 m.
- Medium 2 – At lines RS-I to RS-V, with velocity values: 1.0-1.8 km/s. Thickness of the part: 10-38 m.
- Medium 3 – Defined along all profilesat the footwall of the formation 2, on the basis of velocity values: 2.4-4.0 km/s.

It was concluded that formations 1 and 2 (velocity values: 0.5-1.8 km/s) present unstable and weathered part of the slope, while the formation 3 (2.75-4.00 km/s) is relatively stable and compact rock mass.

Comparing results of electric soundings and seismic refraction measurements, directs us to the following conclusions:
- By seismic refraction method, in domain of lines RS-I and RS-IV, boundary between formations 2 and 3 was not determined (probably because of absence of enough difference in velocity values of the formations).
- Sliding medium is characterized by average resistivity values 271 and 569 Ohmm and average velocity values 1.4 km/s and 2.6 km/s.
- Autochthonous (relatively unmoved medium) is characterised by resistivity values 255 and 365 Ohmm (average values) and 3.75 km/s (average value).

CONCLUSIONS
In domain of engineering geology, geophysical methods are successfully applied not only for: determination of rock characteristics in natural conditions and space anisotropy in hard rock massif, in exploration of subsurface cavities and building material, but also of soil aggressiveness and characteristics of physico-mechanical properties (during geotechnical control of a building construction and quality control of building constructions), etc. Geophysics has also a very important role in landslides exploration. Namely, it is known that, by application of one or several geophysical methods, it is possible to determine: outer landslide and boundaries between blocks within a landslide; depth of sliding surface; groundwater level; direction and velocity of filtration flow; state and physico-mechanical properties of rocks; depth, structure and composition of basement; composition and character of tectonic disturbance zone and zones of karstification; weathering zones of basement rock; landslide regime within a monitoring network. In the paper, beside a concise review and critical analysis of justified choice of each geophysical method within a certain field of landslide exploration, two case studies for landslides exploration at open pits in Yugoslavia (Majdanpek and Vares) are presented.

REFERENCES
Komatina, S. & Komatina, M. (1998) Sophisticated geophysical methods applied in landslides exploration and monitoring. Proc. of ALBA '96 (La Prevenzione delle Catastrofi Idrogeologiche: il Contributo della Ricerca Scientifica), 75-84.
Milanovic, B., Timotijevic, Z. & Milovanovic, Z. (1989) Geophysical exploration at Smreka open pit. Reports of Geophysical Institute, Beograd.
Timotijevic, Z. (1976) Electric exploration of landslide at the open pit of Majdanpek copper mine. Reports of Geophysical Institute, Beograd.

Slope instability hazard evaluation in the Flysch Western Carpathians (Czech Republic)

DR. O. KREJCI, M BIL, Mgr., ZITA JUROVA, Dipl. Eng., Czech Geological Survey, Brno, the Czech Republic, and
JAN RYBAR, Assistant Professor, Academy of Sciences of the Czech Republic, Institute of Rock Structure and Mechanics, Praha, , the Czech Republic

INTRODUCTION

Occurrence of numerous landslides and their reactivation following high total precipitation are known since long ago. During the last centuries hundreds of people died in huge landslides triggered by heavy rainfalls in the Czech Republic. Many casualties were caused mainly by absence of communication means and by lack of suitable rescue technology. Naturally, the best historically documented landslide calamities are known primarily from densely populated areas (Praha and vicinity, Western and Northern Bohemia).

The extreme rainfalls in July 1997 triggered slope movements recorded particularly in the flysch-type sediments of the Western Carpathians and sediments of the Bohemian Cretaceous Basin with Tertiary Neoidic volcanic intrusions and pyroclastic rocks.

THE HISTORY OF THE RESEARCH OF SLOPE MOVEMENTS IN THE CZECH REPUBLIC

The first documented areal landslide on the territory of the Czech Republic is the locality of Chrochovice-Krásný Studenec where the slope movements were observed in 1736. In 1770, 13 landslides were registered in total near Ústí n. Labem and Děčín. First rock falls were registered as soon as in 1132 in Praha Chuchle. The landslide phenomena are characteristic by frequent reactivation of the movements on the same localities. In Praha, for instance, 8 landslides occurred on the same site in 1897-1965. Other examples are known from Ústí nad Labem in 1767-1959 and from Stranné (Žatec) in 1820-1900. During the recent years, frequent rock falls are reported of Cretaceous sandstone blocks in the northern Bohemia (Hřensko, Děčín and surroundings) which permanently endangered the dwelling houses and roads.

Frequent areally extended complex landslides reaching several kilometres form in the region of the Western Carpathian Flysch Belt. The first scientifically documented landslide in the Western Carpathians, which occurred in Hošťálková in 1919, was recorded in 1922. It was a large-scale slide of slope talus of the flysch sediments of 750 m in length and 200 to 350 m in width. The rate of slide movement - recorded by eye witnesses - was 710 m per hour. The landslide destroyed 6 farmer houses and a small lake came into existence at a point where the landslide front blocked the local brook. Actual danger from the slope movements in the Flysch Carpathians was recognised in the course of extensive building of water dams. The constructions of water reservoirs called for a geological research of unusual extent. This applies mainly to the dam on the Stanovnice Brook (a tributary of Vsetínská Bečva River from the Javorníky Mts.) at the village of Karolinka. The geophysical research and borings in the landslide area above glassworks at Karolinka where water infiltration from the Stanovnice water reservoir might be expected revealed an absolutely exceptional depth of the slope movements. Three huge blocks were find out and the landslide shear plane is situated deep in the bedrock, reaching up to 70 m deep in the frontal part of the landslide. The landslide number distribution map with schematic geological situation are shown on the Figure 1.

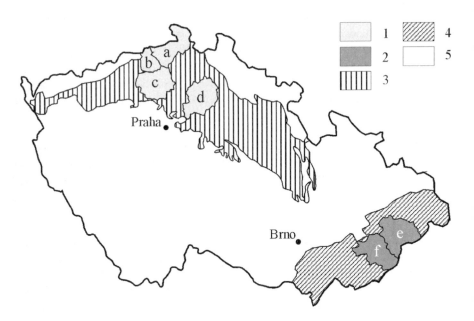

Figure 1. Schematic geological map of the Czech Republic with marked the most jeopardised district areas to geodynamic phenomena. Explanations: (1) district areas with number of the landslides <500 (a-Děčín, b-Ústí nad Labem, c-Litoměřice and d-Mladá Boleslav) (2) district areas with number of the landslides >500 (e-Vsetín and f-Zlín). Geology: (3) The Bohemian Cretaceous Basin and Neoidic volcanic intrusions (4) Flysch Western Carpathians (5) The Bohemian Massif platform.

THE CHARACTER OF THE SLOPE MOVEMENTS IN THE BOHEMIAN MASSIF PLATFORM COVER AREA

In the Bohemian Massif the representatives of the Cretaceous system are occurrences of the Cretaceous transgressing from the Alpine-Carpathian region during Cenomanian and Turonian.

Upper Cretaceous sediments are deposited mostly almost horizontally in a shallow syncline basin with NW to SE axis reach thickness up to 700 m. They influenced the carved-in topography of the northern part of the Czech Republic. The surface morphology of the Czech Cretaceous Basin results from Late Tertiary and Quaternary denudation and accumulation. Their final product consists of structural plains, river terraces, and deep canyon-like valleys, e.g. that of the Labe river.

Saxonian tectonic movements segmented the originally uniform basin to a series of plates uplifted to different elevations forming an interesting landscape. The articulated topography of the Czech Cretaceous Basin is marked by disseminated clean-cut monadnocks mostly formed as denuded fillings of volcanic chimneys and various types of dykes. The distinct elevation of the Labe canyon walls over the valley exceeding 600 m in combination with block disintegration of thick sandstone layers and occurrence of block-shaped unconsolidated sediments of volcanic origin (tuffs and tuffites) are the main reasons for a high tendency to slope movements. These are known mainly as rock falls and less frequent earth or debris flows and avalanche-like slope deformations up to several hundreds of meters in length. The frequency of slope deformations in this region is the highest in the Czech Republic. The individual sliding objects represent isolated blocks or tors with size reaching tens of meters which are close to tumbling down. In 1978 a 1981, for instance, a 23 m tall Neubauer tor collapsed in this region. The landslide phenomena are distributed mainly within the administrative districts Ústí nad Labem, Litoměřice and Mladá Boleslav.

The engineering-geological survey made up to now in a 1.4 km long section of the highest potential instability risk near the Hřensko village in the Labe valley near the German state border in the administrative district Děčín revealed 256 unstable rock objects. At present, the monitoring system here consist of 165 sites for dilatometric measurements of relative displacement among individual blocks of sandstone mass.

THE CHARACTER OF THE SLOPE MOVEMENTS IN THE WESTERN CARPATHIANS FLYSCH BELT AREA

The Mesozoic and Tertiary complexes of the Western Carpathians Flysch Belt are characteristic by mountain and highland relief of mainly erosional and structurally-denudation nature. The flysch complexes are formed by alternating layers of claystones and sandstones. These kinds of rocks are little permeable, which means that their surface zone gets water saturated quickly. The layer bedding and tectonic fracturing also cause formation of shear zones controlling slope movements. Other suitable materials for gravitational movements are loamy-stone, loamy-clay and loamy-sand sediments as well as thick unconsolidated residual mantles of weathered flysch rocks. The extreme precipitation in July 1997 resulted in both new slope failures and the activation of slope movements of earlier origin. Many localities exhibited landslides of enormous destruction force, earthflows being less frequent (rockflows or mudflows), and rockfalls sporadic. The landslide phenomena are distributed mainly within the districts Vsetín (more than 1500) and Zlín (600).

The most extensive activated landslide area in the Flysch at the present can be found in the valley of Bystřička Brook, locality Vaculov-Sedlo (district Vsetín). Length of the landslide area is 4 km, width 1.2 km. There are various types of slope failures which affect both colluvial (slope) sediments and bedrock. Thick landslide accumulations alternate with water logged depressions (existence of a small lake). The open parent rock links up with fissure caves and pseudokarst dolines. Many lonely houses and special-purpose roads were damaged and particular losses were seen on the soil cover in meadows, pastures and forests. The sizeable blocks of bedrock proceed into the streambed of Bystřička Brook and the name-less brook, which are gradually being blocked. Possibilities of protection in such an extensive landslide area are limited.

The total number of landslides in the Flysch Carpathians region only documented to April 2001 amounts to over 3700.

THE SOLUTION TO RECLAIM LANDSLIDE LOSSES
The total damages from activated landslides in 1997 and later reached about 18 millions Ł. The Czech government guaranteed 7 millions Ł from the state budget during the years 1997-2000 and addressed it for the survey and subsequent remediation. The reparation of the damaged state roads and railways demanded following 9 millions Ł. Now in 2001 the financial support from state budget for monitoring and remedial works in landslide areas reaches 1.8 millions Ł per year and the same support is planned up to 2003. Division of the geology of the Ministry of the Environment together with experts of Czech Geological Survey are dealing professionally with this problem in co-operation with other institutes and enterprises.

The complementary research differs as related to individual localities and their topical situations the basic method being the geodetic alignment of the landslide and monitoring of its movements. Inclinometric measurements in boreholes are normally made in larger landslides with drilling works in the most extensive landslide areas reaching into the depth of up to 40 m. The bores further serve to the calibration of geophysical methods and the most commonly used complex of these methods is as follows: georadar with depth engagement of 30 to 40 m, shallow refraction seismic, vertical electrical scanning (VES) and dipole electromagnetic profiling (DEMP). The complementary research is always accompanied by a geotechnical assessment of the locality and a proposal of rescue measures which are normally quite costly works ranging between 20 000 Ł up to over 2 millions Ł (landslides above the railway track near the village of Bystřička, district Vsetín). The alternative solution should consist of an assessment of necessary changes in the area plans of affected villages, prospects of their future habitability, displacements of engineering networks and roads. Another possibility comes in view to leave the extensive landslide areas entirely to their own natural evolution or to convert the affected meadows and pastures into a forest land and to technically stabilise only the landslide areas which represent a real jeopardy to peoples homes and important infrastructure landscape elements.

The special research project is running now to summarise and evaluate as much information as possible about the physiographic settings and initiation mechanisms of landslide phenomena in the model areas. The project was divided into three parts: part 1-documentation of individual landslides, part 2-systematic survey and classification of landslides for new determination of the areas with special conditions of geological structure and part 3-the integrated evaluation of the model areas, information database and conclusions.

Special research project is worked up jointly by several institutions under the co-ordination of the Czech Geological Survey, Praha. To stabilise the landslides and recover from the losses will take a long time. However, each landslide locality has its own specific features which apart from the natural prerequisites also include factors caused by human activities. The registration of landslides is being processed in digital format, the extensive landslide localities are studied in detail in order to bring into life rescue measures that would correspond with the extent of the damaged houses and landscape losses. After the analysis of all available records from the landslides areas, district area maps in a scale of 1:10 000 will provide information about hazardous areas of possible occurrence of landslide phenomena. The maps display recent and intermittent slope deformations, presently quiescent in black, recent active deformations in red colour. Fresh cuts of water flow banks and erosional trenches are indicated also in that. Hydrological and hydrogeological data are in blue. Green colour is used to mark damaged and endangered objects or remedied ones. Documentation of the registered phenomena, written as well as photographic, is an implicit share of the evaluation. The areas are classified in categories by the degree of landslide danger and intensity of its occurrence. Special maps of stability conditions provide primary basis for graphically simple prognostic maps of landslide susceptibility sometimes called also landslide hazard maps. Evaluation of input data (stability, morphometric, geological and hydrogeological conditions) provide means to define borderlines of quasihomogeneous zones of certain specific level of landslide hazard. Colours using a signal head principle differentiate individual zones. Zone I - stable zone is expressed by green, on the other hand the warning red has been claimed for Zone III - instability zone. Zones conditionally exploitable, i. e. zones where potential stability problems cannot be excluded are expressed in amber (Zone II). Each zone contains subzones expressed by hatching which characterise local conditions in geology and geomorphology. Explanations to susceptibility maps are given in tables with short definitions of conditions regarding possible use of subzones as individual construction sites. Regional authorities take steps to implement the landslide hazard maps in the regional and town planning. Characteristic of the areas regarding stability conditions and conditions for the area to be used as construction site of residential and industrial buildings, roads, pipelines are shown on Tables 1, 2.

CONCLUSIONS

The activity of slope movements (mainly landslides) in the Czech Republic is given by the geological control as well as by the character of relief in this area.The impact of area of the Flysch Western Carpathians by landsliding is of the character of small natural hazards. The main impulse were heavy rainfalls in July 1997. To stabilise the slope movements and recover the losses will take a long time. However, each landslide locality has its own specific features which -apart from the natural prerequisites- also include factors caused by human activities. A number of buildings were erected on old landslide terrain and often it was inadequate house extensions that loaded the landslide area. In other cases, one-sided incisions relieved the slope foot and resulted in landslidings. Other causes of slope movement activation in the past are considered to be human activities resulting in the increased surface runoff into critical points such as insensitive deforestation, liquidation of anti-erosion measures on agricultural land and -in contrast- insufficient maintenance of forest reclamation rills draining water off the landslide areas. And it was also some ground formations at road construction works such as one-sided deep undercuts or embankments, or furrows made for telephone cables - improperly situated in the landsliding areas and affecting the landslide movement activity.

Table 1. Characteristic of the area regarding stability conditions.

zone	characteristic of the area regarding stability conditions
I.	*stable areas*
I.1	flat flood plains
I.2	permanently stable slopes with very moderate gradient and flat areas above valleys
II.	*area where slope instability cannot be excluded*
II.3	moderate slopes without verified signs of more serious failure
II.3a	moderate slopes without verified signs of more serious failure (with colluvial deposits > 2m)
II.4	steep slopes without signs of deeper failure
II.4a	steep slopes without signs of deeper failure (with colluvial deposits > 2m)
II.5	slopes deformed by superficial creep movements
III.	*unstable areas*
III.6	slope deformed by landslides and block-type movements in the past time
III.7	slope deformed by present active or dormant slides, as well as by earthflows
III.8	erosion gullies of occasional as well as of permanent small flows
III.9	steep rock slopes and their toes where rockfalls may occur

Table 2. Conditions for the area to be used as a construction site.

zone	conditions for the area to be used as construction site of		
	residental and industruial buildings	roads	pipelines
	conditionally usable areas		
II. 3	suitable for not sophisticated structures, stability failures due to improperly designed earthwork are not excluded (cuttings, cut-offs, embankments, water leakage etc.)	lay-out of the line, as well as earthwork is to be designed with respect to slope stability	slope stability must not be inflicted with deep furrows excavated for long-distance lines
II. 4	unusable for ordinary building, otherwise enormous expenditures must be accepted	suitable only for local roads (e.g. forest roads)	if the lay-out cannot be changed then increased expenditures are to be accepted
II. 5	if chosen as construction site increased expenditures for preventive remedy measures must be considered (e.g. for superficial or deep drainage in the area)	suitable only for local roads, othrewise increased expenditures for preventive remedy measures	lay-out of long-distance lines along the dip
	unsuitable areas		
III. 6	unusuitable for building, acceptable only with enormous expenditures for survey, monitoring and remedy measures	construction possible only with enormously increased expenditures	unsuitable area, in case of absolute necessity lay-out along the dip
III. 7	construction has to be avoided unless stabilisation measures carried out beforhand with successful results proved by monitoring	construction possible with enormously increased expenditures for preventive remedy measures and monitoring	entirely unsuitable area
III. 8	entirely unsuitable area	unsuitable area that can be passed by a bridge	unsuitable area
III. 9	unsuitable area	change of the lay-out is recommended, preventive remedy measures would be too expensive	without limitation under the condition of keeping the lines underground in the area of accumulation

Statistical methods for establishment of landslide hazard maps using GIS

JUNYONG LEE, Seoul National University, Korea
HYEONGDONG PARK, Seoul National University, Korea

INTRODUCTION

Every year, there have been a lot of losses in properties as well as human lives due to landslide and slope failure in Korea. However, through controlling and predicting these aspects, the loss by landslides can be reduced. These days, many researchers focus on the management of natural and artificial slopes. For this reason, the necessity of a landslide hazard map is increasing, because a landslide map can play an important role in decision-making concerning economical sides as well as social sides. Varnes (1984) defined natural hazard as the probability of occurrence of a potentially damaging phenomenon within specified period of time and within a given area. Einstein (1988) proposed comprehensive procedures for landslide management. According to his proposal, state-of-nature map is the first step. State-of-nature map presents data without interpretation of investigators. In other words, this type of map includes the data of site investigation, geologic map, topographical map, precipitation and so on. The next step is danger mapping. Danger map provides possible types or mechanism, such as rockslide, debris flow, and so forth. After this step, a hazard map can be developed. A hazard map contains the data of the probability that such a failure could occur during a given time. This type of map can provide some basic information for risk map and management map that are the final step in landslide mapping. From this point of view, engineering geologists have to establish, at least, a landslide hazard map. Thus, an ideal landslide hazard map should provide the probability of landslide occurrence and basic information to the next step of landslide mapping.

The key point of landslide management is to provide a specific number that helps managers or decision-makers evaluate whether some slopes will be dangerous or not. This specific number can be used not only as a criterion, but also as basic information in calculating economic or social cost. In other words, the manager or decision-maker is able to derive some cost about managing a slope by combining this specific number and other factors related to cost. In this regard, it is crucial to use specific number in slope stability analysis, and a lot of researchers have been using probability as a specific number because probabilities present the degree of instability quantitatively. Thus, in order to derive probabilities, it is necessary to use many different statistical methods.

In this study, many types of slope failures were classified into two categories, and two statistical methods were applied to each category of slope failure to provide a specific number, probability, for a landslide hazard map. These statistical methods include Monte Carlo simulation and logistic regression analysis.

MONTE CARLO SIMULATION
Theory
The Monte Carlo method is a numerical method of solving mathematical problems by the simulation of random variable. Monte Carlo simulation is, originally, a mathematical technique for numerically solving differential equations. It is used mainly in finance for such tasks as pricing derivatives or estimating the value at risk of a portfolio. This method, basically, is executed by many times of repetitions. The repetitions provide a most-likely value and through this most-likely value, a probability can be derived. In this study, this method is applied to Hoek and Bray's simplified method for slope stability.

Data acquisition
Along the highway or railroad, there are so many cutting slopes that have a couple of joint sets. To evaluate the stability of the slopes, investigators measure these joint sets with two methods. One is window method and the other is scanline survey method. In this study, the latter – scanline survey – is used for the collection of the joint data. In the scanline survey, the artificial scanline tends to intersect a specific joint set and scarcely meet with another specific joint set. Thus, this method, usually, induces the bias on the sample measured. To remove the bias, Priest (1985) suggested, a weight factor is used, and the definition of the weight factor is following.

$$ w = \frac{1}{|\cos(\alpha_n - \alpha_s)\cos\beta_n\cos\beta_s + \sin\beta_n\sin\beta_s|} $$

Fisher distribution and Randomly generated joint set
Fisher (1953) assumed that a population of joint orientation values was distributed about some "true" value, and the probability, $P(\theta)$, that an orientation value selected at random form the population makes an angle of between θ and $\theta+d\theta$ with the true orientation is given by

$$ P(\theta) = \eta e^{K\cos\theta} d\theta $$

where K is a constant controlling the shape of the distribution and often referred to as Fisher's constant. η is a variable that ensures the following properties: the value $P(\theta)$ must be proportional to $\sin\theta$ and the sum of all possible values of $P(\theta)$ must be 1. From these properties, η is expressed by following.

$$\eta = \frac{K \sin \theta}{e^K - e^{-K}}$$

Combining the definition of P(θ) vary the equation as following.

$$f(\theta) = \frac{e^{K \cos \theta} K \sin \theta}{e^K - e^{-K}}$$

The parameter K is a measure of the degree if clustering, or preferred orientation, within the population. For a sampling size M and the magnitude of the resultant vector $| r_n |$, Fisher (1953) suggests that k, the estimated K, is given by

$$k = \frac{M - 1}{M - | r_n |}.$$

Through the procedure above, Fisher's constant can be obtained and will be used to generate random joint orientation. If Fisher's constant has a large number, like as 1000, the dispersion from the mean value is very small. On the contrary, low value of Fisher's constant means that the orientations of joints are highly dispersed from the mean value and the data used are not confidential data. In other words, the data generated by very small Fisher's constant tend to show the low relationship with mean value that researchers need to use. Therefore, before the generation of joint set using the Fisher's constant, the test of confidential range of the Fisher's constant.

Fisher (1953) proposed the problem of placing confidence limits on the mean, or resultant vector and showed that the probability Pr(<θ), that the resultant vector makes an angle of less than θ with the true orientation is given by following.

$$\Pr(< \theta) = 1 - (\frac{M \neg | r_n |}{M - | r_n | \cos \theta})^{M-1}$$

The following approximations for Pr(<θ) are valid when M is large

$$\Pr(< \theta) \approx 1 - e^{K | r_n | (\cos \theta - 1)}$$

for which the inverse is $\cos \theta \approx 1 + \dfrac{\ln(1 - \Pr(< \theta))}{K | r_n |}$

The purpose of the test in this study is to know that the least value of Fisher's constant for satisfying the following conditions: A certain percentage (confidence level) of the generated joint orientations lie within about given number of the angular radius.

For a fixed confidence level and angular radius, Fisher's constant is inversely proportional to the magnitude of the resultant vector and expressed by

$K \geq \dfrac{\ln(1 - \Pr(< \theta))}{(\cos \theta - 1) \mid r_n \mid}$. When confidence level is 90% and angular radius is 20°, the

equation can be expressed as $K \geq \dfrac{38.18}{\mid r_n \mid}$. If $\mid r_n \mid$ equals to 20, K should be over

19.09. This means that if K is over 19.09, the 90 % of the generated joint orientation data whose resultant magnitude is 20 lies within 20° of the resultant vector.

If only the mean orientation of discontinuities and their Fisher's constant have been identified, following procedure can generate random joint orientations of each set.

1) The relationship between Fisher's constant and $\Pr(<\theta)$ is $\Pr(< \theta) = 1 - e^{-K(1 - \cos \theta)}$,

and this equation can be changed as $\cos \theta = \dfrac{\ln(1 - \Pr(< \theta))}{K} + 1$.

2) The probability that a vector selected at random makes an angle less than θ from the mean discontinuity normal is randomly generated following uniform distribution.
3) Through the generated $\Pr(<\theta)$ and given K, θ can be calculated by using the second equation in procedure 1).
4) Define α_1, β_1 as trend and plunge of the generated discontinuity normal and α_n, β_n as trend and plunge of mean discontinuity normal.
5) Insert obtained input value and find out the values of α_1, β_1.
6) Iterate procedure 1)~5) as many times as required.

Wedge failure
Hoek and Bray's simplified method (1981) was devised to calculate the factor of safety (FOS) of rock slope about wedge failure. Wedge failure takes place when tetrahedron, which consists of two intersecting discontinuities, slope face and upper face of slope, slides due to gravity. This method has limitation on some aspects. It does not take rotational failure and toppling failure into account. And it, also, does not take tension crack on the upper part of the slope into account. This method, however, has been used for calculating quick FOS with the geometry of the slope because it requires just simple arithmetic calculations. This method contains two procedures. One is the discrimination of forming the wedge due to two intersecting discontinuities given. The other is the calculation of FOS with rock properties, water pressure, cohesion, friction angle and so on.

Application
For the application of this approach, data acquisition was conducted at the face of slopes along the road cut. This area is covered with moderately or highly weathered granitic gneiss where national road was constructed recently. The study slopes are 8 road-cuts along this national road and several hundreds of data were obtained. To analyze these slope stabilities, some following assumptions should be considered. The height of sliding wedge is one-third of slope height and stability of slope is not

affected by water, in other words, only joint orientations and generated block size are taken into consideration.

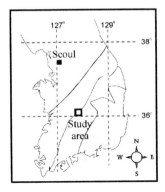

Figure 1. Study area

Table 1 shows Fisher's constant of obtained data and recommended least Fisher's constant for controlled conditions that the 90 % of the generated joint orientation data whose resultant magnitude is 20 lies within 20° of the resultant vector. From the Table 1, the third joint set of the second slope is not proper to use in analysis. Table 2 is the result of Monte Carlo simulation of each slope and each combination of joint set excluding the third joint set of the second slope. According to the result of analysis, slope 4, 5, 6 are relatively stable and slope 6, 8 are very unstable in wedge failure.

Table 1. Fisher's constant of obtained data and recommended least value

	Joint Set1		Joint Set2		Joint Set3		Joint Set4		Joint Set5	
	A*	B**	A	B	A	B	A	B	A	B
Slope1	13.1	2.9	39.6	7.8						
Slope2	12.3	5.1	17.9	4.5	6.5	10.8	14.3	8.2		
Slope3	22.5	4.4	62.7	27.7						
Slope4	17.9	4.1	33.0	5.5						
Slope5	20.6	2.8	63.0	6.1						
Slope6	34.6	4.0	23.1	5.7	5.0	1.8	63.6	12.0	56.5	9.5
Slope7	34.6	5.1	64.5	2.9	106.0	6.4	25.7	8.6		
Slope8	20.1	3.5	69.9	15.6	52.9	3.9				

* : Fisher's constant of obtained data
** : Recommended least Fisher's constant

Table 2. Principal result of each slope

	Probability (%)	Combination of Joint set
Slope1	35.1	J1 & J2
Slope2	31.8	J1 & J2
Slope3	5.7	J2 & J1
Slope4	5.4	J1 & J2
Slope5	8.7	J1 & J2
Slope6	46.8	J3 & J1
	57.2	J3 & J5
Slope7	34.9	J3 & J2
Slope8	36.5	J3 & J1
	51.4	J3 & J2

LOGISTIC REGRESSION ANALYSIS
Theory
In many cases, variables related to landslide can only be described by ordinal variables or nominal variables. In landslide inventory, only nominal variables, such as failure or success, can be found. In order to treat these types of variables, logistic regression analysis can take an advantage over traditional regression analysis. In other words, logistic regression analysis can be used in analysis when dependent variables are categorical variables.

Conceptually, for logistic regression analysis, the logit should be calculated. The logit defined as the logarithm of the ratio of P_a to P_b, where P_a means the probability of presence of the phenomenon, P_b represents the absences of the phenomenon and $P_a + P_b = 1$. Traditional regression analysis for this logit would be followed. Through this procedure, the logistic regression model can be derived and the model is expressed as follows:

$$P_a = \frac{\exp(U_a)}{1 + \exp(U_a)}$$

$$U_a = \beta_0 + \beta_1 X_1 + \beta_2 X_2 + \cdots + \beta_n X_n + e$$

Where,

U_a: a quantity commonly known as the utility function of event a, expressed as a linear combination of a number of explanatory variables, $X_1, X_2, \ldots, X_n$

β_n: the estimated parameter of the variable X_n

e: randomly distributed error term

In the logistic model, the greater value of Ua is, the greater probability for the event to take place. As previously seen, logistic regression analysis is an appropriate technique for dealing with a categorical dependent variables. Thus, this method can be applied to slope stability analysis, because for a certain given explanatory variable, the response to slope may be either 'failure' or 'no failure'. An explanatory variable can be regarded as a triggering factor that might induce a slope failure. In this study, precipitation data are utilized as a triggering factor. This approach concentrated only on the relationship slope failure and precipitation. However, this approach has a great advantage in that the cases -in spite of the precipitation there had not been any slope failure- were taken into account, while the previous studies have addressed only the question of how much precipitation could induce slope failure.

Many researchers prove the fact that landslides are triggered by cumulative precipitation for several days as well as daily maximum precipitation, therefore, this analysis include not only the relationship between daily precipitation and landslide occurrences, but also the relationship between the cumulative precipitation and landslide occurrences.

Application

For this approach, precipitation data of every meteorological observatory of Korea (69 sites), and the landslide history that recorded in Korea for 7 years (1995~2001) were considered in this analysis.

The landslide occurrence data include the only case that was reported in the main daily newspaper, therefore this case of landslide induced the damage of social system or the injury of human lives. In other words, the landslide that was not identified and did not harm properties and human was not included in these data.

The following equations are the result of logistic regression by using statistic package, SAS and Table 3 and Table 4 show the result of logistic regression analysis and estimated probability of landslide with plausible precipitation. According to this result, when the amount of rainfall of a certain region exceeds 300mm in a day, the probability of landslide in the region rises above 80%. To reach the approximate level of probability of previous case, the cumulative precipitation during three days should be over 500mm. On the other hand, in case of analyzing both of two factors simultaneously, landslide is a little more affected by cumulative precipitation for three days than a maximum precipitation in a day.

$$z = 0.0301 \times daily - 7.6028$$
$$z = 0.0188 \times three - 7.9210$$
$$z = 0.0142 \times three + 0.0127 \times daily - 7.9645$$

where, 'three' and 'daily' mean cumulative precipitation for 3 days and daily maximum precipitation respectively.

Table 3. The estimated probability of landslide with each factor

PRECIPITATION (MM)	50	75	100	125	150	175	200
Probability (1 days)	0.22	0.47	1.00	2.10	4.36	8.82	17.04
Probability (3 days)	0.09	0.15	0.24	0.38	0.61	0.97	1.54

PRECIPITATION (MM)	300	400	500	600	700	800
Probability (1 days)	80.65	98.83	99.94			
Probability (3 days)	9.27	40.11	81.44	96.64	99.47	99.92

Table 4. The estimated probability of landslide with both factors

	DAILY MAXIMUM PRECIPITATION									
	0	50	80	100	150	200	300	400	500	700
50	0.1	X	X	X	X	X	X	X	X	X
80	0.2	0.3	X	X	X	X	X	X	X	X
100	0.3	0.4	0.5	X	X	X	X	X	X	X
Cumulative 150	0.6	0.8	1.0	1.9	X	X	X	X	X	X
Precipitation 200	1.1	1.6	2.1	3.8	7.0	X	X	X	X	X
for 3 Days 300	4.4	6.4	8.1	14.2	23.8	52.6	X	X	X	X
400	16.1	22.0	26.6	40.6	56.4	82.1	94.2	X	X	X
500	44.3	53.8	60.0	73.9	84.2	95.0	98.5	99.6	X	X
700	93.2	95.2	96.3	97.9	98.9	99.7	99.9	99.9	99.9	X
1000	99.9	99.9	99.9	99.9	99.9	99.9	99.9	99.9	99.9	99.9

CONCLUSIONS AND SUGGESTIONS

Stochastic methods such as Monte Carlo simulation can provide a reliable probability of slope failure through a large number of repetitions. In addition, controlling the obtained raw data can raise the reliability in result. Moreover, considering the statistical distribution characteristics of other factors will be able to reduce and predict the uncertainty in the given conditions.

Logistic regression analysis has an advantage in providing a quantitative relationship between precipitation and landslide occurrence and probability in case of given rainfall. But, this approach has a limitation that precipitation data and landslide occurrence data used are too regional to consider just points or slopes that have failed. However, more local data and addition of other factors related to landslides can derive better result.

In using GIS layers, it is possible to quantify the degree of effect of each factor and establish landslide hazard maps by manipulation of built layers with derived probabilities. Furthermore, landslide risk maps or management maps can be established, using the economic and social cost.

REFERRENCES

Einstein H.H., 1988. Landslide Risk Assessment Procedure, In: Proc. Fifth International Symposium on Landslides, Lauscanne, Switzerland, A.A. Balkema, Rotterdam, Netherlands, Vol.2, pp.1075-1090.

Fisher, R. (1953) Dispersion on a sphere, Proceedings of the Royal Society of London, A217, pp295-305.

Hoek, E., Bray, J. W. 1981, Rock slope engineering, The institution of mining and metallurgy, 358pp.

Park, H. J., 1999, Risk Analysis of Rock Slope Stability and Stochastic Properties of Discontinuity Parameters in western North Carolina, Ph. D. Thesis, Purdue University, 369pp.

Priest, S. D., Discontinuity Analysis for rock engineering, 1993, Chapman and Hall, 466pp.

Varnes, D.J., 1984, Landslide Hazard Zonation: A review of principles and practices, UNESCO Press, Paris, 63pp.

ssessment of landslide hazards in Iowa, USA

A. LOHNES, University Professor and B.H. Kjartanson, Associate Professor, Iowa
ate University, Ames, Iowa, USA

NTRODUCTION
he objective of this study is to provide Iowa county engineers and highway
aintenance personnel, who are not geotechnical specialists, with procedures that will
low them to efficiently and effectively interpret and repair landslides. Although the
etails of the study are specific to Iowa, it is thought that the approach suggested here
ill be useful in other geologic environments. The study provides methods to identify
cations of potential slope instability thereby avoiding the problem or developing
roactive remediation.

VERVIEW OF IOWA PHYSIOGRAPHIC REGIONS
even physiographic regions have been defined in Iowa (Prior, 1976); however for this
esearch, Iowa was generalized into three upland regions of significantly different
pography and surface geology as shown in Figure 1.

n the north-central portion of the state, referred to as the Des Moines Lobe, glacial till
omprises the uplands. Local relief is generally less than 6 m with deeper valleys along
ajor streams such as the Des Moines River.

he western portions of the state, adjacent to the Missouri River floodplain, have deep
ind-blown loess deposits that form very steep slopes and narrow drainage divides. The
oess in this region has depths up to 60 m with local upland relief in excess of 50 m. The
ource of this aeolian material is the floodplain of the Missouri River, therefore the
hysical characteristics of the material close to the river differ from the loess that is
urther away from the source. This western-most low plasticity soil is referred to as
riable loess.

he remainder of the state is covered with higher plasticity loess of variable thickness,
from 10 m to less than 3 m, overlying glacial till. Local relief varies from 30 m to 9 m
and the slopes are intermediate in slope angle between the Des Moines lobe and the

western Iowa loess hills. The loess in this region is called plastic loess. Alluviu occurs along large streams in all parts of the state.

SCOPE OF LANDSLIDE PROBLEMS IN IOWA

In 1999, a survey of Iowa County engineers was conducted to assess the extent a nature of slope stability problems in the state and to determine successful rep; methods. A questionnaire was prepared and sent to all 99 county engineers. T questions focused on landslides that have occurred since 1993. Sixty questionnair were received, giving a response rate of 61%.

Of the 60 counties that responded to the landslide survey, 80% had some landsli activity and 31% of the counties had more than 11 landslides in the previous six yea Only 17% of the counties had more than 15 slides in this time period. All counties wi high landslide frequency are in the western loess hills or southeastern area of moderate deep loess over till. On a statewide basis, nearly equal amounts of slides occur in cu and in embankments.

Reconnaissance observation of over 50 slides throughout the state indicated bo curvilinear and planar failure surfaces. The majority of slides were relatively shallo less than 2 m deep. The slides were rotational or translational with failure surfac through the toe of the slope. Although these observed slope failures are small, many present a hazard to highway traffic.

The data from the survey were compared with the topographic map of the state and correlation between frequency of landslides and relief was apparent. This resulted in landslide susceptibility map, Figure 2, that categorizes regions of Iowa as either low ris medium risk, or high risk regions for landslides. This map is interpretative and based somewhat incomplete data but it does suggest regions of the state where landslid might be problematic.

DATA BASE FOR SLOPE STABILITY GUIDELINES

In order to develop preliminary, quantitative guidelines for stable slopes in differe geologic materials, it was necessary to obtain strength and unit weight values for th materials. Laboratory triaxial test data used in this analysis included results consolidated, undrained (CU) data with pore pressure measurement and consolidate drained (CD) data for long term, effective stress analyses. Undrained shear strengt were used in total stress, short term analyses.

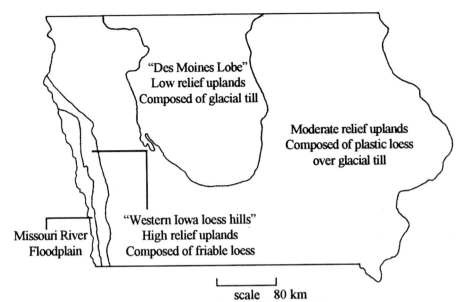

Figure 1. Physiographic regions of Iowa defined for this study.

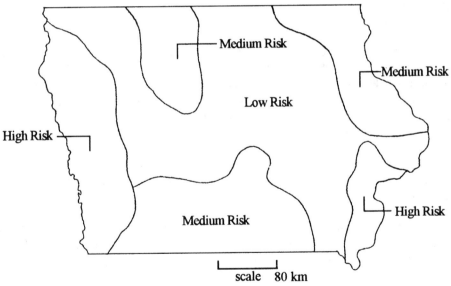

Figure 2. Landslide susceptibility map of Iowa.

Consulting engineering companies, Terracon and CH2M Hill and unpublished these (Olson, 1958; Akyiama, 1964; and Benak, 1967) were the sources of the CD and C data. The undrained shear strength results were compiled from Iowa Department Transportation (DOT) project reports. All of the sources of strength data also include information on unit weights. The data were sorted according to geologic pare materials, and the mean and standard deviation were calculated for cohesion, frictic angle, undrained shear strength, dry unit weight, total unit weight and moisture conte for each geologic material. Means and standard deviations of effective stress cohesio friction angle and dry unit weights for four materials are summarized in Table 1.

Table 1. Average effective stress strength parameters and dry unit weights f Iowa geologic materials. (Std. dev. denotes standard deviations)

Geologic	Cohesion kPa		Friction angle degrees		Dry unit weight kN/m^3	
Material	Mean	Std.dev.	Mean	Std. dev.	Mean	Std.dev.
Glacial till	7.65	5.59	28	1.2	15.1	2.0
Friable loess	5.21	4.00	25	1.4	13.5	0.5
Plastic loess	6.91	4.19	29	4.2	14.3	1.2
Alluvium	2.28	1.90	31	1.3	15.3	0.7

The undrained shear strengths, s_u, and total unit weights are summarized in Table 2. Th data in Tables 1 and 2 show high standard deviations for cohesion and undraine strength for all materials, indicating a high degree of variability in these data.

Table 2. Average undrained shear strengths and total, saturated unit weights f geologic materials in Iowa. (Std. dev. denotes standard deviations)

Geologic	Undrained strength Kpa		Total unit weight kN/m^3	
Material	Mean	Std.dev.	Mean	Std.dev.
Glacial till	30.3	9.3	19.1	1.7
Friable loess	21.9	7.0	18.2	1.1
Plastic loess	31.5	18.4	18.7	1.4
Alluvium	29.9	19.8	19.0	1.1

SLOPE STABILITY ANALYSES

In order to facilitate the recognition of potentially unstable slopes, theoretical curves were developed to provide guidance on stable slope angles and slope heights for different geological materials encountered in Iowa. These curves are not intended for design but only preliminary assessment of stable slope angles and heights. Two analyses were selected to calculate the relationship between stable slope angle and slope height. The Taylor (1948) analysis was used for the undrained condition and the Culmann analysis (Das, 1998) was chosen to represent the slope in a drained condition.

Taylor Analysis

The Taylor (1948) analysis was used to analyze the stability of slopes immediately after construction, before pore pressure equalization and establishment of steady state seepage conditions. As applied here, the failure surface is assumed to be a circular arc that passes through the toe of slope, i.e. the depth factor D = 1, with a safety factor of one. Undrained strength parameters from Table 2 were used in this analysis. Stability number charts (Taylor, 1948) found in virtually all geotechnical engineering books, provide stability numbers, SN, as a function of slope angle, β. The stability number is defined by the following equation:

$$SN = \frac{s_u}{\gamma_t H}$$

where s_u is the undrained strength and γ_t is total unit weight and H is the stable slope height. From the charts where D = 1, theoretical stability numbers can be determined for any slope angle. From the theoretical stability numbers and mean values for s_u and γ_t, the stable height (H) can be determined for the four materials in Table 2. The calculation is then repeated for a variety of slope angles to generate a curve of slope height (H) versus slope angle (β) for each geologic material.

Culmann Analysis

The Culmann analysis was used to analyze slopes in a drained condition; i.e. the analysis is in terms of effective stresses. The equation is based upon the assumption that the failure surface is a plane passing through the toe of the slope. Field observations show that this assumption is approximately valid for high angle slopes, whereas lower-angle slopes tend to fail along a circular arc or a logarithmic spiral. The equation relates slope height, H, to slope angle, β:

$$H = \frac{4c \sin \beta \cos \phi}{\gamma_t [1 - \cos(\beta - \phi)]}$$

where, c = cohesion, ϕ = friction angle, and γ_t = total unit weight of the soil. Mean effective stress strength parameters, shown in Table 1, were used in the Culmann analysis for various slope angles to generate curves of slope height (H) versus slope angle (β) for each geologic material.

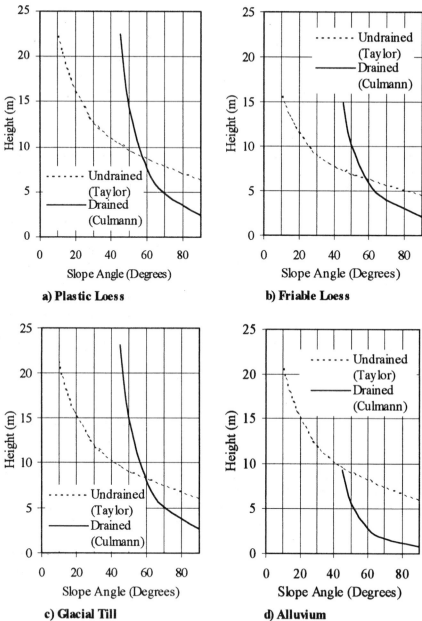

a) Plastic Loess

b) Friable Loess

c) Glacial Till

d) Alluvium

Figure 3 Slope height versus slope angle relationships for various Iowa soils.

.OPE STABILITY GUIDELINES

ιe results for the Taylor analysis and the Culmann analysis are shown in Figure 3 a) rough d). No factors of safety are applied to the results and pore pressures are not cluded in the Culmann analysis. Combinations of slope height and slope angle that ll below the curves represent stable conditions while those above the curves represent stability. These curves are for preliminary assessment of potential stability problems d are not to be used for design.

ONCLUSIONS

ιe objective of this study is to provide Iowa county engineers and highway aintenance personnel, who are not geotechnical specialists, with procedures that will low them to efficiently and effectively interpret and repair landslides.

statewide survey of engineers and a relief map provided data for developing a slope ιbility risk map for the state of Iowa. While this map is interpretive and based on ·mewhat incomplete data, it does suggest regions of the state where landslides might be ·oblematic. Areas of high risk are along the western border and southeastern portion of e state. These high relief regions contain deep to moderately deep loess. The low lief central portion of the state is a low risk area where the soils are glacial till or thin ess over till. The south-central portion of the state, with loess of moderate thickness ver glacial till, is an area of intermediate risk. Reconnaissance observation of over 50 ides throughout the state indicated both curvilinear and planar failure surfaces. The ajority of slides were relatively shallow, less than 2 m deep. The slides were rotational translational with failure surfaces through the toe of the slope. Although these ·served slope failures are small, many do present a hazard to highway traffic.

ιil shear strength data compiled from the Iowa DOT and consulting engineers files ere correlated with geologic parent materials and mean values of shear strength ιrameters and unit weights were computed for glacial till, friable loess, plastic loess ιd local alluvium. Effective stress cohesion and undrained shear strength data have gh standard deviations, indicating a high degree of variability in these data. The shear rength and unit weight data were used in slope stability analyses for both drained and ιdrained conditions to generate curves that relate slope angle to maximum slope height. he Culmann analysis was used for effective stress analyses and Taylor's design charts ·r undrained stability. No factors of safety are applied to the results and pore pressures ·e not included in the Culmann analysis. These curves are intended for a preliminary ·aluation of the stability of slopes comprised of the four geologic materials and are not ·tended for design.

ACKNOWLEDGEMENTS

The research presented in this paper was funded by the Highway Research Board and th
Highway Division, Iowa Department of Transportation (DOT), Ames, Iowa. The
authors wish to thank Bob Stanley of the Soils Design Section, Iowa DOT for his input,
support and providing access to DOT data files. Individuals and firms who contributed
shear strength information are Dennis Whited and Brad Levich, Terracon and Roch
Player, CH2M Hill. Their contribution is gratefully acknowledged. Appreciation is als
extended to former graduate research assistant Norman Chu. The opinions, findings an
conclusions expressed herein are those of the authors and not necessarily those of the
Iowa DOT or Highway Research Board.

REFERENCES CITED

Akiyama, F.M., 1964. Shear Strength Properties of Western Iowa Loess, Master Thesi
Ames, Iowa, Iowa State University.

Benak, J.V., 1967. Engineering Properties of the Late Pleistocene Loess in the Omah
Council Bluff Area, Ph.D. Thesis, University of Nebraska-Omaha.

Das, B.M., 1998 *Principles of Geotechnical Engineering,* PWS Publishing, Boston, 712
pp.

Olson, G.R., 1958. Direct Shear and Consolidation Tests, *Master Thesis*, Ames, Iow
Iowa State University.

Prior, J.C., 1976. "A Regional Guide to Iowa Landforms", *Iowa Geological Surve
Educational Series 3*, Iowa Department of Natural Resources, Geologic
Survey Division, Iowa City, IA 52242.

Taylor, D.W., 1948. *Fundamental of Soil Mechanics*, John Wiley and Sons, New Yor
700pp.

Hazard identification and risk management for road networks

GIORGIO LOLLINO, CNR-IRPI Strada delle Cacce 73, 10135 Torino, Italia
PAOLA ALLEGRA, CNR-IRPI Strada delle Cacce 73, 10135 Torino, Italia
FILOMENA CRISTALDI, CNR-IRPI Strada delle Cacce 73, 10135 Torino, Italia
FRANCO GODONE, CNR-IRPI Strada delle Cacce 73, 10135 Torino, Italia

INTRODUCTION
More and more attention is paid to road safety hence the area of subject matters related to land planning and all the problems caused by hydro-geological instability.

Figure 1. Piedmont and District of Turin Maps

The analysis conduced by *CNR-IRPI* of Turin *and Servizio Difesa del Suolo e di Protezione Civile* District of Turin is the object of the following study (Figure 1). The purposes were those of creating a methodology for the safety of the road networks in order to obtain a cartography about the risks. This cartography should be aimed at the planning of the prevention and the control of the emergency situations.

Priority was given to the roads as these represent the places were the damages caused by river and stream activity mainly occur as well as the dynamics of the slopes. On top of that, on many different occasions the roads proved to be highly unstable because of their slopes (like valley bottoms and stream crossing, as well). Besides, as the lines of communication are extremely complicated infrastructures it is more difficult to protect them in a suitable way, especially if we consider that, in emergency situations, they are the only link-ups with cut-off-areas.

METHODOLOGICAL LAYOUT OF THE MITIGATION OF THE RISK

The methodology planned for the safety of the road network, or at least a reasonable cut in the dangerous situations, is organised according to different phases. This is necessary for a territory analysis because of the existent interactions among the phenomena of instability and road networks both located in the valley bottoms and on the slopes.

A wide retrospective has started thanks to an informatics elaboration system and taking into consideration the historical events of the past with reference to the most damaging natural events occurred and therefore studied. All this has been possible thanks to CNR-IRPI of Turin who has collected al the data. The information refer to landslides and paroxysmal floods in the last five hundred years with more specific reference to the two last centuries.

ISTAT Code /Event Code	
Area Code	
District	
Place /*Stretch*	
Provincial road number	
Progressive km and altitude	
Main hydrographic basin/ *minor*	
Data of phenomenon	
Phenomenon description (*historical data*)	
Phenomenon description (*investigation data*)	
Potential phenomenon description	
Phenomenon typology	
Damage typology	
Antropic or natural causes	
Rainfall 3-15 days before event	
Interventions typology	
Interventions efficacy and efficiency	
Source information and/or investigation data	
Notes	

Figure 2. Card example used for risk evaluation

Figure 2, based upon scientific basis, has the purpose of creating a new instrument which can be easily employed by the operators of the Civil Protection either for a correct handling of usage situations or to plan suitable control system for a timely and effective preventing action. The first column of the card contains all the information about the description of the event and its location. This card can be widely used as there are no restrictions to utilise it in different geo-morphological situations.

The data collected in the card previously described, were then shifted on a digital cartography by employing GIS-Arc View Software, through an exchange of information between Excel and Arc View. The location on the cartographic basis of the cards was performed either with the use of kilometric progressiveness or by following the toponymic about the place where the event occurred.

The upheavals have been represented by provided graphic subject matters, located on the digital cartography. They are drawn in a thick line and with different colours to diversify the river dynamics from the one of the slopes. For each descriptive card of each event, punctiform symbols are employed and connected to those cards which allow to visualise immediately all the information about that particular event. The precision about the exact location belongs to a series of factors among which the information is the most important one. When a map is drawn with a right scale, it will be highly precise; less accurate it is, less specific the type of information will be.

The starting point is constituted by the Regional Technical Map with scale 1:10.000 and 1:5.000, turned from a paper map into a digital one with a scan density of 200 DPI. The image format TIF, according to the UTM geographical system, is provided with information about georeference.

STANDARDS OF THE EVALUATIONS OF RISK AND HAZARD CONNECTED WITH INSTABILITY PHENOMENA OF THE SLOPES AND WITH HYDROGEOLOGICAL EVENTS

This work is meant to explain what we consider hydro-geological risk and the probability for a given phenomenon to occur (GOVI, 1984) and the negative impact that it will have on physical environment and therefore on people and their activities on the other hand if we consider the hazard we mean according to probabilistic terms, the potential instability of a given area, not considering the antropic presence, connected with the typology, the frequency and the intensity of the events which may occur.

The connection existing between hazard and risk is related to the amount of damage caused to the important antropic infrastructures. The risk degree is determined by the social and economic value of the same infrastructure.

There are many difficulties if you want to apply these kind of schematisations to the events of the alpine valleys, as landslide and stream activity. These phenomena have extremely variable features and often they interfere among themselves to give rise to complex phenomena, whose parameters cannot be estimated with precision.

Within the area of the subject matters the rock falls are extremely quick and so they are practically unforeseeable. The rock falls generally occur into very fractured slopes where it is really difficult to identify premonitory signs.

A different approach concern deep landslides, in fact   they involve the rocky substratum; These ones move along sliding plans and their dynamic is slower than the rock falls. Within this situation it is possible to identify premonitory signs before the paroxysmal phase of the event. These phenomena involve sometimes the whole slope and they develop with complex mechanisms causing changing effects. The deep and complex landslides are easy to locate while the whole mass seldom moves completely, so it is very difficult to evaluate the effects on the territory.

Observing processes related to stream activity we have noticed that they are related to heavy pluviometric events (especially autumn and spring) in concomitance with snow melting. Hence there is a direct connection between temperature, precipitation and the different kind of instability events, making it therefore possible to evaluate the value of the critical precipitation (mm) above which the phenomena of instability are more frequent.

NOTES ON GEO-MORPHOLOGY RELATED TO TYPOLOGY, DISTRIBUTION AND REPETITIVITY OF INSTABILITY PHENOMENA WITHIN THE STUDIED AREA

Some hydro-graphic basins of Western Piedmont have been chosen to set up the method because of their geo-morphological and antropical characteristics. In fact these areas represent a wide casuistry of instability phenomena typology. However the whole work covers the whole Piedmont territory. The natural phenomena that have involved these alpine valleys have been divided into two categories: stream and fluvial activities and slopes instabilities.

The processes related to the slopes have been distinguished in terms of their evolution velocity, slow and rapid landslides: rock falls (Germanasca Valley, Perrero), sliding (Germanasca Valley, Praly, Gardiola landslide), soil slip (Chisone Valley, Porte) and complex landslides (Susa Valley, Salbertrand, Serre la Voute landslide).

Stream and fluvial activities have been divided into three typologies (lateral erosion, flood, overflowing) whose effects often add to one another. Debris Flow has been included under the definition of flood dated back from the old historical document already considered.

In total of the considered events, 73% is related to stream and fluvial activity and 27% is related to slope instability; the total number of event-card is 3308 (Figure 3), among these phenomena rock falls are the most dangerous and frequent.

Table 1. Rate distribution of the events typology related to the history data into the Piedmont Valleys

Analysed basins	Cards number	Area of the basin kmq	Slow landslides %	Rapid landslides %	Flood %	Watercourse length km	Lateral erosion %	Overflowing %
Stura di Lanzo	260	600	10,7	13	25,7	96	29,9	19,5
Dora Riparia	625		5,1	8,2	30,2		27,8	28,3
Chisone and Germanasca	512	589	6,2	13,8	37,2	66	19,1	23
Pellice	391	395	2,3	6,6	45,5	56	19,7	25,8
Sangone	42	270	7,1	33,3	9,5	43	45,2	4,8
Dora Baltea and Chiusella	632	406	5,1	20,7	7,9	104	44,9	17,9
Orco and Soana	677	914	1.6	13.6	7.2	83	37.7	13
Malone	169	353	3,5	13,6	7,7	42	55,6	14,8
Total	3308	3527	41.6	122.8	170.9	490	279.9	147.1

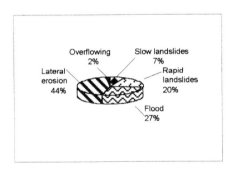

Figure 3. Diagram of the rate distribution of the instability phenomena into the Piedmont Alpine Valleys

PLUVIOMETRIC CHARACTERISTICS OF THE STUDIED AREAS

In order to examine the correlation between the events and the rainfall regime, we have indentified the most significant weather stations which have therefore been considered to obtain major pluviometric characteristics. Some more significant weather stations have been considered to obtain their pluviometric characteristics.

Monthly average rainfall has been calculated for each weather station, then these data have been shifted on specific graphics, where have been represented monthly average rainfall of a specific year characterised by instability phenomena (Figure 4).

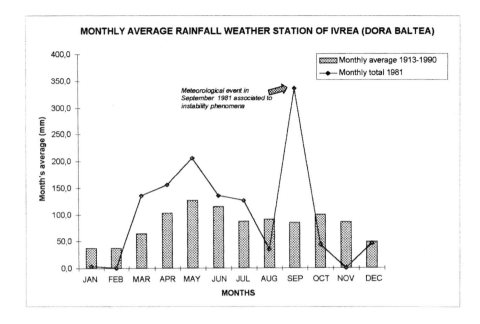

Figure 4. Monthly rainfall average (1913-1990) for Ivrea Station (267m) compared to the total monthly average for the meteorological event in September 1981 associated to instability phenomena

APPLICATION EXAMPLE OF THE METHOD

Through land mapping some roads links have been identified in terms of their relationship with natural events. In this way many places, where hydrogeological risk is high, have been located on the thematic map.

In particular a specific table has been drawn where these elements have been taken into consideration:
-location of the studied area;

-provincial road involved-hydrographic basin;
-typology of natural phenomena
-typology of instability event;
-geolitologic characteristics;
-last interventions executed to defend the provincial road;
-chronological order of the natural events, i.e. repetitiveness of the events;
-traffic related to the particular road stretch.

With reference to these elements, it has been possible to assume a setting based upon the level of the risks of the considered places (low, average, high). This information has been reproduced on the thematic map (Figure 5).

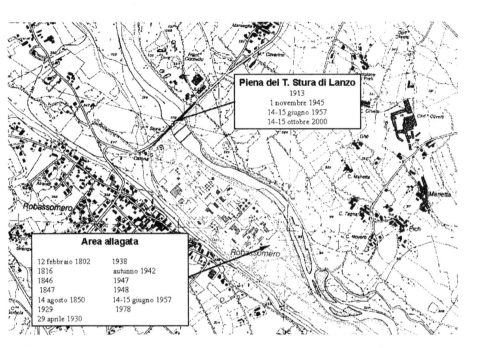

Figure 5. Exctract of risk cartography resulting from historical survey and from site inspection

CONCLUSIONS

This methodology can be considered as a valid instrument to studying every kind of territory. In fact it's suitable both in plain and in mountain areas where there are different kinds of antropic elements.

This method can be used as a basic instrument for land planning studies both for prevention and application works. This possible thanks to a data bank informatic system which can always be updated.

REFERENCES

Allegra P.,Carboneris M, Godone F., Lollino G. (1998) – "Studio della metodologia per una banca dati dei dissesti relativi alla viabilità della Provincia di Torino", Convegno sul rischio idrogeologico, opere di difesa e uso del territorio nel Canavese, Associazione Georisorse e Ambiente, Ivrea 8-9 maggio 1998.

Allegra P.,Carboneris M, Godone F., Lollino G. (1998) – "Studio delle metodologie per la messa in sicurezza di una rete stradale dagli eventi naturali con l'impiego di tecniche GIS", 2ª Conferenza Nazionale della Federazione delle Associazioni Scientifiche per le Informazioni Territoriali e Ambientali, Atti vol.1, Bolzano 24-27 novembre 1998.

Allegra P., Tropeano D., Turconi L. (2001)– *"Enregistrement cartographique des sites routieres soumis aux risques naturels dans la Province de Turin"*, Cemegref – IGAT, Gestion Spatiale des Risques (in stampa).

Bosi C., Dramis F., Gentili B. (1985) – "Carte geomorfologiche di dettaglio ad indirizzo applicativo e catre di stabilitaà a base geomorfologica" Geol. Appl. & Idrogeol., 20 (2).

Caine N. (1980) – "The rainfall intensity –duration control of Shallow landslides and debris flows" Geofisika Annal, 62A.

Canuti P., Casagli N. (1994) – "Considerazioni sulla valutazione del rischio di frana", Atti del Convegno Fenomeni sulla Valutazione del Rischio di Frana, 27 Maggio 1994, Bologna"

Carrara A., Cardinali M., Detti R., Guzzetti F., Pasqui V., Reichenbach P. – "GIS technique and statistical models in evaluating landslide hazard", Earth Surface Processes and landforms, 16.

Cotecchia V. (1978) – "Evoluzione dei versanti, fenomeni franosi e loro controllo (P.F. Conservazione del suolo: S.P. Fenomeni franosi), Mem. Soc. Geol. It., 19.

Govi M. (1984) – The instability processes induced by meteorological events. An approach for hazard evaluation in the Piedmont region (NW) Italy. CNR-PAN meeting, Progress in Mass Movement and Sediment Transport Studies, Problems of Recognition and Prediction, Torino 5-7 dicembre 1984.

Varnes D.J. & IAEG Commission on Landslides (1984) – "Landslide hazard Zonation – a review of principles and practice", UNESCO Paris.

Slope movements in the Majorca Island (Spain). Hazard analysis

DRA. R. M. MATEOS RUIZ. Instituto Geológico y Minero de España. Oficina de Proyectos de Palma de Mallorca. Spain

SUMMARY
In this research, a slope movement hazard analysis at the littoral side of the Serra Tramuntana in the Majorca Island is carried out.

Starting from fieldwork as well as aerial photography analysis, recent and ancient slope movements affecting the studied area were recorded and characterized, distinguishing three main typologies of slope movements: rockfalls, rotation rockslides and complex soilslides.

Different instability conditioning factors have been thoroughly analyzed, such as geology, geomorphology, strength of materials, slopes, hydrogeology and vegetation, mapping each one up to 1:25,000 scale. With the aim of covering the whole studied area at such scale, three subareas have been distinguished, developing the specific maps of every subarea.

The coastal location of the studied area conditions additional determining factors to those regulary analyzed in geohazard studies. The main hydrodinamic aspects, as well as the littoral sedimentation of the Mallorca Serra Norte, have been typified. Shallow inner shelf drilling data recording has allowed the elaboration of 1:25,000 scale mapping of littoral dynamics and sedimentation. This mapping reveals an evident correlation between the existence of important unconsolidated sediment accumulation in the submerged littoral fringe and the presence of large inland slope movements.

A susceptibility analysis has been achieved starting from a land movement inventory mapping and the different instability conditioning factors, according to criteria established by our own methology so-called "The Circles Method". The result is a diverse susceptibility 1:25,000 scale maps, zoning the area with regard to three susceptibility levels: low, medium and high.

Temporal hazard analysis involves a study of the factors starting movement. Four of these factors have been analyzed: rainfall, seismicity, littoral erosion and human

Instability – Planning and Management, Thomas Telford, London, 2002, 339–346

activity. The main starting factor of recent movements is the meteorological one, so that a probability analysis of severe rainfall occurrence in the Serra Tramuntana has been carried out.

The correlation between potentially unstable areas (susceptibility) and severe rainfall occurrence probability allows to gauge the hazard rate, obtaining a space-time response to the future occurrence of slope movements.

INTRODUCTION

The island of Majorca is located in the western Mediterranean sea. With an extension of 3667 Km^2, it has different geographical features, being prominent the "Serra de Tramuntana" mountain range in the northern side of the island (Figure 1). This mountain chain has 90 km length and 15 Km width, with several tops upper than 1000 m height. It is an alpine chain, with geologiocal materials from the lower Triasic to the Quaternary Age, mainly limestone and dolostone. The coastal side is very steep with a sea cliff predominant landscape. This geomorphological pattern in addition to a mediterranean weather (torrential rainfall concentrated in very few days) and a great geological complexity, determine the ocurrence of slope movements.

Throughout the last two centuries several slope movements have taken place in the coastal side, with personal and material damages. The increase of tourism in the island during the last 30 years (nowadays more than 10 million tourists per year) results also in an increase of the risk of landslide and other slope movement types.

The hazard assessment has been carried out in an extense area of 360 Km^2 in the coastal side of the "Serra de Tramuntana".

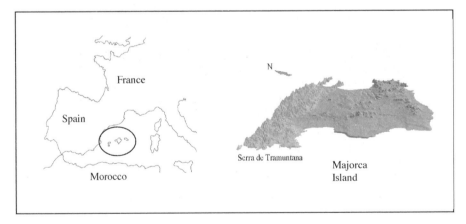

Fig.1. Geographical location of the Balearic islands and the "Serra de Tramuntana" in the Majorca one

TYPES OF SLOPE MOVEMENTS

In the "Serra de Tramuntana" mountain range three types of slope movements have been identified: rock falls, rotational rock slides and complex soil slides.

Rock falls

Rock falls are the most frequent movements in the area. They are related with geological and geomorphological features: fault scarps, karstic valleys, contact of materials with different hardness, etc. They involve masses of hard rock affected by discontinuities. The volume of the fallen blocks can be variable, up to 20 m³. The debri materials which cover the natural slopes come from these rock falls along the Quaternary temps.

Rotational rock slides

Rotational rock slides are the largest movements in the area (Fig.2). They can involve up to 10^9 m³ of rock materials. The materials involved are hard rock ones which are seldom uniform, however, they commonly follow inhomogeneities and discontinuities. The scarp below the crown of these slides is almost vertical and unssuported. Further movements may cause retrogression of the slide into the crown. Usually, the lateral margins of the surface of rupture are high and steep. The scarps of the crown of these movements along the area are aligned parallel to the thrusting fronts. Rotational rock slides in the area are generally multiple movements. They are recent ones, but there is no record of them along the historical age.

Fig. 2. Rotational rock slide in the "Serra de Tramuntana" coast

Complex soil slides

Complex soil slides are very extended along the area. They are movements where the displaced material initially broken by slide movements subsequently begins to flow. They involve fine-grained or weak materials, mostly the alluvial and colluvial sediments of the Cuaternary Age. The flow movements associated to these slides can be earth flow, mud flow and debris flow.

HAZARD ANALYSIS

Hazard assessment tries to estimate the localization, type, magnitude and occurrence of landslides. An ideal landslide hazard map should provide information concerning the spatial and temporal probabilities of all anticipated landslide types within the mapped area (Wu et al, 1996).

Methodology

The methodology to assess the landslide hazard in this area is represented in Fig.3. It could be divided into 2 sections: susceptibility analysis and hazard assessment.

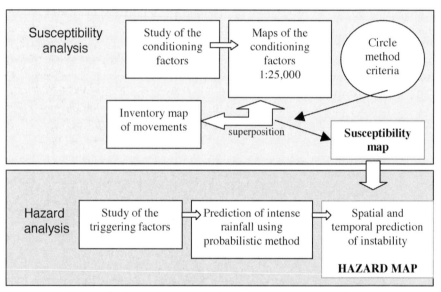

Fig.3. Methodology for hazard assessment

Susceptibilty analysis

Landslide susceptibility assessment requires a detailed knowledge of the processes that are or have been active in the area and of the conditioning factors which determine the spatial distribution of slope instability. The practice of susceptibility zonation requires:

- A detailed inventory of slope instability processes and the study of those processes in relation to their environmental setting.
- The analysis of conditioning factors as well as the representation of the spatial distribution of those factors.

The conditioning factors analyzed in the studied area are geology, hydrogeology, mechanical properties of soil and rock, geomorphology, slope rate, vegetation and land uses. Maps at medium scale (1:25,000) of each determining factor have been designed as well as an inventory map with the localization of unstable areas.

To determine the susceptibility degree along the area, a methodology named "Circle Method" has been developed. It consists of a circle divided into the same sectors as conditioning factors have been analyzed for each type of movement. The aperture of the sector into the circle is indicative of the "weight" of this factor in the instability. For example, in the generation of rockfalls, the importance of factors such as, conditions of water as well as the number and distribution of the joint surfaces in the rock mass, is more relevant than vegetation and lithology ones. At the same time, each sector is divided into several subsector according the legend of their corresponding map. For intance, the slope rate sector of the circle is divided into: <10°, 10° -20°, 20°-35°, 35 – 50° subsectors. The way to place the subsectors into the circle allow us to define the different susceptibility degrees, that is, the susceptibility increase progressively to the centre of the circle. In Fig.4 is represented the circle corresponding to complex soil slides where 3 categories of susceptibility are shown: low, moderate and high

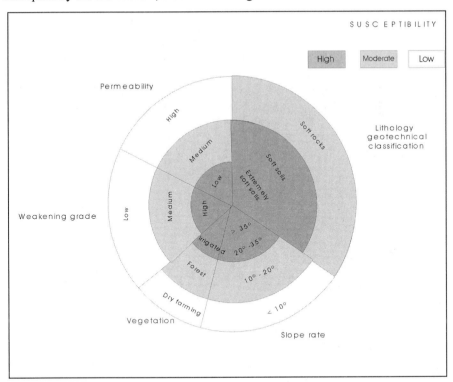

Fig.4. Susceptibility circle corresponding to complex soil slides

By applying the criteria developed in the "Circle Method", a superposition of the inventory map with the conditioning factors have been carried out. According to the convergency of the main factors determining the instability, a susceptibility map escale 1:25,000 is designed which provides information on the spatial probability of occurrence of the mass movements predicted in the area. The maps show, by colour, three degrees of susceptibility: high (unstable areas), medium (relatively stable areas) and low (stable areas) and, by traces, the type of movement expected. In Fig.5 a susceptibility map of the southern area of the "Serra de Tramuntana" is shown.

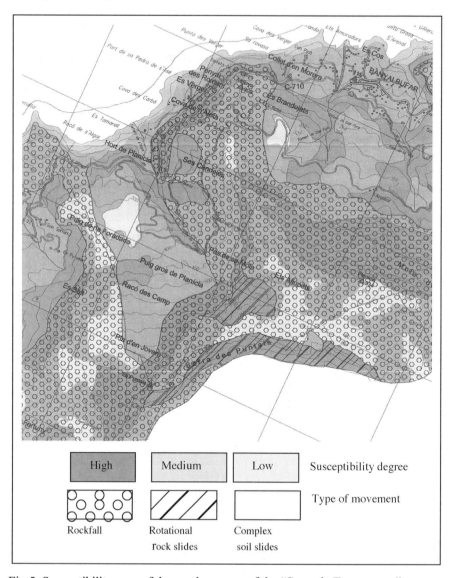

Fig.5. Susceptibility map of the northern area of the "Serra de Tramuntana"

Hazard analysis

Hazard assessment has to give answers to some basic matters such as types of movements, location and time of failure. Forecasting the time of ocurrence of a slope failure is more difficult than assesing the degree of stability because of the uncertainity (Hansen, 1994). The knowledge of triggering factors constitutes the first step to achive this objective.

In the area of study the triggering factors analyzed are intense rainfall, earthquake shaking, coastal erosion and human activity. In areas affected by actual shallow movements, correlations with landslide triggering factors is very useful. In this case, slope movement density and severity depend on both intensity and antecedent rainfall. Seismicity is not relevant in the area and human activity is reduced to slope instability in road talus. We can not underestimate the wave attack at the toe of the coastal cliffs as the main triggering factor of slope failures along the coast, where the height of active cliffs can reach up to 300 m.

Probabilistic analysis of intense rainfall occurrence
The historical landslide record shows that failures are found to be very sensitive to high intensity and short duration rains. A correlation between landslide data and the value of maximun rainfall in 24h has been established. The ocurrence of shallow slope movements, rockfalls and soil slides takes place during the ocurrence of intense rainfall over 130 mm in 24 h. This is the estimated value for time forecasting.

Rainfall data series along the last 50 years have been used in order to identify return periods of intense rainfall that would probably induce slope movements. Several maps with the values of intense rainfall in 24h for different return periods have been drawn. The one corresponding to 100 years shows that the critical value (130mm) of intense rainfall will be reached in almost all the Serra de Tramuntana area.
This probabilistic analysis of intense rainfall data alow us to get a map (Fig.6) with the probability of intense rainfall event in the studied area for a medium time forecasting (25 years).

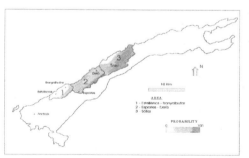

Fig.6. Map of The "Serra de Tramuntana" with the probability of high intense rains (over 130mm in 24 hr) for a 25- year forecasting

The superposition of the map of probability of intense rainfall event and the susceptibility one provides information on the spatial and temporal probability of the mass movement predicted in the area. Hazard zonation mapping a regional scale is carried out taking into account this spatial-temporal analysis (Fig.7).

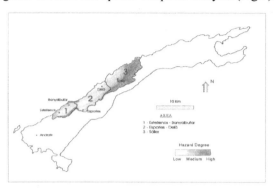

Fig. 7. Regional hazard map

CONCLUSIONS
The main conclusions obtained from this study are he following:
1.   Three types of slope movements have been identified in the studied area: rockfalls, rotational rock slides and complex soil slides. Rockfalls are the most frequent slope failures which have caused several damages along the time. Complex soil landslides are shallow and they are very extended around the area. Both rockfalls and soil landslides have been recorded in the historical time.
2.   Rotational rock slides are the biggest ones, they can put in motion millions of cubic meters of material. They are recent movements because of their morphological features and their correspondence with non consolidated and coarse grained sediments in the shore shelf.
3.   An area of 104 Km^2 (30% of the total surface) has been identified as having a high degree of susceptibility. The susceptibility tends to increase towards NE and SW in the "Serra de Tramuntana".
4.   High intensity and short duration rains trigger mostly shallow slope movements. The probability of intense rainfall event tends to increase towards NE.
5.   According to hazard assesment, the probability of slope movements occurrence tends to increase towards NE as well as towarsd the shore line.

REFERENCES
Hansen, A. 1994. Landslide Hazard Analysis. Slope instability. Edited by Brundsen &Prior. 523-595.
Wu,T., Tang, W. & Einstein, H, 1996. Landslide hazard and risk assessment. Landslides, investigation and mitigation. Special report 247. Transportation Research Board. 106-116.

A sediment budget approach for estimating debris flow hazard and risk: Lantau, Hong Kong

DR R MOORE, Halcrow Group Ltd[1], Birmingham, UK
E M LEE, University of Newcastle upon Tyne[2], UK, and
J S PALMER, Caldex Consultants[2], UK

INTRODUCTION

Quantitative risk assessment involves estimating the probability of landslide events of different magnitude and their consequences (e.g. Cruden and Fell 1997). A recent study of natural terrain hazard and risk (Moore *et al.* 2001) demonstrated the very large uncertainties in quantifying the risk of low frequency, high magnitude events, such as channelised debris flows. The study concluded that in such cases a consequence-based approach is more practical, particularly as a basis for making decisions on hazard and risk mitigation. The paper describes how a combination of geomorphological and engineering geological mapping provided the framework for deriving channelised debris flow event scenarios for a small mountain catchment on Lantau Island, Hong Kong.

The approach involved estimation of a sediment budget for the catchment using detailed site mapping and ground investigation techniques. The whole catchment has been viewed as a cascading sequence of inter-related physical systems (i.e. terrain sub-units). The *outputs* from one system provide the *inputs* to the next in line. Within the catchment, five main stages of landslide activity were recognised. *Initiation* and detachment of material from hillsides; *transport* and delivery of this material into the channel system; *storage* of material within the channel system (and also, in the short-term, on hillsides before delivery to the channels); *entrainment* and runout from the catchment; and *deposition* on a debris fan.

THE SITE

The site is located on the northern slopes of Lantau Island, east of Tung Chung, Hong Kong. It comprises a single very steep catchment (35-40° slopes) characterised by a series of northwest trending drainage lines that join about mid-slope into a single stream course running to the base of the slope. The site is

---

[1] Consultant to Geotechnical Engineering Office, Hong Kong SAR Government
[2] Sub-contractor for detailed field mapping

approximately 400m in length, from mountain crest to the sea, and 100m at its widest narrowing downslope to 40m. The area has a high incidence of recent debris flow activity triggered by rainstorms (Evans *et al*. 1997; Franks 1999).

The catchment has formed in an area of mixed volcanic ash (tuffs) and lava of Upper Jurassic / Lower Cretaceous age assigned to the Repulse Bay Volcanic Group (Hong Kong Geological Survey, 1995). The geological structure consists of a generally south-easterly dipping fabric and a series of generally north-east / south-west trending faults defined along drainage lines.

SITE INVESTIGATION
An integrated programme of engineering geological and geomorphological mapping and subsurface ground investigation (boreholes, probing, trial pits and geophysics) was undertaken in order to define site conditions. A key feature of the terrain is the potential for open hillside landslides (typically shallow debris slides and flows on steep slopes) and channelised debris flows (associated with stream valleys).

Field mapping identified a series of distinct terrain units and sub-units, based on interpretation of slope morphology and surface drainage (Hansen 1984). These include an upper catchment (Terrain Unit 1), comprising a series of convergent incised stream channels (Terrain sub-units 1.1 to 1.4), and a lower catchment (Terrain Unit 2), comprising a single incised stream channel (Terrain sub-unit 2.2). Below Terrain sub-unit 1.3 there is a narrow, sediment filled channel storage zone (Terrain sub-unit 2.1). A large debris fan occurs at the mouth of the catchment (Terrain Unit 3) that extends onto the shoreline. A schematic summary of the distribution of terrain units and the associated natural terrain hazards across the catchment is presented in Figures 1 and 2.

A ground investigation confirmed that the debris fan deposits were probably of debris flow origin and revealed up to 5m of debris forming at least 3 layers of comparable thickness. Given the area and depth of the debris fan, a total volume of 48,000m^3 was estimated assuming a factor of 0.5 to account for depth irregularities and thinning of the debris fan towards the edges. The evidence suggests the fan comprises debris from at least 3 events of comparable magnitude of about 16,000m^3.

LANDSLIDE FEATURES
A total of 75 landslides were identified within the catchment. Most landslide features occur on 35-40° slopes, are *recent* (i.e. vegetated or partly vegetated failures that most likely occurred within the last 50 years; Halcrow 1999) and have involved an initial phase of detachment as a shallow (<3m deep) *debris slide or avalanche* (Evans *et al*. 1998) within weathered rock and/or saprolite. Many slides may have mobilised as hillside *debris flows* (as recorded by Franks 1999, after the 1992/93 rainstorm-triggered landslides in the Tung Chung area), although little or no evidence remains other than elongate, vegetated tracks downslope of the main slide

area. Evidence for *channelised debris flows* is limited to the presence of extensive debris flow deposits at the catchment mouth (debris fan) and the indistinct traces of debris trails on the steep valley sides and stream channels.

A deeply incised area or possible 'relict landslide scar' forms Terrain Sub-unit 1.2 and was thought to have originated from a *major rock avalanche*, with more recent degradation of the scar apparent through relatively shallow debris slides.

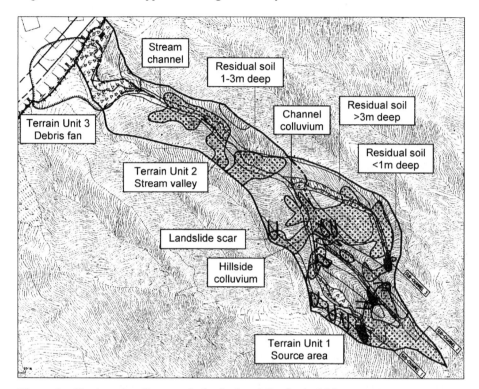

Figure 1 – Engineering Geomorphological and Geological Map

SEDIMENT BUDGET

An estimation of the sediment budget for the catchment (i.e. the balance between sediment *inputs* from landsliding and the *outputs* from the catchment) provides an indication of the current potential sediment supply to the debris fan at the catchment mouth. The potential sediment supply is time-dependent and reflects the rate of weathering and landsliding, and the interval and passage of extreme rainstorm events. A large proportion of the potential sediment inputs remain *stored* within the natural terrain as:

- landslide debris upon hillsides; most landslides within the area have delivered less than 50% of the failed mass to the stream channels and valley stores;
- colluvial and alluvial deposits within channels; as ribbon-like stores along

stream channel banks and beds (often as angular and sub-rounded boulders within a sandy matrix) or broader accumulations in valley floors (valley floor stores).

Estimation of sediment inputs, stores and outputs have been based on detailed field mapping and inferred depths of deposits supported by field observations and sub-surface investigation data where available. The volume of material was estimated for defined landslide scars and as an average volume per metre run for channels.

The debris stores appear to have two contrasting roles in controlling channelised debris flow behaviour in the catchment. Firstly, for the relatively high frequency, shallow open hillside landslides, debris enters the slope or valley stores rather than being directly mobilised as channelised debris flows. Thus, the stores act as a *buffer* between hillside instability and stream valley transport, either as channelised debris flows or in flash floods. Secondly, for low frequency, high magnitude events, debris stored upon hillsides and within valleys provide a *source* of generally unconsolidated sediment that can be entrained and mobilised by channelised debris flows. For example, the Tsing Shan debris flow of 1992 increased its initial failure volume by 200% as it entered a drainage line and entrained bouldery colluvium (King 1992). It follows that the accumulation of unconsolidated debris from numerous shallow hillside landslides within valley stores over time can provide a large volume of sediment capable of being mobilised in a single episodic event.

For this study, it was apparent from site observation that there was limited storage and entrainment potential within the main stream channel (Terrain Sub-unit 2.2); upstream of the debris fan the channel is cut in bedrock, with riffle-pool sequences and occasional waterfalls. Therefore, along the main stream channel, sediment entrainment potential was estimated to be limited to around $2.5\text{m}^3/\text{m}$.

Figure 2 presents a summary sediment budget for the catchment. For each terrain sub-unit the sediment budget was calculated as follows:

- the *inputs* have been estimated as the volume of material supplied from open hillside landslides (i.e. landslide volume less the amount remaining on the hillside) plus the volume of material supplied from upslope terrain sub-units;
- the hillside and valley *storage* has been estimated from field observations (i.e. length × average width × average depth);
- the *output* from each terrain sub-unit is assumed to be the balance between the inputs and storage. Where the inputs are less than the estimated hillside and valley storage, the sediment inputs are assumed to only contribute to the storage (i.e. storage will increase over time). This assumption is likely to be broadly valid as there is little field evidence of sediment throughput in the valley stores i.e. these stores do not appear to be actively supplying material to the stream channels at their downslope margins. Where the inputs are greater than the estimated hillside and valley floor storage, the excess is assumed to be the output from the terrain sub-unit and, hence, an input to the next sub-unit.

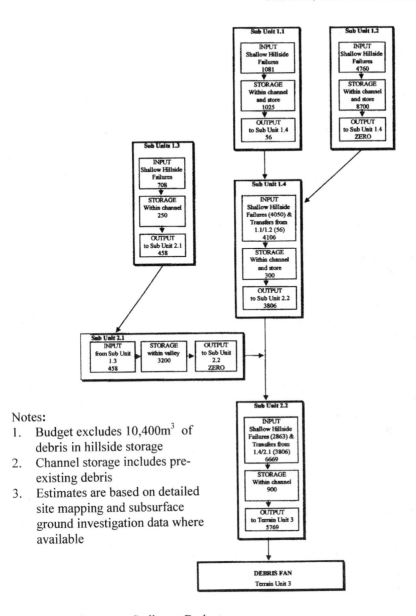

Figure 2 – Summary Sediment Budget

In this way, the contemporary catchment yield was estimated to be about 5,770m^3 i.e. the supply to the debris fan. This figure represents the difference between the inputs from the current suite of open hillside landslides and the catchment storage. The recorded landslide features have contributed around 13,500m^3 of sediment to the catchment (a further 10,400m³ remains in storage upon the hillside within landslide scars), of which 7,730m³ has added to the channel storage and 5,770m³ has been transported to the debris fan by channelised debris flows and flash floods.

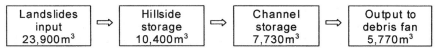

| Landslides input 23,900m³ | ⇒ | Hillside storage 10,400m³ | ⇒ | Channel storage 7,730m³ | ⇒ | Output to debris fan 5,770m³ |

Clearly, a different suite of landslide features would generate different sediment inputs and catchment yield. It is possible that the current suite of landslides represents the product of landslide activity over a relatively limited timescale (e.g. 50 years; after Evans and King, 1998). If this is the case, the distribution and scale of landslide activity over the last 50 years or so, and the resultant catchment yield, is the product of a unique series of triggering events and hillside responses. Different triggering events over the next 50 years may produce a different pattern of landsliding, storage and catchment yield.

CHANNELISED DEBRIS FLOW SCENARIOS: EVENT MAGNITUDE

The sediment budget approach can provide an indication of the maximum potential channelised debris flow event that could be expected to occur within the catchment. For example, the 'best estimate' potential channelised debris flow event with an expected frequency of 1:50 years, was estimated by defining a maximum potential open hillside landslide for each terrain sub-unit (i.e. source area) and entrainment of colluvium from the existing channel storage areas (Table 1). Field evidence indicates that the largest existing open hillside landslide has an estimated volume of 1,800m³. It was assumed that events of a comparable magnitude could be triggered simultaneously in each of the 3 sub-catchment source areas, providing an input of 5,400m³ of sediment. Consideration was given to the potential initiation of a large-scale landslide (rock slide) but field observation did not reveal any evidence of potential hillside failures of this magnitude, which was therefore discounted.

A series of channel sections were defined from each potential source landslide along the stream course highlighting channel storage areas from which colluvium can be entrained by the channelised debris flow. It was assumed that all debris from channel stores could be mobilised and transported down the channel to a point where deposition takes place. Based on existing knowledge of channelised debris flow behaviour in Hong Kong, debris entrainment has been observed on channel gradients above 22° whilst deposition occurs where gradients reduce to 15° or less. An 'initial deposition point' was therefore defined by the 15° break in slope contour, below which channel gradients are generally less than 15°.

For channel sections steeper than 22°, it has been assumed that 2.5m³ of debris per metre length of stream channel may be entrained. This assumption accounts for the potential entrainment of loose colluvium from the steep side-slopes adjacent to the stream channel. In some locations, the potential channelised debris flow passes over colluvium previously deposited in stream channels. If the colluvium is within a channel at gradients of 22° or above, it has been assumed that all material within the channel could be entrained. For example, a 5m deep deposit of colluvium in a 7.2m wide stream channel could yield 36m³/m length of channel: 7.2m × 5m × 1m = 36m³. From field observation, one section of stream channel (channel section.8) has a gradient of less than 15°, where the channelised debris flows might deposit debris. However, in this analysis it was assumed that all the debris passes through this short, low-angle channel section.

Table 1 – Best Estimate Potential Channelised Debris Flow Event

Terrain Sub-units and landforms	Volume m³
Terrain Sub-unit 1.2	1,800
Maximum Open Hillside Landslide Source Event	
Channel Section 1 (35°)	
Entrainment of colluvium = 64 m × (2 m × 2 m/tan 35)	367
Channel Section 2 (17°)	
No entrainment due to low gradient = 135 m × 0 m³/m	0
Channel Section 3 (27° – 30°)	
Entrainment = 195 m × 2.5 m³/m	488
Terrain Sub-unit 1.1 and 1.4	
Maximum Open Hillside Landslide Source Event	1,800
Channel Section 4 (28°)	
Entrainment of colluvium = 163 m × (1m × 1m/tan 35) = 1.43m³/m Since 1.43 < 2.5, use 2.5m³/m	408
Channel Section 5 (25° – 36°)	
Entrainment = 157 m × 2.5 m³/m	393
Terrain Sub-units 1.3 & 2.1	
Maximum Open Hillside Landslide Source Event	1,800
Channel Section 6 (30°-50°)	
Entrainment of colluvium = 185 m × (2 m × 2 m/tan 35)	1,057
Channel Section 7 (20°)	
No entrainment due to low gradient = 65 m × 0 m³/m	0
Channel Section 8 (12°)	
No entrainment due to low gradient = 75 m × 0 m³/m	0
No deposition	0
Terrain Sub-unit 2.2; V-shaped stream channel below confluence	
Channel Section 9 (18° – 20°)	
No entrainment due to low gradient = 300 m × 0 m³/m	0
Channel Section 10 (25°)	
Entrainment = 54 m × 2.5 m³/m	135
TOTAL POTENTIAL OUTPUT TO TERRAIN UNIT 3	8,248m³

The results indicate a 'best estimate' potential channelised debris flow event of about 8,250m³. This compares with a possible 'worst-case' potential channelised debris flow event of about 24,000m³ (see Section 5) that assumes widespread initiation of new open hillside landslides and removal of all stored debris within existing landslide scars and channels. The likelihood of the 'worst-case' scenario is considered extremely remote with an expected frequency well in excess of 1:200 years. Indeed, an event of this magnitude has not been recorded in historical times. Several observations are, however, worthy of note:

- The existing landslides within the catchment are recent, many of which were triggered by an extreme rainstorm in 1992/93; the estimated sediment budget for this event (16,000m^3) supports the view that the theoretical 'worst-case' scenario involving mobilisation of up to 24,000m^3 is extremely remote;
- Extreme rainstorms of different intensities, frequency and storm-paths could generate a very different pattern and catchment response;
- Evidence from the debris fan of at least 3 events suggests that channelised debris flows are extremely rare and that there could be a threshold storage capacity and yield of the catchment of about 16,000m^3;
- In this case, high-magnitude channelised debris flows are only likely to be triggered by the rare occurrence of multiple open hillside landslides coupled with delivery of debris into fast flowing stream channels that increase the mobility and entrainment of landslide debris down the mountain stream course.

ACKNOWLEDGEMENTS

The work reported in this paper formed part of the Natural Terrain Hazard and Risk Area Studies and is published with the permission of the Director of Civil Engineering and the Head of the Geotechnical Engineering Office, Civil Engineering Department, Hong Kong SAR Government.

REFERENCES

Cruden D and Fell R (eds.) 1997. *Landslide Risk Assessment*. Balkema.

Evans NC, Huang SW and King JP 1997. The Natural Terrain Landslide Study: Phase III. *GEO Special Project Report* SPR 5/97.

Evans NC and King JP 1998. The Natural Terrain Landslide Study: Debris Avalanche Susceptibility. *GEO Technical Note* TN 1/98.

Franks CAM 1999. Characteristics of some rainfall induced landslides on natural slopes, Lantau, Hong Kong. *QJEG*, 32;247-259

Halcrow Group Ltd. 1999. Natural Terrain Hazard and Risk Area Studies: Mount Johnston North and Tung Chung East. *Final Report* to the Geotechnical Engineering Office, Hong Kong Government.

Hansen A 1984. Engineering geomorphology: the application of an evolutionary model of Hong Kong's terrain. *Z. Geomorph*. NF, 51; 39-50.

Hong Kong Geological Survey, 1995. Geology of Lantau District. Hong Kong *Geological Survey Memoir* No. 6. Hong Kong Government.

King J P 1996. Tsing Shan Debris Flood. *GEO Special Report* SPR 7/99.

Moore R, Hencher SR and Evans NC 2001 An approach for area and site-specific natural terrain hazard and risk assessment, Hong Kong. *Geotechnical Engineering Meeting Society's Needs: Proceedings of the 14th South-east Asian Geotechnical Conference, Hong Kong, December 2001, pp155-160.*

The management of risk on the Chalk cliffs at Brighton, UK

J. S. PALMER, CalDex Consultants Ltd, Hertfordshire, UK
A. R. CLARK, High Point Rendel, London, UK
D. CLIFFE, High Point Rendel, London, UK
M. EADE, Brighton & Hove City Council, Brighton, UK

ABSTRACT

The cliff frontage within the urban area of Brighton extends for some 5km to the east of the city centre from the Marina to Saltdean. With growing concern about the long-term stability of the cliffs, and increasing awareness of public safety and the potential future effects of climate change, a cliff management strategy has been implemented including surveys to assess the risk and hazards that may arise from instability in the cliffs.

The paper will describe the details of the geology and engineering geology of the site area, and the techniques used to establish and manage the risk. The unprecedented rainfall experienced throughout the UK during the winter of 2000/2001 had a significant impact on the cliff and caused a series of major falls. The paper will also discuss these events with respect to the risk assessment and management strategy.

INTRODUCTION

The 5 km of cliffs between Brighton Marina and Saltdean on the UK South Coast (Figure 1) consists of near vertical Upper Chalk, some 25-40m high. The toe of these old sea cliffs has been protected since the 1930's by the construction of the coastal defences comprising a seawall and a groyne system along the entire section; the cliffs were re-profiled to a uniform angle of about 70° at this time. A relatively narrow promenade lies at the toe of the cliffs. This is an extremely popular attraction for both the local population and visiting holidaymakers. Up to 2700 visitors per day may use the area in the peak periods as a leisure area and as an access to the beach. In addition to the facilities at the base of the cliff, the main A259 south coast road runs along the top of the cliff carrying up to 28,000 vehicles a day.

Brighton & Hove City Council is the owner of the cliff and the coast protection authority, and under their health and safety duty, a hazard and risk assessment has been carried out for the whole area. This has involved identification of the hazards of cliff falls and evaluation of relative risk to the public and property. This process has also allowed the identification of remedial measures to be identified and prioritised. The survey addressed the lithologies and geological structure visible in the cliff together with human issues of property and public proximity to potential hazards. The evaluation of risk is ongoing to account for new data and changing cliff conditions as they occur, including repeat inspections undertaken every 2 years

The cliffs had been largely stable since the construction of sea defences. Whilst isolated rockfalls had been experienced since the sea defence construction, there was no historical precedent for the extensive instability that affected the cliffs during the winter of 2000/2001.

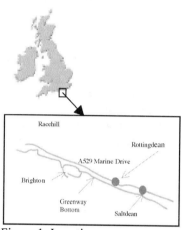

Figure 1, Location

Figure 2, "Hanging" Dry Valley

GEOLOGY
Three principle geological deposits exposed within the cliffs are:

Residual soil and dry valley deposits
Head deposits are present along the entire cliff top area representing a residual soil. In addition, Head or Coombe deposits occupy dry valleys of post-Pleistocene origin that occur intermittently along the cliff section, many of which are observed *"hanging"* along the cliff top (Figure 2). The Coombe Deposits are typically Chalky loam or Chalk rubble in extreme cases formed of a clast supported interlocking flint gravel in a sandy clay matrix.

Chalk
The most widespread deposit exposed along the whole cliff section between Brighton Marina and Saltdean is the Cretaceous Upper Chalk. This is approximately horizontally thickly bedded and is typically traversed by 4 major sets of angled joints (typically 50°/South; 70°/Southwest; 70°/Southeast; sub-parallel to face). The Chalk is generally well structured, although weathering of the upper portion is

present throughout the cliff line, and this becomes pervasive over the whole cliff height where the cliff line is breached by dry valleys. The combination of jointed near vertical cliffs with varying degrees of weathering and exposure to the elements, is responsible for past rock falls and a slow but progressive retreat of the cliff line.

Of particular importance to stability is the presence of joints dipping out of the cliff face at typically 50° which create basal release surfaces for plane failures, and 70° orthogonal joints dipping sub-perpendicular to the face that create lateral release surfaces for such events. The back scars of significant historic events formed along these joints are apparent intermittently along the cliff line, although with the exception of a single landslide that resulted from unloading due to cliff excavations (Corbett 1990), there was no recorded incidence of such instability since the construction of the sea defences. Nevertheless, the risk assessment identified this as a potential failure mode.

Historically instability would have occurred in a *"bottom up"* fashion arising from undercutting due to exposure of the cliff to marine action. Where the sea defences are effective, the instability tends to originate within the upper portion of the cliff (where weathering is more pronounced since the construction of the sea defences) and propagate downwards. In this manner it may be appreciated that whilst the construction of the sea defences has slowed cliff regression, it is not effective in completely arresting instability.

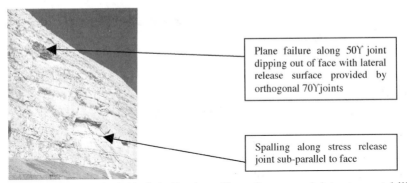

Plane failure along 50Y joint dipping out of face with lateral release surface provided by orthogonal 70Y joints

Spalling along stress release joint sub-parallel to face

Figure 3.   Typical Chalk indicating effect of common joint sets on stability

Raised Beach Deposits

Of particular geological interest is the area at the west end of the cliff line (inland from the Marina) that represents mixed Coombe and chalk debris deposits infilling an abandoned Pleistocene cliff line, orientated approximately perpendicular to the current cliff (Figure 4). The importance of this exposure is recognised in its designation as a Site of Special Scientific Interest (SSSI), and is described in detail in Mortimer 1997. For the purposes of the paper, the various infill deposits are referred to collectively as *"Raised Beach Deposits"*. The raised beach at Black Rock is reported to be one of several in Britain indicating a mean sea level of about 7m above that of the present day (Castleden, 1996).

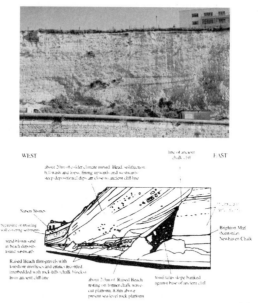

Mortimer 1997

Figure 4.   "Raised Beach Area", Black Rock

RISK ASSESSMENT

In order to carry out a risk assessment it is necessary to establish the hazards that exist and the vulnerability of the public and property to those hazards. There is little documented data prior to 2000/01 concerning the timing, frequency and magnitude of past landslide events, and evidence of rockfalls that pre-date 1930's will have been partially or fully removed by the re-profiling works.

Evidence for large-scale historical instability can, however, be observed intermittently along the entire Chalk cliff section, and this provided information from which judgements on the scale of historic failures were made. Such judgements cannot so easily be made as to the timing of past landslide events or whether the instability occurred as a single event or as the result of a progression of small failures. Recent landslide events (2000/01) provide the best data on which future judgements on the probable scale and location of landslides may be made.

Because of the lack of historic data, and difficulties in the precise definition of the nature of past failures, it has not been possible to complete a rigorous quantitative risk assessment for the study area. Therefore, a qualitative assessment based on observations and expert judgement was undertaken. Details of the cliff condition were recorded including description of lithology, weathering and geological structure, vicinity of properties to the cliff top or bottom, condition of the sea wall, and indications of cliff top instability, etc. The collection of this data indicated hazards of various type and scale affecting the cliff line. Not unexpectedly, this included potential instability ranging from individual rock falls, to large-scale *en masse* slope instability, and erosion (including scour on the cliff top and on the cliff face).

The assessment concluded similarly to work on chalk cliffs elsewhere in the UK, that there was likely to be precursory indicators to large-scale failure events (e.g. isolated rock falls may occur without warning, whilst large-scale failures are usually preceded by development of tension cracks). This assumption has required significant re-thinking in light of the landslide events of 2000/01.

RISK RATING SYSTEM

Arising from information gathered in this fashion, a management strategy for the cliff line was developed, together with a comprehensive schedule of prioritised remedial works.

The identification of hazards which cause a risk to the public and property involves the assessment in probabilistic terms of the likely location of the public, vehicles and structures in the study area which could be affected by landslide and rock fall impact. In this respect, the temporal and spatial distribution of the public is significant. For example, with respect to risk to the individual, a person walking along the seaward side of a wide section of the promenade, well away from the cliff base, is exposed to a lower risk from rockfalls than a person walking directly at the foot of the cliff. The risk may also be increased where the population is concentrated; most notably on the Marina reclamation and at access paths. The approach to the management of the risk in either of these two examples is different. The first example may be addressed by encouraging the public away from the toe of the cliff (i.e. the hazard remains the same, but the risk is reduced). With respect to the second example remedial works may be undertaken to contain rockfalls (i.e. the risk is reduced by addressing the hazard).

The level of risk can also be assessed by consideration of the magnitude of the hazard and frequency of occurrence. A small isolated block failure or release of a flint nodule may occur at any time, without any warning. The risk to a member of the public being struck by such an event is not considered to be high because:
- the frequency of failures is apparently low;
- the size of the block or nodule is typically small (generally < 0.1 m^3) and;
- the average density of people on the promenade is low.

The probability of a member of the public being struck by a rock is approximately proportional to the size of the fall; and there is a greater probability of minor injury compared with serious injury. The time of the day and the time of year will also affect this probability, as the density of people on the promenade is not constant.

On the basis of the above it was possible to construct a risk rating system based on frequency and extent of potential instability at any location along the cliff line, and exposure of the public or properties to these hazards. The general risk rating system is summarised in Table 1.

Table 1, Risk Rating System for the Brighton Cliffs

Risk Rating	Description of Risk	Consequence of Risk	Remediation Requirement*
1	**Very Low Risk:** to the public and property in areas where concrete structures cover cliff face (such as access steps and retaining structures within dry valley sections)	Negligible risk from cliff falls.	Routine maintenance of retaining structures only
2	**Low risk:** to the public and property from individual boulder and rock falls (including flint nodule falls) up to $0.2m^3$ in volume. Distance of cliff top edge to road >10m.	Small chance of public and property being hit by rock fall. Low risk sections require periodic visual inspections in order to check for deterioration of the cliff.	No immediate requirement of remediation other than monitoring.
3	**Medium risk:** as per risk rating 2; and including major joint sets which daylight on cliff face forming potential rock failures greater than $0.2m^3$; potential for rolling or bouncing during rock fall; overhanging vegetation fall and localised overhanging pipes and weathered cliff top materials. Distance of cliff top edge to road <10m.	Moderate chance of public and property being affected by rock fall/debris. Cliff hazards can be defined but further deterioration is required prior to the probable onset of instability. Remediation is not considered necessary in the near future; these sections require periodic visual inspections in order to monitor further deterioration	Remediation required within c.2-5 years to reduce risk
4	**High Risk:** as per risk rating 3; and including the widening of joints leading to block and wedge falls in the near future; precarious overhangs of cliff top Chalk and foundations (including brickwork, drain pipes, fence footings); recent failure scars along cliff face and/or spurs between scars of historic major failures in areas of wide promenade; Distance of cliff top edge to road <8m.	These sections present cliff hazards that can occur in the near future and are of a risk to the public but are not considered as high a priority in terms of remedial action compared with the Very High Risk sections. Provision for remedial engineering works should be included for in future budget planning. Regular monitoring and inspections should be undertaken until remediation is implemented.	Remediation required within c.2 years to reduce risk.
5	**Very High Risk:** Conditions generally as risk rating 4, but located at areas of population concentration (such as access paths), and areas where people are encouraged towards the base of the cliff (such as benches placed at the foot of the cliff, or areas of narrow promenade). Distance of the cliff top edge to road <5m.	These sections comprise the greatest risk to the public in the near future and are priority areas for remedial action. The state of instability is such that immediate provision is required for remedial engineering works.	Urgent attention required to reduce risk

INSTABILITY
Historic Events
Records of past instability along the cliff line are sparse. The most complete record of a significant failure is that documented by Corbett (1990). Historic maps provide an indication of the past cliff recession rates (Figure 5.), and assists in highlighting specific locations along the cliff line where recession occurs at an accelerated rate, and implicitly may be more prone to instability than the average cliff line.

The obvious spatial discrepancy between the recession rates remains unexplained but is likely to relate to hydrogeological conditions near the raised beach area. This may include locally high perched water tables, perhaps arising from the lower hydraulic conductivity in the raised beach deposits than in the Chalk. Also, the major dry valleys are perched high in the cliff in the west of the area, whilst in areas of slow retreat the base of dry valleys is at or below sea level. Lithological variation has been proposed to explain the variation in recession rates, however, there is no visible evidence of this.

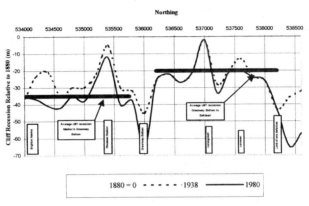

Figure 5, Cliff recession indicated from historic maps

Recent Events
The true significance of the increased recession rates west of Greenway Bottom became dramatically apparent following the unprecedented rainfall of 2000/01. Some 15 significant failures occurred within a 300m section of the cliff land ward of the Marina. The extent of these failures was without recorded precedent, although the importance of the Chalk structure and weathering in controlling the geometry of the events was consistent with evidence of past instability recorded in the cliff survey. Instability also affected the raised beach area, with the largest single event releasing 2000m^3 of material from these deposits (Figure 6).

Irrespective of the geological factors that define the location and occurrence of the failures, the presence of groundwater resulting from the exceptionally high rainfall must be contributory to all of the failures. This is evidenced by the coincidence of so many events within such a concentrated time span. It appears likely that any influence of groundwater is localised due to perched groundwater or storm water falling directly onto the cliff face and environs.

DISCUSSION
The instability affecting the cliff line followed the extreme climatic conditions of 2000/01. From historic data, it was evident that past cliff recession was higher in the

Marina area than elsewhere. Further, the geometry of the large-scale events was broadly as indicated from the relict back scars of landslides that predated the construction of the sea defences. The risk assessment drew attention to the increased probability of instability with time (due to weathering), but without precedent landslide data since the construction of the sea defences, the risk assessment could not conclude a high probability of significant landslides occurring in the Marina area. The scale and frequency of landslides was not, therefore, predictable from the risk assessment, and while remedial activities were defined within the cliff strategy, such were of limited extent reflecting the general longstanding stability. It was only following the events of 2000/2001 that full intrusive investigations and implementation of extensive stabilisation measures was instigated.

The case study has indicated the benefits of completing site surveys and interpretation of the mechanisms of potential failure. However, it has demonstrated the limitations of assigning absolute risk with respect to predicting the incidence and timing of landslides in the absence of precedent landslide data. Of equal significance, the methods adopted in this case study are common to the assessment of risk along cliff lines elsewhere. It is clear from the example of Brighton during 2000/2001 that these methods may not be reliable in cases of extreme climatic conditions. It remains to be seen if our predictions of risk become generally unreliable in light of future climate change, and to what extent we will have to re-evaluate the methods used to predict such risk.

Figure 6, Failure within the raised beach deposits

REFERENCES

Mortimer R. N. (1997). The Chalk of Sussex and Kent. Guide No. 57. Geologists Association.

Castleden R. (1996). Classic Landforms of the Sussex Coast. The Geographical Association

Corbett B O (1990). Slip in the chalk cliffs at Brighton Chalk. Thomas Telford, London.

Dynamic analysis of recent destructive debris flows and debris avalanches in pyroclastic deposits, Campania region, Italy

PAOLA REVELLINO, Faculty of Science, University of Sannio, Benevento, Italy
OLDRICH HUNGR, Department of Earth and Ocean Sciences, University of British Columbia, Vancouver, Canada
FRANCESCO M. GUADAGNO, Faculty of Science, University of Sannio, Benevento, Italy
STEPHEN G. EVANS, Geological Survey of Canada, Ottawa, Canada

INTRODUCTION

Numerous debris avalanches and debris flows have occurred on steep slopes mantled by pyroclastic deposits in the Campania region of south-central Italy in recent years, some with disastrous impact (Guadagno, 1991; Del Prete et al. 1998). The most recent events took place in two clusters - the Sarno/Quindici events of 5 - 6 May 1998 and the Cervinara landslides of 15 – 16 December 1999 (Fiorillo et al., 2001). Altogether, over 100 major individual landslides took place. The landslides were remarkably similar to each other in terms of style, and geological and topographic setting. It therefore seemed possible that their dynamic behaviour could be simulated by using a single model, with a limited range of input parameters. Such a model, if successfully calibrated, could become a useful tool for predicting the behaviour of future landslides of the same type in similar settings in the Campania region. A systematic program of back-analysis was therefore undertaken, with the aim of calibrating an existing dynamic model against a number of landslide cases from this region.

RECENT EVENTS IN THE CAMPANIA REGION
Geological and Geomorphological setting

The geology of the Campanian Apennines is characterised by the presence of calcareous and dolomitic monocline ridges, which derive from Plio-Pleistocene tectonic activity (Ippolito et al., 1975). During periods of volcanic activity of the Somma-Vesuvius and the Phlegraean Fields, air fall deposits mantled this rugged relief being deposited on the steep slopes of the bedrock surface.

In general, the pyroclastic mantle is of varying thickness up to several metres, that gradually decreases towards the top of the slopes. The air fall deposits in the date from volcanic eruptions of the last 22,000 years (Rolandi et al., 1998).

The thickness of the pyroclastic mantle, formed by airfall deposits and soil horizons, was controlled by the exposure in relation to the prevailing wind directions and by surface morphology and slope angles. The pyroclastic deposits covered surface irregularities, creating a smooth topography; although steps can be visible where limestone horizons form vertical faces with typical sub-vertical cornices (Guadagno, 1999).

The presence of pyroclastic deposits on slopes with angles as steep as 40-50 degrees is difficult to explain and points to a rather high friction angle for the material, although possible cementing and the presence of dense vegetation may also play a role.

The fertility of the pyroclastic soil has allowed the growth of dense vegetation on the slopes. The land has been used, since the times of the Romans, for the cultivation of chestnut and filbert nut trees. During the last few decades, many trackways have been created for access to these plantations, which have interrupted the continuity of the pyroclastic coverage of the steep slopes through deep cuts.

Events at Sarno-Quindici
On 5 – 6 May 1998, after prolonged rainfall, a large number of landslides occurred in the area of the towns of Sarno and Quindici, situated at or near the base of the slope of Pizzo d'Alvano. As much as 1 M m^3 of slope materials were displaced in the landslides which caused the death of 161 people. The landslides can be classified as debris avalanches and debris flows (Hungr et al., 2001). They initiate as shallow initial debris slides, but expand by incorporating material from the slope and water to become fluid debris avalanches. Some become channelized and develop into debris flows.

The initial movements took place in the steepest parts of the slopes, on surfaces inclined of 40°-50°, often at the heads of gullies; the location of the instabilities appear to have been controlled by the presence of cliffs and the artificial trackways crossing the slopes (Guadagno, 1999). Over 100 initial failures combined to form more than twenty long debris flows with sufficient mobility to reach built-up areas near the base of the slopes. Shortly after initiation, the landslides acquired considerable velocity, eroding pyroclastic horizons and colluvial soils (0.5-2 m thick), from the sides and base of the paths. As a result of the progressive increase in width, the upper reaches of many of the paths exhibit triangular shape in plan, typical of debris avalanche scars.

The materials appears to have been subjected to liquefaction by means of rapid undrained loading (cf. Johnson, 1984; Sassa et al., 1998). The channelized motion in mid-reaches of the slopes facilitated the incorporation of surface water. Some of the events were transformed into hyperconcentrated streamflow in the distal reaches of the deposition areas (Pierson and Costa, 1987).

Events at Cervinara

In December 1999, debris flows occurred on the northern face of the Mt. Cornito, in the Partenio Range, in the same geomorphological environment as the events of Sarno and Quindici. One event, in particular, was large enough to reach the village of Cervinara and cause the death of six people (Fiorillo et al., 2001). This event involved a volume of approximately 30.000 m^3 of material. The initial failure was a soil slide in the cut slope of a trackway, on surface inclined at about 40 degrees. The slide evolved into a debris avalanche, eroding the full thickness of the pyroclastic cover, and at the base of the slope was transformed into a high velocity debris flow.

TYPICAL BEHAVIOUR OF THE DEBRIS AVALANCHES AND FLOWS

From the above descriptions, the typical behaviour of debris flow and avalanches in this area can be summarized as follows:

i) the failures initiate as shallow translational slides on steep slopes, controlled by minor topographic details and man-made trackways. The slides appear to accelerate suddenly and incorporate material during motion down the slopes, often widening the path.

ii) at middle or lower slope levels, some of the avalanches enter confined gullies or steep stream channels, narrowing considerably, but still eroding the pyroclastic and colluvial cover from the slope. Velocities may reach up to 20 m/sec near the toe of the slopes, as deduced on the ground of observational characters and testimonies.

iii) on leaving the gullies, the flows spread out over depositional fans or aprons. They continue moving at relatively high velocities, destroying vegetation and houses, but neither eroding nor depositing. The velocities decrease in the distal parts of the paths and deposition occurs on gentle slopes. The deposits are often reworked by large quantities of flowing water and hyper-concentrated flows ("debris floods") develop downstream of the main deposition zones.

THE DYNAMIC MODEL

Numerical modelling of the debris avalanches has been carried out using "DAN" (Dynamic Analysis), developed by Hungr (1995), based on an explicit Lagrangian solution of the equations of unsteady non-uniform flow in a shallow open channel. The solution includes an open rheological kernel, so that a variety of rheologies can be implemented. The program has been modified in two ways: 1) the source volume is assumed to consist of a slab of constant depth; 2) downslope from the source area, the debris avalanche is assumed to erode a constant thickness of material from the path.

The rheological relationship selected for the analysis is the Voellmy (1955) model, modified by Hungr (1995), which consists of a frictional component and a turbulence parameter:

$$\tau = \gamma H (\cos\alpha + \frac{a_c}{g})\mu + \gamma \frac{v^2}{\xi}$$
(1)

Here, τ is the resisting stress at the base of the flow, α is the slope angle, γ is the unit weight of the flowing material (approx. 20 kN/m^3), H is the flow depth and μ the dynamic friction coefficient of the material, a_c is the centrifugal acceleration resulting from the vertical curvature of the flow path and ξ is a turbulence coefficient with dimensions of m/s^2. A justification for use of the Voellmy model for analysing debris avalanches has been discussed by Ayotte and Hungr (2000).

BACK-ANALYSES

All of the larger debris flows/avalanches from the two study areas (as show in Del Prete et al., 1998 and Fiorillo at al., 2001) have been selected for analysis, resulting in 15 cases (12 from Sarno-Quindici and 3 from Cervinara). In preparing the data, the flow paths have first been outlined, based on 1:5.000 scale topographic maps with 5 m contour interval, and on airphotos. Some of the paths had to be simplified by combining the widths of both tributary and distributary branches. This procedure contains an implicit assumption that the slides occurred instantaneously. While this in not likely to have been the case, the resulting errors are not serious, as one can consider that adjacent branches act as neighbouring stream channels. In any case, the simplification was necessary only for a few of the analyses. Where the distal margins of the true avalanches came to deposit in drop-shaped flow ends, the model was assumed to be able to travel on at constant width. For purposes of comparison, the actual toes of the debris avalanche deposits were assumed to lie at the distal limits of strong damage as interpreted from post-event aerial photographs. Areas covered by hyperconcentrated flow deposits downslope from the main deposits were not included in the runout measurements.

All of the initiating debris slides were assumed to have a uniform thickness of 1.5 m, measured perpendicular to the slope. The erosion depths downslope from the source areas was also assumed to be 1.5m. The downstream limit of the erosion zones was assumed to be placed at the toe of the slopes (slope less than 20°) and at the mouths of confined gullies or channels.

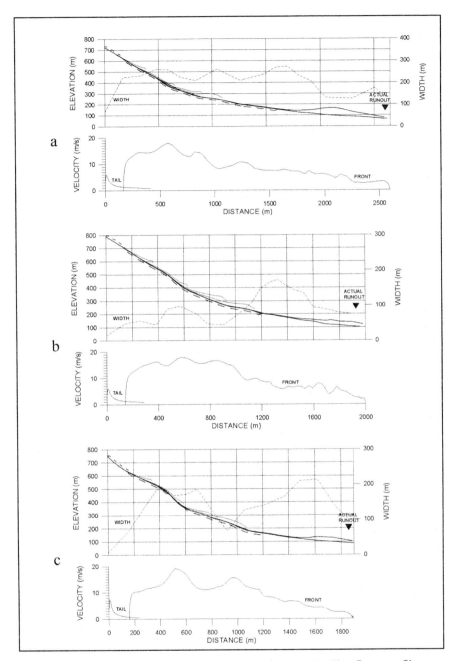

Figure 1. Three examples of back-analyses using DAN. The flow profiles are plotted at 10 second intervals. All normal depths (flow depths and erosion depths) are exaggerated 20 times. The letters a, b and c correspond to landslides 5, 6 and 7 in Fig. 3.

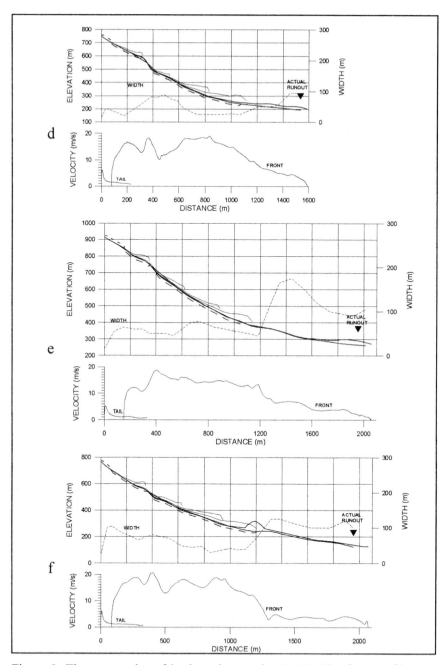

Figure 2. Three examples of back-analyses using DAN. The flow profiles are plotted at 10 second intervals. All normal depths (flow depths and erosion depths) are exaggerated 20 times. The letters d, e and f correspond to landslides 9, 10 and 14 in Fig. 3.

Systematic back analyses of the 15 cases were carried out. At first, a combination of rheological parameters was found (by trial and error), that provided good correspondence for most of the events, in terms of overall runout distance, velocity of motion and thickness of deposits. Examples of six individual analyses are shown in Figures 1 and 2.

Further detailed back-analyses were then carried out for those events which were not satisfactorily simulated by the constant parameters. For these, alternative friction coefficients were found to duplicate the observed actual runout distance. In this way, we obtain both a set of parameters capable of predicting "average " mobility, and limiting parameters applicable to some of the more mobile cases.

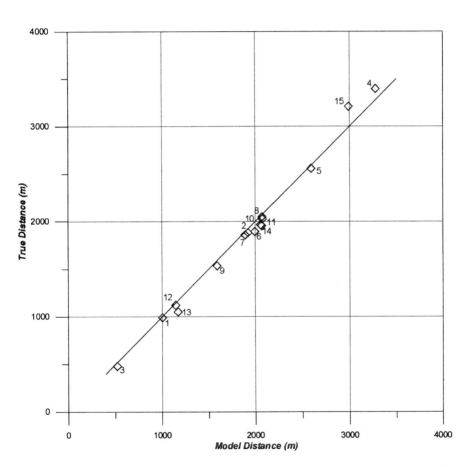

Figure 3. A comparison of simulated runout distances, obtained by DAN with a single pair of Voellmy parameters (friction coefficient μ = 0.07, turbulence coefficient ξ = 200 m/sec^2) and actual runout distances for the 15 cases analysed.

RESULTS AND CONCLUSIONS

By trial and error, it was determined that most of the landslides can be simulated using the Voellmy model, with a friction coefficient of 0.07 and a turbulence coefficient of 200 m/sec^2. All of the cases in Figure 1 and 2 have been analysed with the same pair of parameters. In all cases, maximum velocities of about 20 m/sec were obtained near the mouths of the confined part of the paths, as observed in the real cases. The model also satisfactorily simulated the lack of deposition in the proximal parts of the fans and the gradual deceleration and even distribution of deposits in the distal parts.

Figure 3 shows a summary of all 15 simulations, all of which have been obtained with the same pair of rheological parameters. In 14 out of the 15 cases, the model produced a maximum displacement within 100m of the true value, usually erring on the safe side. Serious under-prediction occurred only in one case (Slide 15). This debris flow followed an artificially reinforced concrete channel. The unusually strong degree of confinement, smoothness of the channel and lack of obstruction probably account for the unusual degree of mobility encountered in this case.

Figure 4 shows a histogram of friction angles that would be required to achieve a precise simulation of the observed displacements. The turbulence coefficient was kept constant at 200 m/sec^2 in all these cases. A value of the friction coefficient between 0.07 and 0.10 can provid precise simulation of the observed displacement for 80% of the cases.

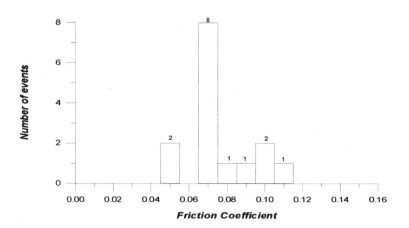

Figure 4. A histogram of friction coefficients required for exact simulation of the runout distances, with a turbulence coefficient $\xi = 200$ m/sec^2.

The simple dynamic model, as applied here, is capable of remarkably accurate simulation of the observed behaviour of the debris avalanches and debris flows, with closely constrained selection of input parameters. The model could be used to produce *quite reliable* first-order predictions of runout of potential slides, provided that the locations and sizes of the initiating failures and paths can be estimated. Such predictions could serve to map maximum potential runouts from specific landslide sources, as well as velocities and discharges required for the design of protective measures.

REFERENCES

Ayotte D. & Hungr O. 2000. Calibration of a runout prediction model for debris-flows and avalanches. Debris-Flows Hazards Mitiagation: Mechanics, Prediction, and Assessment, G.Wieczorek & Naeser (eds). Balkema, Rotterdam, pp. 505-514.

Del Prete M., Guadagno F.M., Hawkins B. 1998. Preliminary report on the landslides of 5 May 1998, Campania, southern Italy. Bull. Eng. Geol. Env.,57, pp. 113-129.

Fiorillo F., Guadagno F.M., Aquino S., De Blasio A. 2001. The December 1999 Cervinara landslides: further debris flows in the pyroclastic deposits of Campania (southern Italy). Bull. Eng. Geol. Env., 60 (3), pp 171-184 .

Guadagno F.M. 1991. Debris flows in the Campanian volcaniclastic soil. Slope Stability Engineering. Thomas Telford, London,1991, pp. 109-114.

Guadagno F.M. 1999. The landslides of 5th May 1998 in Campania, Southern Italy: natural disasters or also man-induced phenomena? Journal of Nepal Geological Society, 22, pp. 463-470.

Hungr O. 1995. A model for the runout analysis of rapid flow slides, debris flows, and avalanches – Can. Geotech. J. 32, pp. 610-623.

Hungr, O., Evans, S.G., Bovis, M., and Hutchinson, J.N. 2001. Review of the classification of landslides of the flow type. Environmental and Engineering Geoscience, VII, No. 3 (in press).

Ippolito F., D'Argenio B., Pescatore T.S., Scandone P. 1975. Structural-stratigraphic inits and tectonic framework of Southern Apennines. Geology of Italy. Tripoli, pp. 317-328.

Johnson A.M. 1984. Debris flows. In Brudsen D. and Prior D.E. (eds). Slope Instabbility. Wiley, London, pp. 257-361.

Pierson T.C., Costa J.E. 1987. A rheological classification of sub-aerial sediment-water flows. Geol. Soc. Am. Rev. Eng. Geol. 7, pp. 1-12.

Rolandi G., Bartollini F., Cozzolino G., Esposito N., Sannino D. 2000. Sull'origine delle coltri piroclastiche presenti sul versante occidentale del Pizzo d'Alvano (Sarno-Campania). Quaderni di Geologia Applicata, 7-1, pp. 37-47.

Sassa K. 1998. The mechanism starting liquefied landslides and debris flows. Proceedings, IV Int. Symp. On Landslides, Toronto, 2, pp. 349-354.

Voellmy, A. 1955. Uber die Zerstorungskraft von Lawmen. Schwerzerische Bauzertung, 73, 212-285.

Groundwater composition specified by weathering degree of the ground and trial estimation of the next occurrence of slope disaster

DR. H. SAKAI, Railway Technical Research Institute, Tokyo, Japan
DR. H. TARUMI, Railway Technical Research Institute, Tokyo, Japan, and
S. HIRAIWA, Kyushu Railway, Fukuoka, Japan

INTRODUCTION

The groundwater composition is basically dependent on the sort of layer that the groundwater passes through. Moreover, the concentration of inorganic ion contained in the groundwater would be changeable with the weathering degree of the layer. This is because the ion-exchange reaction continuously performed where groundwater and soil particles contact each other results in exchanging ions between them. In this case, the larger the surface of soil particles becomes, the faster the ion-exchange reaction must be. Therefore, the reaction rate is effectively controlled by the change in the area where groundwater comes into contact with the soil particles. This suggests that the ion concentration in groundwater where the ground is deeply weathered should be higher than that in the ground not weathered. Thus, it is easily estimated that the ions can indicate the weathering degree of the ground.

Based on this expectation, the compositions of groundwater collected at a number of spots located on the ground in the coast in Japan which simply consists of the same sort of layer or volcanic tuff were compared with one another. Additionally, the record of disasters taking place at the spots where groundwater was sampled was thoroughly looked into. As a result, it was found that the groundwater composition is satisfactorily classified. All the spots in the area investigated in this paper were generally equal to one another in the ion concentration except at some spots where the concentration is higher. More specifically, at the spots where the higher concentration appeared in groundwater, slope disasters, namely collapse or sliding of slope, occurred although the spots, whether their ion concentrations were high or not, are broadly and homogeneously covered with the same volcanic tuff. This fact means that the groundwater composition would be a useful indicator to express the weathering degree of ground.

On the other hand, a spot was found where no record on disasters was available but some inorganic ions with extremely higher concentrations were found in groundwater in the investigation. It would also be estimated that the spot has already become unstable but is not weathered well enough to collapse by falling into unbalance. When the observation of slopes at the spots, which would probably fall in the future, was under way with much expectation and great interest, the slopes thoroughly slipped as estimated in advance.

INVESTIGATION
Outline of the investigated site

The plain section of the area on which the work has been performed in this paper is illustrated in Figure 1. At this site, groundwater was sampled and the detailed record on disasters taking place on the slope was also carefully traced. The area is just on the coast located in Kyushu, Japan, and the spots, where the groundwater composition as well as the weathering state of the slope surface were observed, lined up right on the coast for the length of 10 km. Basically, this district including the spots is homogeneously, broadly, deeply, and simply covered with volcanic tuff as the surface layer, but the bed rock consists of two layers. Therefore, the upper layer, which is thin, is rhyolite, and lower pyroxene-andesite. Along the coast, the ground continuously extends without changes in the sort and combination of layers, and the gradient of the slope just looking the sea is 30 to 40 degrees. Right behind the slope, a plain widely extends at the elevation of 200 through 300 m.

This area has repeatedly suffered from a lot of disasters on the slope, especially in the season with much rainfall. To be remarked, significantly large-scale debris flows unexpectedly hit mainly the spots L and M in 1993, which are shown in Figure 1, thoroughly devastating

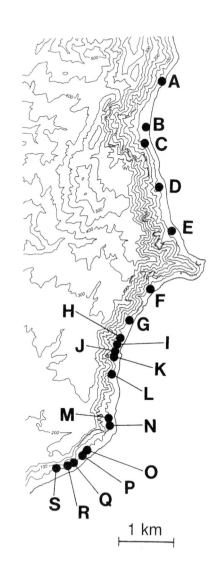

Figure 1   Plain section of the site in this work

railway tracks as well as a trunk line right below the slop just beside the coast.   On the other hand, quite a few other spots were furthermore attacked by a lot of debris flows at the same time leaving the slope slipped on a smaller scale.   Three severe storms repeatedly attacked the area with the recorded maximum hourly rainfalls of 70, 96, and 75 mm respectively during the first, second, and third wave.   Just for two months when rainfall waves visited over and over again, 2,176 mm of rainfall was recorded in total even though the average of rainfall observed in this area is only 2,237 mm.

Sampling of groundwater

As pointed out in Figure 1, the steep slope in the area faces to the sea with a large gap in elevation, which is a few hundred meters per approximately 500 m in the horizontal distance.   This means that it is expectable to easily find groundwater seep ing out at the foot of the plateau.   As a matter of fact, groundwater was actually available over there as spring water and surface water right after flowing underground.   Beside those categories of water, groundwater pumped up at wells is also obtainable.   At the end, in 1995, groundwater was collected at 19 spots found on the coast in the map given in Figure 1.   The information on the ground water as well as the manner to sample at all the spots is listed in Table 1.   Almost all of the samples of groundwater collected were regarded as fresh groundwater right after coming out from the ground even if groundwater at some spots would contain surface water.   This is because the slope is extremely steep with the gradient being

Table 1   Information on groundwater sampled and slipping of slopes

Spot	Feature of the place where groundwater was sampled[a]	Recent record on disasters
A	Drainage	
B	Dam	
C	Dam	Slipped in 1998
D	Water pipe	
E	Drainage	
F	Dam	
G	Drainage	
H	Dam	Already slipped
I	Dam	Already slipped
J	Natural spring	
K	Drainage	
L	Slope surface	
M	Dam	Already slipped
N	Dam	Already slipped
O	Well (60 m depth)	
P	Slope surface	Already slipped
Q	Natural spring	Already slipped
R	Well (70 m depth)	Already slipped
S	Rock slope	Slipped in 1999

[a] Sampled on August 7th and 8th, 1995.

at 30 to 40 degrees and the distance between the top and the foot of the slope being less than 500 m.   Moreover, no river, even brook and stream, runs or creeps down from the plateau on the surface of the slope in the area.   Therefore, it is exceptionally hard to expect foreign water such as surface water and rainwater on the slope and the edge of the plateau behind the slope.   This means that the samples are mainly groundwater without being mixed with surface water but just resulting in the fundamental characteristic of groundwater in chemical values, if the groundwater originated from the ground under the slope were a little bit diluted with foreign water.

Determination of chemical composition of groundwater
Groundwater samples were collected in 100 mL polyethylene bottles at all the spots indicated in Figure 1.   The bottles were immediately capped thoroughly with a middle stopper without leaving space inside the bottles by completely filling up with the groundwater to be sampled when collecting.   Moreover, the bottles capped further with screw stoppers outside were completely sealed with a paraffin-coated polyethylene film, and then kept in the dark and cool until when the samples were subjected to determination in a laboratory.   Right before the analysis of the samples, they were filtered with a membrane filter with the pore size of 0.45 μm.   After that, 10 μL of the sample was quantitatively injected into an ion chromatograph through an injection valve so that the ions contained in the samples can be determined with the peak areas given by the sample signals.   Thus, alkaline metals, alkaline earth metals, halides, and oxo-aions in the samples were accurately determined with the absolute working curves.

RESULTS
Groundwater composition
The concentrations of inorganic ions measured by taking the means described in the Investigation Section are illustrated in Figure 2.   As for single-charged cations, the concentrations of sodium in the groundwater samples collected at spots C, H, I, M, N, P, Q, and S are higher than those at other spots.   Furthermore, potassium concentrations in the samples at spots C, H, J, K, M, N, R, and S are also higher than those at other spots.   Regarding double-charged cations, magnesium concentrations in the groundwater samples obtained at spots C and M are higher than those at all spots except C and M.   In a similar manner, calcium concentrations in the samples at spots C, H, and M are higher than those at any spots excluding spots C, H, and M. At any rate, the concentrations of all the cations contained in groundwater collected at spots C, H, and M are at least absolutely higher than those at other spots (in the case of magnesium, the concentration only at spot H is not high).   Among them, the concentrations of all the cations at spot C are significantly higher.

Additionally, as to anions, chloride concentrations in the groundwater samples obtained at spots C, M, N, P, Q, and S are higher than those at other spots. Moreover, sulfate concentrations in the samples at spots C, M, N, P, Q, and S are

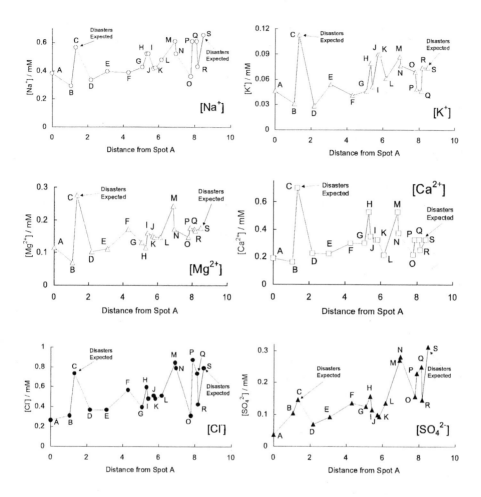

Figure 2   Concentrations of inorganic ions contained in the groundwater sampled at all the spots illustrated in Figure 1.

higher than those at other spots as well.   Shortly, the concentrations of anions contained in groundwater at spots C, M, N, P, and S are definitely higher than those at all other spots.

Slope stability
As listed in Table 1, the stabilities of the slopes close to or just on the spots where groundwater was collected as samples are shown.   It is not necessarily accurate and quantitative to determine the degree of weathering on the surface of the slopes by observing with naked eyes and touching by hands in a traditional manner.   So that the weathering degree of the slopes can semi-quantitatively be evaluated but simply classified, the stabilities of the slopes were sorted just into two stages dependent on

the progress of weathering, in other words, only already slipped or not yet slipped. In any case, slopes with the experience of slipping are to be regarded as those with lower stability. On the other hand, slopes without any experience of slipping are to be estimated as sturdy slopes or those still far from losing their balance. Thus, the slopes near spots H, I, M, N, P, Q, and R are, if not fragile, must be disturbed leaving the soil particles' structure decomposed or deformed. Nevertheless, the rest of the slopes located at the spots mentioned right above, that is to say, slopes near spots A, B, C, D, E, F, G, J, K, L, O, and S are still sturdy as of the date when the investigation was done in August 1995.

In the meantime, the slopes at spots H, I, M, N, P, Q, and R already slipped or slid before the groundwater composition was observed. However, no record on ground disasters are available with spots C and S. What is the difference in geotechnical characteristic between them is that the inorganic ion concentrations in groundwater at spots C and S are both still absolutely high without any experience of collapsing while the concentrations at spots H, I, M, N, P, Q, and R are also high with concrete records on slope disasters. Therefore, it is effortlessly expected that the slopes at spots C and S would be in the same condition as that sharing the slopes at spots H, I, M, N, P, Q, and R by comparing the groundwater compositions with one another.

By the by, the slopes close to spots C and S classified as slopes with the balance expectedly falling slipped in 1998 and 1999 respectively resulting in more than 15 m^3 of ground collapsing while periodical watch of the slopes had been taken with care even after the observation of the groundwater composition in the site in 1995.

DISCUSSION
What controls groundwater composition
Basically, when the ground constantly ranges with a simple structure without changes in the sorts of layers in a site, it can be estimated that the composition of groundwater seeping out from the site remains constant wherever the groundwater is collected throughout the site. Because, the same rainfall in the chemical composition is provided in the site in case the site area is not so large. However, the composition of groundwater usually depends on the spots where the groundwater is collected although the site surrounding the spots is constantly covered with the same layers.[1,2] Therefore, groundwater samples different in composition are commonly collected even in the same site with the same rainfall available in the composition. This fact suggests that the groundwater composition be contributed by the weathering degree of slopes where the groundwater passes. This is because there is nothing except the difference in the weathering degree of the slopes on the route of groundwater originated from the same rainwater.

As already reported,[2] the groundwater composition is extremely easily and drastically changed before and right when the ground is displaced, and some of inorganic ions contained in the groundwater coming out from the landslide area points out the ground behavior caused by landslide, namely ground displacement to be generated in the immediate future. In short, the changes in groundwater composition will effectively indicate the next occurrence of ground displacement with success actually prior to the event. What this means is that the micro-scale deformation of soil in the internal ground at deep places is generated to result in micro and local collapse. Thus, owing to the incident, the surface area of soil particles is immediately increased apparently, leaving the field to be subjected to ion-exchange reaction enlarged. On the other hand, ion-exchange reaction is continuously taking place on the surface of soil particles coming in contact with groundwater. Therefore, the numbers of ions exchanged on the surface of the soil particles just depend on the surface area under the ion-exchange reaction. Anyhow, the groundwater composition expressed with the concentrations of ions given through the ion-exchange reaction is controlled by the surface area of the soil particles in the ground. Furthermore, the area can be increased resulting from the deformation or displacement of the ground. Ended up with the event, the groundwater composition is governed by the particle sizes of the soil which is in fact reduced by the ground displacement when the ground gets well weathered losing the balance. As readily associated with, the soil particle sizes can easily be related to the weathering degree of the ground. Finally, the groundwater composition will be available as an indicator or a scale suggesting the progress in development of ground weathering.

CONCLUSION

The slopes at spots C and M slipped as expected in advance with the groundwater compositions which are similar to those at the spots with experience of slipping. As mentioned in Discussion Section, the groundwater composition is useful not only to evaluate the current weathering degree of the ground , in other words, the slope stability, but also to estimate the further occurrence of slope slipping as well as the progress in decreasing in the stability by weathering.

On way or the other, the knowledge of the balance of the ground which is obtainable from the groundwater composition gives practical suggestions on ground disasters. Especially in the instance when the ground simply extends with the same structure and sorts of layers, the groundwater composition will successfully bring beneficial information on the disasters to be caused in the future with brief work on the observation of groundwater composition. All the efforts to be done in completing the investigation above mentioned are only to take action to collect samples in plastic bottles for determination.

REFERENCES
(1) Sakai, H.; Murata, O.; Tarumi, T. *Proceedings of the 7th International Symposium on Landslides*; 1996; Vol. 2, p 867.
(2) Sakai, H.; Tarumi, H. *Proceedings of the 8th International Symposium on Landslides*; 2000; Vol. 3, p 1289.

The detailed hazard map of road slopes in Japan

Y. SASAKI, DR N. DOBREV, and DR Y. WAKIZAKA, Public Works Research Institute, Japan

INTRODUCTION
70% of Japan is mountains, and the land is surrounded in 30,000 km coastline. The geology belongs to the mobile belt, and there are many earthquakes and volcanoes. The climate belongs to the monsoon, and the heavy rain frequently arises. Therefore, Japan has suffered many slope disasters every year. The number of slope disasters in Japan shows about 900 cases per year with several hundreds lives lost.

The total extension of the roads in Japan is 973,838 km. The cost for the disaster prevention is enormous. Road administrator must select the most appropriate combination of hard and soft countermeasures. Hazard map becomes an important tool for estimating the cost and transmitting the risk of the road to administrators and users.

Especially, Detailed Hazard Map (DHM) aims to express the danger caused by small-size slope disasters like shallow landslides, rockslides, rockfalls. It is very important because such disasters occupy 67% from the all slope disasters in Japan, while debris flows and big landslides occupy only 14% and 19% respectively. As the shallow landslides are generally several-tens meters in width, the scale of DHM is required to be 1:5,000 or 1:2,500 at least.

In this paper, the hazard evaluation system of the road slope in Japan and the present state of the DHM are introduced first of all. And the next, the DHM for road slope by GIS are described as a case history.

STATE OF THE DETAILED HAZARD MAPS OF ROAD SLOPES IN JAPAN
There are various DHM of road slope disasters in Japan. We introduce three main DHM.
The detailed engineering geological maps have been made in a part of national road in Japan by road administration offices. The Disaster Prevention Geological Map (DPGM) is shown in Fig.1. The geology and landform interpretation data are layered over the topographical map in scale 1:5,000. Passage regulation zone in the heavy rain, past landslides and disasters, expected slope movement forms, and risk ranks by the inspection for road disaster prevention are described additionally.

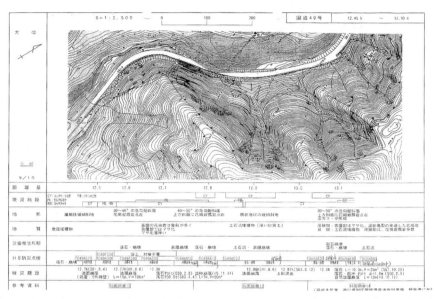

Fig.1 Disaster prevention geological map (Iwaki National Road Construction Office)

The Land Condition Map (LCM), in scale 1:5,000, is made by Geographical Survey Institute (Fig. 2). This map shows the geomorphic elements that are connected with the slope disaster along a road. Also, the map informs about the slope movement type, risk rank, and desirable planned investigations.

The Disaster Prevention Card (DPC, Fig.3) is made by the Road Disaster Prevention Inspection (RDPI) (Ministry of Construction, 1998). RDPI is the synthetic inspection that is done in every 5 years. 424,400 points were investigated in the last time in 1996, the DPC were made on 145,500 points that cannot construct the countermeasures immediately although the importance in the disaster prevention is high.

DPGM, LCM, and DPC are used as basic maps for the selection of the countermeasure techniques as well as reference

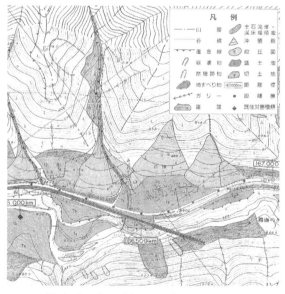

Fig.2 Land condition map (GSI)

materials of main monitoring position in the daily check. However, there are some problems. The first problem is that the data are rough and maldistributed. The detailed data have been limited to the very narrow region, and many disasters occur at points without data. The second problem is the lack of some basic data, for example, the detailed topographic map. Such data shortage is based on budgetary and technical limitation. The data, which must be continuously improved in future from long-term viewpoint, are the following: 1) Detailed topographic map (1:2,500 –1,000), 2) Detailed DEM (5~2m), 3) High precision aerial photograph (1:5,000 –2,500) and oblique photograph, 4) Aerial photograph interpretation map and landform interpretation map, 5) Past collapse data, 6) Geological data, topsoil depth, ground water, etc.

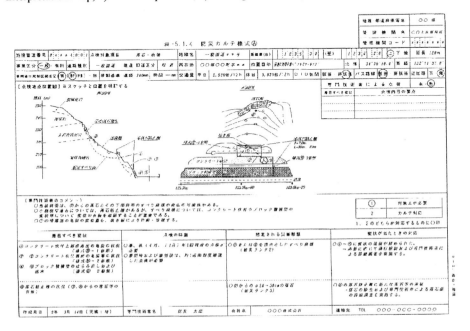

Fig.3 Disaster prevention card (Ministry of Construction,1998)

LEVEL OF THE DETAILED HAZARD MAP
Trying to resolve this data shortage in the future will take time and budget. So, it is necessary to construct hazard map in proportion to the improvement level of the data. The levels of the hazard map are as follows.
   Level 1: Inventory Map (Engineering geological Map etc.)
   Level 2: Susceptibility Map (Unstable Source Map and Fragility Map)
   Level 3: Hazard Map (in a narrow sense)
   Level 4: Risk Map.

Inventory Map- provides information on the physical-mechanical properties of soft and hard rocks, foundation conditions, groundwater conditions, topographical properties, and landslides.

Susceptibility Map- provides information on stability of slope. This map could divide into two levels. The first is Unstable Source Map that shows unstable zone and slope movement type such as rockfall. The second is Fragility Map that shows fragility or stability against expected specific triggers such as a rainfall that the amount is 300 mm/day.

Hazard Map- provides information on hazardous area by slope movements that include not only unstable source but also expected debris distribution.

Risk Map- provides information on a risk of disaster, which include economic and life risk.

In the road without basic data, it is the prior settlement to improve Inventory Map. It is the Digital Evaluation Model (DEM), landforms interpretation figure, geological maps, past collapse data, etc. to be important in Inventory Map. The Inventory Map has the function of the rough hazard map if results of the road disaster prevention inspection are added. The greater part of Japanese road is Inventory Map level.

In the road with DEM data and landform interpretation data, it is possible to make Susceptibility Map. The accuracy of Susceptibility Map depends on the availability of the past collapse data. The quantitative Fragility Map could be automatically described using the sufficient number of past collapse data and DEM. The Unstable Source Map is mainly made in the site that has no/insufficient past collapse data.

In the road with Susceptibility Map, it is possible to make Hazard Map. It is difficult to make Hazard Map automatically, because we have to estimate not only instability but also the expected size of landslide, expected debris distribution, debris volume on the road, effectiveness of countermeasure facilities, etc. Therefore, in present technology level, the preparation of Risk Map is also very difficult.

CASE STUDY
Here, some detailed Inventory Maps and Susceptibility Maps by GIS are introduced. These case studies are continuous at present. Hazard Map and Risk Map will be also made in future.

Detailed Inventory Map and Rockfall Hazard Map of National Road No.49
Detailed hazard mapping of a dangerous slope located near to Iwaki City, Fukushima Pref., has already begun as a detailed study on the dangerous slopes along National Road 49.

The geology of the slope is represented by Mesozoic granites and Quaternary loose deposits. The granites are fresh at the medium levels of the slope, and weathered at lower and higher levels. Below the watershed, granites are affected by spherical weathering. Some dangerous granite scarps are protected by retaining walls, anchors, metal rope net and sprayed concrete.

The Quaternary deposits are represented by topsoil, colluvium, deluvium and debris flow deposits. The *topsoil* is thin (less than 0.5 m) and it covers larger part of the slope area. The *colluvium* includes all fragmentary loose and incoherent materials, which are accumulated by rockfalling. These deposits are accumulated at foots of the scarps as well as at some sections of beds of the dales. *Debris flow deposits* fill the beds of dales containing different types of materials: mainly coarse fragments, organic soil, gruss and vegetation. At the upper levels of the slope, particle size of these materials decreases, and these materials are composed mainly by redeposited topsoil. The older Quaternary slope deposits, which cover the lowest oblique parts of the slope, we determine as *deluvium*.

Hazard phenomena
During the field studies, we established the following geohazard phenomena: rockfalls, slope failures, debris flows and soil creep.

The *rockfall phenomenon* is the most dangerous one. It affects an important part of the granite scarps. The fallen granite fragments create additional hazard of secondary (rolling) movement on the slope. We divided the rockfall deposits into recent and old. Recent ones include those that contain relatively fresh fragments. *Slope failures* are not largely distributed. Nominally, they are shallow, and their depth does not exceed 1 m. *Soil creep* affects the upper surface topsoil zone. Many inclined and curved trees as well as wavy soil surface can be seen over a large part of this area.

Rockfall hazard assessment
The rockfall hazard includes the site of rockfall source, the falling place, the transportation trace, and the runout line. In the world practice, the predictions of the possible path of a given rolling fragment is based on the local cases happened in the past (Evans & Hungr, 1993). We classified the rock scarps using two criteria: 1) availability of fresh traces of rockfall, and 2) main crack systems of each outcrop. We divided two types rock scarps: with high and low level of rockfall hazard. We accepted the high level includes fresh traces of rockfall phenomena and/or two and more dangerous crack systems; low – only one dangerous crack system or lack of such system and without any traces from previous rockfalling.

Rockfall hazard map
To compose the map of rockfall hazard, we applied the isopleth method of geohazard assessment already depicted by Wright et al. (1974), DeGraff & Canuti (1988), and others. The chosen method is based on the available cases and expresses the present stability slope conditions. This method is used mainly for landslide hazard maps. The aim of this map is to predict the locations that are most likely to be affected by rockfall occurrence, and to be outlined. The method represents the percentage of affected area by slope movements. We used an inventory map in scale 1:2,500. The final map is in the same scale. Its reduced copy is shown in Fig.4. In our calculations, we involved the data of dangerous scarps in granite outcrops, and the rockfall deposit areas. Old colluvial talus deposits in the Southern part of the studied area were omitted due to the protected outcrop above them.

Geohazard mapping through GIS
The field work data was digitalized through ArcView GIS module. That gives us the possibility to apply various models for estimation of hazard probability including the isopleth

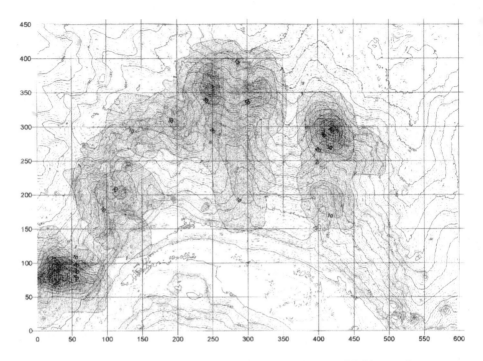

Fig.4 Rockfall hazard map. The map represents probability to rockfall impact in percentage steps 1%, 10%, 20%, 30%, 40%, and 50%, according to DeGraf & Canuti (1988)

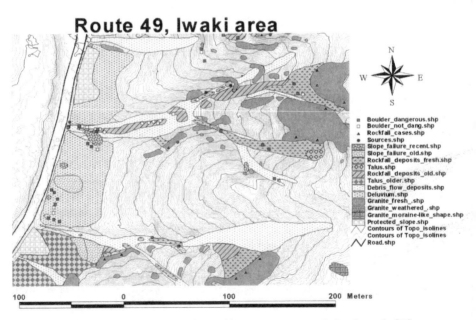

Fig.5 A detail of the geohazard inventory map made by through GIS.

method.   We use a topographic digital (vector) base map in scale 1:1,000.   Now   whole obtained field mapping data is included and an inventory hazard map is made (Fig. 5). Additional data, which are related to the calculation of slope stability as the soil test data, are planned to be established and included.

Detailed Fragility Map of Hiroshima

Fragility was estimated and mapped by multiple regression analysis using topographic data in the 1999 Hiroshima rainstorm disaster.

Method

Explanatory variables are micro-topography division, i.e. crest slope, valley-head slope, side slope, watershed slope, foot-slope, convex break of slope, terrace, etc. by aerial photograph, and topographic data by 10 m DEM, i.e. a slope azimuth, altitude, inclination, convexity ratio, and relative height.   The purpose variable is collapse/no-collapse in the 1999 Hiroshima rainstorm disaster.

Result

The relationship between estimated value by regression formula and collapse rate was compared, and the expected collapse rate was mapped (Fig.6). This map also means fragility map at a future heavy rain that is similar to the 1999 Hiroshima rainstorm. The high correlation was recognized between this rate and convexity ratio, inclination, landform division, azimuths, distances from convex break line. While,

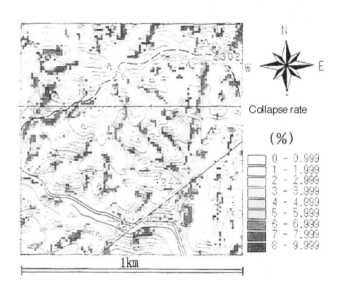

Collapse rate

(%)

```
0 - 0.999
1 - 1.999
2 - 2.999
3 - 3.999
4 - 4.999
5 - 5.999
6 - 6.999
7 - 7.999
8 - 9.999
```

1km

Fig.6   Expected collapse rate map (fragility map)
          in the future rainfall that is similar to the 1999 Hiroshima
          Rainstorm.

the correlation was low for altitude. The maximum collapse rate was about 8%, which distributed mainly in valley-head slope. The maximum collapse rate in the relationship between collapse rate and slope inclination was only 1.3% at over 50 degrees,   and those of other explanatory variables were also similar levels. Consequently, the multiple regression analysis enables to heighten about 6 times at the estimation accuracy of the fragility.

Field confirmation of topsoil depth

Though the collapse rate was improved by the multiple regression analysis, it was about 8% at most. The reason is that the data is only the landform. Ideally, 100% of collapse rate is required to the hazard map. Then we confirmed the effect of topsoil in the field, because topsoil depth is another important factor for shallow landslides. Topsoil depth distribution was measured in steep slopes by a rod penetration test that was developed originally.

The result showed shallow landslides in the 1999 Hiroshima rainstorm have occurred only in the area which topsoil was deep. This fact means the accuracy of the fragility map would improve if the topsoil depth is added in explanatory variables.

CONCLUSION

We introduced various DHM for the disaster prevention of road slopes in Japan. Though the basic data are insufficient, we tried to develop the quantitative hazard map, which enable to inform road administrators and users of the risk certainly and to help administrators to make suitable plan of countermeasures for each road in association with users and researchers.

REFERENCES

DeGraff, J.V., P.Canuti. 1988. Using isopleth mapping to evaluate landslide activity in relation to agricultural practices. Bull. IAEG, No.38, Paris: 61-70.

Evans, S.G., O.Hungr. 1993. The assessment of rockfall hazard at the base of talus slopes. Can. Geotechnics J. 30, 620-636.

Ministry of Construction of Japan. 1998. Manual for filling and the application of disaster prevention card, 47 (in Japanese).

Wright, R., R.Campbell, T.Nilson. 1974. Preparation and use of isopleth maps of landslide deposits. Geology, 2, 483-485.

Risk analysis of potentially damaging rainfall events

ORESTE TERRANOVA, CNR-IRPI, Via Cavour, 87030 Rende (CS), Italy,
E-mail terranov@irpi.cs.cnr.it

ABSTRACT AND PREMISES
Rainfall is a natural phenomena that often, directly or indirectly, causes considerable damages to human activities. Damage extent and frequency are related to rainfall amount totalized in a time depending on the physical environment and physical process that we are considering.

With regard to landslides the time concerned depends, apart from the influence of other factors, on the thickness of the materials that are involved, or that are possibly involved; in particular deep landslides can be caused by remarkably high rainfall values accumulated in periods that are often superior to the hydrologic year. In the same way highly intense rainfall can cause superficial flows in periods as short as one hour or some fraction of an hour.

Even though in the last decades there have been many contributions to the study of the relationship between rainfall and landslides [Campbell, 1975; Govi & Sorzana, 1980; Cancelli & Nova, 1985; Cascini & Versace, 1988; Clarizia et al., 1998], applying instruments or an exhaustive reorganisation of the subject are not available yet. The complexity of the analysed process, the lack and heterogeneity of direct observations and the variety of climatic contexts make the study hard to progress. Then it is necessary here to use on one hand generally available data and on the other to limit enquires to the single landslide typologies.

This essay describes a completely general statistic-probabilistic method, even though it is referred exclusively to superficial landslides, that enables us to locate potentially damaging rainfalls through the study of historical series of daily rainfalls. An application of the methodology is illustrated for the Southern Ionic area of Calabria that has become more interesting following the storm that occurred on 8th-10th September 2000 [SIMI, 2000]. The storm culminated in the tragic destruction of the camping area situated in the low flow riverbed of the torrent Beltrame. Moreover a comparison between this last event and potentially graver past situations is given here.

METHODOLOGIC ASPECTS AND PRELIMINARY STATEMENTS

From a generic historical series of daily rainfalls, sequences of consecutive not null rainfall values can be gathered. Each sequence, necessarily preceded and followed by at least one null value, is assumed to be a single pluviometrical event without more precise meteorological observations.

Pluviometrical conditions can be resumed in two parameter typologies related to their capacity to cause superficial landslides. The first typology is particularly related to the generic event as it has been identified -it tries to resume magnitude and shape and it has a bigger and more direct influence on landslides trigger. The second typology tries to define the antecedent state of soil imbibition through the quantity of rain that has fallen before the event as well.

Some parameters, between all the possible parameters that could be ascribed to the two typologies, are shown below:

First typology	Second typology
a) Total rain P_{ev} b) Maximum intensity I_{max} c) Medium intensity I_{med} d) Duration D_{ev} e) Position of peak f) Ratio $r = I_{med} / I_{max}$	α) Rain Σ_k totalized in k days before rain peak β) Time lapse Δ_{ev-ev} between the event under discussion and the previous event χ) Total rain Π_{ev-pr} of the event previous to the examined one.

Firstly we have examined just P_{ev}, I_{max}, I_{med}. parameters of the first typology for whose statistic-probabilistic treatment is used a simplified "Peaks over threshold" method that is included in the rare events theory, usually used to define the connection between the number of events and their probability beginning from Poisson researches. Main interest in fact is given to the knowledge of the number of events that exceed established threshold values, relatively to the values assumed from a certain parameter. In short the risk R, or probability to exceed, that the generic parameter x gets over S has been studied in an empirical way changing the value of S threshold in the relationship:

$$R(x>S) = 1 - P(x \leq S) = 1 - \int_{-\infty}^{s} p(x)\,dx\,,$$

This study has been made counting the events exceeding S and avoiding to use functions of theoretic probability distribution to interpret their magnitude series.

The methodology, shortly illustrated in the following pages, is generally interesting from an hydrological point of view as it's useful for various natural phenomena and some of the analysed aspects belong to hydrological sector of great importance. This methodology is new though then, to avoid loads from probabilistic implications that anyway we should use, in this stage we preferred to use empirical relationships.

The specific aim is to underline the peculiar severity of rainfalls as potential trigger of superficial landslides through, at the beginning, P_{tot}, I_{max}, I_{med} parameters and then through the other parameters.

The elaboration has been made turning rainfall and threshold values into adimensional on the basis of the normal rainy day value (GPN) of each rain gauge station. This value is given by the ratio between medium annual rainfalls and medium annual number of rainy days. We obtain in this way an improvement in comparability between different stations during the elaboration.

Later, assumed a conveniently acceptable risk level, or the relative return period T=1/R, the events presenting parameters values superior to this risk threshold are identified. The number of these events depends on the accepted risk and consequently is not too high for practical application cases. Events that are not important for the research are rejected while a detailed exam can be conducted on a few events left.

Another selection can be made to identify events characterised by simultaneous overcoming of the risk prefixed for more parameters. This kind of events is called Severe Events (SevEv) compared to P_{ev}, I_{max}, to P_{ev}, I_{med}, to I_{max}, I_{med} or, more, to P_{tot}, I_{max}, I_{med}. Obviously the return period of these events is superior to the prefixed one and it is unknown, as the three variables P_{tot}, I_{max}, I_{med} are not mutually independent from each other. It is right to talk about a number N of presence of this kind of SevEv in the number N of examined years.

APPLICATION
As we said this work is illustrated with reference to the area called Southern Ionic Calabria (fig.1). In this area the rainfall system is typically seasonal with a low number of rainfall episodes concentrated between the months of October and March

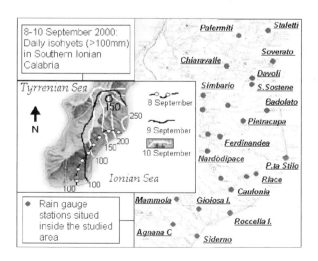

Fig. 1- Area interested by the 8-10 September 2000 event and location of the rain gauge stations

and with medium yearly rainfall, from 650 to 1700 mm more or less, that increase with the raising of altitude from sea level (fig.2). Soil geotechnic characteristics and geo-lithological and morphological conditions that we find make this area ideal for frequent superficial landslides phenomena in large areas; these aspects are better described in the researches made in this area by Cosenza CNR-IRPI's researchers after 8th–10th September ?000's event [Antronico et al., 2001].

In the interested area 24 daily rainfall series are available (fig.1 and tab.I) with a duration from 10 to 76 years located on variable altitudes from 5 to 1420 m slm.

Once the pluviometrical series are turned into adimensional with respect to GPN local values (tab.I), we have extracted the single pluviometrical events to calculate P_{ev}, I_{max}, I_{med} parameters to confront them with threshold values related to GPN too; particularly we assumed as P_{ev} threshold values variable from 0 to 35·GPN, as Imax threshold values variable from 0 to 25·GPN and as Imed threshold values variable from 0 to 10·GPN.

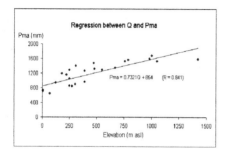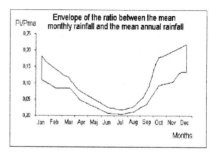

Fig.2 – Simple regional relationships: a) P_{ma} as a function of elevation; b) mean monthly rainfall (P_i) as a function of P_{ma} and of the month

The number of events that exceed threshold values is illustrated on Gunbel's cartogram on fig.3 together with relative frequencies. The number of events is related to the number of years of observation of each series. From the graphics we can see that with the raise of threshold we have the typical and well known "dog-leg effect" or "separation phenomenon" [Potter, 1958; Matalas et al, 1975, Rossi et al., 1984] due to "outlier" that states the low adaptability of the exponential probabilistic law, if not in the case of high threshold values [Madsen et al., 1995]. For each parameter under discussion anyway the shapes of the curves related to the different stations are very similar and, to interpret this shape, have been proposed [Rosbjerg, 1987; Cruise & Arora, 1990] probabilistic laws as GEV law, generalised Pareto's law, Weibull's law, etc.

Tab. I – Rain gauges characteristics and threshold values

STAZIONE	2000 Year	Elevation m s.l.m.	N	P_{ma} (mm)	NGP	GPN (mm)	S_1	S_2	S_3
AGNANA CALABRA		180	65	1188,9	103,2	11,5	322,6	138,2	69,1
BADOLATO		250	38	1052,9	97,8	10,8	301,4	129,2	64,6
CAULONIA	Y	275	56	847,8	92,4	9,2	256,9	110,1	55,1
CHIARAVALLE C.	Y	550	75	1285,5	120,9	10,6	297,7	127,6	63,8
DAVOLI		390	10	1266	77,9	16,3	455,0	195,0	97,5
FERDINANDEA	Y	1050	31	1531	143,5	10,7	298,7	128,0	64,0
GIOIOSA IONICA	Y	125	65	936,1	110,3	8,5	237,6	101,8	50,9
MAMMOLA		250	56	1279,6	114,7	11,2	312,4	133,9	66,9
MAMMONE		981	21	1588,7	138,7	11,5	320,7	137,5	68,7
MONTE PECORARO		1420	10	1581	77,9	20,3	568,3	243,5	121,8
NARDODIPACE		670	57	1342,8	87,3	15,4	430,7	184,6	92,3
PALERMITI	Y	480	67	1313,9	111,8	11,8	329,1	141,0	70,5
PIETRACUPA		1000	12	1682,4	120,4	14,0	391,3	167,7	83,8
PLACANICA		250	45	854,6	97,3	8,8	245,9	105,4	52,7
PUNTA STILO	Y	70	68	638,9	79,2	8,1	225,9	96,8	48,4
RIACE		304	55	902,2	96,3	9,4	262,3	112,4	56,2
ROCCELLA IONICA	Y	5	48	732,2	93,6	7,8	219,0	93,9	46,9
S. NICOLA DI CAUL.		225	22	1153,8	98,4	11,7	328,3	140,7	70,4
SAN SOSTENE	Y	475	43	1480,6	124	11,9	334,3	143,3	71,6
SERRA SAN BRUNO	Y	790	76	1564,7	155,3	10,1	282,1	120,9	60,5
SIDERNO MARINA		7	49	708,2	97,1	7,3	204,2	87,5	43,8
SIMBARIO		760	51	1524,6	117,1	13,0	364,5	156,2	78,1
SOVERATO MARINA	Y	6	49	735,7	96,2	7,6	214,1	91,8	45,9
STALETTI'		390	47	958,5	91,1	10,5	294,6	126,3	63,1

N.B.: Y = lenght (in years) of historic records; P_{ma} = mean annual rainfall; NGP = mean annual number of rainy days; GPN = normal rainy day; S_i = threshold values for the generic i parameter

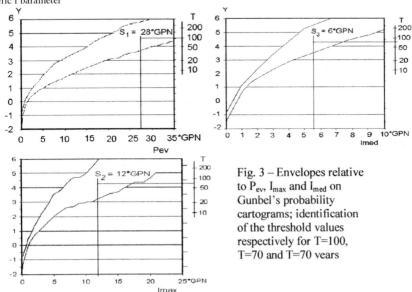

Fig. 3 – Envelopes relative to P_{ev}, I_{max} and I_{med} on Gunbel's probability cartograms; identification of the threshold values respectively for T=100, T=70 and T=70 years

In this phase of the work we do not consider this important matter because the actual interest is to calculate threshold values of P_{ev}, I_{max}, I_{med} parameters related to an established risk R. Thresholds determination can be practical made in a graphic way; the thresholds, symbolically indicated as $S_1 = S(P_{ev}, R_1)$, $S_2 = S(I_{max}, R_2)$, $S_3 = S(I_{med}, R_3)$, are dependent only by the studied parameter and by the fixed risk. The events exceeding S_i (i=1,2,3) can then be identified. These events are later on named $SevEv_1$, $SevEv_2$, $SevEv_3$; their number is reported in tab. II, for each rain gauge station, for approximatively: T = 100 corresponding to $S_1 = 28 \cdot GPN$, T= 70 corresponding to $S_2 = 12 \cdot GPN$ and T=70 corresponding to $S_2 = 6 \cdot GPN$.

Tab. II – Number of severe events exceeding one ($N°S_i$) or two ($N°S_{i,j}$) or three ($N°S_{1,2,3}$) threshold conditions

STAZIONE	S_1	S_2	S_3	$N°S_1$	$N°S_2$	$N°S_3$	$N°S_{1,2}$	$N°S_{1,3}$	$N°S_{2,3}$	$N°S_{1,2,3}$
AGNANA CALABR	322,6	138,2	69,1	10	11	26	5	4	4	2
BADOLATO	301,4	129,2	64,6	7	12	16	6	5	10	5
CAULONIA	256,9	110,1	55,1	5	12	17	4	3	6	3
CHIARAVALLE C.	297,7	127,6	63,8	31	46	26	19	9	14	9
DAVOLI	455,0	195,0	97,5	1	1	2	1	0	0	0
FERDINANDEA	298,7	128,0	64,0	12	13	7	8	4	6	4
GIOIOSA IONICA	237,6	101,8	50,9	10	24	22	8	7	13	7
MAMMOLA	312,4	133,9	66,9	9	16	15	6	3	9	3
MAMMONE	320,7	137,5	68,7	11	24	16	10	6	9	6
MONTE PECORAR	568,3	243,5	121,8	0	0	0	0	0	0	0
NARDODIPACE	430,7	184,6	92,3	4	3	12	1	2	0	0
PALERMITI	329,1	141,0	70,5	8	26	27	5	3	13	3
PIETRACUPA	391,3	1677	83,8	6	10	3	6	1	2	1
PLACANICA	245,9	105,4	52,7	3	12	11	1	2	4	1
PUNTA STILO	225,9	96,8	48,4	7	15	18	4	3	9	3
RIACE	262,3	112,4	56,2	6	17	28	5	5	11	5
ROCCELLA IONIC	219,0	93,9	46,9	7	14	21	5	5	10	4
S. NICOLA DI CAU	328,3	140,7	70,4	3	6	4	1	2	2	1
SAN SOSTENE	334,3	143,3	71,6	13	23	12	8	5	8	4
SERRA SAN BRU	282,1	120,9	60,5	44	50	21	26	6	12	6
SIDERNO MARIN/	204,2	87,5	43,8	7	19	20	7	5	11	5
SIMBARIO	364,5	156,2	78,1	4	9	11	4	3	5	3
SOVERATO MARI	214,1	91,8	45,9	6	17	15	5	4	8	4
STALETTI'	294,6	126,3	63,1	5	6	7	3	0	2	0

Particular attention for superficial landslides trigger has to be given to events that belong to two $SevEv_i$ sets at the same time, or to all the three sets of $SevEv_i$, for i=1, 2, 3; in the first case the corresponding events have been indicated as $SevEv_{1,2}$, $SevEv_{1,3}$ or $SevEv_{2,3}$, in the second as $SevEv_{1,2,3}$ (tab. II).

Adopting the shown method and criteria for the series for which are known the data relative to the year 2000, it resulted that the event of 8th-10th September 2000 belongs to $SevEv_{1,2,3}$. In the tab. III are reassumed the typical P_{ev}, I_{max}, I_{med}

parameters of the events that overcome those relative to the event of 8th–10th September 2000. Moreover for this few events are shown the remaining parameters too.

We observed that the 8th –10th September 2000 event has assumed considerable gravity characteristics, as it belongs to $SevEv_{1,2,3}$ set and it has a risk corresponding to a return time superior to 50-100 years for every single P_{ev}, I_{max}, I_{med} parameter. We have to consider that the risk that a generic event belongs to SevEv1,2,3 set is not known even though it is obviously bigger or the same as the established risk; in fact the examined parameters are not mutually independent statistic variables. In conclusion, even though we can identify the number of $SevEv_{1,2,3}$ related to the number of years taken as sample, we need more researches about this statistic dependence structure.

Tab. III – Parameters of the severe events heavier than the 8-10 September 2000 one

STAZIONE	Year	Day	P_{ev}	I_{max}	I_{med}	D_{ev}	P_p	r	Σ_{10}	Σ_{20}	Σ_{30}	$\varnothing_{ev-ev}$	Π_{ev-pr}
CHIARAVALL	2000	252	562,2	250,2	140,6	4	3	0,56	561,8	561,8	561,8	2	0,2
	1951	288	1046,0	436,1	174,3	6	4	0,40	1180,0	1206,0	1253,4	4	13,8
PALERMITI	2000	252	306,6	114,2	102,2	3	2	0,89	225,6	225,6	230,8	2	0,2
	1935	327	348,6	332,2	116,2	3	1	0,35	1006,8	1025,0	1404,7	3	46,6
	1972	356	611,5	270,8	122,3	5	2	0,45	411,3	442,0	533,4	4	108,0
	1972	364	565,5	433,4	80,8	7	5	0,19	871,4	1283	1293	2	2,7
S. SOSTENE	2000	252	449,2	241,3	112,3	4	3	0,47	483,0	483,0	483,0	3	11,2
	1951	289	1195,5	417,1	239,1	5	3	0,57	967	1159	1179,2	2	2,6

Interesting observations can be made from the location of SevEv in the hydrological year, in fact we can see (fig.4) that they happen almost exclusively in the months from October to December and then the 8th–10th September 2000's event is "anomalous" for this aspect too.

In the end, using the frequent contemporaneity for the SevEv happening in the 18 stations, even with different importance for P_{ev}, I_{max}, I_{med} parameters, we can draw up spatio-temporal evolution maps of each SevEv and then verity the presence of eventual similarities between the graver past situations.

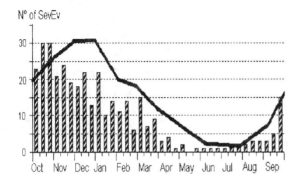

Fig. 4 – Distribution in time of $SevEv_i$ (i=1,2,3) number (bars) compared with the qualitative pattern of rainfall averaged on the whole area (black line)

CONCLUSIONS AND RESEARCH PROSPECTS

A method has been proposed to select pluviometrical events that assume bigger superficial landslides trigger potentialities on the basis of the most suitable parameters. Possible applications of this method are clear both in forecasting model ambits and alarm systems in real-time, both to other natural erosive process phenomena and floods in basins of extents proportional to the daily time scale.

The illustrated proposals enable us to adopt, in an objective way, threshold values of the characteristic parameters of pluviometrical events on the basis of risk levels prefixed at will.

A regional approach to pluviometrical events analysis has been suggested, on one hand with regard to the magnitude expressed by P_{ev}, I_{max}, I_{med} parameters, on the other with regard to the number of events. We can find a lot of common points with the study of the maximum yearly values of high intensity and short duration rainfalls.

From fig.3's graphic we can observe that with the raising of the value on the abscissa there's an increase in dispersion of the patterns parameter-frequency relative to each rain gauge station. This bigger dispersion is due to increased doubts caused by the diminution of the dimension taken as sample as far as threshold values increase. To reduce this uncertainty more precise regional studies are necessary to explain the variability from station to station.

It has been observed that SevEv regards more particularly a limited period of rainy months (fig. 4).

The illustrated application has permitted underlining of the gravity characteristics of 8th–10th September 2000's event in relation to the past ones that have produced more damage. In particular (tab. III) just three rain gauge stations have recorded three events, relative to one or more of them, heavier than the 8th-10th September 2000's one.

Possible developments basically regard: 1) the extension of this study to other territorial ambits, 2) the arrangement of relations in a regional kind of analysis, 3) the determination of the dependence structure between the random variables and 4) the identification of their theoretic frequency distributions.

REFERENCES CITED

ANTRONICO L., GULLÀ G. & TERRANOVA O. (2001) *Dissesti sui versanti e processi in alveo in due aree della Calabria ionica meridionale a seguito dell'evento pluviometrico dell'8-10 settembre 2000.* Proc. of the Conf. "Il dissesto idrogeologico: inventario e prospettive". Acc. Naz. dei Lincei, 5th June. Rome (Italy)

CASCINI L. & VERSACE P. (1988) *Relationship between rainfall and lanslide in a gneissic cover.* Proc. 5th Int. Symp. On Landslide. Lausanne, 10th-15th July 1988. 1:565-570

CANCELLI, A. & NOVA, R. (1985) *Landslides in a soil debris cover triggered by rainstorms in Valtellina (central Alps – Italy).* Proc. 4th Int. Conf. And Field Workshop on landslides, Tokyo, 1985: 267-272

CAMPBELL, R. H. (1975) *Soil slips, debris flows, and rainstorms in the Santa Monica Mountains and vicinity, Southern California.* U. S. Geol. Survey Prof. Paper 851

CLARIZIA M., GULLÀ, G. & SORBINO, G. (1988) *Sui meccanismi di innesco dei "soil slips".* Proc. Int. Conf. Prevention of hydrological hazard: the role of scientific research. Alba (CN), 5th-7th Nov. 1996, Torino, Italy, 585-587

CRUISE, J.F. & ARORA, K. (1990) *A hydroclimatic application strategy for the Poisson partial duration model.* Water Resour. Bull. 26(3), 431-442

GOVI,M. & SORZANA, P.F. (1980) *Landslide susceptibility as a function of critical rainfall amount in Piedmont basin (North-Western Italy).* Studia Geomorph. Carpatho-Balcanica, 14, 43-61

MADSEN, H., ROSBJERG, D. & HARREMOËS, P. (1995) *Application of Bayesian approach in regional analisys of extreme rainfall.* Stoch. Hydrol. and Hydraul., 9(1), 77-88

MATALAS, N.C., SLACK, J.R. & WALLIS, J.R. (1975) *Regional skew in search of a parent.* Wat. Resour. Res., 11(6), 815-826

POTTER, W.D. (1958) *Upper and lower frequency curves for peak rates of runoff.* EOS. Trans. AGU, 39, 100-105

SIMI (2000) *Rapporto sull'evento dell'8-10 settembre 2000. Caratteristiche pluviometriche dell'evento.* Serv. Idr. e Mareogr. Italiano – Sezione di Catanzaro

ROSSI, F., FIORENTINO, M. & VERSACE, P. (1984) *Two component estreme value distribution for flood frequency analisys.* Water Resour. Res., 20(7), 847-856

ROSBJERG, D. (1987) *Partial duration series with log-normal distributed peak values.* In: Hydrologic frequency modeling, Edited by V.P. Sing, 117-129

An example of risk assessment from British Columbia, Canada

D.F. VANDINE, VanDine Geological Engineering Limited, Victoria, BC, Canada
P. JORDAN, Ministry of Forests, Nelson, BC, Canada
D.C. BOYER, Ministry of Water, Land and Air Protection, Nelson, BC, Canada

INTRODUCTION

Perry Ridge is located in the West Kootenay region of British Columbia (BC). The relatively flat-topped ridge is approximately 25 km long, 9 km wide and ranges in elevation from 500 m to 2100 m. Most of its uplands is Crown land. Privately owned land extends along the base of, and in places part way up, the east side of Perry Ridge. Much of the private land has been cleared for pasture or cultivation, or has been logged, and numerous residences, with associated surface water supply intakes, are located along the base of the ridge.

Most forest land in BC is owned by the Crown, and the BC Ministry of Forests (MOF) manages the forest resource and licenses it to forest companies. MOF plans to harvest timber from a portion of the Crown land on the ridge. Local residents are concerned about the resulting risks to life and limb, property, and water supply from slope instability and changes in the hydrologic regime and as a result of logging and associated logging roads.

MOF has undertaken a comprehensive planning process for forest development on Perry Ridge that significantly exceeded the requirements of the current BC Forest Practices Code (the Code). This planning process included several mapping projects which addressed geologic and hydrologic hazards, both on the ridge and the adjacent populated valley bottom. The authors were retained to use these hazard studies, and other information, to carry out an overview geologic and hydrologic assessment of existing risks to life and limb, property and water supply, and anticipated risks from road building and logging on Crown land as proposed in a Total Chance Plan (TCP) – a hypothetical forest engineering plan. The resulting document was the Perry Ridge Risk Assessment (Boyer, Jordan and VanDine, 1999).

This is the first time in BC that such an overview risk assessment has been carried out as part of the forest development planning process. The purpose of this paper is not to present the results of the risk assessment, but to describe the method.

In the past two decades, a number of natural hazard risk assessments have been carried out in BC, primarily for transportation, hydroelectric development and municipal planning. The theoretical basis of these studies, and some examples, were reviewed by Morgan (1991), Morgan et al (1992) and Cave (1992). For this assessment, the authors attempted to use definitions consistent with past studies, with due consideration to the variety of hazards involved and the differences in scale of the past studies.

Risk is the product of the probability, or likelihood, of a hazardous event occurring, and consequence, or effect or potential effect, of that event. Consequence is a combination of the elements at risk and the severity of the hazardous event on those elements at risk. For hazardous events that occur frequently in the same location, such as floods and snow avalanches, and where historical data exists, quantitative risk assessments may be appropriate. For most forest development planning purposes, however, hazardous events are relatively infrequent and/or the historical data is lacking. To accommodate this difficulty, the authors carried out a "consensual qualitative risk assessment" based on available information and empirical evidence, combined with the experience and judgement of the authors. The consensual aspect of this risk assessment reduced biases and provided a balance of three individual's qualitative judgements.

BACKGROUND DATA AND METHODOLOGY
The study area covers approximately 76 km^2, composed of 69 km^2 of Crown land and 7 km^2 of private land along the east side of Perry Ridge.

No forest development has taken place on Crown land although construction of an access road has begun. Most of the ridge was burned by large fires 80-100 years ago, and timber in the burned areas is now nearing maturity. The area has a moist climate and is heavily forested. Most creeks flow year-round and are dominated by spring snowmelt.

In general the level of geomorphic activity on most of the ridge is low. Landslides and other sediment sources are scarce. Many creeks, however, are deeply incised in bedrock, and their channels and fans have debris flow deposits. In the valley bottom, there are extensive and deep glacial deposits that are subject to a variety of potentially hazardous processes, including slumps, slides, and sinkholes.

Hydrologic Units/Zones of Influence/Forest Development Areas/ECA Index
To relate the risks to the geography, the study area was divided into 16 watershed units and 16 face units between the watersheds, a total of 32 *hydrologic units*. The lowermost point of the watershed units was located at the furthest upstream water intake. The lower boundary of the face units was assumed to be the topographic break between the Perry Ridge slope and the valley bottom, which is also typically the boundary between forest and agricultural land.

The approximate areas downslope of the hydrologic units that could potentially be affected by surface water, groundwater, and landslide runout were delineated and referred to as *zones of influence*. They were identified using 1:10,000 scale, 10 m contour maps; 1:5000 scale, 2 m contour valley bottom floodplain maps; and 1998 1:15,000 scale air photos. Groundwater flow was inferred from surface contours. The zones of influence between adjacent watershed and face units commonly overlapped.

The TCP prepared by D.S. Spencer Forestry Consulting Ltd (1998) extended the traditional hypothetical forest engineering plan by considering some non-timber resources. The TCP identified over 300 hypothetical cutblocks and over 100 km of logging roads. For the purpose of this risk assessment, the authors divided the study area into 18 *forest development areas* based primarily on independent access for development.

Equivalent Clearcut Area (ECA) is a forest hydrology concept used to estimate the per cent of a watershed that is equivalent to having been clearcut. ECA considers the silvicultural system (clearcut or partial cut), harvesting system, and amount and density of regeneration (Forest Practices Code of BC, 1995). For each hydrologic unit, the an *ECA Index* was estimated, as an indicator of the hydrologic effect of the existing condition and proposed logging. The following assumptions were made: 30% of the timber on Crown land would be logged during the initial 30 years of development; between 40% and 80% of the private land had been logged; logging would not be concentrated in any one hydrologic unit; logging a given volume of timber would have the same effect whether clearcut or selectively logged; and reserve areas identified in the TCP would not be logged.

Geological Hazards

Key documents used for the hazard assessment included: Detailed Terrain Stability Map at 1:20,000 scale (Wehr, 1985; updated by S. Chatwin, 1998b); Stream Channel Assessments at 1:10,000 scale (S. Chatwin, 1998a); Geological Hazards Mapping on the private land at 1:25,000 scale (Apex, 1998); Floodplain Mapping along the adjacent Slocan River at 1:5,000 scale; and 1:15,000 scale 1998 air photos. Using this information the geological hazardous events were grouped into 9 different types, in 3 categories:

- Primarily water related events: peak flow/flood; sediment yield
- Primarily slope/channel related events: debris slide into a stream; debris flow down a stream; debris flood/avulsion along stream channel; open slope landslide (any type of landslide that occurs on an open slope above the valley bottom and does not enter a stream); snow avalanche
- Primarily valley bottom related events: valley bottom landslide; sinkhole development.

The existing probability of each of the 9 event types was rated High, Moderate, Low, Very Low or None based on the occurrence of an event, independent of its magnitude. One of the challenges was to establish a common hazard rating system that could be applied to an entire hydrologic unit for 9 quite different event types.

The anticipated hazard with logging on Crown land as proposed by the TCP was then assessed, based on the existing hazard plus an inferred effect due to, or linkage with, forest development. Linkage was subjectively inferred from the TCP, including the proposed road system, harvesting system, estimated ECA Index after logging, and location of roads and logging with respect to hazard polygons that were higher rated. Linkage was relatively rated as Low, Moderate and High. Assumptions included the 30% removal of timber in development areas, as discussed above, Code standard road building and logging, and overlapping adjacent zones of influence. The effects of present and/or future land use activities on the private land were not considered.

An example of existing hazards, inferred linkage to forest development, and anticipated hazards after logging for one hydrologic unit, Watson Creek, is shown in Table 1. To illustrate the concept of linkage using this example: the linkage of "valley bottom landslide" to forest development is Low because the proposed development does not occur near High hazard polygons. The linkage of "sediment yield" is High because there are seven proposed road crossings of the creek. Low, Moderate and High linkages increase the hazard class by zero, one and two classes respectively.

Table 1: Hazards summary for Watson Creek

No.	Event type	Existing hazard	Linkage[1]	Hazard with TCP
	Primarily water related events			
1	Peak flow/flood	Moderate	Moderate	High
2	Sediment yield	Very Low	High	Moderate
	Primarily slope/channel related events			
3	Debris slide into stream	Moderate	Moderate	High
4	Debris flow down stream	Moderate	Moderate	High
5	Debris flood/avulsion along stream	Low	Moderate	Moderate
6	Open slope landslide	Low	Low	Low
7	Snow avalanche	None		None
	Primarily valley bottom related events			
8	Valley bottom landslide	High	Low	High
9	Sinkhole	None		None

[1]Linkage with forest development

Elements at Risk and Consequences

Elements at risk considered were: life and limb, property and water supply. Residences and other buildings were located from high-resolution satellite photos, 1998 1;15,000 scale air photos and 1:5000 scale floodplain maps; roads, utilities, and agricultural land were located from forest cover maps; and water supply intake locations were located from the Ministry of Environment (MOE) water rights maps.

This mapping was accompanied by several assumptions and limitations. The intent was to have a uniform level of information across the study area and within each zone of influence, rather than to accurately map each element at risk. Only existing elements at risk were considered.

Consequences were rated High, Moderate, and Low based on the elements at risk and the inferred severity of the type of events to which those elements at risk are, or could be, exposed. The relative ratings were not intended to be compared among the three groupings of elements at risk. Some examples of consequences are:

- Consequence to life and limb: High - death; Moderate - serious injury; Low - minor injury.
- Consequence to property: High - destruction of multiple residences; Moderate - destruction of single residence, or damage to multiple residences; Low - damage to single residence.
- Consequence to water supply: High - destruction of multiple water intakes, or very high increase in turbidity; Moderate - destruction of a single water intake, or high increase in turbidity; Low - damage to a single water intakes, or moderate or low increase in turbidity.

RISK ASSESSMENT

The risk assessment for each of the 32 hydrologic units were carried out with the help of three 3 X 4 risk matrices: one relating to life and limb, one relating to property and one relating to water supply. The risk matrices for Watson Creek are shown in Table 2.

The numbers in Table 2, correspond to the 9 types of events (see Table 1 column 1). The location of the numbers indicate the existing risk rating (no logging on Crown land). Event numbers appearing in Table 1 but not in Table 2 indicate that consequences were assessed to be Very Low.

The anticipated risks with logging on Crown land were then assessed and entered in the matrices. It was assumed that the consequences would not change after logging, only the hazards, and the amount the hazards would change was related to the linkage with forest development. With a Low linkage, the hazard would remain the same; with a Moderate linkage, the hazard would increase by one hazard rating; and with a High linkage, the hazard would increase by two hazard ratings.

The changes to the existing hazard, and therefore risk, are shown graphically by the locations of the arrow heads in Table 2. As for consequences, relative risks were not to be compared among the three groupings of elements at risk.

The relative risk ratings were applied conservatively across an entire hydrologic zone of influence, and do not necessarily refer to an individual element of risk within that zone.

Table 2: Risk matrices for Watson Creek

Hazard	Consequence to Life and Limb			Consequence to Property			Consequence to Water Supply		
	High	Mod	Low	High	Mod	Low	High	Mod	Low
High	VH	H ▲	M	VH	H	M ▲ ▲	VH	H ▲ ▲ ▲	M
Mod	H	4 M	L	H	M	1 4 ▲	H	1 3 4	L
Low	M	L	VL	M	L	5 VL	M	5	6 ▲ VL
Very Low	L	VL		L	VL		L	2 VL	

Where Very High (VH), High (H), and Moderate (M) risks were identified, either existing or anticipated with logging, and based on the information contained in the background documents and the authors' experience, alternatives to the logging or road building were suggested and recommendations made, to reduce the effects of the proposed forest development on life and limb, property and water supply. The effect of the alternatives on the feasibility (economic or otherwise) of the development plan was not considered. The difference between alternatives and recommendations, was that alternatives were suggestions to be considered during subsequent planning, while the recommendations should be carried out, in many cases prior to logging, regardless of the details of the final forest development plan.

Examples of suggested alternatives are: eliminate logging in specific areas; limit the ECA in specific areas (equivalent to reducing the rate of cut); add riparian reserve zones; modify road locations; eliminate specific stream crossings; eliminate specific roads and instead consider skyline logging, helicopter logging, or no logging.

Examples of recommendations are: extend geological hazard mapping in specific areas; carry out stream channel assessments on specific streams not already assessed; determine stream courses or watershed boundaries in some locations where they are poorly mapped; carry out Terrain Stability Field Assessments to address specific hazards; pay special attention to the location, design and construction of specific logging roads and creek crossings; design and construct specific logging roads to higher drainage control standards than normal; minimize ground disturbance associated with skidding in specific areas. Most recommendations were aimed at doing further geotechnical studies, or detailed planning, design or operations in excess of minimum Code standards, to reduce identified hazards.

This risk assessment did not address the acceptability of risk. Thresholds or criteria of risk acceptability should be established by government, which must incorporate appropriate socio-economic and environmental factors into its decision making.

CONCLUSIONS AND DISCUSSION

Was the process useful? Is it applicable to other forest planning areas? And considering that this was the first such study done for forestry purposes, should any major changes in the methodology be made for future studies?

Having a risk assessment such as this done during the overview stage of planning helps identify major constraints, and helps set priorities for further mapping and information gathering. In considering possible future studies elsewhere, it should be noted that the information used in this risk assessment exceeded what is normally available at this stage of planning, especially the stream channel assessments and hazard mapping on private land. The methodology used, therefore, may be more applicable to studies for smaller, high risk areas, such as populated alluvial fans, carried out at a more detailed planning stage. At this later planning stage, field work would be an important component of the study and would be targeted at specific proposed roads and harvesting areas. Future studies at the overview stage are likely to be simpler, commensurate with the information available, however, there is a limit to how general such a study can be. There must be sufficient baseline information to be sure that moderate or high hazard areas are not overlooked.

The authors found the consensual approach, using a 3-member panel, to be useful in dealing with the inevitable lack of quantitative hazard data. The process, however, was time-consuming and expensive. Future risk assessments in other areas, however, are likely to have less available information. Considering the large areas of land with which forestry planning deals, it is unlikely that the routine collection of more detailed geologic and hydrologic information would be affordable for most planning studies.

The project did not receive much support from local residents. The public, and even some forestry and engineering professionals with an interest in the process, had difficulty understanding and/or accepting the concepts of probability on which risk assessment is based, and the limitations of such a study. Some sectors of the public had a very low tolerance for any risks associated with proposed development, and that they did not understand the concept that damage to a single property, which to them personally is of very high consequence if it is their own property, might be considered only a Moderate consequence in the context of a larger planning unit. This is not unexpected and it is inevitable in any study that attempts to keep citizens informed about risks that affect them locally.

ACKNOWLEDGEMNTS

The authors would like to acknowledge the assistance and encouragement of the BC MOF, Arrow Forest District, especially Mr Pat Field and Ms Leah Malkinson; members of the Perry Ridge Local Resource Use Planning Table; and the residents of Perry Ridge

REFERENCES

Apex Geoscience Consultants Ltd. 1998. Geological hazards mapping of the Slocan Valley, Phase I. Prepared for MOF Arrow Forest District, Castlegar, BC.

Boyer, D., Jordan P. and VanDine, D.F. 1999. Perry Ridge Risk Assessment. Prepared for the Perry Ridge Local Resource Use Plan Table, MOF Arrow Forest District, Castlegar, BC, including 3 addenda.

Cave, P.W. 1992. Natural hazards, risk assessment and land use planning in British Columbia: progress and problems. *In* Geotechnique and Natural Hazards, Symposium sponsored by Vancouver Geotechnical Society and Canadian Geotechnical Society, May 6-9, 1992. Bitech Publishers, Vancouver, BC.

S. Chatwin Geoscience Ltd. 1998a. Perry Ridge stream channel survey. Prepared for MOF Arrow Forest District, Castlegar, BC.

S. Chatwin Geoscience Ltd. 1998b. Upgrade of TSIL C mapping and surface soil erosion hazard mapping for the east side of Perry Ridge. Prepared for MOF Arrow Forest District, Castlegar, BC.

Forest Practices Code of BC. 1995. Interior watershed assessment procedure guidebook (IWAP). BC MOF and BC MOE.

Morgan, G.C. 1991. Quantification of risks from slope hazards. *In* Geologic Hazards in British Columbia, Proceedings of the Geologic Hazards '91 Workshop, Feb. 20-21, 1992, Victoria, BC British Columbia Geological Survey Branch, Open File 1992-15.

Morgan, G.C, Rawlings, G.E, and Sobkowicz, J.C. 1992. Evaluating total risk to communities from large debris flows. *In* Geotechnique and Natural Hazards, *op cit.*

D.S. Spencer Forestry Consulting Ltd. 1998. Engineering plan report, Perry Ridge total chance plan. Prepared for MOF Arrow Forest District, Castlegar, BC.

Wehr, R. 1985. Terrain survey: Perry Ridge, preliminary assessment. Report and maps prepared for Slocan Forest Products, Slocan, BC.

National map data base on landslide prerequisites in clay and silt areas - development of prototype

L. VIBERG, Swedish Geotechnical Institute, Linköping, Sweden
J. FALLSVIK, Swedish Geotechnical Institute, Linköping, Sweden
K. JOHANSSON, Swedish Geological Survey, Uppsala, Sweden

DEFINITIONS

The prototype described in this paper is based on a map database. We distinguish between the paper map and the digital (map) database. The paper map is only a printout of the database. All information is gathered in the database. It is easy to mix these definitions because we are used to think of the traditional map as holder of all information.

We also make a distinction between mapping and classification in this paper. Mapping is the physical process of localisation and determination of geological, topographic and other features in the landscape. When we use the term classification in this paper we mean the mental process of grouping mapped landscape properties into units with similar stability prerequisites for landslide.

The database furnish only information on landslides in clay and silt, Figure 1. Landslides in frictional soils or rock is not considered.

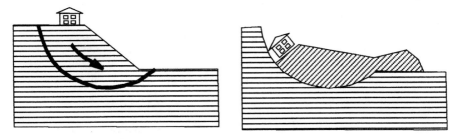

Figure 1. Typical landslide features in clay and silt.

BACKGROUND AND SCOPE OF STUDY

In Sweden insufficient stability in clay and silt slopes is a threat in many built up areas. Therefore a national survey investigation of the landslide hazard in built areas in clay and silt soils has been going on since nineteen eighties. However there is a demand for landslide hazard maps outside built up areas as a tool e g for city and infrastructure planning, for planning and executing of rescue actions and for judgement of landslide hazard for existing constructions outside built up areas.

Swedish geotechnical institute, SGI, was commissioned by the Ministry of Environment to develop a prototype survey landslide hazard map. (The term "map" has later been changed to database for reason mentioned in the DEFINITION article). This work was carried out in co-operation with the Swedish Geological Survey, SGU, Lantmäteriet, LMV (Swedish land surveying authority) and Swedish Rescue Service, SRV. A reference group with representatives for city and county planning, city rescue service and geotechnics was also connected to the project.

The work comprised of development of a theme database, judgement of the area to be mapped, proposal of organisation for the work, judgement of cost and proposal of financing. The prototype database was to be tested in an area.

It was decided early in the development work that the map should be digitally processed by GIS-method. Thus the major part of the work was to develop a database and an algoritm that could combine soil and topographical data into stability zones.

NEEDS AND BENEFITS

Swedish society annual cost for landslide damage and remedies is several hundred millions SEK. The cost/benefit ratio for prevention measures versus damage is 1/10 to 1/100 (Räddningsverket, 1996)

There has been a need for development of a landslide hazard mapping in digital format for GIS applications.

It obviously beneficial if information on landslide hazard is available and easily accessible in early planning phases. The database format makes possible to combine landslide information with other themes like flooding.

LEVELS OF INFORMATION ON LANDSLIDE HAZARD

The information on slope stability may roughly be divided into three detail levels depending on available information, Table 1, where level 1 is most general and level 3 is most detailed. The described development work concerns level 1.

The described database in this paper belongs to detail level 1.

LEVEL OF DETAIL	TYPE OF INFORMATION	REQUIRED DATA
1	General prerequisites for landslide	Geology and topography
2	Safety factor	Data for stability calculation (shear strength, pore water pressure, profiles etc)
3	Landslide risk (probability x consequence)	Data for calculation of probability of landslide and consequences of landslide

Table 1. Detail levels of slope stability information.

STABILITY CLASSIFICATION MODEL

The stability classification model used is a slightly modified model based on the model used by Swedish Rescue Service (Räddningsverket, 1997). The terrain is subdivided into stability zones based on soil and topographical data, Table 2 and Figure 1. The inclination criteria 1:10 is based on an inventory of occurred landslides in Sweden, Norway and Canada (Inganäs & Viberg, 1979). No initial landslide was found to have occurred in slopes with less inclination than 1:10. This means that many slopes in zone I have required stability, but this has to be controlled.

DEVELOPMENT OF DATABASE

The development work was carried out in a river valley in northern Sweden in Sundsvall municipality. The database is based on soil map data and topographical elevation data in digital format.

The soil data was furnished by Swedish Geological Survey, SGU. The source data was a soil database, which had to be reclassified to fit the soil classes for the stability zonation, Table 2 and Figure 2. As the original soil data was in digital database format the reclassification work was rather easily done.

The topographical data source was a set of standard contour level isolines in paper format furnished by Lantmäteriet, LMV. The lines were digitised and the elevation data was the base for the creation of an elevation terrain model.

STABILITY ZONE	CRITERIA		STABILITY CONDITION	RECOMMEN-DATIONS FOR PLANNING
	Soil type	Incli-nation		
I **ORANGE**	Clay and silt with or without cover of other soils	> 1:10	Prerequisites for initial landslide	Special attention to landslide hazard
	For clay and silt bordering water the zone width >= 50 m	All inclina-tions		Attention to erosion hazard
II **YELLOW**	Clay and silt with or without cover of other soils	< 1:10	No prerequisites for initial landslide Areas adjoining zone I may be affected by landslide	Empirically based judgement of slope stability by geotechnician is normally required Attention to erosion hazard
III **GREEN**	Sand on till, gravel, cobbles, boulders or rock Till, gravel, cobbles, boulders or rock	All inclina-tions	No prerequisites for initial landslide In steep terrain flow slide may develop	Attention to landslide hazard steep terrain Attention to erosion hazard Attention to vibration activities

Table 2. Subdivision into stability zones. Modified after Räddningsverket (1997)

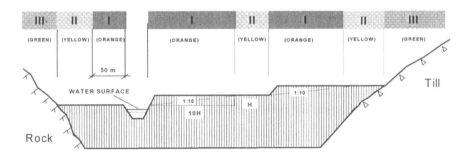

Figure 1. Principle section through clay and silt showing stability zones I, II and III. Räddningsverket (1997). (Original colours are indicated).

The soil and height data were fed into a GIS software (ArcInfo/ArcView). The GIS-classification was made in raster format with a cell side length of 5 m. An algorithm was designed to process the data and classify the terrain into stability zones I, II and III.

Three different algorithms had to be developed before satisfaction was reached. The two first algorithms were far too slow. It is a simple problem, but the number of calculations may be extremely high even for a moderately large calculation study area, in our case about 1 km^2. The problem with the high number of calculations was solved by an algorithm that compared a few cells at a time and "excavated" down successively to the inclination line 1:10.

The database is named "survey map database on landslide prerequisites in clay and silt soils". A printout from the database in grey tones is shown in Figure 3.

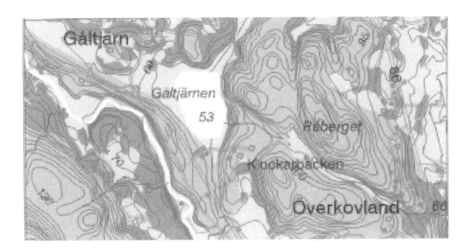

Figure 3. Example of printout from database on landslide prerequisites in clay and silt soils (Swedish Geotechnical Institute, 2001). Lower picture illustrates the principle of stability zonations. (Real map in colours).

END USER OPINIONS

The reference group has given their opinions on the use of the database. Example of opinions are: good planning base for municipality and county planning, important with clear understandable description of content and use, good tool for risk planning and for total defence crisis planning, important that data can be combined with planning data, e g flooding data.

PROPOSAL FOR PRODUCTION

Geographical scope

The working group suggests that the classification should comprise all municipalities where prerequisites for landslides in clay and silt is at hand, that is areas with high and moderate frequency of landslide scars according to Figure 4. This corresponds to about 300 map sheets 25x25 km in the scale of 1:50 000.

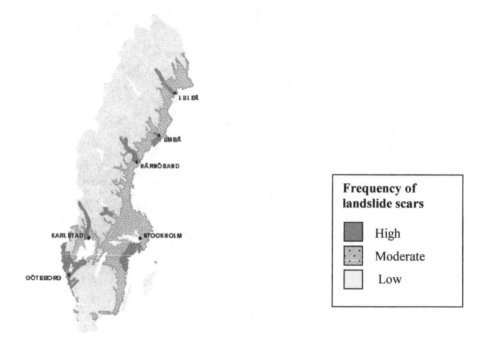

Figure 4. Generalised map on landslide frequency in Sweden. (From Swedish Geological Survey, homepage www.sgu.se)

In order to give a priority list of the regions the ranking should be based on current and future construction and planning activities together with the stability situation in the region. The working group has made a preliminary ranking list mainly based on stability conditions.

Availability of soil and elevation data in Sweden

The availability of soil and elevation databases is a crucial factor for the databse production. Soil data is available in different levels: Database is available, database must be supplemented ether by soil strata data or detailed geological mapping.

Level data in the form of contour isolines are available over all areas of interest. These have to be digitised and labelled with their elevation number.

Organisation

Swedish geotechnical institute is proposed to be responsible for the production in cooperation with SGU, LMV and SRV. These authorities provide SGI with data necessary to do the classification into stability zones. LMV is supposed to store and distribute the information, Figure 5. The distribution may be carried out in different ways. In Figure 5 an Internet distribution solution is outlined.

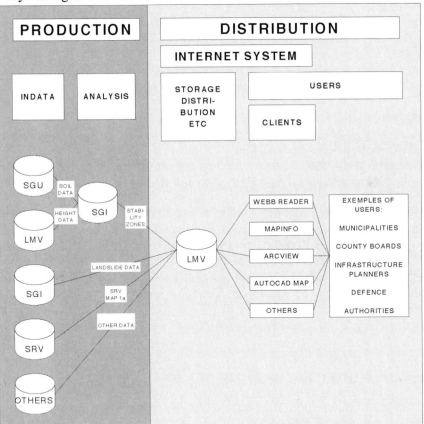

Figure 5. Proposed organisation and distribution of   national database on prerequisites for landslides in clay and silt. (From Swedish Geotechnical Institute, 2001).

Cost
The production cost is roughly estimated to about 60 millions SEK in year 2000 money value. Administration and distribution cost is estimated to 25 000 SEK. The working group has proposed that the production should be financed by government.

Production time
The production time is depending on available financial resources. Production of the database is recommended to be carried out in 10 years.

CONCLUSIONS
It is shown that it is possible to produce a national database on landslide prerequisites in clay and silt areas in Sweden.

A prototype database on landslide prerequisites for landslides in silt and clay is developed and

Necessary data on elevation and soils exits or can be completed. The soil data has to be supplemented in some places in Sweden.

Cost, financing, time table and organisation has been proposed.

REFERENCES
Inganäs, J, & Viberg, L (1979). Inventering av lerskred i Sverige. Nordiska geoteknikermötet 8, Esbo, Finland, May 1979. Föredrag och artiklar, s 549-556.
Räddningsverket (1996). Lönar det sig att förebygga skred ? Rappport Räddningstjänstavdelningen R53-151/96. (Swedish Rescue Service. Does it pay off to mitigate landslides ?)
Swedish Geotechnical Institute (2001). Nationell översiktlig kartdatabas över skredförutsättningar i ler- och siltjordar. Utveckling av databasprototyp och förslag till produktion. SGI Dnr 1-0005-0399. (National survey map database on landslide prerequisites in clay and silt soils. Development of data base prototype and proposal for production). (In cooperation with Swedish Geological Survey, Lantmäteriet, Swedish Rescue Service. Report to Swedish government, Ministry of the Environment).

A modelling tool for predicting coastal cliff recession and analysing cliff management options

DR M.J. WALKDEN, Department of Civil Engineering, University of Bristol, UK
DR J.W. HALL, Department of Civil Engineering, University of Bristol, UK
E.M. LEE, Department of Marine Science and Coastal Management, University of Newcastle Upon Tyne, UK

INTRODUCTION
Because of the risk of land loss, and the potential importance of eroding cliff sites in larger scale morphodynamics, soft cliff dynamics are of interest to coastal landowners and managers, so tools are needed that can predict future recession. The most common method of predicting future shoreline positions is to calculate historic rates and to assume that they will persist. A variety of algorithms exist for this purpose (Crowell et al, 1997, Dolan et al, 1991). When recession becomes strongly non-linear, perhaps when the eroding shore is responding to changing coast protection strategies, simple statistical analysis is less reliable. In order to deal with more complex situations, particularly those that are not represented in the historic record, it is necessary to represent the processes involved in shore recession.

LITERATURE REVIEW
Clearly erosion rate varies with wave forces (F) and the inverse of material strength (R), but it is not obvious which hydrodynamic force and type of material strength are most important (for a discussion of this problem see Sumanura, 1992). Recently Mano & Suzuki (1999) represented R by Young's modulus of the soft cliff material and F by wave energy at the break point, whilst Wilcock et al, (1998) considered a ratio of wave pressure to cohesive strength. Kamphuis (1987) provided an expression for the erosive potential of breaking waves, based on an analysis of the wave power in the breaker zone, the rate of energy dissipation and the energy contained in each breaking wave. The material strength was represented by a simple coefficient that was found through calibration. In further investigations Kamphuis, et al, (1990) studied the initiation of erosion of a range of clays in a laboratory. They found that erosion was strongly associated with discontinuities in the material and with the presence of sand, which was mixed into the water. The importance of the fissures within the material implies that laboratory tests should be conducted on undisturbed clay samples. Skafel & Bishop (1994) conducted laboratory experiments in which they installed large slabs of undisturbed glacial till into a wave flume. They generated pseudo-random waves and studied the magnitude and

distribution of erosion under a variety of conditions, including different water levels, breaker types and degrees of sand cover. This study probably represents the most realistic laboratory representation of an eroding till foreshore. Damgaard *et al*, (1998) conducted physical model tests on near vertical soft sand cliffs fronted by a mobile beach. As the cliff collapsed it provided material to the beach. They provided an expression linking offshore wave climate to erosion rates through wave run-up.

The form of the shore profile is another factor that has a strong influence on recession. In laboratory experiments Sumamura (1976) showed how the erosion of an initially near-vertical soft cliff reduced as its profile adapted to the prevailing wave conditions. The importance of the cliff toe level and the severity of wave runup has been recognised, modelled and compared to observations of erosion (Shih et al, 1994, Kirk *et al*, 2000). Meadowcroft *et al*, (1999) produced the first process based probabilistic model of an eroding soft cliff system. This included a vertical cliff fronted by a deep beach. Beach motion was calculated every time-step using a one-line model approach, and the erosion of the cliff toe was obtained from the empirical function developed by Damgaard *et al*. This model did not include a platform, so the profile form was essentially determined at the start of the model run. Wave heights, material strength, and ground water levels were all represented with probability distributions.

THE STUDY SITE
The model development has been informed by observations made at The Naze peninsula, which is on the Essex coast of the UK. These cliffs are oriented approximately North-South, and are composed of London Clay overlain with Waltonian Red Crag (for a fuller description of the site geology and geomorhpology see Daley & Balsan, 1999 and Flory *et al*, 2002). The interface between the two geological layers is at or above the mean high water level of spring tides (MHWS), so the cliff toe and platform are composed entirely of clay. There is a distinct discontinuity, a rapid change in surface gradient, between the lower cliff and shore platform. This discontinuity appears to arise for hydrodynamic reasons since there is no noticeable change in the clay. The tidal cycle seems to be particularly important since the discontinuity appears to be a little above MHWS. It follows that wave set-up may also be significant. Despite the differences in slope, and the resulting wave forces, the shore profile appears to be relatively constant through time, i.e. the retreat rates across the profile and at the cliff toe are equal.

A sandy beach intermittently covers the platform and is occasionally thick enough to protect the underlying platform. Erosion of the foreshore occurs through a variety of processes. The presence of sand and a reasonably energetic wave climate implies abrasion due to corrasion by sand particles saltating or being carried within turbulent water. The bedding planes in the foreshore appear to be approximately horizontal and there is evidence of small thin sheets of clay being lifted out of the foreshore. This might be caused by high wave impact pressures within fissures at the bedding planes, or low pressures above induced within breaking wave turbulence. The platform very close to the cliff toe shows evidence of the removal of small lens-shaped chips, apparently plucked out by plunging waves.

MODEL DEVELOPMENT

It was clear from the literature review that some fundamental problems still need to be solved before a complete model of erosion processes could be prepared. In particular, appropriate forms and values of F and R needed to be established. The approach used by Kamphuis seemed the most appropriate for the problem, where wave forces were represented with a simple function based on an analysis of the surf zone, and material strength represented by a single coefficient of resistance. Adopting such an approach in a model means that average recession rate becomes a fixed property rather than an emergent one. This raises questions about how model validity can be established. If, regardless of model content, the average recession rate is fixed during calibration, how will it be possible to check how well the model is representing the involved processes? This issue was resolved by allowing differential erosion cross-shore and along-shore, so that variation about the average could be observed. Also, no initial shoreline was assumed other than a deep vertical face, so that the model would form the cliff and platform. Thus, though the average recession rate would be fixed through calibration, differential recession and shoreline shape would be emergent. These emergent properties could then be compared to the subject site, and a decision reached on model performance.

A flow chart of the model that was developed is shown in Figure 1. It has been called cliffSCAPE (Soft Cliff And Platform Erosion). It describes a two-dimensional shore section, which can be made quasi-3D by using a series of such sections and

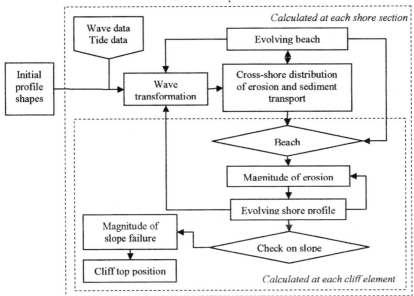

Figure 1. Flow-chart representation of cliffSCAPE.

allowing interaction between them. Model time-steps are typically one tidal period or one hour. At every time-step, wave and tide data are sampled from files. The waves are shoaled, refracted and diffracted over a topography and around obstructions that represent, in a simplified way, site bathymetry and coastal structures. A cross-shore distribution of longshore sediment transport is calculated using the breaker depth combined with tidal changes in water level and an assumed distribution under static conditions. A similar approach provides a cross-shore erosion distribution function. Potential sediment transport at each shore section is calculated using a CERC type equation. The cross-shore distribution of drift is compared to the actual beach width to discern what proportion of the potential transport actually occurs. At the same time the magnitude of erosion at different elevations across the shore is calculated using the duration of submersion during the tidal cycle, location within the surf zone and local slope. These were combined with the expression of Kamphuis and a distribution of erosion under static conditions obtained from Skafel & Bishop. The beach is considered to protect the profile when it is of adequate thickness. Slope stability is represented, to a first approximation, using a probabilistic model that calculates the probability and magnitude of slope failure from the angle of the cliff face. This angle is made more severe by the wave-induced erosion and reduced by the slope failure events. It can be seen in Figure 1 that the model includes several feedback paths, as current beach and clay profile conditions influence future wave transformation and erosion patterns.

MODEL OUTPUT

CliffSCAPE is undergoing development, so a fully calibrated model of the Naze is not yet available. The future scenarios presented in this section should be regarded as demonstrations of how the model can be applied rather than indicating future evolution of The Naze.

The Naze was first represented as an infinitely long, two-dimensional straight cliff. Appropriate wave and tide conditions were used and the level of the cliff top was defined. A rate of sea level rise of 2 mm/Annum was used to represent background eustatic and isostatic effects. CliffSCAPE has been designed to evolve a shore profile but some initial condition still has to be specified. A vertical cliff face was chosen as this shape is far from any equilibrium condition. A beach was considered inappropriate for such an initial shoreline and was not included. The model was run and the emerging profile was observed. Once the retreat rate had equalised over the whole profile the simulation was halted. It should be noted that a process of iteration was necessary at this stage in order to achieve an appropriate strength parameter and recession rate for the site. Figure 2 shows the shape that emerged over 1000 years. The lines represent the profile at 100 year stages. Also shown are mean levels of the extremes of spring tides. It can be seen that the initial erosion is concentrated between the tidal limits. In deeper water the slopes become very gentle and shoreline retreat is ultimately dominated by rising sea-levels.

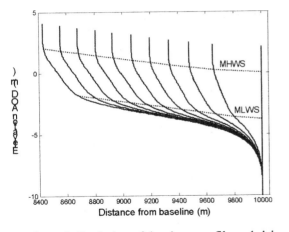

Figure 2. Evolution of the shore profile and rising sea-levels over 1000 years

During the second stage of modelling a quasi 3D representation of the Naze was constructed using eight shore sections. Each of these was initially identical to the final profile shown in Figure 2, but with different relative offsets to approximate the plan shape of the Naze. A beach was introduced and appropriate CERC constants were calibrated. The introduction of the beach changed the equilibrium form so the model was run for a period to equalise the retreat rate over the heights of the profiles. There was longshore variation in the shore profiles that emerged. The main differences were a general predicted decrease in steepness and increase in recession rate towards the Northern point of the Naze. The same trend is observed on site and seems to be due to reduced beach protection in this region. One of the resulting profiles is shown in Figure 3, with a measured profile at the corresponding location on The Naze.

Figure 3 shows the two profiles within 200 m of the cliff toe. In this region the profile shapes match very well. The differences that can be observed may be due to cross-shore dynamics of the beach, which are not represented in the model. Beyond

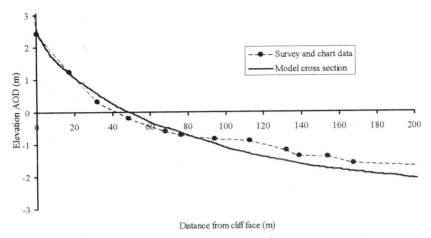

Figure 3. Comparison between measured and predicted shore profiles.

Average rate of recession (m/A)

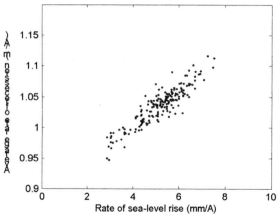

Figure 4. Comparison between rates of sea-level rise and recession

200 m the model seems to overpredict profile depths by up to 0.6 m. It should be noted that survey data in this region is extremely sparse so it is not currently possible to make a detailed comparison.

Having established a shape for the Naze model that was both similar to the site and showed equalised retreat rates, i.e. comprised of equilibrium profiles, it was possible to begin investigating future shoreline response to different scenarios. The model was used to investigate how such a site might respond to future, uncertain, rates of sea level rise (SLR). The model was run in a probabilistic manner, using Monte-Carlo simulation, during which a new SLR was sampled for each realisation. Using estimates obtained from the UK Environment Agency it was decided to represent SLR due to global warming as a normal distribution with a mean of 4.2 mm and a standard deviation of 1 mm/A. This was assumed to exist in addition to the 1 mm/A increase due to isostatic change. The results of 300 simulations can be seen in Figure 4, which compares SLR with total recession of a point at the cliff toe. The results show, as expected a strong positive relationship between rate of sea-level rise and rate of recession. The data is rather scattered; this seems to occur because of the feedback loops within the model. Small differences in the system conditions can build into larger differences through positive feedback. At the same time these fluctuations are bounded, providing evidence of larger scale negative feedback.

To demonstrate how cliffSCAPE may be used to investigate shoreline response to structures, three long groynes were introduced to the Naze model. It can be seen in Figure 5 that the two southern structures behaved as expected and protected the areas of the cliff closest to

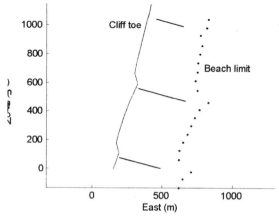

Figure 5. Cliff-line deformation behind groynes.

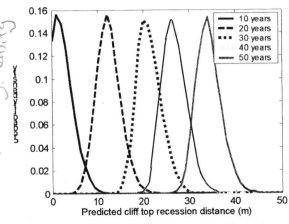

Figure 6. Probabilistic predictions of recession of a point on the cliff top

them. They did this partly by providing shelter from wave attack, and partly by retaining a substantial beach. The seaward extent of the beach is also shown. Between these structures both the line of the cliff toe and the beach have formed bays. The most Northerly of the groynes has been less effective. This appears to be due to it being the shortest of the three and most exposed to large waves approaching from the North.

The predictions of cliff toe recession have been combined with a simple probabilistic model of landsliding to generate predictions of cliff top recession. Analysis of cliff angle at the Naze indicates that the toe erosion increases the cliff angle up to a maximum stable value, at which point a major landslide event returns the cliff to a lower angle. Both the upper and lower bounds on the cliff angle are roughly stable, for a given cliff height, though they naturally show some scatter. By representing this scatter as normal probability distributions, a simple probabilistic model of landsliding was established. Results from a combined run of the shore platform model and the probabilistic landslide model are shown in Figure 6.

DISCUSSION AND CONCLUSIONS
CliffSCAPE is a new tool for predicting future recession of soft coastal cliffs. It describes the main elements of the coast/cliff system including wave transformation, sediment transport, shore erosion and cliff stability. It can be used to represent and investigate inherent uncertainties because it has been developed for probabilistic application. CliffSCAPE requires calibration against historic data in order to establish a material strength coefficient, but differential erosion and shore shape are emergent properties and have been shown to closely reproduce conditions observed at the Naze site.

Whilst the landsliding aspects of the model will always need to be developed on a site-specific basis, the shore platform model is readily transferable to many coastal sites and is hardly more difficult to set up than a conventional one-line beach model.

REFERENCES

Crowell, M., Douglas, B. and Leatherman, S. (1997), "On forecasting future U.S. shoreline positions: a test of algorithms". Journal of Coastal Research, 13(4), pp. 1245-1255.

Daley, B. & Balsan, P. (1999), "British Tertiary Stratigraphy". Part of the Joint Nature Conservation Committee Series (15). ISBN 1 86107 469 7.

Damgaard, J.S., (1998), "Towards improved modelling of recession of dry areas: development of a predictive model framework", HR Wallingford Report SR514.

Dolan, R., Fenster, M. & Holme, S., (1991), "Temporal analysis of shoreline recession and accretion". Journal of Coastal Research, 7(3), pp. 723-744.

Flory, R., Nash, D., Lee, E.M., Hall, J., Walkden, M., & Hrachowitz, M. (2002), "The application of landslide modelling techniques for prediction of soft coastal cliff recession". Proc. Conf. Instability – planning & management. Ventnor, U.K.

Kamphuis, J. (1987), "Recession rates of glacial till bluffs". Journal of Waterway, Port, Coastal and Ocean Engineering, 113(1), pp. 60-73.

Kamphuis, J., Gaskin, P. & Hoogendoorn, E. (1990), "Erosion tests on four intact Ontario clays". Can. Geotech. J. 27, pp. 692-696.

Kirk, R.M., Komar, P.D., Allan, J.C. & Stephenson, W.J. (2000), "Shoreline erosion on Lake Hawea, New Zealand, caused by high lake levels and storm-wave runup". Journal of Coastal Research, 16(2), pp. 346-356.

Mano, A. & Susuki, S. (1999). "Erosion characteristics of sea cliffs on the Fukushima coast". Coastal Engineering Journal, 41(1), pp. 43-63.

Meadowcroft, I.C., Hall, J.W., Lee, E.M., & Milheiro-Oliveira, P., (1999), "Coastal cliff recession: development and application of prediction methods". HR Wallingford Report SR528.

Shih, S.M., Komar, P.D., Tillotson, K.J., McDougal, W.G. & Ruggiero, P., (1994), "Wave run-up and sea-cliff erosion". Proc. 23rd Int. Conf. Coastal Engineers, Kobe, pp. 2170-2184.

Skafel, M.G., & Bishop, C.T., (1994), "Flume experiments on the erosion of till shores by waves". Coastal Engineering 23, pp 329-348.

Sunamura, T. (1976), "Feedback relationship in wave erosion of laboratory rocky coast". Journal of Geology, (84), pp.427-437.

Sunamura, T., (1992), "The Geomorphology of Rocky Coasts". Chichester: Wiley. ISBN 0-471-91775-3.

Wilcock, P.R., Miller, D.S., Shea, R.H., & Kerkin, R.T., (1998), "Frequency of effective wave activity and the recession of coastal bluffs: Calvert Cliffs, Maryland". Journal of Coastal Research, 14(1), pp. 256-268.

Overview geo-hazard assessment of China by GIS

K. YIN, PROF., Engineering Faculty, China University of Geosciences, Wuhan, China
Y. LIU, DR., Ministry of National Land Resources, Beijing, China
L. ZHANG, PROF., Institute of Land Resource and Economy of China, Beijing, China
Y. WU, DR., Engineering Faculty, China University of Geosciences, Wuhan, China
L. ZHU, Engineering Faculty, China University of Geosciences, Wuhan, China

ABSTRACT

In recent years, China has suffered serious geological disasters mostly of slope movements due to complex geology, geomorphology, unusual weather conditions, and large-scale land explorations during economic development. According to geological hazard investigations organized by the Ministry of National Land Resources, there are 400 towns and more than 10,000 villages under the threat of these geo-hazards. A project entitled "National geo-hazard risk assessment" was supported by the Ministry from 2000 to 2001 which aims to evaluate the overview geo-hazard potentials, vulnerabilities of lives and structures, and risks in conterminous China at the scale of 1:6,000,000. This is the first overview geo-hazard potential map of China, and will take the role for geo-hazard management by the administrations.

INTRODUCTION

As a result of the complex geology, geomorphology, climate, and recent high economic development, China is one of the countries seriously influenced by geological disasters in the world. In recent years, the losses from geological hazards especially from landslides, rock falls, and debris flows have increased rapidly due to the triggering by massive infrastructure constructions in the process of economic growing and unusual weather conditions. According to statistics, there are more than 200 casualties and 2 to 3 billion US dollars both direct and indirect economic losses per year in the last two decades. In the most destructive year of 1998, heavy rain fall swept across large parts of China during July and August, which caused more than 180,000 events including landslides, rockfalls, and debris flows, and as a consequence 1157 people died, more than 10,000 wounded, about 500,000 houses damaged. Total direct economic losses were estimated at around 3.5 billion US dollars. According to geological hazard investigations organized by the Ministry of

National Land Resources, there are 400 towns and more than 10,000 villages under the threat of these geo-hazards. The project of "National geo-hazard risk assessment" was supported by the Ministry to evaluate the overview geo-hazard potentials, vulnerabilities and risks in China at the scale of 1:6,000,000.

As a part of the project, this paper presents the details of geohazard potential assessment concerning main procedures and methodologies used in geo-hazard assessment with regarding to slope hazard, which is strongly based on a nation wide geo-hazard database consisting of about ten thousand individual or regional slope hazard investigations and their consequent losses in last three decades. In terms of the software of MapGIS, all factor-maps are digitized at same scale and superposed respectively with geo-hazard distribution map to produce quantitative potential map according to Information Model (Yin & Yan 1988, Yan 1989). Then all factor-potential maps are superposed together to generate a comprehensive predictive map in which different hazard levels are indicated based on the sum value of Information.

GEOMORPHOLOGY, GEOLOGY AND CLIMATE

China has 9.6 million square kilometres of territory and almost 1.3 billion people. It is located in eastern Asia. Neotectonic movement of Pacific, Euro-Asian and Indian plates controls the geomorphologic feature of the conterminous China's territory. Southwest is the Qinghai-Tibet plateau which is 5000 metres above sea level on average, the highest in the world. The chain of mountain trends mainly east-westwards. In the middle, a series of innercontinent crust faults of north-south direction dominate the landform of parallel deep valleys. Earthquakes and triggered landslides are the serious geological hazards in the region. The altitude ranges between 1000 and 4500 metres above sea level. The coastal Pacific in eastern China is mainly composed of plains, such as the deltas of the Yellow River and the Yangtze River. Some middle high mountains consist of igneous rock and volcanic rock along the coast of the southeast. Precipitation varies greatly from the coast to far west. Annual precipitation in the southeast is above 1400 mm, 800mm to 1400 mm in the central area and less than 400 mm in the northwest. The smallest annual precipitation in the desert of the far northwest is below 50 mm. Most geo-hazards (landslides, rock falls and debris flows) are normally triggered by continuous and abnormal rainfall during summer and autumn seasons. But more and more geo-hazards relating to human activities have become significant environmental problems during the recent economic development of China.

HAZARD ZONATION
Principle of information model for geo-hazard assessment
In geo-hazard assessment, any kind of phenomenon is influenced by many factors. Different factors have different importance or weight among the whole contribution. How to evaluate the weight becomes the key point in quantitative methods. Susceptibility grading or other similar methods (Brabb 1984, Einstein 1988, Carrara

1983, Ross 1998) are a good approach to geo-hazard mapping. Based on the definition of information by Shannon (1948), which presents the reduction of the uncertainty of a random phenomenon, Yin (1988), Yan (1989) and Wu (2000) have developed the methodology to determine the weight, in which each factor contributes a certain amount of information for predicting a geo-hazard potential. Assuming geo-hazard phenomenon is y and influential variable is x_i, information value contributed by variable x_i is mathematically expressed as

$$I(y, x_i) = \log_2 \frac{p(y/x_i)}{p(y)} \tag{1}$$

where, $p(y/x_i)$ is conditional probability of geo-hazard occurrence with the existence of x_i, $p(y)$ is average probability of geo-hazard in the whole area. The bigger the information value $I(y,x_i)$ is, the more favourable for geo-hazard the factor x_i is.

From statistics, it is possible to estimate $p(y/x_i)$ and $p(y)$ in terms of the ratios among the area affected by geo-hazard in whole region, the area affected by geo-hazard with factor x_i, and the area of the region. Equation (1) can be rewritten as

$$I(y,x_i) = \log_2 \frac{S_{0i}/S_i}{S_0/S} \tag{2}$$

where, S_{0i} is the area affected by geo-hazard with factor x_i, S_i is the area of factor x_i existence, S_0 is geo-hazard area in the region, S is total area of the region.

Because of the characteristics of mathematical addition of information, it is easy to build a comprehensive linear model concerning the co-effect by group of factors. Formula (3) presents the principle of the model in which influential factors X_i and their categories (X_{ij}) affect geo-hazard potential through coefficient I_i. I_i is the information value that contributed by the factor X_i. Mostly, each factor contains categories. For instance, the factor of slope may be divided into different gradients, strata may also include various rock or soil formations, and so on. In real operation of assessment, studied area is normally divided into units. Each unit is then computed by the sum of information value of different categories.

$$I = I_{1j} X_{1j} + I_{2j} X_{2j} + \ldots + I_{nj} X_{nj} \tag{3}$$

The value of I above zero in any unit means that the combination of all factors gives useful information for reducing the uncertainty of geo-hazard prediction.

For simplicity, grids are usually divided into well defined shaped such as squares or polygons. Coefficients I_{ij} are calculated based on statistical analysis through the superposition of geo-hazard deposit map respectively with other factor maps. With the development of GIS technique, renovations have been made in order to make more precise considerations for natural boundaries like geological limit, topographical boundary or other irregular limits possible, which makes more reasonable for superposition of multiple map layers (Fig.1).

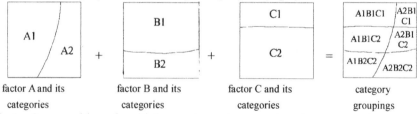

| factor A and its categories | factor B and its categories | factor C and its categories | category groupings |

Fig.1 Superposition of multiple map layers to generate category groupings

Based on the principle of equation (2) and the advantages of GIS mapping, it is possible to compute the information value contributed by each categories through the superposition between geo-hazard map and factor maps.

Factors and categories
To select proper factors is the key for modeling. A suitable factor group can lead to high accuracy of the result. On one hand, geo-hazard is controlled by environmental factors such as geological formation, geomorphology, slope gradient, seismic background, geological structure mode, and so on. These factors, dominating the genetic environment of geo-hazards can be treated as stable through the time scale of human activity. On the other hand, geo-hazard is also affected by external factors such as climatic condition, earthquake, or human activities; these are changeable. So it is necessary to consider both environmental and triggering factors as the variables in modeling. Based on the geological, geomorphologic, seismic, climatic conditions and human activities within the territory of China, four environmental and three triggering factors are selected for modeling. 35 categories are subdivided according to their basic behaviours (Table 1).

Procedures of geo-hazard mapping by GIS
1. Digitize basic maps
According to selected factors and the geo-hazard distribution map, four environmental factor maps and three triggering factor maps are compiled at the same scale of 1:6,000,000. Furthermore, these maps are digitized through GIS operation (see Fig. 2-4).

2. Superpose maps
Digitized geo-hazard distribution maps generated from the database are respectively superposed with both environmental and triggering maps to create quantitative factor information maps. Information values of all 35 categories are automatically computed by the superposition by GIS (see Table 1). The bigger the information value is, the more favourable to geo-hazard the category is. Theoretically, negative information means unfavorable action for reducing uncertainty.

Table 1 Factors, their categories and information

Factor (X_i)		Category (X_{ij})	I_{ij}(bit)	
Lithologic formation group		quaternary soil	-4.24	Environment factors
		loess	1.32	
		clastic sedimentary rock	0.63	
		carbonate rock	0.89	
		volcanic rock	-0.46	
		igneous rock	-0.83	
		week metamorphic rock	0.84	
		hard metamorphic rock	-0.55	
Geomor-phology	altitude	very high mountain >5500m	-1.65	
		high mountain 3000~5500m	0.48	
		mid-high mountain 1000~3000m	2.10	
		middle mountain 800~1000m	0.48	
		low mountain 500~800m	-0.26	
		hill 100~500m	-0.73	
		plain and basin <100m	-1.60	
	slope	very steep	1.84	
		steep	0.08	
		moderate	-0.66	
		gentle	-2.23	
		flat	-2.15	
Intensity of earthquake micro-zonation		≥IX	1.81	
		VIII~IX	0.71	
		VII~VIII	0.61	
		VI~VII	-0.32	
		<VI	-1.79	
Annual average precipitation		>1600mm	-0.64	Triggering factors
		1600~1200mm	0.21	
		1200~800mm	1.89	
		800~400mm	0.33	
		<400mm	-3.03	
Abnormal weather		heavy rain center	0.13	
		snow melting center	0.30	
Intensity of human activity		strong	0.45	
		moderate	0.09	
		slight	0.02	

3. Group quantitative information maps
Total 7 individual information maps are further superposed to generate a sum information map which presents comprehensive contributions by all factors and their categories.

4. Grade geo-hazard
Based on the map from procedure 3, a geo-hazard grading map which indicates different hazard levels can be compiled in terms of total quantitative information values. Through analysing the distribution of the sum information and the

geo-hazard distribution map, four hazard levels are finally defined: high hazard (I>1.0 bit), middle hazard (0~1.0 bit), low hazard (-1.0~0 bit), and very low hazard (<-1.0 bit) (Fig. 5).

CONCLUSIONS

Based on the information analysis and geo-hazard grading map of China, the following can be mainly concluded:

1. Slope hazards in China are significantly dominated by complex geology, seismic background, high and steep topography. Clastic rock formations, carbonate rock systems, week metamorphic rocks such as schist, slate, and loess are the most favourable geological environments. Seismic intensity above VII corresponding to earthquake acceleration around 0.1g can normally trigger slope failures.

2. Meanwhile, external factors such as heavy rainfall or human activity played important roles in recent geo-hazard events.

3. High hazard regions are mainly located in southwest China and along the upper stream and middle stream of the Yangtze, middle hazard regions are mostly in the central and the southeast. Large parts of the north show low and very low hazard levels.

4. This first overview geo-hazard zonation map will further serve for land planning, geological hazard management, or risk assessment by administration agencies.

ACKNOWLEDGEMENT

This paper is sponsored by the project "National geo-hazard risk assessment" of the Ministry of National Land Resources and by the Project 40072084 of National Natural Science Foundation of China.

REFERENCES

Brabb E. E., 1984, Innovation approaches to landslide hazard and risk mapping, Proc. of 4th ISL, Toronto, pp307-323.

Carrara, 1983, Multivariate models for landslide hazard evaluation, Math. Geol.,(15), pp403-426.

Einstein H., 1998, Landslide risk assessment procedure, Landslide, Balkema, Vol.2, pp1075-1090.

Ross M., 1998, Landslide susceptibility mapping using two matrix assessment approach, Geohazards in engineering geology, Geological society engineering geology special publication No.15, London, pp247-264.

Yan T., Wu F & Yin K., 1989, Static and dynamic regularity of landslide and space-time prognosis of slope instability, Earth science, Journal of China university of geosciences (in Chinese), Vol.14, pp117-133.

Yin K., Yan T., 1988, Statistical prediction models for slope instability of metamorphosed rocks, Landslide, pp1269-1272.

WU Y., YIN K. & Y. Liu, 2000, Information analysis system for landslide hazard zonation, Landslides in research, theory and practice, Thomas Telford, Vol.3, pp1593-1598.

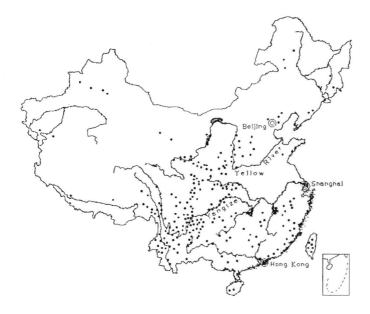

Fig. 2   Distribution map of major geo-hazard in China

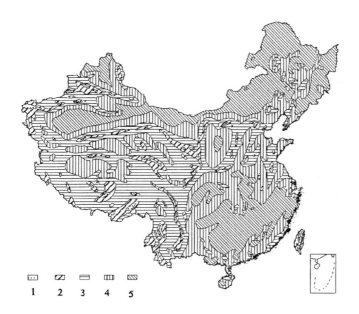

Fig.3   seismic intensity map of China
1-- ≥IX, 2--VIII～IX, 3--VII～VIII, 4--VI～VII, 5--＜VI

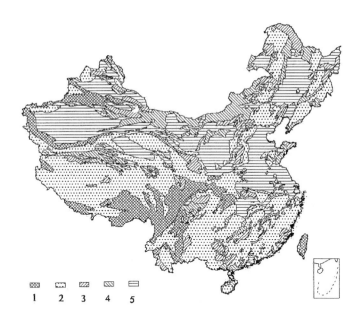

Fig.4   Slope map of China
1--very steep, 2--steep, 3--moderate, 4--gentle, 5--flat

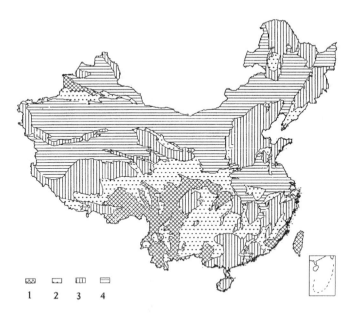

Fig. 5   Geohazard grading map of China
1--high, 2--middle, 3--low, 4--very low

Session 4:

Handling information relating to unstable land

Handling, storage and dissemination of instability data and information, including monitoring equipment data, innovations and providing information on unstable land to local residents and businesses.

Landscape planning and its cartographical support

MS. N. ALEXEENKO, Cartography and Geoinformatics Dept., Moscow State University, Russia

In modern Russian economical conditions the importance of administrative regions increases and administrative regions turn into main territorial objects of management and planning. For the best organization of regional authorities activity in social, economical and nature protecting areas the planning of landscape is necessary. Landscape planning is a new procedure for both Russian science and economics. That's why neither science-methodical nor economical base for developing of large-scale works doesn't exist.

Environmental pollution, wastes of natural resources and their consequences (it's just the same as in the case of illness) are better to prevent than to cure. It is possible in case if nature-protecting activity will be started on the stage of planning. It is necessary to project and build the future not only to predict it. This idea is not new even for our country.

Territory planning had already existed in Russia in soviet time. It can be characterized in the following way:
- Priority of economical interests;
- Architects were the main developers of region planning;
- Subordinate role of nature protecting method (not every regional plan included the part of nature protecting)
- The medium scale planning prevailed – the majority of maps made in scale 1: 200 000 or 1:100 000.

Consequences of such territorial planning in country's scale are well-known: turning Volga –river into range of reservoirs with stagnant water and as a result the fate of Caspian herd of sturgeons; poisoning of Azov sea by herbicides from Cuban rice fields; ecological disaster in Aral region. The result of regional and numerous local planning are less known, but they have created ecological unfortunate background in Russia.

Instructions and methodological recommendations on how to make schemes and projects of territory planning had been changing periodically. But in whole the list of main maps schemes and of project of regional planning doesn't differ from those maps, which were enumerated in one of the first methodical works about territory planning which was published in 1941.

Nowadays the actuality of developing of methods of landscape planning in Russian conditions is clear. Practical importance of that work is determined by role of landscape planning in solving of geo-ecological problems on different territories.

Poor cartographical studying and lack of information could characterize many Russian developing regions. All this facts make the landscape planning realization more difficult.

Special attention must be paid to regions with good perspectives of economical development and unknown ecological conditions. That's why it is useful to enrich methodological and practical experience of both traditional and computer cartography for landscape planning realizing. It is good also to use the remote sensing data for mapping as a basis of landscape planning. Recently both the administrative bodies and new private industries have to revalue the perspectives and re-plan social- economical development.

Now it's necessary to take into consideration the interests and specifics of regions in exact details. Ecological component also is important. Modern projecting must be territorially differentiated; it has to include all natural and social-economical conditions for every region and its inner differences. The main aim is to create the effective mechanism of interaction between society and nature instead of "direct administration" which acted together with inactive laws on nature protection.

The analysis of modern tendencies in nature guarding and regional development shows, that current situation in Russia is favorable for ecologically oriented territory planning:

1. Ecological culture is increased. Economical profit has not abstract meaning.
2. The process of redistribution of competences between towns and regions is being done.
3. For all economical units has appeared the opportunity to involve in world economical relations through the direct contacts with foreign partners. So it is necessary to sequence the normative base of nature recourse using with international standards.
4. The number of conflicts increases due to resources, territories and infringing of citizens' rights on living in ecologically favorable land, so the mechanism of sequence of interests and prevention of conflicts doesn't work properly.

But together with positive moments there are some serious obstacles:

1. In fact perfect regional development conceptions doesn't exist (it's not clear is it necessary to support undeveloped regions, borrowing resources

from the developed regions and by this way making them weaker or to support developed regions and to get much resources for Federation developing).
2. Lacking of normative documents for all kinds of planning.
3. Organizational scarce and lacking of skilled staff. The most "anti-ecological" production gives the biggest profit – this paradox also influences the situation.

Landscape planning is firstly - combination of measures which must be used to create such activity organization of the society in certain landscape which could provide stable nature protection and saving of main function of these landscapes as a system of life support; secondly – that is a communication process: subjects of nature protecting and хозяйствующей деятельности on the territory of planning are involved into that process. Communicative aspect provides revelation of interests of nature guarders problems of using of natural resources and conflicts solving. That process also provides developing of the plan of activities and measures.

Usually landscape planning includes 3 levels: district (oblast'), region (rayon) and the lowest territorial unit. Landscape planning includes:
- developing of landscape programs of territory developing;
- producing of frame landscape plan (scale 1:200 000 1:100 000);
- large-scale landscape mapmaking (scale 1:25 000 and larger);
- developing (or systemizing and coordinating) of normative documents on realization of landscape plans and control

Composition of landscape planning contains 5 stages:
- **inventory** – obtaining and generalization of available information about the environment of that territory, its social-economical conditions, structure and specifics of land using, revelation of the main conflicts of natural using in the context of analysis of ecological problem of that territory.
- evaluation of the natural resources and potential of the planning territory in terms of importance and sensitiveness as well as estimation of land use character
- developing of conceptual aims for the recourses usage (for the each natural component)
- developing of the integrated program of territorial using
- main activities program developing

This paper concerns only **inventory** stage. Our research main objectives are the following: to find out the suitable set of information data for that stage and to detect the maps list for information block; determination of bases and units for mapping; elaboration of demands for maps sequence in block and recommendations on the maps contents.

The fist task for landscape planning conduction is choosing of information sources for information gathering about the territory, information analysis, grounds of demands. The statistical, cartographical and remote sensed data have been represented widely among these sources. Additionally field research and survey data as well as textual sources can be used. Also it is possible to use mental maps and the respondents might be of different age, social and educational level.

The main results of that stage are inventory maps in scale 1:100 000 and the list of main problems of the territory planning. The content and information fullness of maps (scale 1:100 000) have to reflect current "status quo" of the environment and the peculiarities of land's using.
Inventory maps have to contain the following information:

1. *Modern land using.* In that item information about administrative arrangement of the territory: structure of land using with characteristics in different categories of the land, manmade territory violation is presented. That information may be presented on the landing arrangement schemes, on forest-using, water-using maps, on the maps of territories governmentally protected, geomorphological maps with the exogenous processes characteristics, erosion and inclination angle breach. It is good to add maps of agricultural specialization of the territory, melioration maps and maps with information about the lands productiveness.
2. *Climate and air condition.* Besides the maps with average year values of the different climate components (temperature, precipitation, winds), maps fixing pollutions are necessary.
3. *Soils.* Soil maps (kinds and types of soils), maps of current processes (degradation, deformation, erosion), the maps of soils pollution must be included.
4. *Water environment.* There must be hydrological map (with traditional set of characteristics), energy resources, pollutions, and hydrographical constructions.
5. *Vegetation.* There must be maps of vegetation, geo-botanical, and forest-taxation. Maps about changing of areas and content of species, Red books of different levels and the places of medical grasses growing must be represented in this item.
6. *Landscapes.* Landscape map. Map of the landscape changing.
7. *Population.* Dynamics of mortality and birth rate, migration. Distribution on the territory. Health Babies mortality (up to 1 year age), the level of illness, the level of disabled people. Number of working citizens. Information on the unemployment. Serves on the needs of the society, attitudes towards land using, emigration, possible ways of economic development, appraisement of the level of life.
8. *Infrastructure.* The level of living conditions (information on transport security, the most important service types). Cultural centers, possibility to get education, the criminal situation.

9. *Industry and agriculture*. Structure. Ecologically dangerous production. Traditional types of agricultural activities of inhabitants. Subsidies and fees. Provision with energy and with local material in different branches. Environment pollution and cleaning systems.

Such list of information can be varied because it depends on the natural features of the territory.

The most important task of inventory stage is revelation of interests of nature users and analysis of the conflicts in this sphere. That is why it's necessary to consult people engaged into this work and spread information about tasks and procedures of work by mass media, open debates and other means even on the stage of the landscape planning.

Map of conflict situations is the most important on this stage. To attract population to rational nature using it is necessary to establish relations, to create the situation of cooperation. Such map had never been created earlier; the situation was analyzed in the text only.

The aim of that map is not to show the decision of such problem, but to make first step in that direction, to reveal the areas of the conflict, to make typology of them, to determine reasons of escalation, to fix the participants. Cartographical localization of territory doesn't cause any difficulties. In territory aspect conflicts can be differ on point, line and area ones. The second important characteristic of the conflict situation in nature using is a time aspect. It must be reflect on a map of conflict situation. In that aspect we can outline the following situations: long-time and short-time, non-interrupted and impulsive, seasonable and all over the year, 1 hour and 24 hours.

The following situation is possible: in territory aspect the interests of nature users can cross and in the time aspect they don't. It's possible to call the situation conflict in this case?

The next characteristic of conflict situation is intensivity and tense. It's apparent that conflict must be ranged in that case. No doubt it's necessary to characterize the conflict on extent of complexity: there are mono-conflicts, pare conflicts, multy-conflicts.

Looking through certain conflict situation it is pleasant to know if the conditions of the verging territory influence upon the condition of the conflict and is there mutual influence? What does determine the limits in fixation of areas in conflict situation? Does conflict exist already? Is it going to reveal or is it potentially possible?

Necessity of development of scientifically proved territory planning works on the detailed level exists in our country and it has to develop according to ideas of settle development. Territories where landscape planning probably will take place are different in geographical position; they are different in the level of social-economy development and in conflicts in sphere "environment - man". Landscape planning cartographically supported in a proper way could help to solve these problems. In order to solve conflict in a concrete situation could be developed a special program but that task can be solved easier using general methodical recommendations.

Planning mitigation works in a large-scale landslide periodically affecting an inhabited area (Central Italy)

MACEO-GIOVANNI ANGELI, National Research Council, Perugia, Italy
FABRIZIO PONTONI, Geoequipe Consulting, Camerino, Italy
FRANCO PONTONI, Geoequipe Consulting, Camerino, Italy
STEFANO LEONORI, Geoequipe Consulting, Camerino, Italy
STEFANO LEOPERDI, Councilor, Macerata Province, Italy

ABSTRACT

This paper is the result of more general research aimed to study the evolution of landslides affecting inhabited areas in the Marche Region (Central Italy). In this region, landslides of different sizes and types are widespread; this is due to the presence of unfavourable geological and geomorphological conditions. The research deals with the evolution of instability phenomena and examines the causes of their activation (or reactivation) such as human activities and natural events. After the earthquake that shocked the whole area of Central Italy, causing loss of human lives and major damage to buildings, the area was included in the general reconstruction programme financed by the Central Government, both for the buildings and for land instability. Taking also into account the past climate fluctuations (since XVIth century) and the precipitation data recorded in the area during the last century, the large-scale and ancient landslide affecting the village of Sant'Agata Feltria has been investigated in detail and the general design for the main control works has been worked out.

INTRODUCTION

The village of S. Agata Feltria is located in the north-western end of the Marche region, at the boundary with the Romagna region (Fig. 1). The territory is typically hilly, with altitudes ranging between 265 m asl and 930 m asl. The geology is characterised by terrains belonging to the Val Marecchia complex and the Umbria-Marche Sequence outcrops (ANGELI & Alii, 1996).

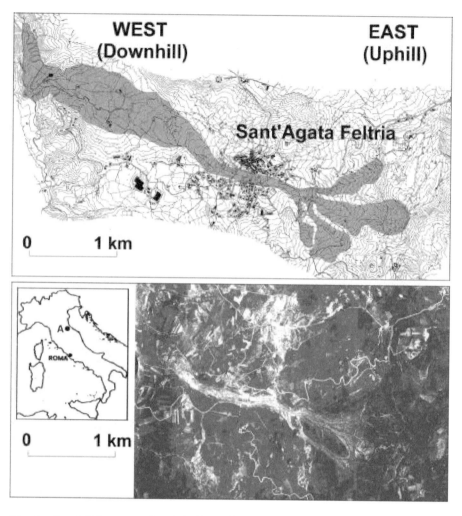

Fig. 1 – Landslide map and aerial photo of the event that occurred in March 1934 (at that time the remobilisation of the mass stopped just 1 km downhill the village) with a location map of the village (A).

The chaothic complex of the Val Marecchia (emplaced by gravity during the lower Pliocene) is made up of allochthonous Palaeogene scaly multicoloured clays (Liguridi) with Oligocene sandstones and pelitic sandstones of the Monte Senario Formation (Epiliguridi). The Umbria-Marche succession in this area is represented by Messinian units underlying the Val Marecchia complex. The Messinian units are formed by interbedded sandstones and marly clays (Marnoso-Arenacea and Ghioli di Letto Formation) and by microcrystalline gypsum with intercalated bituminous clays and solphide-bearing dolomitic limestones (Gessoso-Solfifera Formation).

The landslide affecting the village of Sant'Agata Feltria (Fig. 1) is very ancient and its reactivation has been recorded at least twelve times starting from XVIth century. It presents itself in the form of a mudslide 5.4 km long, which crosses the inhabited area endangering many buildings and roads. The last important landslide event occurred at the end of the winter of 1934, after a very critical rainfall season. Major damage to the buildings and to the infrastructure induced the Central Government to require the Air Force to make investigations with the most sophisticated instrumentation available at that time. Aerial photos were taken just after the event (Fig. 1) in order to record the size and the shape of the landslide. Thanks to these photos we are now able to assess no doubt the maximum extent of a new possible event.

The oldest well documented event (1561 AD) took place during a period of severe climate involving the whole Europe, well-known as "Little Ice Age", which is the closest strong climate change which lasted some centuries. It started approximately, depending on the different latitudes, in 1400-1500 and terminated at the end of 1800 (Fig. 2).

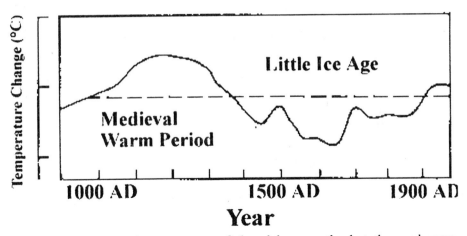

Fig. 2 – The mean air temperature of the globe over the last thousand years (Houghton *et al.*, 1990)

It was a very cold period, with abundant precipitation, which strongly affected the surviving conditions of many people around the world. Historical research, geomorphological analyses and recent geotechnical investigations allowed a detailed reconstruction of the evolution of the phenomenon, outlining its possible causes and attempting a risk zonation. In particular, in the upper part of the landslide, a deep-seated gravitational phenomenon affecting a very thick sandstone plateau (marked by the presence of large trenches and many ponds or small lakes) has been recognised as the main source area of the landslide.

The reactivation of the mudslide generally coincided with particularly rainy periods, even though a connection with the strong earthquakes that affect the area cannot be excluded.

The above recent geotechnical investigations, included in the general reconstruction programme financed by the Central Government, besides having pointed out the hydro-geological features of the phenomenon, allowed preparation of a better plan for some large-scale mitigation control measures.

THE CASE HISTORY

The main mass movement (Fig. 1) involves part of the village of Sant'Agata Feltria, dividing it into two parts, as clearly shown by the damage to the buildings during the critical event of March 1934 (Fig. 3). This mass movement, whose reactivation has been recorded 12 times starting from XVIth century, is 5.4 km long and has about 665 m of vertical descent (from about 930 m asl down to 265 m asl), with an average slope angle of about 12°. It is up to 1.6 km wide and involves more than 50 million cubic metres of terrain. For its large size and for having hit the village many times in the past, this phenomenon has been the object of detailed investigations, especially after the strong earthquake of 1997.

Historical research, geomorphological analysis and geotechnical investigations have allowed a detailed reconstruction of the evolution and the mechanism of the phenomenon.

Fig. 3 – Damage to the buildings during the critical landslide event of March 1934.

The first documented period of activity dates back to 1561, when the movement had already assumed its present shape and direction, even though a notary act dated 1485 reports about a "ruined area" in the neighbouring of the castle. It must be taken into account that the second half of XVIth century was characterised by very severe climatic conditions, with deposition of thick snow cover and intense rainfall, coinciding with a colder phase of the "Little Ice Age" (Fig. 2). The second documented activity period occurred in 1604, again characterised by a particularly severe climate. The recording of particularly late dates for the starting period of vintages have been used to recognise unusual and severe climatic conditions.

Further movements occurred during 1644 and 1647, when the climate, even though progressively getting better, was still characterised by intense precipitation. During the second half of XVIIth century climatic conditions improved and consequently no landslides have been documented in the area.

Starting from the beginning of XVIIIth century, the climate turned cold and humid, as testified by many documents related to soil instability. In particular, in 1714, 1723, 1743, 1748, 1750, 1753, 1754, 1756, 1758, 1772, 1773 and 1781 indications of landslide activity have been found. In 1782, a strong earthquake affected the area damaging some buildings, but no report exists of remobilization of the landslide.

Then the landslide reactivated during the winter 1801-1802 and in 1803. In 1815 a partial remobilization took place. Successive events of remobilization happened in 1834, 1839, 1840, 1844, 1845 and 1846. Starting from 1848 the climate improved again, and, therefore, for several years no landslides took place.

In March 1934, after a prolonged and severe rainfall period, the mudslide experienced a strong reactivation.

THE LANDSLIDE MECHANISM
The mechanism of the landslide and the climatic conditions that cause instability can be outlined as follows. It has been found that in the mountains, which are uphill of the village (Fig. 1), the Oligocene sandstones "floating" over the scaly clays seem to be affected by deep-seated gravitational movements. These deformations gave origin to steps and trenches mainly trending N-S or NNE-SSW which disconnected the huge lithoid slab (the plateau) into minor parts. These parts continued to supply the mudslide with material. In fact, large portions of the plateau (more than 200 metres thick, as detected by means of deep boreholes) tend to detach from the top, forming very large and deep rotational landslides (Fig. 4), favouring the formation of small lakes and ponds in the counter-sloping areas. These landslides, formed by very large blocks of sandstones, go to overload the plastic clays below, causing undrained shear strength conditions in them. As a result different lobes of clays tend to move and converge toward a unique channel of mudslide (Fig. 1).

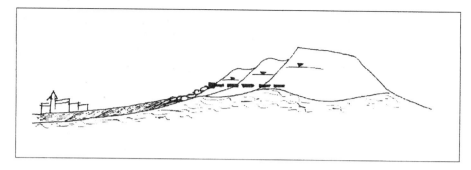

Fig. 4 – Schematic cross section in the landslide uphill area with indication of the deep drainage system.

The clay mudslide is not very thick (only a few metres), but it proceeds very quickly downhill, creating other overloads to the approaching clayey strata found along its path. At the very end a continuous flow of soil takes place and only after reaching the natural drainage at its base and in the sandstone portions already collapsed uphill (the engine of the landslide) does the mudslide stop.

The uphill sandstone plateau corresponds to a huge natural water tank (Fig. 4). The measurements taken in the boreholes drilled at the top, in the framework of the overall monitoring system installed all through the mudslide, give a very high water table. Under these conditions a continuous groundwater flow supplies the uphill landslide bodies, maintaining them at the limit of equilibrium.

Looking through the rainfall recordings it is quite evident that the landslide movements mostly coincided with periods of long-lasting and intense rainfall (Fig. 5, 6). The event of 1934 demonstrated that the cumulative rainfall over 8 months can cause the sudden remobilization of the landslide.

In fact, we have observed that starting from August 1933 until March 1934 a very large amount of rainfall occurred (1200 mm). The analysis proceeded taking into account all the series of rainfall accumulated over 8 months (from August to March) during the period 1926-1962. But, in this same period at least 3 other similar values of cumulative rainfall were recorded, as clearly shown in figure 5. The point was that no important landslide events renewed in correspondence with these periods. Under these conditions the analysis of each critical situation (shown in figure 5) continued taking into account the temporal distribution of the values of the monthly rainfall over the same period of 8 months starting from August and stopping in March (Fig. 6). In this way we found out that the temporal trend, in the 8 months preceding the 1934 landslide, shows a shape of the curve which is very different from the other situations analysed.

Moreover, the analysis of the air temperatures (over the same 8 month period) recorded in 1933-34 showed a very mild winter. In fact, the temperatures observed at a meteorological station, located nearby the village, but at a higher elevation (812 m. asl), gave values always above 0°C. This definitely implies that the overall precipitation in that period was only rainfall and also that, in total absence of frozen soil surface, all this water could easily and continuously infiltrate at depth.

Therefore, we can conclude that all the climate recordings preceding the 1934 landslide event were very unusual and very favourable to induce critical groundwater conditions in the soil mass.

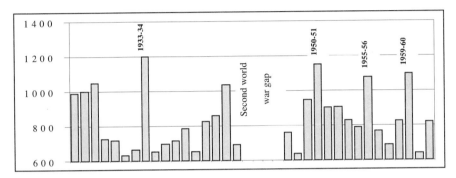

Fig. 5 – Series of the rainfall accumulated over 8 months (from August to March) in the period 1926-1962.

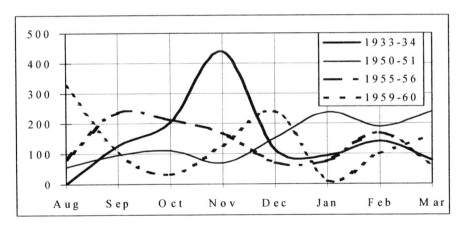

Fig. 6 – Monthly rainfall trend in the most critical periods shown in figure 5; the 1933-34 trend seems to be responsible for the great landslide event of March 1934.

FINAL REMARKS
Due to the huge mass of the landslide analysed and the complexity of the mechanism, the strategy for the control works was mainly directed to the mitigation

of the phenomena at work, due to the impossibility of carrying out a **definitive** solution to the problem.

In detail, the analysis of the rainfall series gave us a simple but effective tool to design control works in so large an area.

The analysis of rainfall series showed that the distribution over selected periods of 8 months (eg. preceding the landslide which occurred in March 1934) was very dangerous and capable of very rapidly filling up the uphill natural tank (the sandstone plateau). The rapidity of this last process definitely prevented the possibility of dissipating the water pressures that can arise inside the landslide mass. This lack in the natural drainage has proved to be crucial for the instability conditions of the whole landslide.

Therefore, the principle of the project was the design of both a deep and a superficial drainage system capable of intercepting the groundwater in the uphill plateau, in order to drastically reduce the possibility of the water flowing freely downhill. It will be achieved by means of very long tubular drains (Fig. 4) and trenches. The water will be completely collected without penetrating into the downhill landslide body anymore.

The main objective is to stabilise the large portions of the plateau, already detached from the top and hanging on the plastic clays (therefore maintaining dangerous stability conditions for the downhill village of Sant'Agata Feltria), while at the same time stopping the possibility of triggering a chain reaction, propagating the instability downhill with the well known mechanism of the undrained shear strength. This objective will be reached not only by blocking the dangerous groundwater flow from the sandstone plateau, but also by maintaining the water table much lower throughout the year, even in the presence of adverse climatic conditions.

No clear direct connection seems to exist with earthquakes, even though the area is considerably active from a seismic point of view. However, seismic shocks probably contributed to the general instability of the area and, most of all, can be considered among the most important factors for emphasising the fragmentation of the huge lithoid slab located at the top of the slope.

REFERENCES

ANGELI M.-G., BISCI C., BURATTINI F., DRAMIS F., LEOPERDI S., PONTONI F., PONTONI F. "Two Examples of Large-Scale Landslides Affecting Built-Up Areas in the Marche Region (Central Italy)" Quaderni di Geologia Applicata, vol. 1, pp.131-140, 1996.

HOUGHTON, J.T., JENKINS, G.J. AND EPHRAUMS, J.J. (Eds.). 1990. Climate Change: The IPCC Scientific Assessment. Cambridge University Press, Cambridge, UK.

Ground-based radar interferometry: a novel technique for monitoring unstable slopes and cliffs

C. ATZENI[1], P. CANUTI[2], N. CASAGLI[2], D. LEVA[3], G. LUZI[1], S. MORETTI[2], M. PIERACCINI[1], A. J. SIEBER[3], D. TARCHI[3]
(1) Department of Electronics and Telecommunications, Univ. of Firenze, Italy
(2) Earth Sciences Department, University of Firenze, Italy
(3) European Commission, Joint Research Centre, ISIS, TDP Unit, Ispra, Italy

ABSTRACT

This paper describes a new remote-sensing technique for monitoring landslides and unstable slopes; it is based on a synthetic aperture radar (SAR) interferometry and it is fully implemented by using portable ground-based instrumentation. The technique has been tested on a large rockslide in the Italian Alps producing multi-temporal deformation maps of high resolution and accuracy. The results have been validated through independent measurements obtained from geotechnical monitoring instrumentation.

INTRODUCTION

Monitoring of slopes and dormant landslides is necessary to prevent disasters in unstable areas, like mountain and hilly ranges or costal cliffs. Existing methods of geotechnical monitoring are based on networks of sensors which must be installed directly on the unstable areas, with serious problems of maintenance and effectiveness. The potential of remote-sensing techniques, based on *synthetic aperture radar* (SAR) interferometry from satellite platforms, has recently been explored (Fruneau et al., 1996; Singhroy et al., 1998; Rott et al., 1999; Refice et al., 2000), but has given reliable results only for large landslides on gently inclined slopes and with moderate displacement rates. The use of SAR ground-based sensors permits the avoidance of most of the problems associated with satellites, guaranteeing the necessary flexibility and adaptability to each specific case.

INSTRUMENTATION

A portable device for ground-based SAR interferometry, called LISA (*Linear SAR*), was developed by the Joint Research Centre of the European Commission, specifically for applications in the field (Rudolf et al., 1999). It is able to provide 17 GHz measurements with a 2.8 m synthetic aperture.

Radar data are collected by a couple of antennas mounted on a linear track that slides them horizontally, forming a synthetic aperture (Fig.1). Coherent SAR processing converts the raw radar data into an image which contains, for every pixel, information on the phase depending on the target-sensor distance. The accurate measure of the relative displacements of each pixel, in the time span between two acquisitions and in the direction of the line of sight between the antenna and the target, can be directly achieved by repeating the measurement from exactly the same position (zero baseline). Assuming that the dielectric properties in the observed scene remain constant in the interval between the two acquisitions, the phase difference between corresponding pixels in the two images can be directly related to ground displacements (Zebker & Goldstein, 1986), which can be assessed with an accuracy of a small fraction of the wavelength.

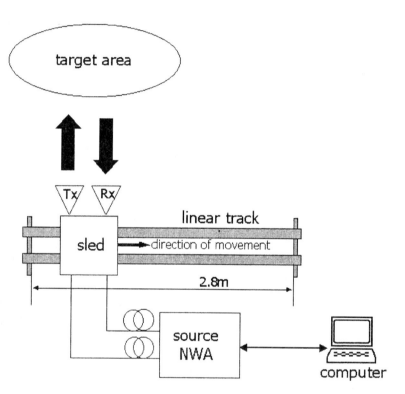

Figure 1. Scheme of the hardware components. Tx = transmitting antenna; Rx = receiving antenna

The microwave component of the system is composed of a continuous-wave-stepped-frequency (CW-SF) radar, based on a network analyser which includes a signal source between 30 kHz to 6 GHz. A coherent conversion module of about 17 GHz is employed for measures at higher frequencies. Observations are typically carried out in the frequency band 16.80 – 16.88 GHz with steps of 50kHz and VV polarisation. The transmitted power was approximately 25 dbM (300 mW). The antenna synthesis is obtained by moving a motorised sled, hosting the antennas and other microwave components, at steps of 7 mm along a 2.8 m long straight track controlled by a linear positioner. The system must be installed in a stable site in front of the target area, which must be completely visible, at a distance preferably lower than 2 km.

DATA ACQUISITION AND PROCESSING

A measurement campaign was carried out in the period between 25/07/2000 and 2/8/2000 on the Ruinon landslide in the Italian Alps, to test the system capability and to evaluate the accuracy in field conditions. Ruinon is a 30 million m^3 active rock slide that, since 1987, is kept under constant monitoring for early warning purposes as it represents a major threat to human lives and socio-economic activities. The landslide (Fig.2) is located in the lower portion of a SW facing slope (about 240°N) with an average inclination of about 35°, on the hydrographic right of the Frodolfo stream. The landslide is characterised by two main scarps, oriented WNW-ESE located at an elevation of about 2100 m a.s.l. (Upper Scarp), and 1900 m a.s.l. (Lower Scarp).

Figure 2. Picture of the Ruinon landslide from the opposite slope where the radar was installed

The LISA was rigidly fixed on a stable support (Fig.3) on the slope facing the landslide, at an average distance of 1.3 km. In these conditions the expected spatial resolution of the images is approximately 5 x 5 m with an expected precision of displacement detection of 0.75 mm. Image acquisitions were repeated at time intervals of ca. 35 minutes from the same position. A total of 124 images were collected, arranged in a couple of continuous sequences (Tab.1). In order to eliminate disturbances due to moving objects on the target scene (such as wind-blown vegetation), three images were averaged to produce a coherent final image. For this reason the effective time interval between image couples used to form the interferograms is ca. 1h and 45 min. Interferograms, showing the phase difference (related to ground displacement) between the two images for each pixel, are obtained from the cross-correlation between averaged image couples.

Figure 3. Arrangement of the instrumentation in the field operational conditions

Sequence	Starting time	Total duration	No. of images	No. of interferograms
1st	27/07/2000 14:34	15hrs 55min	28	9
2nd	31/07/2000 00:44	55hrs 14 min	94	29

Table 1. Details on the sequence obtained for the Ruinon landslide

The last interferograms of both sequences are shown in Fig.4 and Fig.5 respectively. The interferograms are georeferenced through a high definition digital elevation model (5x5m) of the slope with a local coordinate system with the origin placed in the central point of the radar image. The spatial resolution of the image is the same as the digital elevation model (5x5 m). The grey scale directly expresses the phase difference in millimetres, which in turn corresponds to the effective ground movements in the direction of the sensor. For each sequence, the first averaged image is taken as a reference and interferograms are produced comparing the successive images with the first, in order to obtain cumulated deformation maps relative to the initial condition. Positive values indicate the movements towards the observer. Pixels with a low coherence (due to irregular superficial movements as those linked to vegetation) have been masked.

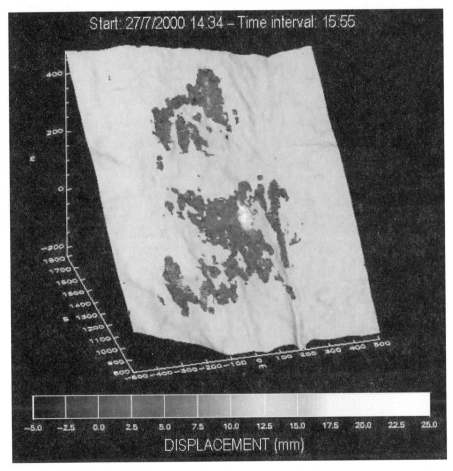

Figure 4. Detail of the last interferogram of the first sequence

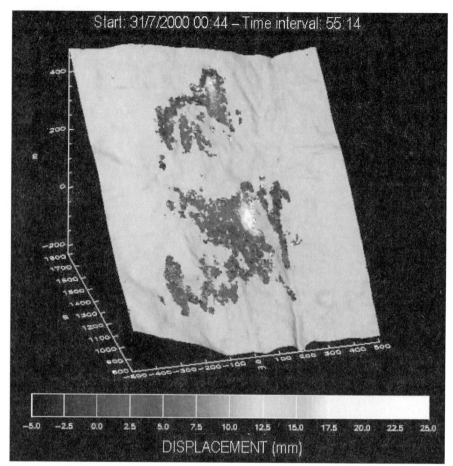

Figure 5. Detail of the last interferogram of the second sequence

The interpretation of the interferograms, highlights the presence of relevant displacements concentrated in well-defined areas within the landslide, with two relative maxima just beneath the two main scarps. Significant displacements are evident at the Lower Scarp in both the sequences, where an area of about 100x100 m with a sub-circular shape shows a maximum total displacement of about 21 mm in the first sequence (average displacement rate of 1.32 mm/h). Relevant movements occur also on the Upper Scarp, affecting an irregular area (about 50x80 m) elongated in the dip direction of the slope with a maximum displacement measured at about 12 mm (average displacement rate of 0.75 mm/h). The interferograms shows also some distributed displacements of low magnitude, but the total duration of the sequence is too low to allow this evolution to be followed in detail.

These results have been validated by using independent measurements obtained from the Geological Monitoring Centre of the Lombardia Region. The comparison between total displacements measured by a network of automatic extensometers, recording data every 4 hrs, and by the radar in the corresponding pixels, shows a satisfactory agreement. Considering the total time interval, the difference between measurements is lower than 2 mm, whereas discrepancies within this interval are limited within 5 mm.

These small discrepancies between the radar and the extensometers can be explained by taking into account that radar displacements are "averaged" both in space and in time, since they refer to a square pixel of 5x5 m and to a time interval of 35 minutes, necessary for a single image acquisition. Conversely extensometers provide single-point, instantaneous measurements. Moreover, small errors in the localisation of the extensometer position or in the assessment of the angle between the extensometer wire cables and the sight line of the radar can affect the comparison between the two data series. Finally extensometer readings are affected by cyclic effects (within a period of 24 hours) linked to daily temperature variations that produce the greater discrepancies compared to radar data.

CONCLUSIONS

The application of ground-based radar interferometry to the Ruinon rockslide has confirmed the interesting perspectives of this technique as a tool for monitoring unstable slopes. The portable SAR device employed in the field campaign has demonstrated its manageability and adaptability even to rough field conditions. The operational parameters, such as the frequency band, the synthetic aperture length and the time interval between acquisition, permitted the detection of ground movements with a precision of 0.75 and an accuracy of some millimetres.

The technique can obtain distributed information over wide areas by assessing the entire displacement field of the less vegetated portions of the slope. Its remote sensing nature allows inaccessible zones and dangerous landslide sectors with high displacement rates to be kept under constant control. These characteristics represent the main innovation in the realm of slope monitoring, with evident improvements compared to GPS or other traditional methods of geotechnical monitoring, which are all based on single-point measurements on benchmarks and sensors directly positioned on the area to be controlled.

Still more evident are the advantages with respect to spaceborne SAR interferometry, since the use of ground based sensors not only permits a higher spatial resolution and a better accuracy, but it also provides a wide flexibility in the choice of the operational parameters to be adapted to a variety of ground conditions (such as steep slopes and narrow valleys) where satellite observations are strongly limited. In particular, the customized interval between measurements makes the investigation of a wide spectrum of landslide velocities possible and ensures reliable results even for fast moving landslides that, obviously, raise the main problems in civil protection and instability of buildings.

ACKNOWLEDGEMENTS

The research on the applicability of ground-based SAR for landslide monitoring is funded by the Italian Space Agency (ASI) and by the National Research Council Group for Hydro-Geological Disaster Prevention (CNR-GNDCI) supported by the Italian Agency of Civil Protection. The application on the Ruinon landslide was carried out in the framework of a research contract between the Lombardia Regional Administration and the University of Firenze *"Testing ground-based SAR interferometry for monitoring the Ruinon landslide in Valfurva"*.

CNR-GNDCI Publication n.2453

REFERENCES

Atzeni, C., Canuti, P., Casagli, N., Leva, D., Luzi, G., Moretti, S., Pieraccini, M., Sieber, A.J. and Tarchi, D. 2001a. A portable device for landslide monitoring using radar interferometry. Landslide News International Newsletter. Disaster Prevention Research Institute, Kyoto (Japan), in press.

Fruneau, B., Achache, J., Delancourt, C. 1996. Observation and modelling of the Saint- Etienne-de-Tinee landslide using SAR interferometry. Tectophysics, 265, pp. 181-190.

Refice, A., Bovenga, F., Wasowski, J. and Guerriero, L. 2000, "Use of InSAR Data for Landslide Monitoring: A Case Study from Southern Italy," IGARSS 2000, Hawaii, 2504-2506.

Rott, H., Scheuchel, B. and Siegel, A. 1999. Monitoring very slow slope movements by means of SAR interferometry: A case study from mass waste above a reservior in the Otztal Alps, Austria. Geophysical Research Letters, vol. 26, no.11, pp.1629-1632.

Rudolf, H., Leva, D., Tarchi D. and Sieber, A.J. 1999. A mobile and versatile SAR system. Proc. IGARSS 99, pp. 595-594.

Singhroy, V., Mattar, K. and Gray, L. 1998. Landslide characterization in Canada using interferometric SAR and combined SAR and TM images. Advances in Space Research vol 2(3), pp. 465-476.

Zebker, H.A. & Goldstein, R.M. 1986. Topographic mapping from interferometric Synthetic Aperture Radar observations. Journal Geophysical Research, vol. 91, pp. 4993-4999.

Early warning of landsliding using weather station data

DR. M. E. BARTON , Civil & Environmental Engineering Department, Southampton University, UK

EARLY WARNING OF LANDSLIDING

Early warning of landsliding initiated by changes in pore pressures can be obtained by three kinds of methods as follows.

- Monitoring the ground movements.
- Monitoring the in-situ pore water pressures.
- By prediction from meteorological observations.

These three methods will give varying degrees of early warning. In the case of the ground movements, the warning is immediate in the sense that the movements are actually taking place. The in-situ pore pressures provide additional time in so far as the rise of pore pressure will be at a rate determined by the permeability of the ground and must reach the critical level for ground movement to occur. By contrast, using weather station data to predict the changes in pore pressures can give much longer advance warning of potential landsliding and hence much longer for appropriate action to be taken. The time lag between the meteorological data and initiation of landsliding will greatly depend on ground conditions and the complexity of the landslide but potentially this period will be several orders of magnitude longer than with either of the other kinds of methods. It is surprising, therefore, that more use has *not* been made of it in areas much at risk from, and likely to be greatly affected by, landsliding.

MODELS RELATING METEOROLOGICAL DATA TO LANDSLIDING

A comprehensive review of the many ways in which meteorological data can be used to predict landsliding and groundwater pressures has been given by Polemio & Petrucci 2000. The following is offered as a simpler and more succinct scheme of classification. The scheme applies to landslides initiated by pore pressure changes: landsliding resulting from toe erosion is excluded.

1. **Empirical models to predict landslide triggering.**

   (a) Simple and direct models relating rainfall duration and intensity to landslide initiation. These work in situations where the ground is highly permeable and readily responding to infiltration as for instance with coarse grained debris slides. Examples are given by Caine 1980 and Keefer et al 1987.

Caine gives the threshold for triggering debris slides as:-

$$I = 14.82 \ D^{-0.39}$$

where I = intensity in mm / hour and D = duration in hours.

Because of the rapid response, the warning time during a rainfall event may be short but when combined with a meteorological forecast, then the warning time can extend to at least several days.

(b)   Models incorporating cumulative and/or antecedent rainfall. Where the landslide mass has some storage capacity which must be filled before pore pressures can reach critical levels, then the antecedent infiltration will be an important factor. The antecedent rainfall is incorporated into the model using a decay function to allow for drainage of the stored water. An example is given by Zaruba & Mencl 1969 for deep seated landslides in Bohemia where the mean rainfall of three antecedent years was taken into account. The model of Crozier 1999 uses soil water balance and empirically correlates this with landslide triggering. The influence of antecedent conditions in these models implies that time is required for critical pore pressures to be obtained - time which contributes positively to the early warning time.

2. Models to predict changes in pore pressure.

(a)   Simple empirical models to relate meteorological data to pore pressure. As with the models of 1(a), these rely on a simple direct correlation. Pore pressure response in a fissured metamorphic rock mass to rainfall is shown by Harper 1975 in Fig. 1. Predicted pore pressures could be used subsequently for feeding into a slope stability analysis but this is not part of the model in a strict sense.

(b)   Models predicting pore pressure changes by using the soil water balance. These refine the above simple models by determining the effective rainfall and respective rates of infiltration and drainage so as to calculate the amount of recharge in the zone of interest. An example is given in Barton & Thomson 1986 and illustrated in Fig. 2. Models of this type are semi-empirical in that they require adjustment to allow for local variations in soil properties by making use of previous sets of data. Two years data were used for the Barton & Thomson model, using the first year to refine the model. This model gives a satisfactory fit to the second year's data as shown in Fig. 2. A similar type of model is given by Sangrey et al 1984 who also indicated its potentiality for predicting landsliding.

3. Models which combine the meteorological conditions with stability analysis of the slope.

This may be accomplished by simplifying the factors, treating them as mainly empirical, or, if the data and understanding of the relationships allow, by using a full analysis and treating the slope as a complete system. An example of the former is the model of Fell et al 1991 who use piezometric data predicted from a soil water balance model in association with a statistical correlation of piezometric levels with landsliding. A less empirical model is that of Anderson & Howes 1985 which combines observations on infiltration into unsaturated soil with an infinite slope analysis for slopes in Hong Kong. A truly, complete model will require full knowledge of all the recharge processes and would combine the resulting predicted pore pressures with analyses which allow for all the likely modes of landsliding. Models of this latter kind are not yet available.

REQUIREMENTS FOR A COMPLETE, METEOROLOGICAL BASED EARLY WARNING SYSTEM

The requirements will be considered in relation to large landslide complexes which are protected from toe erosion. An example would be the Undercliff landslide complex at Ventnor where sufficient protection exists to discount any toe erosion (McInnes 1996). The requirements include firstly, knowledge of all the sources of seepage reaching potential slip surfaces and, secondly, knowledge of all the potential modes of slope failure.

Sources of seepage reaching potential slip surfaces

The current models predicting pore pressure responses are for clearly defined aquifers. In the Barton & Thomson model, the aquifer is the Plateau Gravel overlying the Barton Clay in the Christchurch Bay cliffs. This aquifer is a major source of seepage to the unstable undercliffs developed on the Barton Clay outcrop and its recharge can therefore be considered as potentially promoting further landsliding. However, complex slopes will have many potential sources of seepage which can be listed as follows.

I.   Direct infiltration from the effective rainfall, controlled by the antecedent conditions , the variation in permeability and amount of fissuring.

II. Deep percolation controlled by the regional hydrogeology and very significantly influenced by the antecedent meteorological conditions.

III. Equilibration of previously unloaded clay strata: a process that can extend over many years but at rates which respond to the meteorological conditions as shown by Vaughan and Walbanke 1973.

IV. Artificial sources such as broken pipes and   inadvertent flooding: such sources need to be investigated and recorded since their influence will not otherwise show up in the analysis of the meteorology and soil water balance models.

V.   Redistribution of groundwater resulting from slope movements: the movements may not initially be classed as serious in themselves but their influence may extend to altering the seepage regime with long term consequences.

Potential modes of failure in complex landslides

No distinction is drawn in the following list as to whether these are first time movements or re-activated slips. In a large landslide complex the latter is more likely but the former is also possible, especially with weakened strata and rising pore pressures. The most likely potential modes can be listed as follows.

I. Compound type landslides (as defined by Skempton & Hutchinson 1969).
II. Lobate mudslides.
III. Scarp failure with toppling and spalling.
IV. Debris slides.

Setting up the system

The complete early warning system will incorporate slope stability analyses for the various modes of instability with input from the sources of recharge given above. It will be necessary to install a network of piezometers and construct a secure local weather station, both using automatic recording. A lengthy period of research is necessary, both in order to research the site investigation fundamentals and to refine the models by using the initial data derived from the weather station and the network of piezometers. The research and operational stages for setting up a complete meteorological based early warning system are shown in Fig. 3.

CONCLUSIONS

- Success has been achieved with simple models relating meteorological data with changes in groundwater pressures and triggering of landsliding, especially where there tends to be a single mode of instability resulting from a particular set of ground conditions.
- Large landslide complexes, protected from toe erosion, will require a lengthy period of research to relate the pore pressure changes to the various modes of instability and to use pore pressure and meteorological data to refine models which can be used as effective methods of early warning of landsliding.
- The major benefit derived from setting up systems of this kind will be that of obtaining greatly increased warning times - both from knowledge of the influence of antecedent conditions and the effective use of meteorological forecasts.
- Climate changes resulting from global warming will be accompanied by a greater range of extreme conditions which can be expected to have a significant influence on groundwater pressures. Having a meteorological based early warning system will be an important way in which areas at risk will be able to cope with these conditions.

REFERENCES

Anderson, M.G. & Howes, S. 1985 Development and application of a combined soil water-slope stability model. *Quart. Journ. Engineering Geology.* 18, 225-236
Barton, M.E. & Thomson, R.I. 1986 A model for predicting groundwater level response to meteorological changes. In *Cripps, J.C. et al (Eds), Groundwater in Engineering Geology.* Eng. Geol. Special Pub No. 3, 299-311.
Caine, N. 1980 The rainfall intensity-duration control of shallow landslides and debris flows. *Geografiska Annaler* 62A, 1-2, 23-27.
Crozier, M.J. 1999 Prediction of rainfall-triggered landslides: a test of the antecedent water status model. *Earth Surface Processes & Landforms* 24, 825-833

Fell, R.; Chapman, T.G. & Maguire, P.K.   1991   A model for prediction of piezometric levels in landslides.   In *Chandler, R,J. (Ed), Slope Stability Engineering: Developments & Applications.* T.Telford, London. 37-42.

Harper, T.R. 1975 The transient groundwater pressure response to rainfall and the prediction of rock slope instability. *Int. Journ. Rock Mechanics, Mining Science & Geomechanics Abstracts.* 12, 175-179.

Keefer, D.K. et al   1987   Real-time landslide warning during heavy rainfall. Science 238, 921-925.

McInnes, R.G. 1996 A review of coastal landslide management on the Isle of Wight, UK   In *Senneset, K. (Ed) Landslides.* Proc. 7th Int. Symp. Landslides, Trondheim, Norway. Balkema, Rotterdam 1, 301-307.

Polemio, M. & Petrucci, O. 2000. Rainfall as a landslide triggering factor: an overview of recent international research. In *Bromhead, E. et al (eds) Landslides.* Proc. 8th Int. Symp. Landslides, Cardiff. Telford, London 1219-1226

Sangrey, D.A.; Harrop-Williams, W. & Kleiber, J.A.   1984 Predicting ground-water response to precipitation. Journ. Geotech. Eng., ASCE 110, 957-975

Skempton, A.W. & Hutchinson, J.N. 1969   Stability of natural slopes & embankment foundations. *Proc. 7th Int. Conf. on Soil Mechs. & Found. Eng., Mexico.* State of the Art Volume, 291-340.

Vaughan, P.R. & Walbanke, H.J. 1973 Pore pressure changes and the delayed failure of cutting slopes in over-consolidated clay. *Geotechnique.* 23, 531-539.

Zaruba, Q. and Mencl, V. 1969 *Landslides and their Control.* Elsevier, Amsterdam

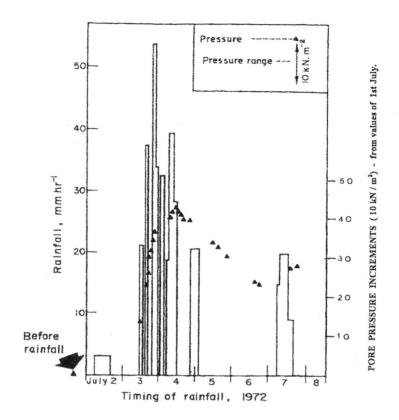

FIG. 1 Pore pressure response to rainfall in a fissured rock mass (Harper 1975).

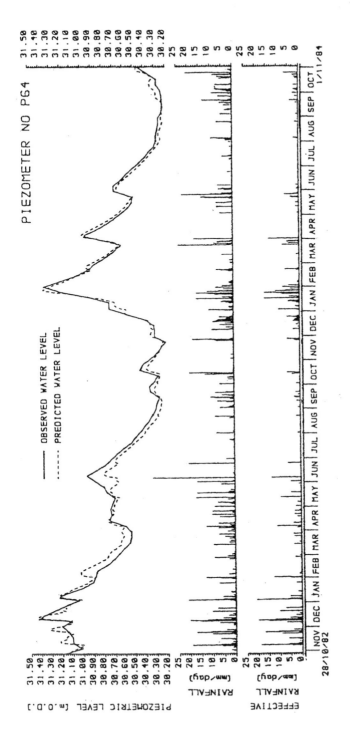

FIG. 2   Observed and predicted groundwater levels in the Plateau Gravel overlying stiff fissured clay at a Christchurch Bay cliff top site (from Barton and Thomson 1986).

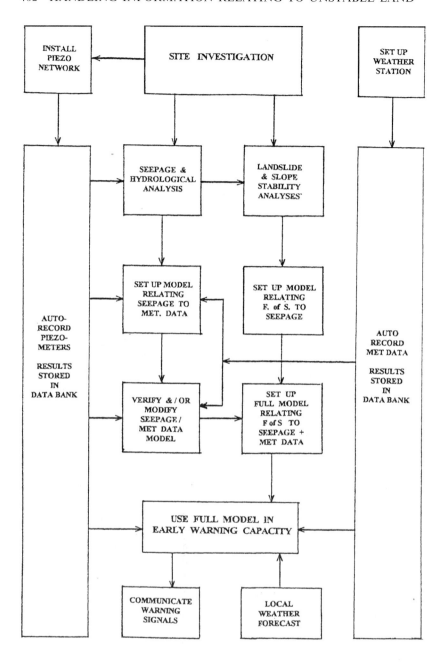

FIG. 3 Network showing the research and operational stages required to set up a meteorological based early warning system for landslide complexes.

Landslide warning and emergency planning systems in West Dorset, England

K. COLE, West Dorset District Council, UK
G. M. DAVIS, West Dorset District Council, UK

BACKGROUND

West Dorset is underlain by ancient landslide systems formed in Lower Lias clays and Cretaceous sands, which may be reactivated by adverse weather conditions or human activity. Instability problems are particularly acute in the coastal areas, where slopes are oversteep as a result of present-day and historical marine erosion. Much of the town of Lyme Regis is constructed on coastal landslide systems which have caused considerable disruption and damage to property. Landslide movements within the town are typically of the slow and creeping type which may cause damage through cumulative effects over many years. However, from time to time there have been much more rapid landslide events which have severely damaged or destroyed property over periods of a few days or even hours.

West Dorset District Council is developing coast protection slope stabilisation schemes to address the problems of coastal in stability in the urban area of Lyme Regis (Clark et al. 2000, Cole et al. 2002). However, due to the scale and complexity of the instability problems, any engineering schemes to reduce the risk of damaging landsliding will take several years to implement. Hence there is some risk that serious landsliding could take place in the interim, before stabilisation works are put into effect.

GROUND MONITORING NETWORK

A ground monitoring network was set up in 1996 as part of a series of studies into coast erosion and landslipping (Fort et al. 2000, Davis et al. 2002). The main purpose of the monitoring system is to assist in the establishment of ground models and landslide mechanisms and to provide information for the design of the remedial works. It comprises some 240 individual monitoring points including ground surface markers, borehole inclinometers and piezometers. The system is monitored routinely in order to obtain ongoing information on landslide behaviour. Monitoring over several years has established that much of the coastal area of Lyme Regis undergoes some degree of ground movement each year, of order of a few millimetres or centimetres, and that much of this movement occurs during the winter and early spring when ground water levels are at their highest.

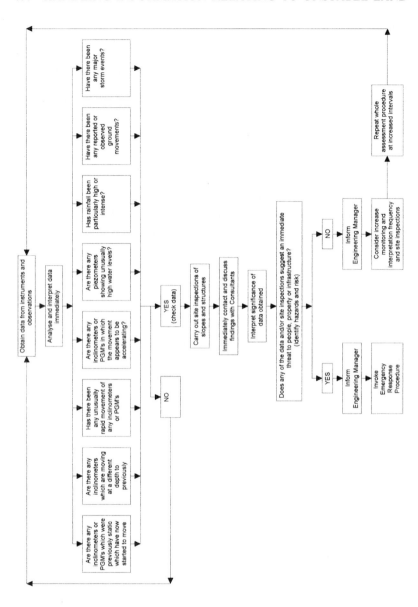

Figure1. Flowchart for interpretation of slope monitoring data.

A standard procedure is used for the interpretation of monitoring data and field observations (Fig. 1). In periods when ground movements are taking place or might be expected to start, for example during a period of wet weather, the frequencies of monitoring observations and, particularly those of inclinometers, are increased, so that any accelerations in ground movement may be tracked. The inclinometers provide the best information in which to observe changes in rate of ground movement as they can detect subsurface displacements as small as a few millimetres which are about an order of magnitude more precise than ground marker data.

Occasionally, instruments within certain elements of a landslide system record continuing increases in rates of ground movement. In these instances, a judgement must be made as to whether the trends in the data indicate the possibility of an escalation of movement to a magnitude that would cause serious damage to property or infrastructure. To assist in this, ground movement data are plotted as rate of movement against time, which identifies trends in acceleration (Fig. 2). The data may also be plotted as inverse rate of movement against time, where a zero inverse rate represents a very large rate of movement - extrapolation of an accelerating trend to the x axis thus gives an indication of at what date movements could become very rapid and potentially damaging. Account is taken of the behaviour of adjacent instruments, visual observations of ground movement in the area and the current and forecast weather conditions. In those exceptional cases where continuation of movements in the same trend would have led to serious damage in the immediate future, West Dorset District Council has invoked a landslide emergency response procedure.

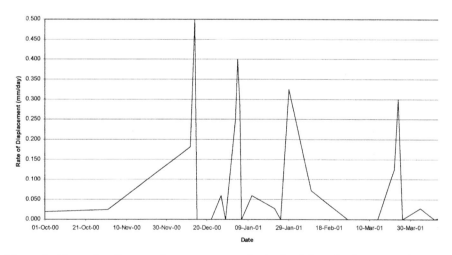

Figure 2. Example of a rate of movement graph obtained from inclinometer data

LANDSLIDE WARNING AND EMERGENCY PLANNING PROCEDURE

When potentially-damaging ground movements are considered to be imminent, landslide warnings are issued in the areas at risk and the first stage of the local Civil Emergencies Plan is implemented. A three-tier system of warnings is employed, analogous to the flood warning system used by the UK Environment Agency:

Level 1: Landslip Watch – detailed observation of areas where monitoring has indicated a landslip is imminently possible

Level 2: Landslip Warning – informing occupants in areas where landslips are expected to occur and may affect their properties

Level 3: Severe Landslip Warning – informing occupants where landslips will occur and will affect their properties

Residents and businesses under threat are informed by news releases to the media, leaflet drops to premises (Fig. 3) and, where appropriate, by personal visits. The public are encouraged to maintain vigilance and to report instances of ground movement so that this information may be used in the ongoing assessment of landside risk. Under the Local Civil Emergencies Plan, the emergency services, public utilities and others are put on alert when a Landslip Watch: Level 1 is declared, and briefed on the possible extent and consequences of a landslide event, so that they may prepare for the worst.

LANDSLIDE MOVEMENTS IN THE WEST DORSET AREA IN WINTER 2000-2001

Exceptionally wet weather during autumn and winter 2000-2001 led to widespread landside activity on the West Dorset coast. Major reactivation took place of the large landslide systems in the coastal cliffs in the Lyme Regis area, including the Stonebarrow and BlackVen/Spittles systems to the east of the town. Significant ground movements also occurred within the urban area of the town of Lyme Regis to the extent that three landslide warnings were issued, two to Level 1 and one to Level 2. An example summary chronology of a landslide emergency that occurred in December 2001 is given in Table 1. In the event, serious damage to property did not take place, the ground movements coming to a halt suddenly after a peak in the rate of movement. This sudden cessation of movement would seem to be related to spells of relatively dry weather, the landslide requiring a continuation of heavy rainfall to maintain an accelerating trend.

IMPORTANT: LANDSLIP WATCH
Help us to help you

As you will be aware, West Dorset District Council has been monitoring water levels and ground movements across Lyme Regis for some time. The work has prompted concerns about the potential for a landslip in this area following the recent wet weather.

As part of the Council's commitment to keeping residents informed a three-tier landslip warning system - similar to that employed by the Environment Agency in relation to flooding - is now being employed by the District Council.

Level 1: Landslip Watch – detailed observation of areas where monitoring has indicated a landslip is imminently possible

Level 2: Landslip Warning – informing occupants in areas where landslips are expected to occur and may affect their properties

Level 3: Severe Landslip Warning – informing occupants where landslips will occur and will affect their properties

The Council has moved to Level 1 (Landslip Watch) in the area covered by Lower Cobb Road, Upper Cobb Road and Langmoor Gardens.

Engineers will be monitoring ground movements very closely over the coming days and weeks. **Householders are asked to report any new cracks in structures and gardens as a matter of urgency on 01297 445051 during office hours, or on 01305 250365 outside office hours.**

A pamphlet advising residents what they can do to help minimise the risk of landslips went to every home in Lyme Regis in mid October. A copy is attached to this leaflet.

The District Council's Civil Emergencies Plan has been produced to help manage the response to emergency incidents, and includes a section on responding to a serious landslip in Lyme Regis. Officers are now following the initial phase of this plan as a precaution.

The development of Lyme Regis has been strongly influenced by coastal erosion and landslipping. The spectacular coastal scenery around the town has formed as a result of continuous erosion and landslipping of local cliffs. The sea walls and structures such as the Cobb protect the town against marine erosion.

However, parts of the town are built on ancient landslides which may, from time to time, become reactivated resulting in disruption and damage to property.

West Dorset District Council is proposing £20 million of works over seven years to protect Lyme Regis from the ravages of the sea and tackle coastal instability problems.

Figure 3. An example of a landslide warning leaflet.

Thursday 7 - Tuesday 12	Inclinometer observations indicate significant accelerations of ground movement at Upper Cobb Road, Lower Cobb Road and Langmoor Gardens.
Wednesday 13	Fresh cracking observed in Cobb Road and lower Lister Gardens. Analysis of data indicates damaging ground movements could take place within next few days. Stage 1 of Civil Emergencies Plan invoked. **Level 1- Landslide Watch - declared for Upper Cobb Road, Lower Cobb Road and Langmoor Gardens.**
Thursday 14	Civil Emergencies meeting held at Dorchester with Police, Coastguard, Ambulance Service, Fire Brigade and Council Emergency Planners. Leaflets dropped in landslide watch area. News release issued to media Further inclinometer observations indicate continuing acceleration. Ground movements rate reaches 0.5mm per day in Langmoor Gardens, indicating imminent destructive movement. **Langmoor Gardens and Bay Hotel area raised to Level 2 - Landslide Warning.** Occupiers informed personally.
Friday 15	Observations in Langmoor Gardens morning and afternoon indicate movements ceased. **Langmoor Gardens and Bay Hotel area lowered to Level 1- Landslide Watch.** Occupiers informed personally. Transco commences installing emergency gas shut-off valves at Cobb Road and Marine Parade
Saturday 16	Dorset County Council seal cracks in Cobb Road with bitumen.
Saturday 16 - Monday 18	Further instrument observations indicate movements ceased in Cobb Road and Langmoor Gardens area
Wednesday 20	Instrument observations confirm no further movement. Weather forecast dry for next few days. **Landslide Watch stood down.** News release issued to media.
Thursday 21	Stand down leaflets dropped.

Table 1. Summary chronology of landslide emergency that took place in Lyme Regis in December 2000.

DISCUSSION

A decision on whether or not to invoke the landslide emergency procedure requires careful judgement, balancing the likelihood and consequences of destructive landsliding with the potential angst amongst the public caused by issuing the warnings and the burden placed on emergency services and others in preparing for an emergency which may not actually occur. The fact that very large scale landslipping occurred in the coastal cliffs adjacent to the urban areas of the town, shortly after the landslide warnings in winter 2000-2001, gave additional confidence that the issuing of the warnings (from a risk point of view) was justified. The overall response from members of the public has been positive, with many, whilst extremely concerned about the possibility of landslides occurring, taking some comfort that the local authority was acting in their best interests.

The limitations of the landslide warning system should be borne in mind. It offers no guarantees that all destructive landslide events will be foreseen and warnings issued in advance. In particular, the monitoring system on which the landslide warning system is based was not designed for such a purpose. It relies on very labour intensive field monitoring, particularly the inclinometers which require heavy equipment to be carried to the boreholes, and with each instrument taking about an hour on average to read. Frequent observations of many holes at times of landslide activity is particularly heavy on the District Council's resources. The quality of the datasets and ease of interpretation depends very much on the frequency of the observations.

CONCLUSIONS
The setting up of a ground monitoring network for the town of Lyme Regis has enabled ground movement and water levels to be routinely monitored. The information obtained is being used to enable engineering schemes to be designed with confidence to alleviate the instability problems of the town. The information had also been found useful in assisting the District Council on making the judgement calls required under its onerous duty of providing an emergency planning response for the community. The District Council has set up a system of landslide warnings to inform the community and to place the emergency services and public utilities in a state of awareness, and the ground monitoring system set up within the urban coastal slopes informs the process.

ACKNOWLEDGEMENTS
The paper is published by permission of West Dorset District Council. The principal consultant is High-Point Rendel. The project is funded by DEFRA. The role of West Dorset District Council's communications team, led by Mr Colin Wood, and of Mr Rob Tapply in the monitoring the 2000-2001 landslide emergencies is acknowledged.

REFERENCES
Clark A.R., Fort D.S. and Davis G.M. 2000. The strategy, management and investigation of coastal landslides at Lyme Regis, Dorset, UK. *Landslides in research, theory and practice* 1, 279-286. Thomas Telford, London.
Cole K., Davis G.M., Fort D.S. and Clark A.R. 2002. Managing coastal instability - a holistic approach. Proceedings of the International Conference on Instability - Planning and Management. Thomas Telford, London.

Davis G.M., Fort D.S. and Tapply, R.J. 2002. Handling the data from a large landslide monitoring network at Lyme Regis, Dorset, UK. Proceedings of the International Conference on Instability - Planning and Management. Thomas Telford, London.

Fort D.S., Clark A.R., Savage, D.T. and Davis, G.M. 2000. Instrumentation and monitoring of the coastal landslides at Lyme Regis, Dorset, UK. *Landslides in research, theory and practice* 2, 573-578. Thomas Telford, London.

Handling the data from a large landslide monitoring network at Lyme Regis, Dorset, UK

G. M. DAVIS, West Dorset District Council, UK
D. S. FORT, High-Point Rendel, London, UK
R. J. TAPPLY, West Dorset District Council, UK

INTRODUCTION
The town of Lyme Regis, on the southern coast of England, is located on ancient coastal landslide systems in Jurassic mudstones and Cretaceous sands, covering an area of some 75 hectares. The landslides undergo slow, creeping-type movements, the cumulative effects from which cause considerable damage to properties and infrastructure. The town also suffers from rapid destructive landslide events from time to time. As part of a project to gain an understanding of the landslide mechanisms operating in the town (Clark et al. 2000, Cole et al. 2002, Brunsden 2002), and ultimately to implement ground stabilisation and coast protection schemes, a landslide monitoring system has been established. This comprises some 240 individual monitoring points and has been in operation since 1997, generating considerable amounts of data.

The principal purposes of the monitoring system are:
1. to determine the location and distribution of landslide movements in the town and flanking areas;
2. to assist in the construction of ground models (Sellwood et al. 2000) and establishment of landslide mechanisms by providing information on the depth of shear zones and vectors of ground movement;
3. to provide information for the design of remedial works;
4. to form a basis for comparison of remedial measures and to determine their effectiveness;
5. to establish patterns of evolution of the landslide systems over several years to assist in the prediction of their behaviour in the future.

Although not a principal objective in its original design, the monitoring network is also used as a type of early warning system, to give prior indication of possible destructive landslide events (Cole and Davis, 2002).

This paper describes how data are collected from the various types of monitoring instrument and then processed and interpreted for dissemination and use.

Instability – Planning and Management, Thomas Telford, London, 2002, 471–478

DATA AQUISITION

The two principal groups of parameter monitored are ground movements and groundwater pressures. These are supplemented by the recording of local rainfall, the rate of discharge of groundwater from drains and the deformation of structures located within the landslide systems. The different types of instrument or method employed and the type of data generated are summarised in Table 1. Further information on the instrumentation and its installation is given in Fort et al. (2000).

The monitoring regime is kept flexible so it may be adjusted to meet changing circumstances. For example, monitoring frequency may be substantially increased at times of rapid ground movement, new instruments or monitoring points may be installed to meet particular requirements for data or, in some instances, monitoring may be discontinued. The monitoring frequency is increased during wetter periods or after significant rainfall when groundwater levels are likely to be elevated and rates of ground movement greater.

Ground Surface Movement

A global positioning satellite (GPS) network is used to gain general ground movement data over the extent of the coastal landslide area to a nominal precision of +/- 7mm in plan and elevation. Coordinates are verified by using two separate static GPS receivers (Fort et al. 2000). Ground markers are positioned to cover the principal individual units and some are placed outside the known coastal landslide systems to act as controls. For local landslide units which are particularly active, arrays of closely-spaced pegs are set up and the coordinates of each peg observed more frequently using a conventional total station linked to base points in the GPS network. High precision levelling (+/- 1mm) to a local datum is used in some areas in order to gain information on subtle changes in ground level that would not be identified using GPS alone.

Subsurface Movement

Biaxial inclinometers are used to provide most of the data on subsurface deformation. In areas of relatively fast moving landslides, where an inclinometer installation would be quickly destroyed, slip indicators are used to provide information on the depth of shearing at relatively low cost. From time to time, a mandrel is dropped down piezometer tubing in slow moving landslide zones to supplement the data on the depths to shear zones obtained from the inclinometers.

Structural Deformation

Certain structures which are being affected by ground movements are monitored by keeping observations of crack widths and the tilt of structural members.

Method	Parameter measured	Description	No. of points	Typical monitoring frequency
Global Positioning Satellite surveying	xyz coordinates of ground markers	The OS National Grid coordinates of ground surface markers are determined with differential GPS, using two static base receivers	72	3 months
Digital levelling	Elevation of ground markers	The elevation of surface markers relative to a local datum is established by surveying with a digital level	47	Monthly
Slip indicators	Depth of shear zone	Mandrels of different sizes and lengths are lowered into piezometer tubing to detect the approximate level of deformation within a borehole	As required	
Borehole inclinometers	Depth of shear zone, displacement	Subsurface deformations measured using a biaxial inclinometer probe inserted into keyway tubing an a borehole	31	Monthly
Total station survey of peg array	xyz coordinates of markers	Survey of peg array to OS NG coordinates using conventional total station	8	Every 1-3 months
Manually read standpipe piezometers	Groundwater pressure	Standpipe piezometers with Casagrande-type tips read manually with a dipmeter	21	Fortnightly
Automated standpipe piezometers	Groundwater pressure	Standpipe piezometers as above, but monitored automatically using pressure transducers and data loggers	18	Every 12 hours
Pneumatic piezometers	Groundwater pressure	Pneumatic piezometers installed in boreholes	16	Weekly
Visual observations	Various	Recording of visually-obvious deformations and failure in the ground and structures	As required	
Crack width monitoring	Crack width	Crack width measurement with digital or vernier callipers	8	Monthly
Tilt measurement	Angle of tilt of structural member	Measured using digital inclinometer, similar in appearance to large sprit level	12	Monthly
Drain discharge rate	Discharge rate	Record is made of time taken to fill a receptacle of known volume	7	Weekly
Rainfall	Daily rainfall	Automatic logging rain gauge	1	Daily

Table 1. Monitoring methods employed at Lyme Regis

Groundwater

The principal type of instrument used for monitoring groundwater levels is the standpipe piezometer. Many of these are automatically monitored using battery-powered data loggers, which are programmed, typically, to record water levels every 12 hours. More frequent readings may be taken, for example in tidally-influenced areas. Pneumatic piezometers are employed where a rapid instrument response is required, for example where groundwater pressures are required shortly after the installation of the instrument or to monitor the response of drainage trials.

Rainfall

Local rainfall is recorded daily using an automatic rain gauge.

Drain Discharge Rate

The rate of water discharge from critical drains is measured by recording the time required to fill a container of known volume. This included the measurement of flows from sub-horizontal drilled drainage trials.

PROCESSING

The principal processing tool is a conventional spreadsheet package. Modern spreadsheets have been found to be well suited to the efficient handling of the large data sets involved and have the additional advantages of allowing easy visualisation of the data through graphs and providing a standard computer file format which facilitates the dissemination of the data amongst interested parties. Some of the simple data measured, for example crack widths and drain discharge rate, do not require much processing and may be recorded on the spreadsheet and plotted on a graph to identify any trends. More complex data require further consideration, as described below. Most of the data are processed and interpreted immediately after they have been collected so that any critical information, such as the onset of fresh ground movements, can be identified quickly.

Surface Marker Data

The data from the GPS and total station survey equipment are transferred to the spreadsheet, via floppy disc, as matrices of eastings, northings and elevations. Any settlement or heave of the ground surface is calculated by a simple subtraction of elevations recorded at different times; the rate of settlement or heave may also be calculated as an additional column in the spreadsheet. The change in the easting and northing of a marker may be used to calculate the magnitude and rate of horizontal movement using Pythagoras's Theorem. The changes in easting and northing are also used to calculate the direction of movement of the marker in plan - this is not straightforward because there is no universal formula which allows this for all quadrants of the compass, hence 'IF' statements are required in the spreadsheet for the general case. The 'angle of dip' of the marker in the vertical plane is calculated from the magnitude of horizontal movement and the change in level (Fig. 1). The parameters calculated from the raw coordinates are summarised in Table 2. In general, the calculations are carried out for each epoch between successive observations and also for the epoch for the period between the initial and latest observation (Fig. 2).

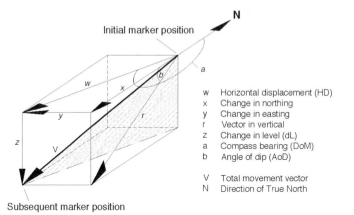

Figure 1. Diagram illustrating parameters which may be calculated from changes in coordinates of a surface ground marker

	GPS surveying	Total station survey of peg array	Precision levelling	Inclinometer	Slip indicator
Depth of shear zone				✓	✓
Magnitude and rate of settlement/heave	✓	✓	✓		
Vector in plan and rate of horizontal movement	✓	✓		✓	
Vector in vertical	✓	✓			
Total vector and rate of movement in 3 dimensions	✓	✓			

Table 2. Parameters obtainable from different methods of ground movement monitoring

Inclinometer Data

Field data from the inclinometers are processed initially using the specialist software appropriate to the particular equipment to generate ground displacement profiles in the two orthogonal directions. This identifies the depth of shear zones. The displacement in the two directions, over a specified depth including the shear zone, is then fed into a conventional spreadsheet, which calculates the magnitude and direction of the movement relative to True North in plan by means of trigonometric formulae. The mean rate of movement for each instrument over a particular epoch between two observations is also calculated.

Marker Ref.	Nov-00 - Jan-01				Jan-01 - May-01				Totals				Epoch
	dL	HD	DoM	AoD	dL	HD	DoM	AoD	dL	HD	DoM	AoD	
02A	-18	5	143	-74	6	7	315	40	0	5	180	0	Feb 97 - May 01
02B	*-3*	*13*	*189*	*-13*	-11	11	0	-45	*-11*	*14*	*124*	*-37*	Feb 97 - May 01
02C	7	7	153	46	5	8	346	31	-8	17	225	-25	Feb 97 - May 01
02D	-7	5	158	-52	-5	6	171	-39	-1	12	180	-5	Feb 97 - May 01
02E	*18*	*4*	*180*	*77*	*20*	*10*	*204*	*64*	*50*	*23*	*205*	*65*	Feb 97 - May 01
02F	7	6	315	51	-5	4	76	-50	9	3	180	72	Feb 97 - May 01
02G	*-10*	*15*	*191*	*-33*	-1	7	164	-8	*-22*	*73*	*189*	*-17*	Feb 97 - May 01
02H	*-5*	*21*	*209*	*-14*	*-4*	*10*	*209*	*-21*	*-1*	*85*	*204*	*-1*	Feb 97 - May 01
02J	-7	2	270	-74	12	7	225	59	8	17	208	25	Feb 97 - May 01

Figure 2. Extract from a spreadsheet summarising ground marker movements. The bold-italic cells denote data interpreted as being genuine landslide movements; see Figure 1 for key.

Groundwater Data

Processing groundwater data is relatively straightforward. The raw data (in the form of depth below ground level in the case of standpipe piezometers or metres head of water in the case of the pneumatic piezometers) are entered into the spreadsheet and then converted into depth below ground level, metres head of water at the piezometer tip and elevation relative to Ordnance Datum, by means of simple arithmetic formulae in the spreadsheet. The data from groups of piezometers may then be displayed on the same graph as level against time. The automatic loggers are uploaded using a laptop computer and the data transferred directly to the spreadsheet via a floppy disc.

Rainfall Data

Rainfall is plotted as bar charts of daily total and as antecedent rainfall for a range of antecedence periods. Each year, statistics are prepared comparing that year's rainfall with that from previous years, data of which is available since 1868.

INTERPRETATION

When carrying out interpretations of the data, account is taken of limits of precision. For example, to be confident that a ground marker is indicating actual landslide movement, its apparent change in position must be greater than the error bounds inherent in the observation method. The interpreter should also be aware of the possibility of extraneous influences on the instrumentation. For example standpipe piezometers may become flooded by surface water run off from time to time, or a particular GPS observation may be less accurate than usual because of interference to the satellite signal by tree foliage during the summer. 'Sense' checks are applied to the processed data, for example the data for a marker which was apparently moving up hill rather than downhill would be scrutinised for causes of error, whist keeping an open mind on the possible landslide mechanisms which might allow this to occur. It is also important to consider long term trends in the data, both to confirm that a phenomenon is taking place and to identify any anomalies in the trends which could be indicative of a change in landslide behaviour.

In order to assist in interpretation, ground movement vectors from the different sources are regularly plotted on summary plans (Fig. 3) to which are added notes on the visual descriptions of ground movement. Ground movement vectors in the vertical and piezometric levels are plotted on geological sections through the landside systems. In this way, the data from the different types of monitoring method may be compared. Great confidence in the integrity of the data is gained if data obtained from different methods at the same location, for example inclinometers and ground markers, are concordant with one another.

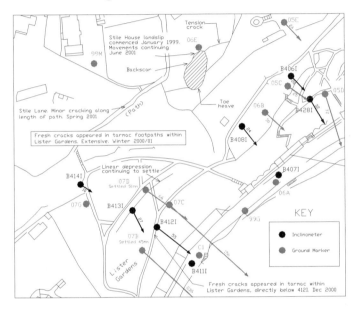

Figure 3. Extract from a ground movement summary plan.

The spreadsheets offer a flexible way of comparing different types of data to assist in gaining an understanding of landslide processes. For example, piezometric levels may be plotted on the same graph as rainfall, or the rate of ground movement for a particular marker.

In maximising the benefit of such monitoring, it is important that the geological ground model and landslide systems are well understood and therefore appropriate geomorphological and geotechnical investigations are a pre-requisite of optimising a monitoring programme. Monitoring data add confidence to the accuracy of the ground models and provides the necessary information to develop appropriate stabilisation measures (Fort and Clark, 2002).

REPORTING AND DISSEMINATION
The monitoring data are presented and interpreted in the overall context of the coastal landslide systems in periodic monitoring summary reports, which comment on the behaviour of each of the instruments or monitoring points and give recommendations for further monitoring or changes to the monitoring regime. Information from the reports and the associated spreadsheets are made available to interested parties, such as consulting engineers working on stability problems in the town, or to local residents in landslide areas.

CONCLUSIONS
The paper highlights the wide range of monitoring techniques used to provide essential data necessary to manage and mitigate the coastal instability problems at Lyme Regis. However for such a monitoring programme to be effective, appropriate resources in terms of investment, mantime and commitment are required. To illustrate this, about 150 man-days are required per year to monitor, process, interpret and action the programme. However, in view of the significant coastal instability problems of Lyme Regis, this is seen to be a necessary procedure.

ACKNOWLEDGEMENTS
The paper is published by permission of West Dorset District Council. The principal consultant is High-Point Rendel. The project is funded by DEFRA.

REFERENCES
Brunsden D. 2002. The Fifth Glossop Lecture - Geomorphological roulette for engineers and planners: some insights into an old game. Quarterly Journal of Engineering Geology and Hydrogeology, 34.
Clark A.R., Fort D.S. and Davis G.M. 2000. The strategy, management and investigation of coastal landslides at Lyme Regis, Dorset, UK. *Landslides in research, theory and practice* 1, 279-286. Thomas Telford, London.
Cole K. and Davis G.M. 2002. Landslide warning and emergency planning systems in West Dorset, England. Proceedings of the International Conference on Instability - Planning and Management. Thomas Telford, London
Cole K., Davis G.M., Fort D.S. and Clark A.R. 2002. Managing coastal instability - a holistic approach. Proceedings of the International Conference on Instability - Planning and Management. Thomas Telford, London.
Fort D.S. and Clark A.R. 2002. The monitoring of coastal landslides: a management tool. Proceedings of the International Conference on Instability - Planning and Management. Thomas Telford, London.
Fort D.S., Clark A.R., Savage, D.T. and Davis, G.M. 2000. Instrumentation and monitoring of the coastal landslides at Lyme Regis, Dorset, UK. *Landslides in research, theory and practice* 2, 573-578. Thomas Telford, London.
Sellwood M., Davis G.M., Brunsden D. and Moore, R. 2000. Ground models for the coastal landslides at Lyme Regis, Dorset, UK. *Landslides in research, theory and practice* 3, 1361-1366. Thomas Telford, London.

The monitoring of coastal landslides: a management tool

D. S. FORT and A. R. CLARK, High-Point Rendel, London, UK

INTRODUCTION
The determination and quantification of risks associated with potentially unstable coastal landslides, especially in developed urban areas, forms an important aspect in managing the coastline. This paper presents an overview of the use of geotechnical monitoring within the ground investigation process as part of the evaluation of the risks associated with coastal instability. From this, the development of monitoring response procedures and risk limitation is discussed. Examples from a number of sites around the UK coastline including Lyme Regis, Barton-on-Sea, at Ventnor and Afton Down on the Isle of Wight, and Scarborough are described.

The development and establishment of geotechnical monitoring systems should be seen as an integral part of the process of evaluating the risks posed by cliff recession and coastal landsliding (Clark et al, 1996). There is a range of settings where such systems could be of considerable benefit in reducing the threat to public safety and property, either to detect signs of *pre-failure movements* on the cliff top or the *reactivation* of pre-existing coastal landslides which may or may not have coast protection structures.

Coastal cliffs can be conveniently categorised in terms of the role of monitoring into:
- **protected cliffs** in urban areas where monitoring systems can provide advance warning of potentially damaging events, e.g. Lyme Regis, Ventnor and Scarborough.
- **unprotected cliffs** where investment in coast protection and/or slope stabilisation is either uneconomic, not technically feasible or environmentally unacceptable. Here monitoring systems could provide a solution to the medium and long term protection of public safety, e.g. Blackgang on the Isle of Wight (Moore et al, 1998).
- **unprotected cliffs** where coast protection and/or slope stabilisation is justified, monitoring and early warning could provide *interim safety cover* until the scheme has been constructed and the threat to public safety reduced, e.g. Afton Down and Holbeck Hall (Clark and Guest, 1994).

MONITORING STRATEGY & SYSTEMS
As part of the normal ground investigation process (Clark et al, 1996(a)), it is common UK practice to monitor the behaviour of landslide instability for a number of reasons, including:

- the determination of the landslide mechanism and geometry;
- identification of the extent of the problem, i.e. rate, direction, position and magnitude of movement;
- to determine the groundwater conditions at the site;
- to determine the interrelationship of ground movement with groundwater level, pore-water pressure and rainfall;
- to establish early warning of movement events in advance of a major failure, i.e. limit risk to property and life;
- to determine the rate of development of a landslide to allow the programming of stabilisation works or further monitoring requirements;
- to monitor the area of instability during and post construction/implementation of remedial measures.

Many methods of instrumentation and monitoring can be used to obtain the above information, some of which are listed below:
- Surface movement
    - comparison of Ordnance Survey bench mark elevation surveys from different dates (e.g. Moore et al,1991)
    - ground survey (conventional survey or GPS methods) (e.g. Fort et al, 2000)
    - photogrammetry (e.g. Moore et al 1991; Brunsden and Chandler 1996)
    - survey pins and peg lines
    - extensometers, strainmeters, tiltmeters, settlement cells etc
    - direct measurement e.g. across tension cracks
    - visual inspections
- Sub surface movement
    - borehole inclinometers
    - slip indicators and mandrels in standpipes
- Groundwater
    - hydrogeological studies
    - measurement of groundwater/ pore pressure using piezometers
    - rainfall data collection – historic and current

Methods and their application are described in the literature (e.g. Dunnicliffe 1993).

CASE HISTORIES
Examples where instrumentation and monitoring have been successfully used for the management of UK coastal landslides are given below:

Lyme Regis, Dorset
Lyme Regis is a coastal town of 3,500 inhabitants which rises to 14,000 in the holiday season. The area has a history of coastal erosion and instability and is affected by landslides covering an area of 75 hectares. As part of the Lyme Regis Environmental Improvement Studies, funded jointly by West Dorset District Council and DEFRA, a long-term monitoring programme has been set up to provide

quantitative information on the location, rate, magnitude and distribution of landslide movements and groundwater variations across the town. This information compliments the geotechnical ground models that have been developed for the town and will form part of the rationale for the design of any future stabilisation measures that may be undertaken. The monitoring also contributes to the early warning of potential landslide instability.

Techniques used on the reactivated landslides at Lyme Regis include monitoring of inclinometers, piezometers, ground movement (using conventional and GPS survey techniques), visual inspections and climate/rainfall, (Fort et al, 2000; Davis et al, 2002). The short term aims of the monitoring programme included the collection of precise geotechnical data to more clearly determine the nature of the complex multiple landslides systems at Lyme Regis, (Sellwood et al, 2000). This required the determination of the position, the direction, rate and magnitude of ground movement. In areas of relatively rapid active movement, the surface survey markers provide both vertical and horizontal vectors, and rates of movement at the ground surface, whilst the inclinometers provide precise measurement of depth and direction (in the horizontal plane) and rate of movement along sliding surfaces. From this information it has been possible to build up a picture of the distribution of ground movement and also to confirm the mechanisms and kinematics of slope instability by examining the vector movement and depth of movement at each individual landslide unit within the multiple landslide complex. The monitoring of the many standpipe piezometers installed at various depths and geological horizons in the landslide system enabled an understanding of the complex multilayered groundwater regime to be made.

Another objective of the geotechnical monitoring is to establish the interrelationship of ground movement with piezometric levels and rainfall, with the ultimate aim being the forecasting of ground movement based on antecedent rainfall and groundwater levels. A weather station has been installed to monitor rainfall, whilst historical daily rainfall records are available for the area since 1868. Rainfall data (daily, weekly or antecedent etc) is routinely compared to the response in groundwater level and rates of ground movement.

Barton-on-Sea, Hampshire
The majority of the coastal slopes along the Barton frontage have historically been subjected to marine erosion and for many years have undergone phases of investigation, landslide stabilisation measures and coast protection. Nevertheless there has been a continuation of slope instability at the site. To determine the nature of the landslides that are operating in the undercliff and the significance of certain controlling lithological boundaries within the Eocene Barton Beds on the formation and development of the various landslide systems, a comprehensive ground investigation, monitoring and geomorphological mapping programme was undertaken.

Monitoring has been used very successfully to confirm landslide mechanisms, rates and position of slip movement and is used as an indicator of developing movement. Techniques used include GPS surveying, monitoring of inclinometers, slip indicators and piezometers, rainfall measurement and slope inspections (Fort et al, 2000). In addition, a series of geomorphological and ground behaviour maps were prepared that enabled the development of the landslides to be determined. Various phases of ground investigation including boreholes and static piezocone penetration tests to depths of up to 35.5m were used to correlate the stratigraphy and landslide systems within the undercliff. The results of these investigations, coupled with the monitoring programme, have enabled the mechanisms of slope failure to be more clearly identified and a management strategy to be developed.

Results from monitoring show clear relationships of ground movement with the onset of the wet winter period. A comparison of antecedent monthly rainfall with ground movement has been made for the period between 1994 and 2001 for those monitoring points that show significant seasonal trends in movement. Accelerating ground movements in October/November generally appear to commence when monthly rainfall for the average current and preceding months (antecedent 2-month average) is in excess of about 80mm/month. Once high rates of movement have commenced during the winter period, monthly rainfall has to reduce to relatively low levels for rates of ground movement to reduce back to summer levels with average monthly rainfall typically being about 40mm to 50mm (Figure 1).

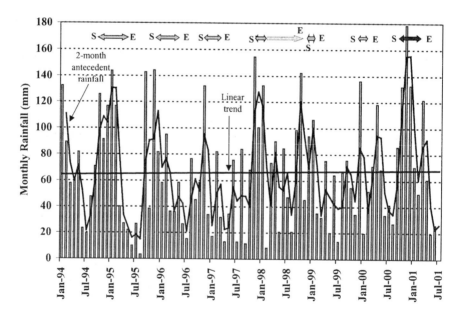

Figure 1   Relationship of start (S) and end (E) of ground movement with rainfall at Barton

Undercliff Landslide, Isle of Wight

The Undercliff forms one of the largest developed landslides in Europe and comprises an assemblage of ancient landslides which has evolved over the last 10,000 years. It is located on the south-eastern facing slopes of the Island and extends for a distance of some 12km from Dunnose in the east to Blackgang in the west. The Undercliff area is widely populated with small towns having developed at Ventnor, Bonchurch, St Lawrence and Niton, (Moore et al, 1995 & 1998). It is unusual in Britain for a town of Ventnor's size to be sited on a marginally stable landslide complex and as such presents a unique set of geotechnical problems.

As part of Isle of Wight Council's Landslide Management Strategy for the Undercliff, an automatic monitoring network was installed at various locations including Ventnor, Castlehaven and Blackgang to provide early warning in areas where ground movement could lead to the disruption of service and infrastructure. Data recorded by the monitoring network is also used to provide a forecast of ground behaviour conditions for the succeeding month.

Ventnor Wastewater Scheme, Isle of Wight.

An example of the use of ground monitoring within the Isle of Wight Undercliff relates to the development by Southern Water PLC of the Ventnor wastewater scheme comprising pumphouse structures, transfer stations and pipeline routes through the town.

A detailed geomorphological, geological and geotechnical investigation was commissioned to provide the necessary design parameters to ensure the integrity of the scheme affected by ground movements. Rates of ground movement had been estimated using various methods ranging from bench mark surveys, extensometers and photogrammetry (Moore et al, 1991). The ground investigation for the works included 36 boreholes drilled to depths of between 15m and 100m below ground level and 48 trial pits. Inclinometers were installed in 10 boreholes to depths down to 92m below ground level whilst 27 standpipe piezometers were used to investigate the local hydrogeology.

The monitoring of the borehole inclinometers since 1992 confirmed movement on the basal shear surface of the Undercliff landslide at Ventnor to be predominantly on the Lower Greensand Sandrock 2d layer at a depth of about -10mOD. Minor movement was also recorded at even greater depth on the Sandrock 2b layer. Monitoring over a number of years confirmed an annual trend of deep-seated movement of 2-3mm/month during the drier summer periods increasing to over 30mm/month during the worst wet seasons. This information, together with the identification of the surface expression of such movement in the town through both instrumentation and geological/geomorphological mapping provided the necessary geotechnical information required to design the wastewater works to accommodate such movement.

Afton Down, Isle of Wight

At Afton Down on the south-west coast of the Isle of Wight the main A3055 road edges along the crest of near vertical chalk cliffs up to 70m in height. In places the road is within 10m from the edge of the cliff top, (Barton and McInnes, 1988). Due to the important environmental designations of the area, coast protection of the chalk cliffs to reduce toe erosion is not appropriate. However, it is important to the economy of the Isle of Wight that the safety of the important tourist route is maintained. In order to ensure this a monitoring system was installed in the cliff top to provide an early warning of ground instability that could adversely affect the safety of the road. A system of surface extensometers, tiltmeters and inclinometers have been installed, part of which has been linked to automatic recording instruments that are connected to road closure traffic lights. When the instruments trigger pre-set thresholds of movement a recorded message is sent to both the local police and the Council and the road is automatically closed.

Scarborough, N Yorkshire

The town of Scarborough located on the NE coast of UK is partly developed on the coastal slopes comprising Jurassic strata overlain by Glacial Deposits. Instability of these coastal slopes is not uncommon with perhaps the most recent example being the failure of the coastal slopes below Holbeck Hotel in 1993, (Clark and Guest 1994). Here there was concern that the landslide would continue to extend inland affecting an increasing number of properties and related infrastructure and a system of monitoring comprising conventional survey and then remotely read tiltmeters provided an early warning system in advance of the stabilisation works. Many additional areas of ground instability were identified during the strategy study undertaken for the Scarborough coastline (High-Point Rendel 1999). To provide a management framework to allow the coastal authority to adequately plan and resource the necessary stabilisation works a programme of investigation and monitoring has been set up. This has included detailed geomorphological mapping of the 7km length of coastline supplemented where necessary by ground investigation, the installation of borehole inclinometers and standpipe piezometers, and surface surveying and inspections.

MANAGEMENT STRATEGY

An important aspect of the monitoring regime described in the examples above is the development of a management response strategy. It is essential that a response procedure be in place both to collect and rapidly process the data being collected and for the possible situation where significant movements are observed which could pose a threat to the safety of the public and property. The response strategy needs to consider the following issues:

- thresholds of significant movement
- an evaluation of the likely consequences of continued movement
- communication procedures & nominated individuals' responsibilities
- actions to be taken in the event of movements
- clearly defined response pathways.

Appropriate resources in both staff time and expertise need to be provided to allow such a strategy to be implemented. Without the provision of appropriate resources to process and interpret data, a monitoring strategy provides little benefit.

MONITORING RESPONSE PROCEDURE & EARLY WARNING SYSTEM

Monitoring Response Procedures (MRP) and Early Warning Systems (EWS) need to be developed for sites of instability where there is a perceived risk to public safety or property. Through an MRP an EWS may enable the owner or emergency services to alert people at risk from imminent events and make advanced preparations to lessen the impact of the event.

Appropriate methods can include simple frequent visual observations, regular monitoring of instrumentation, surveying methods or sophisticated remote monitoring and alarm systems (Clark et al, 1996). With all such systems it is important that the mechanisms of instability are fully understood so that the implications of the data collected by the EWS is correctly interpreted.

In setting up an EWS it is important to consider and address the following:
- type and location of instrumentation equipment
- methods used to collect and process the data (real time monitoring systems, continuous recording data loggers, remote interrogation by modem etc.)
- alarm threshold levels
- method of warning (alarms/telepagers)
- landowner and public awareness
- management procedures
- MRP in response to event/alarm
- availability of sufficient and appropriate resources

Such MRP/EWS were successfully developed for Lyme Regis (Clark et al, 2000; Cole and Davis, 2002).

SUMMARY & CONCLUSIONS

An appropriate geotechnical monitoring strategy provides a valuable, if not essential, management tool in the identification and management of risk associated with coastal instability. A selection of case histories has been presented which illustrates this. However in order for this to be undertaken a large commitment is required both in terms of investment and appropriate staffing resources.

ACKNOWLEDGEMENTS

The Authors would like to thank West Dorset District Council, New Forest District Council, Isle of Wight Council, Scarborough Borough Council and Southern Water PLC and their technical staff for granting permission for using the various case histories to illustrate the paper. In addition, thanks are expressed to the many individuals within High-Point Rendel who have been involved in the studies.

REFERENCES

Barton M.E. and McInnes R.G. 1988. Experience with a tiltmeter-based early warning system in the Isle of Wight. Proc. 5th Int. Symp. on Landslides, Lausanne, Switzerland, July 1988. pp 379-382, Balkema, Rotterdam.

Brunsden D. and Chandler J.H. 1996. Development of an episodic landform change model based upon the Black Ven mudslide, 1946-1995. In: M G Anderson and S. M. Brooks (eds.) Advances in Hillslope Processes, 2, pp 869-896.

Clark A.R., Fort D.S. and Davis G.M. 2000. The Strategy, Management and Investigation of Coastal Landslides at Lyme Regis, Dorset, UK. *Landslides in research, theory and practice* 1, pp 278-286. Thomas Telford, London.

Clark A.R. and Guest S. 1994. Holbeck landslide: coast protection and cliff stabilisation. Proc. MAFF Conf. of River & Coastal Engineers, Loughborough.

Clark A.R., Lee E.M. and Moore R. 1996(a). Landslide investigation and management in Great Britain: A guide for planners and developers. Report prepared by Rendel Geotechnics for the DoE. HMSO.

Clark A.R., Moore R. and Palmer J.S. 1996. Slope monitoring and early warning systems: application to coastal landslides on the south and east coast of England, UK. In Proc. of 7th Int. Symp. on Landslides, Trondheim, Norway.

Cole K. and Davis G.M. 2002. Landslide Warning and Emergency Planning Systems in West Dorset, England. International Conference on Instability – Planning and Management. Ventnor, Isle of Wight, May 2002.

Davis G., Fort D.S. and Tapply R. 2002. Handling the data from a large slope monitoring network at Lyme Regis, Dorset, UK. International Conference on Instability – Planning and Management. Ventnor, Isle of Wight, May 2002.

Dunnicliffe J. 1993. Geotechnical instrumentation for monitoring field performance. John Wiley & sons Ltd.

Fort D.S., Clark A.R. and Cliffe D.G. 2000. The Investigation and Monitoring of Coastal Landslides at Barton-on-Sea, Hampshire, UK. *Landslides in research, theory and practice* 2, pp 567-572. Thomas Telford, London.

Fort D.S., Clark A.R., Savage D.T. and Davis G. M. 2000. Instrumentation and Monitoring of the Coastal Landslides at Lyme Regis, Dorset, UK. *Landslides in research, theory and practice* 2, pp 573-578. Thomas Telford, London.

Moore R., Clark A.R. and Lee E.M. 1998. Coastal cliff behaviour and management: Blackgang, Isle of Wight. In: *Geohazards and Engineering*, eds. J.G. Maund, & M. Eddleston, Geological Society Special Publication 15: pp 49-59.

Moore R., Lee E.M. and Clark A.R. 1995. The Undercliff of the Isle of Wight. A review of ground behaviour. Report prepared by Rendel Geotechnics.

Moore R., Lee E.M. and Noton N.H. 1991. The distribution, frequency and magnitude of ground movements at Ventnor, Isle of Wight. In: R.J. Chandler (ed) Slope stability engineering: developments and applications, pp 231-236. Thomas Telford, London.

Sellwood M., Davis G., Brunsden D. and Moore, R. 2000. Ground models for the coastal landslides at Lyme Regis, Dorset, UK. *Landslides in research, theory and practice* 3, pp 1361-1366. Thomas Telford, London.

Borehole correlation difficulties in unstable coastal slopes between Cowes and Gurnard in the Isle of Wight, UK

W.HODGES, BSc, CEng, MICE, FGS, Fareham, UK

ABSTRACT

To assist developers and their consultants in the planning and execution of ground investigations in the potentially unstable land lying within the coastal strip between Cowes and Gurnard, Isle of Wight, a data bank of archive boreholes is presented. Even where boreholes are closely spaced difficulties should be anticipated in trying to understand the soil succession. These are believed to be largely due to intermittent landslipping over very many years but other geological processes of superficial origin, including the formation of gulls and dip-and-strike fault structures, may contribute to the difficulties. Interpretation of the available borehole data is in the author's opinion made a little more difficult by apparent geological inconsistencies in a slope stability report recently published by the Council. A further difficulty is the denial of sight of important geotechnical data known to be held by Transco and Southern Water. A plea is made for the Council to support more strongly the collection and availability of local geotechnical reports.

BOREHOLE DATA BANK

A borehole data bank of site locations is presented in Table 1. Detailed geotechnical data were obtained from the British Geological Survey, from commercial sources, through a study of local authority records and by other means.

Sixty-three sites are listed in Table 1. Their locations are plotted in Fig.1. Most lie within the area under consideration. One hundred and two boreholes, eighteen of them up to 20 or 30 m deep with near-continuous 100 mm dia. driven core sampling, provided a total of 1995 m of core logging. Three of the logs were for deep wells down to 133 m (Whitaker 1910).

There are two sites within the coastal strip where detailed information from major construction sites has not yet been published. One is the coastal slope rising up from the Transco Gas Valve Compound in Prince's Esplanade (48000 96330). The other is the Southern Water cofferdam site at Woodvale (47780 96070). Data from these two sites, when eventually released, could be invaluable in understanding the irregular occurrence of rock strata within the coastal strip and might explain the apparent inconsistencies in the Council's recently published slope stability report.

Table 1. Site locations

Ref.	Location	NGR	Date	BH	Depth m
1	The Briary Well	SZ 48310 96480	1901?	1	65
2	Baring Drive	SZ 48986 96460	1967c	*	
3	Baring Road	SZ 48910 96429	1967c	*	
4	Baring Road	SZ 48726 96453**	1974	2	9 & 11
5	Egypt Hill	SZ 48550 96505	1984	1	20
6	Cliff Road	SZ 48824 96488	1988	1	15
7	Baring Road	SZ 48828 96474	1988	1	16
8	Queen's Road	SZ 48885 96615**	1991	2	15 & 15
9	Cliff Road	SZ 48826 96509**	1992	2	19 & 18
10	Cliff Road	SZ 48694 96511**	1998	3	20, 15 & 25
11	Cliff Road	SZ 48770 96475	1998	1	20
12	Egypt Hill	SZ 48536 96448**	1998	2	35 & 22
13	Egypt Hill	SZ 48554 96538	1999	1	19
14	Lammas Close	SZ 48384 96294**	1999	2	10 & 20
15	Baring Road	SZ 48807 96439	1999	1	16
16	Cliff Road	SZ 48795 96489	1999	1	8
17	Baring Road	SZ 48783 96450	2000	1	30
18	Baring Road	SZ 48858 96440	2000	1	18
19	Cliff Road	SZ 48781 96489	2001	1	25
20	Queen's Road	SZ 48755 96578**	2001	2	16 & 13
21	Egypt Hill	SZ 48554 96471	2001	1	26
22	Crossfield Ave.	SZ 48890 96026	1984	1	40
23	Prince's Esp.	SZ 47920 96245**	1975?	2	10 & 15
24	Prince's Esp.	SZ 47965 96210	1975?	1	15
25	Prince's Esp.	SZ 48015 96165	1975?	1	15
26	Woodvale	SZ 47780 96083	1984	1	8
27	Woodvale	SZ 47781 96067	1997	1	8
28	Woodvale	SZ 47789 96021	1997	1	10
29	Woodvale	SZ 47830 96000**	1991	3	9, 15 & 15
30	Woodvale	SZ 47800 95955	1997	1	10
31	Woodvale	SZ 47806 95906	1997	1	15
32	Woodvale	SZ 47710 95870**	1999	2	10 & 15
33	Worseley Rd.	SZ 47711 95670	1996	1	20
34	Worseley Rd.	SZ 47760 95640**	1999	3	10, 15 & 20
35	Gurnard Cliff	SZ 47625 95585	1996	1	20
36	Woodvale Well	SZ 48140 95830	1885?	1	133
37	Debourne Manor Dr.	SZ 48410 95680**	2000	2	15 & 15
38	Broadfields Well	SZ 49040 95190	1845	1	133
39	Broadfields WTW	SZ 49060 95190**	1986	4	9, 9, 10 & 10
40	Briary Lodge	SZ 48305 96530	1996	1	8
41	Egypt Esp.	SZ 48590 96600**	1997	2	12 & 15
42	Egypt Point	SZ 48655 96650	1980s	*	?
43	Queen's Rd.	SZ 49160 96585**	1993	2	12 & 15
44	Royal Yacht Squ.	SZ 49335 96545**	1997	3	6, 16 & 20
45	Baring Rd.	SZ 49263 96466	2000	1	30
46	Castle Rd.	SZ 49440 96320**	1999	3	10, 15 & 20
47	Bars Hill	SZ 49523 96318	2000	1	12
48	Griffin House	SZ 49450 96040**	1989	3	7, 15 & 20
49	Denmark Rd.	SZ 49508 95967	1998	1	13

(contd.)

Table 1. Site locations

(contd.)

Ref.	Location	NGR	Date	BH	Depth m
50	Denmark Rd.	SZ 49454 95933	1998	1	19
51	Griffin House	SZ 49433 95981	1998	1	20
52	Griffin House	SZ 49474 96030	1998	1	20
53	Ancasta Marine	SZ 49720 96030**	1986	3	15, 15 & 16
54	Birmingham Rd.	SZ 49735 95850	1978	1	10
55	Vict.Rd./Granville Rd.	SZ 49310 95780**	1991/92	3	10, 12 & 12
56	Fire Station, Vict.Rd.	SZ 49330 95640	1984	1	34
57	Bernard Rd.	SZ 49680 95680**	1976/98	6	10, 15, 15, 15, 11 & 25
58	Medina Rd.	SZ 49930 95690	1985	1	25
59	Medina River	SZ 50010 95740	1985	1	25
60	Medina River	SZ 50090 95840	1985	1	25
61	Albany Rd. (E.Cowes)	SZ 50210 95920	1985	1	25
62	Sam. White Shipyard	SZ 49955 95390**	1973	4	22, 24, 30 & 30
63	Fairey Marine	SZ 49940 95335**	1984/85	5	17, 18, 19 20 & 25
				102 No.	1995 m

* ..... no boreholes carried out
**..... approx. centre of site

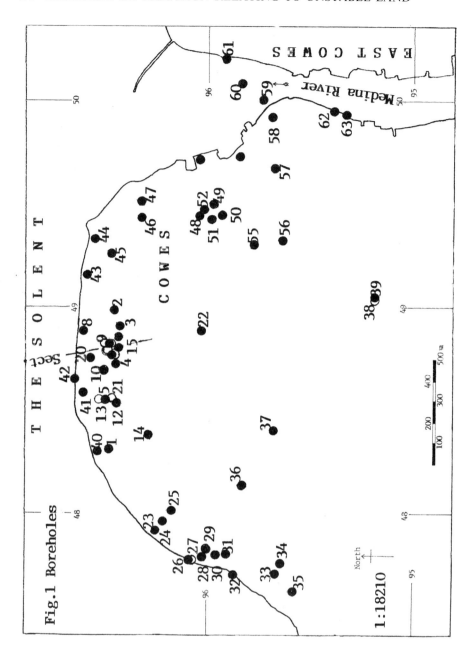

Fig.1 Boreholes

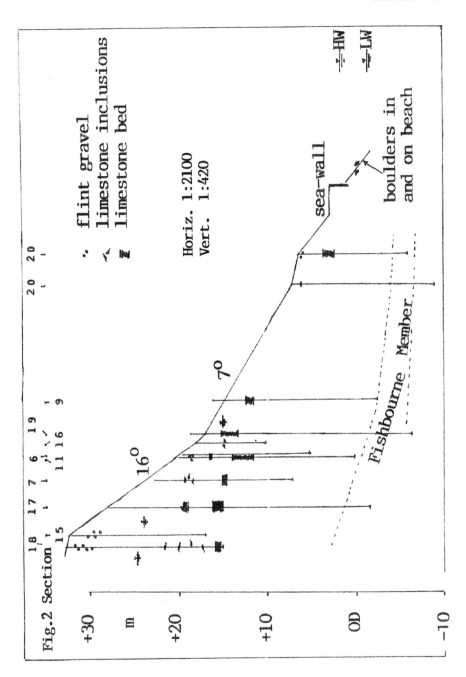

Fig.2 Section

All the core samples from the eighteen boreholes were extruded, split vertically and examined in detail. This 'fabric' examination revealed some visual characteristics which we observed in all boreholes sited in apparently undisturbed ground near the top of the coastal slope. These visual characteristics are described below:-

(i) The Bembridge Limestone Formation generally consisted of two layers of limestone separated by stiff lightly coloured clay. The lower half of the clay between the two limestone layers in many boreholes contained small (mm) specks or zones which produced distinct parallel orange brown smears when cut through. The total thickness of the formation in eleven boreholes ranged between 4.2 and 5.9 m with an average of 4.95 m. (Borehole locations 12, 14, 17, 22, 45, 46 & 48.)

(ii) A metre or so below the lower limestone, in many of the boreholes, the red mottled zone within the stiff clay of the Osborne Marls Member was encountered.

(iii) The Osborne Member in the coastal zone overlay the stiff clay of the Fishbourne Member. The Fishbourne Member could be recognised by the characteristic thin horizontal layers or bands of small (10-20 mm) crushed shells.

In disturbed areas some of the characteristics outlined in (i) & (ii) were missing. For example, at the western end of Cliff Road only a single limestone layer was encountered in the highest of the three boreholes at site No.10 (Fig.4c). In Bars Hill, about half-way down the slope below Castle Road, no *in situ* strata were found down to at least 12.5 m below ground level. BH47, Fig.4d.

SECTION
A section through the centre of the coastal strip is shown in Fig.2.

The ground level along the section line drops from about + 33 m OD at Baring Road to about +19 m OD at Cliff Road, an average slope of about 16 degrees. Below Cliff Road the slope decreases to 7 degrees before reaching Queen's Road and The Esplanade at the bottom of the coastal slope. The level along The Esplanade is +2.70 m OD.

This section goes through eleven site investigation borehole locations between Baring Road and Queen's Road. These boreholes are spread out over a length of 165 m. They are offset from the section line by an average of only 24 m. In spite of this reasonable coverage it is not possible to make any reliable correlation between boreholes. This difficulty is believed to be due to past landslipping. Only eight of the eleven boreholes showed any significant limestone, sometimes as gravel- and cobble-size fragments, and only two found the 'normal' Bembridge Limestone

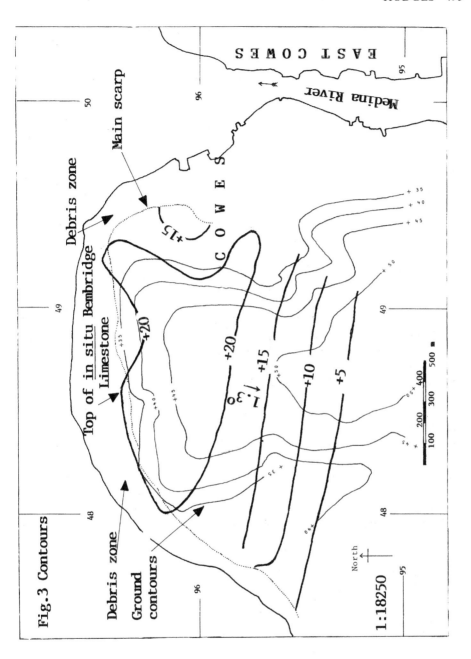

Fig.3 Contours

Main scarp

Debris zone

Top of in situ Bembridge Limestone

Debris zone

Ground contours

1:18250

sequence (No.17, Fig.4b, and No.11). In the other three boreholes no significant limestone was encountered. An analysis of this section would seem to confirm that the coastal slope here overlies an ancient debris zone from past landslipping. Other sections through the coastal slopes exhibited similar features.

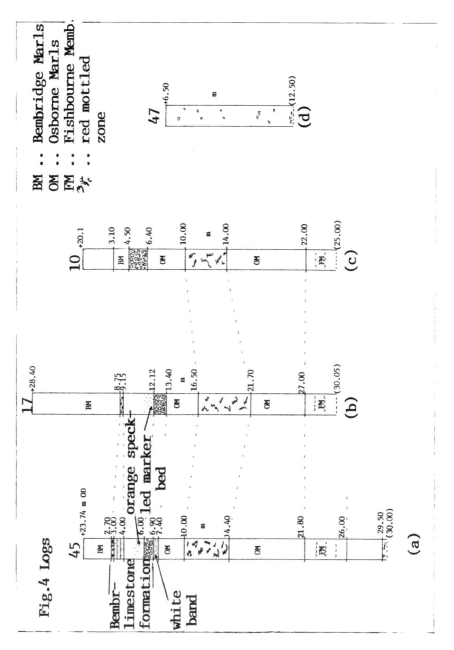

Fig.4 Logs

BM : Bembridge Marls
OM : Osborne Marls
FM : Fishbourne Memb.
≋ : red mottled zone

The Bembridge Limestone Formation would appear to be fairly flat between Broadfields and Baring Road with a reduced level for the top of the formation at least +20 m OD. (Fig.3, Contours). Between Baring Road and the coast any limestone found in boreholes appeared to be part of ancient landslip debris, the original Bembridge Limestone Formation having been broken up or displaced by landslipping or other geological processes. [The origins of these other processes in Southern Britain according to Kellaway (1972) are problematical but landslipping is clearly involved along the northern coast of the Isle of Wight.]

The Halcrow Group report gave the Bembridge Limestone as below sea-level in the four model sections (their Fig.5) although the accompanying Geological Map (their Fig.4) shows the Bembridge Limestone outcropping *within* the coastal slope.

CONCLUSIONS

1. The geology in the coastal slopes cannot be easily simplified into model sections.

2. Small sites on the sloping ground below the main scarp feature cannot be adequately assessed in isolation of the surrounding data.

3. Variations in the borehole data indicate that interpretation of the subsoil conditions in the coastal strip may prove unreliable if geotechnical information from surrounding areas does not exist or exists but is not readily available.

4. Identification of the geological formations and an estimation of whether the strata are *in situ* or displaced by past movements are of critical importance in trying to understand the intermittent instability of the sloping ground along the coastal strip. The proposed lithostratigraphical classification by Insole and Daley (1985) has been adopted in the review.

5. The Bembridge Limestone Formation is fairly flat under the high ground inland of the coastal slopes between Cowes and Gurnard with a reduced level at least +20 m OD. This is in contrast to the model sections in the Halcrow report where the Bembridge Limestone is said to be 'below sea-level'.

6. The difficulties correlating strata between boreholes in the coastal strip is put down to past landslipping and other superficial geological processes. Any limestone encountered in shallow boreholes (10-30 m) in the coastal strip is probably part of ancient landslip debris and not representative of the *in situ* Bembridge Limestone Formation as recognised by Insole and Daley.

7. The collection and availability of commercial geotechnical data would appear to be in need of improvement. Some of the inconsistencies and anomalous data encountered along the coastal strip could perhaps be explained by an examination of the data held by Transco and Southern Water, at present withheld from public scrutiny. Perhaps the Council may be able to offer some help in this regard. Developers and their consultants would be in a much better position to comply with the Council's guidelines if all local commercial data could be made freely available.

REFERENCES

Bristow, H.W. (1862). The geology of the Isle of Wight. Memoirs of the Geological Survey. HMSO. 72 & 77.

Bristow, H.W., Reid, C., Strahan, A. (1889). 2nd Edition of 1862 Memoirs. 158, 160-167.

Colenutt, G.W. (1893a). The Bembridge Limestone of the Isle of Wight. Hampshire Field Club. Vol.2. 174 & 175.

Daley, B., Balson, P. (1999?). British Tertiary Stratigraphy. British Geological Survey. Joint nature conservation committee.

DoE (1990). Planning Policy Guidance: Development on Unstable Land. PPG14. HMSO.

-------(1992). Planning and Policy Guidance: Coastal Planning. PPG20. HMSO.

-------(1996). Planning and Policy Guidance: Development on Unstable Land: Landslides and Planning. PPG14 (Annex 1). HMSO.

DETR (2000). Planning Policy Guidance. Note 14. Development on Unstable Land. Annex 2: Subsidence and Planning. Consultation paper. HMSO.

Dyer, K.R. (1975). The buried channels of 'The Solent River', Southern England. Institute of Oceanographic Sciences, Taunton.

Forbes, E. (1856). Tertiary fluvio-marine formation of the Isle of Wight. HMSO. 55 & 62.

Halcrow Group Limited (2000). Cowes to Gurnard slope stability study. Ground behaviour assessment. Isle of Wight Council. 26 & 27. Fig.4 & 5.

Hutchinson, J.N. (1965). A reconnaissance of coastal landslides in the Isle of Wight. Building Research Station. Note EN11/65.

--------- (2001). Talking Point. Quaternary Geology and Geomorphology. British Geotechnical Association. Pub. Ground Engineering.

Insole, A., Daley, B. (1985). A revision of the lithostratigraphical nomenclature of the late Eocene and Early Ologocene strata of the Hampshire Basin, Southern England. Tertiary Research. Leiden 1985.

Kellaway, G.A. (1972). Development of non-diastrophic Pleistocene structures in relation to climate and physical relief in Britain. Inter. Geological Congress. 24th Session. Section 12 Quaternary Geology. Montreal. 136-146.

West, I.M. (1980). Geology of The Solent estuarine system. NERC. Series C. No.99?. 6-19.

Whitaker, W., Mill, H.R., Matthews, W., Thresh, J.C. The water supply of Hampshire. Memoirs of the Geological Survey. HMSO. 164-166.

Session 5:

Instability, planning and the natural environment

Environmental issues have increasing importance with respect to instability investigation, management and remedial measures, particularly in relation to reconciling socio-economic needs with environmental legislation and biodiversity requirements.

Instability in an area park: naturalistic protection and safeguarding tourism-economic productivity

DR M.C. BOROCCI, architect, Ancona, Italy,
DR V.M.E. FRUZZETTI, engineer, Ancona, Italy,
DR M.E. MOBBILI, architect, Ancona, Italy,
DR. A. PACCAPELO, geologist, Ancona, Italy, and
DR L. POLONARA, geologist, Ancona, Italy

GEOLOGY-GEOMORPHOLOGY

The area studied forms part of the Conero Park, a section of coastline in the central Italian Marches region where wave-cutting of cliffs is in progress. The promontory of Monte Conero, 572 metres high, is an asymmetrical structure running NW-SE. The seaward sides are very steep, almost vertical, with the typical shape of cliff formations, with sinuous and fragmented forms.

Figure 1. View from the sea of the Monte Conero Park cliff

The central core of the promontory consists of a carbonatic formations with outcrops of the older geolithological units of the Umbrian-Marches Succession. These older structures date from the Cretaceous (Maiolica, Marna a Fucoidi, and Scaglia Bianca) to the Eocene (Scaglia Rossa), and the Oligocene (Scaglia Cinerea). The formations are structured in an anticline, with faults and a large number of unconformities.

The geomorphology of the Conero coastline is complex, with many rock falls as well as other types of movements such as slides, soil slips and debris flows. The instability of the slopes is due to a variety of factors such as rainwater flows, wave action and the neotectonic activity of the Monte Conero ridge. The rock fall accumulations are partly transported by the tides along the adjoining beaches. To a large extent, the evolution of the gravitational phenomena described constitutes a sort of natural nourishment for the beaches.

The geological-structural conditions are favourable for the triggering of large landslides with a rotational factor, which have always characterised and influenced the evolution of the Conero coastline's landscape. The fairly steep shapes and the rock types present are the key factors behind the processes of denudation of the slopes. Moreover, the specific weather and climate conditions and factors related to the sea conditions tend to encourage the reactivation of dormant and/or inactive landslides in the form of real debris flows. At present, these movements are occurring mainly around beaches which have formed at the foot of the areas where these gravitational falls take place. In the case of the "Due Sorelle" bay, the lower part of the slope contains a number of debris flows which discharge debris with large particle size, mainly limestone gravel, along the beach.

The morphodynamic processes described are especially active in periods of heavy rainfall and their evolution is also closely linked to the high degree of fracturing of the Monte Conero ridge. In periods of especially heavy rainfall, gravitational creep phenomena occur as the terrain which forms the sheets becomes waterlogged.

Figure 2. Due Sorelle Bay

The study of the area was performed through careful photo-interpretation analysis aimed at identifying the factors which have caused the changes in the landscape. This yielded valuable information which might represent the guidelines for a programme of measures to improve and upgrade the zones affected by gravitational instabilities. The examination of the aerial photographs, from surveys conducted in 1955 and 2000, clearly revealed the profound change in land use. In 1955 the coastal slopes were used by farmers and shepherds. The agricultural practices (see section between S. Michele and Sassi Neri Bay), involving traditional methods, not only ensured stabilisation of the slope by growing crops on terraces, but also guaranteed proper, effective control of the surface waters. The many main ditches which received the water from the drainage channels of the individual cultivated fields underwent regular seasonal maintenance (cleaning and reshaping). The resulting good drainage capacity allowed the rainwater to flow to the sea quickly, reducing seepage into the subsoil. The comparison between the aerial photographs from the two flights revealed that in the '50s landslides were less frequent and affected a smaller area than today. Another feature observed was the effect of the waves. The coastline has retreated a few metres from 1995 to 2000. Moreover, the wave action, especially intense during high tides, has removed material from the feet of several dormant and/or inactive landslides, reactivating them. Examples include the Sassi Neri landslide at Sirolo and the one between the quay beach and the Mezzavalle bay at Portonovo.

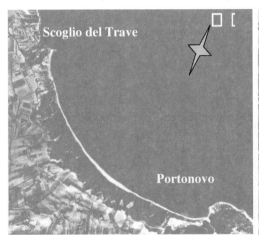

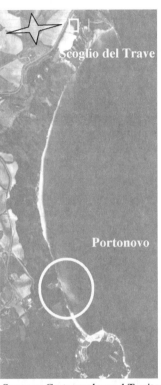

Figure 3 and 4. Comparison of the aerial photographs taken in 1955 (top) and 2000 (right) of the shore between Scoglio del trave and Portonovo. The retreat of the coastline due to marine erosion, leading to cutting into the foot of dormant landslides which have thus been reactivated (white circle) is clearly visible. The analysis of the aerial surveys revealed that compared to the past, the surface water drainage system along the coastal slope of the bay has been obliterated and in some points eliminated

Sources: Cartography and Territorial Information Dept-Marches Region

Recently (1997), in the case of Sassi Neri, stabilisation operations have been carried out using low-impact methods to safeguard the landscape, conserve the biodiversities, protect the autochthonous species and allow the continued use of the area by holidaymakers. The instability of the S. Michele - Sassi Neri Bay coastal slope is reflected in rock slides with a rotational factor and soil slips. In order to safeguard human activities within the landslide body, surface channels were constructed to lessen the effects of the surface waters. This considerably reduced the triggering of mud flow. A coast-hugging reef was constructed at the foot of the cliff to combat the wave action; submerged in the beach, the reef consists of gravel and pebbles obtained from the materials generated by the nourishment process.

The artificial reef has been very successful in improving the general stability of the area; it lies in summer between the beach and the cliff, is semi-submerged by the sediments and blends perfectly into the natural context of the Park, without interfering with the natural dynamic of the cliff. Measures along the slope ought to aim to restore the primitive conditions. A historical analysis of the morphology clearly reveals that the frequency of hazardous events has increased as a result of the cessation of cultivation. This has led mainly to an obliteration or, in some zones, the elimination of the water drainage system at the lowest level, leading to denudation of the slopes due to run-off water. Moreover, the growing of crops on terraces stabilised the clayey slopes, preventing the occurrence of gravitational instabilities.

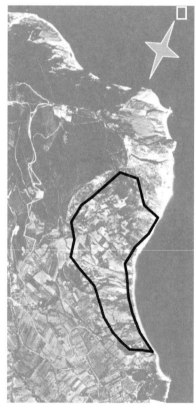

Figures 5 and 6.Comparison between the aerial photographs taken in 1955 (left) and 2000 (top), highlighting the profound change in land use (enclosed area) and the retreat of the coastline along the Sassi Neri-San Michele bay as a consequence of human factors and marine erosion.

Sources: Cartography and Territorial Information Dept. - Marches Region

WARNING SIGNS, MONITORING AND CONTROL

In most cases it is possible to recognise a series of warning signs which, with varying degrees of obviousness and significance, allow the recognition and definition of a zone with geomorphologic instability, or at least identification of the potentially unstable area within which more in-depth surveys need to be conducted using specific geotechnical monitoring techniques. When the preparatory stage of the instability phenomena is long enough, the surveys must be conducted systematically, in order to verify the ways in which the processes are evolving and pick up any tendency for movements to accelerate. For the purposes of prevention, the checks and monitoring procedures to be performed refer primarily to all the factors which can be identified as initial signs of instability.

Rock falls occur suddenly from walls of rock or on very steep slopes. The conditions which may indicate that falls of material are probable must be identified through monitoring of the degree of fracturing of the rocks, the geometrical position of the fractures and the way in which they intersect with each other. Hazardous situations arise when criss-crossed systems of cracks break down the mass of rock into several parts, cutting out a large number of blocks which rest on fracture or stratification surfaces which slope downward, towards the outside of the slope. The presence of individual boulders which have already broken away from the wall and are lying at its feet is symptomatic in these cases. Springs of water flowing from the base of the outcrop indicate that the fractures run continuously through the body of the rock; in this case the pressure of the water and the disintegrating action of the freeze-thaw cycle may lead to large rock falls and therefore to serious hazards. These monitoring operations, which often result in measures intended to allow controlled use of the zones concerned, and in some cases in a ban on access, should be repeated regularly and across a wide radius. However, the situations requiring a detailed geostructural survey aimed at identifying any blocks requiring scaling or

Figure 7. View from above of the Sassi Neri-S. Michele landslide.

nailing and tying operations must also be recognised. As well as rock falls, in mainly clay sections of cliff retroactive landslides with a rotational factor, soil slips and plastic deformations also occur. Examples include Rupe Sermosi and the Gigli landslide at Numana, S.Michele-Sassi Neri at Sirolo, and Mezzavalle at Portonovo di Ancona.

Slides are movements with flat or curved slip surfaces affecting slopes of average or low gradient. The initial stage of a flat or rotational slide features the opening of more or less continuous cracks concentrated in the top of the slope, often of considerable depth. Locally, in the middle and lower parts of the slope, areas of the ground will be seen to swell, sometimes in association with intermittent outflows of water. However, a careful, systematic analysis can identify growing fracture sizes, variations in water outflows, the appearance of new cracks and deformations on any nearby man-built structures, and the gradual, slow tilting or sliding of tall trees. The accentuation of these phenomena must be detected by means of regular on-site inspections which may use even quick methods such as checking the alignment of tall trees in the unstable area which can be observed from a fixed point in a stable area, capable of providing useful information about the trends of the gravitational movements.

Soil slips are phenomena occurring on slopes of varying gradient, with movements which persist even in zones where the slope is very gentle. The initial warning signs consist of gentle undulations in the ground, which gradually become a series of swollen and sunken areas. In more advanced stages these forms tend to be arranged around arcs of a circle running across the direction of movement, and temporary subparallel cracks appear in the land in motion. The moving mass may subdivide into several lobes moving at different speeds. Phenomena apparently similar to those described above but evolving much more quickly occur further to fluidification of the waterlogged surface soil during heavy rains, and affect areas without efficient woodland cover. These phenomena occur a very short time after the rainfall which triggered them and the only warning sign for potential instabilities of this kind is sometimes the fact that similar slips have already occurred in the same area in the past.

THE PARK PLAN. FEATURES OF THE COASTAL ECONOMY
This zone, with a large number of areas of hydrogeological instability, and with the congestion of the city of Ancona immediately to the north and intensive tourist-industry development along the coastline to the south, is protected by the Plan of the Conero Park, established in 1987 but only actually operational since 1991.

The Conero Park occupies an area of about six thousand hectares lying in four municipal areas of the province of Ancona: Ancona, Camerano, Sirolo and Numana; its main feature is the presence along the Adriatic coastline of sheer cliffs which make its use by the tourism industry difficult. The Park's perimeter runs through an extremely complex area of regional importance featuring urban spread, a high concentration of infrastructures and varying degrees of anthropic pressure; from this point of view the Park is in direct relationship with the context of Ancona to the north and the Aspio basin to the west, where the area's largest concentrations of industrial and business activities are located.

The aim of the Plan is to upgrade the zone from the point of view of the natural environment and the protection of tourism and economic productivity, by taking direct action on the territory to deal with its hydrogeological instability. Along the Ancona coastline the direct economic activities (i.e. those directly dependant on the sea) and indirect activities, which have strong commercial links to the coastal activities as such, mainly comprise fishing, shipbuilding, the transportation of tourists, the construction and renovation of tourist accommodation, and agriculture and holiday farms.

The problem of the sustainability of the economic processes in the Conero area derives from the physical limitations on the transformation of the natural features into an urban context and thus the fact that it is impossible to adopt a policy of growth. The most heavily urbanised areas show problems of restoring the equilibrium of the cycles and the procedures for differentiating and distributing the provision of tourism services and the activities and functions present.

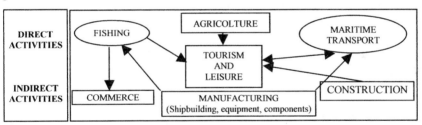

Table 1. Direct and indirect coastal economic activitie Source: Quaderni del Parco, University of Ancona, Economics Department, September 2000.

The Park's problem in relation to one if its main resources, the coastline, is that of defining sustainable use capable of combining economic growth and protection-upgrading in the light of the highly seasonal nature and extremely high spatial concentration of the main businesses and resources in the tourist industry. There is a need to differentiate tourism across different times of year by offering tourists products and itineraries linked to the different types of users, local and occasional (environment, sports, local cuisine, etc.). Demand in the tourism sector is evolving and becoming more and more differentiated and segmented; the sector must therefore increase its ability to specialise its services and offer a wide range of specific packages. Another clear sign of imbalance is the lack of livestock farms integrated with arable activities. Far from favouring a positive environmental impact, this contributes to a deterioration of soil quality and a substantial disintegration of the geological and hydrogeological equilibriums. Another topic of great interest on the larger scale is that of the quality of the landscape through which the various routes through the Park pass. These routes are key to the different ways in which the Park can be used (on foot, by car, by bicycle) and upgrading the network of routes across the agricultural landscape can on the one hand extend and differentiate the alternatives for use of the Park and on the other reduce congestion in the areas under greatest stress from pressure of tourist numbers. To achieve this, the Park could implement a general system for conservation of the common goods (goods and services of interest to the community such as monuments, water drainage channels, meadows, walks and rides through the most attractive environments, etc.) and for management of tourism in a way compatible with the characteristics of a protected area, through improvement of the services and facilities in terms of use and profitability.

Only integrated, all-inclusive upgrading of the area will be able to activate this economic potential by involving all the sectors in a widespread, diversified way, encouraging the use of local products (guided excursions and tours, information and promotion of these typical local assets, accessibility of locations) and by reducing the impact on the natural resources (minimisation of energy consumption, reduction of waste production and effluents and atmospheric emissions). All this must also consider the many danger factors linked to the intensive geological activity of the cliffline, which must always be borne firmly in mind if it is not to constitute a disturbance to the other activities planned.

INSTABILITY AND POTENTIAL FOR USE: AN INTEGRATED PROPOSAL

The need is therefore for a plan of measures with the main objective of generating in-depth knowledge of the dynamics of the cliffline, which identifies the indicators for the sustainability of the tourist industry on the one hand, and on the other the indicators linked to the possibilities for use of the coastline with regard to the many points of hydrogeological instability. Drawing on the experience of the Marches region and other regions of Italy (Tuscany, Liguria, etc.) with regard to the state of resources, an outline is suggested for the construction of the sustainability indicators for the area under consideration.

In conclusion, what should be stressed is the need for integration between the different aspects of this complex situation (instability, coastal erosion, dangers, possibility of use, potential and economy of the park, etc.) in order to allow the bodies responsible to identify the performance objects for the planning and scheduling of measures intended to maintain and improve environmental quality, while at the same time guaranteeing the economic funding derived from tourism. The complexity of the topics dealt with can be grouped into a series of measures which could constitute the basic lines of action:

- an improved and more qualified provision of recreational and leisure time activities; the construction of technological networks and the provision of filtering plants; the re-qualification of the area from the beach and a functional and architectural reorganisation of the objects and equipment for touristic exploitation of the beach (identification of stable areas for building sanitary facilities and characteristic beach restaurants with lightweight wooden structures, equipped with suitable toilets);

- the demolition of damaged structures;

- the restoration of vehicle access and pathways; recovery of degraded vegetation and gradual reclamation of the natural character of the woods;

- the strengthening of projects regarding regulation and drainage of surface water using ecological engineering techniques;

- beach nourishment;

- monitoring and supervision;

- awareness training for tour operators regarding certain aspects of monitoring and maintenance of actions taken;

- banning all motor boats access to the bay.

The need is therefore stressed for a plan of action which sees as its main objective a profound process of awareness regarding the actual dynamics of the cliffs. This must start from the identification of indicators representing the sustainability of tourism on the one hand, and on the other hand those concerning the possibility of exploiting the coast in reference to the numerous hydrogeological instabilities.

The planning and development of a coast protection scheme in an environmentally sensitive area at Castlehaven, Isle of Wight

A R CLARK, High-Point Rendel, London, UK
C V STORM, High-Point Rendel, London, UK
D S FORT, High-Point Rendel, London, UK
R G M^cINNES, Isle of Wight Council, UK

INTRODUCTION

Castlehaven is a fishing hamlet at Reeth Bay located on the south coast of the Isle of Wight. The area forms a large bay backed by steep cliffs above which are coastal slopes that form part of the Niton Undercliff (Figure 1). These slopes are well developed with large Victorian villas and cottages, many of which have substantial grounds and gardens (Plate 1). The area also has a well developed infrastructure and includes the main east/west road through the Undercliff, the A3055.

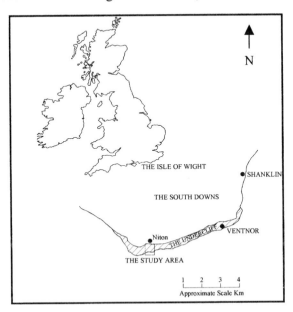

Figure 1. Location of Castlehaven, Isle of Wight, UK

Instability – Planning and Management, Thomas Telford, London, 2002, 509–518

The Niton Undercliff forms part of a much larger undercliff which covers the southern coastal fringe of the Island and which is a post glacial landslide complex some 12km in length and 500m wide. Coastal erosion of the soft cliffs and the consequent coastal retreat has the effect of reactivating the older landslides such that movement may be experienced significant distances inland from the coast. As a result the properties and infrastructure in the area show evidence of structural damage to varying degrees depending on their distance from the coast and their form of construction. This movement and loss of land has been ongoing for centuries. A recent major landslide event occurred in response to the wet winter of 1994/95 resulting in accelerated erosion and significant damage to property up to 400m inland from the coast. (Plate 2).

Plate 1. Aerial photograph of Castlehaven.

The proposals for addressing the coastal erosion and instability problems consist of a range of engineering measures. The selection of remedial measures however, and indeed the opportunity to implement them has been significantly influenced by the consideration of the important environmental issues, which are relevant to the site. The site is designated an Area of Outstanding Natural Beauty (AONB) and is adjacent to a marine area now designated a candidate Special Area of Conservation (cSAC). The installation of measures to arrest slope instability to protect property and public safety are diametrically opposite to the environmental interests that wish to maintain instability to retain the habitats regarded as of particular value.

It is within this context that the scheme has been developed through the feasibility, planning and local inquiry stages to a successful conclusion to allow a scheme to proceed to detailed design and construction. This position has been reached by cooperation, collaboration and compromise between the client, the scheme consultants and English Nature (EN), the UK Governments' advisors on environmental matters.

Plate 2. Landslide damage at Castlehaven.

GEOLOGICAL BACKGROUND

The geology of the area consists of a series of Cretaceous rocks including principally Upper Greensand overlying the Gault Clay (Refs 1&2). This in turn overlies weekly cemented lower Cretaceous sandstone sequences of Carstone and Sandrock which are exposed as vertical 8-15m high unprotected coastal cliffs surrounding the beach at Reeth Bay. The Upper Greensand forms the inland scarp and limit of the Undercliff landslide complex approximately 500m inland from the shore. Marine processes since post glacial times have progressively eroded cliffs at the toe of the landslide complex causing coastal recession and reactivation of the instability. Historic maps data and local history indicates that up to 40m recession has occurred since 1862 and nearby offshore subseabed geophysical data indicates that the landslide complex extended up to at least a further 1.5km offshore (Ref. 3).

In addition to coastal erosion, groundwater hydrogeology plays a significant role in the long term stability of the area. The geology comprises essentially an interbedded sequence of relatively high and low permeability strata. The permeable Upper Greensand forms much of the hinterland but downward drainage of the groundwater is restricted by the underlying overconsolidated Gault Clay and Passage Beds aquiclude which forces drainage along dip towards the rear scarp cliff of the landslide (Figure 2). The geological junction is obscured by the landslide debris and thus groundwater infiltrates the landslide complex debris and emerges as a series of springs, sinks, ponds, seepages and surface streams; underground streams have also been reported. In addition to natural groundwater the landslide is further surcharged by artificial drainage from uncontrolled surface soakaways, highway drainage and foul sewage from septic tanks and cesspools from all the properties in the area.

Thus the stability of the area is controlled by marine erosion and coastal recession of the weak rock sequence exposed at the toe of the slope and by groundwater. The resulting recession progressively undermines the overlying landslide complex which in turn rests on the Gault Clay, which is notorious for its instability and is locally known as the "Blue Slipper". In addition, the whole sequence of both intact geology and landslide debris are surcharged by both groundwater and artificial water sources.

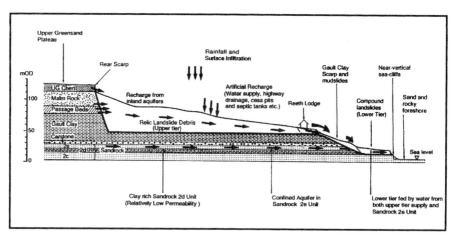

Figure 2. Cross section through the Niton Undercliff showing principle sources of groundwater.

ENVIRONMENTAL ISSUES

The site area is within an Area of Outstanding Natural Beauty and the cliffs are within a locally designated Site of Interest for Nature Conservation (SINC) and the shore below low water mark is within the South Wight Maritime cSAC. Although Castlehaven is not a designated conservation area, it has considerable heritage value as a site of a number of Victorian villas in their own landscaped gardens, several of architectural or historic interest.

The principal ecological interest of the site is the unstable south-facing soft rock cliff habitat, which is of value for a diversity of rare invertebrates, especially solitary bees and wasps. This environment is favourable due to a combination of mild climate, overall habitat diversity and, as a consequence of landslip activity, the availability of bare ground of varying composition in which many of these species excavate nests.

The 1988 Environment Impact Assessment Regulations introduced the requirement for certain planning applications to be accompanied by an environmental statement (ES). It was decided, at an early stage, that the Castlehaven proposals would be treated as a development of a significant scale and nature in a sensitive environment and that would warrant preparation of an ES.

The most significant environmental issues, identified in the scoping report and examined in the ES, were: marine and terrestrial ecology; landscape (visual impact); material assets (property and infrastructure) and heritage (principally listed buildings and historic landscape). Effects on the first and second were generally considered adverse, and mitigation measures were proposed. On the third and fourth, the effects were considered beneficial in that the works would help to protect the material assets and heritage of the developed parts of Castlehaven.

SCHEME PROPOSALS

If left unchecked the progressive loss of both housing development and infrastructure caused by the landslide reactivation would lead to combined direct and consequential losses of approximately £18 million at 2001 prices. Consequently an initial engineering scheme was prepared to solve the problem. Having considered many alternatives a preferred scheme, which met the required cost benefit criteria comprised a rock armour toe protection plus slope stabilisation works consisting of both deep and shallow drainage systems and slope reprofiling. Notwithstanding the engineering and economic feasibility of the preferred scheme, the scheme was subjected to rigorous testing with respect to environmental impact and extensive studies were undertaken during the scheme development culminating in an environmental statement.

MITIGATION MEASURES

At Castlehaven a 600m length of the cliffs would have been affected in the initial scheme proposal by localised slope regrading and extensive drainage works, including some habitat damage during construction and, in the longer term, by the progressive loss of bare ground as stabilised slopes became vegetated. Extensive areas of neighbouring cliffs would, however, retain these characteristics.

Mitigation measures were developed as part of the scheme proposals. The ecological advice was that since there was such an abundance of bare ground of varying texture, slope and aspect in the vicinity it was unlikely that the effect of the scheme on this habitat would significantly limit the population of rare species. Possibly of more importance would be the loss of foraging area through succession of scrub and woodland. The principal mitigation proposal therefore, was for a land management plan to reduce the levels of scrub and encourage the development of flower-rich grassland. Longer term monitoring was also proposed, so that the management regime could be adjusted in the light of experience and areas of bare ground created, if necessary.

A second potential ecological impact concerned the effects of delivering rock armour by sea on the coastal reefs. The reef sheltering Reeth Bay, partly within the maritime cSAC, supports luxuriant algal communities of very high conservation value, although the bay itself, around the foot of the cliffs, is mostly sandy and of little ecological interest. Fortunately, a gap in the reef opposite Castlehaven, with a sandy bottom, enabled a route for barges to be identified that would not affect the nature conservation interest of the shore or cSAC. The only other environmental impact that raised concern was the potential loss of the natural appearance of the cliff slope landscape, although this had to be balanced against the damage that severe instability could have on the developed landscape heritage of Castlehaven itself.

THE UK PLANNING CONSENT PROCESS

A report on the proposed scope of the environmental statement was prepared in December 1996 and circulated to the Local Planning Authority, statutory consultees and other bodies. It became apparent that EN and the local Wildlife Trust were inclined to oppose the protection of this section of coast on the basis of its environmental sensitivity and at this stage did not fully engage in the environmental assessment process. Conversely the residents, directly affected by the problems of land instability on their property and businesses were understandably keen to see a coast protection scheme implemented as soon as possible.

In order for a coast protection scheme to proceed, consents, notifications or licences are necessary under the Town and Country Planning Act, the Coast Protection Act and the Food and Environmental Protection Act, thus involving the Local Planning

Authority the Department of Transport, Local Government and Regions (DLTR) marine division and Department for Environment Food and Rural Affairs (DEFRA formerly MAFF). The outline scheme was submitted to the Planning Authority for approval in January 1998 and EN submitted a full written objection in March 1998.

Despite a number of meetings with EN to discuss scheme modification and mitigation measures during the spring of 1998, agreement in principle could not be reached and EN requested the then DETR (now DTLR) to call the application in for the Secretary of State to determine. Thus the application was taken out of the Local Planning Authority's hands by the DETR in April 1998. This could have led to a public inquiry but, after due consideration by the Government, it was returned to the Local Authority for determination locally and consequently planning permission was granted, with conditions, in September 1998. There were nine planning conditions. Three were specifically imposed "in order to preserve as far as possible the nature conservation interest of the area". They covered the detail of the construction of the drainage works in the cliff, the delivery of rock armour by sea and a land management plan for the site and adjacent land.

The UK procedures for coast protection scheme implementation require a detailed "Engineers Appraisal Report" to be submitted to DEFRA confirming the technical, economic and environmental viability of the scheme to accompany the application for approval and grant aid. This must include planning permission and the agreement of English Nature as the Governments' environmental advisor. Following the granting of planning permission English Nature and the Hampshire and Isle of Wight Naturalist Trust still maintained their objections. Consequently at this stage DEFRA was unable to approve the scheme for grant aid until all objections had been withdrawn.

POST APPLICATION NEGOTIATIONS
The Council's aim was to deal with the concerns expressed by EN through scheme mitigation and the development of the nature conservation planning conditions into some form of legally binding agreement. Meanwhile, there was increasing concern on the part of the residents that an urgently needed scheme was in their view being held up unreasonably. It also appeared at this stage that a draft UK Biodiversity Action Plan was in the course of preparation by EN with the target of seeking to retain and, where possible increase the amount of maritime cliff and slope habitats unaffected by coastal defence. At this point the EN objection became not only to the 'loss' of the habitat for solitary bees and wasps, but also conflict with the emerging UK Biodiversity Action Plan. In addition the possibility of an "Appropriate Assessment" under the EC Habitats Directive Regulations being needed in relation to the offshore reefs in the cSAC was raised by EN.

Further discussions and site meetings were held with EN, joined by the Wildlife Trust, early in 2000 which eventually led to a basis on which EN felt able to withdraw its opposition. Mitigation measures involved a minor reduction in the length of the protected sea cliff, the omission of additional surface drainage on the seaward Gault Clay slope and the omission of localised slope regrading. On the basis of the proposed inland deep drainage system, such measures were not deemed to compromise the technical viability of the scheme. In addition, there was a general firming up of the arrangements for, and extent of, the land management proposals. These were finally agreed as amendments to the planning permission in January 2001.

Unfortunately this was insufficient mitigation for the Wildlife Trust to withdraw their objection. They regarded any measures to 'stabilise' the coast as being in conflict with its features of importance and considered the proposed mitigation measures questionable in terms of effectiveness and achievability. The failure to have this objection withdrawn without rendering the scheme ineffective finally caused DEFRA to decide to hold a Local Inquiry.

THE LOCAL INQUIRY
Local Inquiries under the provisions of the Coast Protection Act are relatively unusual, although this is one of the three regularly used means of hearing appeals against refusal of planning applications (the others are written representations and a full public inquiry). The advantage is that a Local Inquiry is less formal than a full public inquiry in that legal representation is not normally required, formal cross-examination does not take place and all parties or persons are encouraged to participate, under the chairmanship of DEFRA. It is thus more conducive for residents and local groups.

The hearing was held on 1st August 2001. The Council, the Wildlife Trust and EN had circulated statements of case beforehand and presented them to the hearing. Each party accompanied the Inquiry Chairman on a site visit before the Inquiry opened. At the Inquiry each presented its case and there was an opportunity for questions. Local residents also presented a statement and put questions. The Inquiry concluded with each party being afforded and opportunity to make a short summary statement.

The outcome was that the scheme was subsequently approved by DEFRA in December 2001 with conditions regarding the need to implement the mitigating measures.

CONCLUSIONS
At Castlehaven there was an urgent need as a result of coastal landsliding for works to protect public safety and property and infrastructure worth many millions of

pounds. The outcome in this case has been extensive delay and high public costs, exacerbated by considerable financial costs and extensive anguish imposed on residents.

It is clearly regrettable that the current system allowed significant delay to the decision on a relatively urgent scheme which did not affect a site of designated national or international importance. There seems to be three areas where the current system could be improved:

- at least four separate consents under three different government acts and involving three government or local government decision-making departments are required, each providing a separate opportunity for a determined objector.
- the advisory role of single-interest government agencies has been allowed a status that extends beyond advice;
- there does not appear to be a working mechanism to assess the validity and weight to be given to objections by single interest groups, taking into account the broader considerations in the decision making process, within a reasonable timescale.

The issue covering single-interest government agencies and other bodies is perhaps more difficult to resolve, especially outside a co-ordinated consents process. Obviously, in the case of Castlehaven, there was a strongly held opinion on the part of EN and the Wildlife Trust, that any form of coast protection or cliff stabilisation would destroy the nature conservation interest of this part of the coast and that no such scheme should be approved. It was equally obvious that others, that is the Council as Coast Protection Authority and the residents, held just as firmly to an opposing view, supported in this case by the Shoreline Management Plan.

In the case of English Nature an eventual spirit of co-operation and understanding enabled a workable solution to be achieved through acceptable mitigation measures. In the case of non-statutory single interest group objectors, the current system allowed them the authority to object and delay but without having the responsibility for the consequences.

ACKNOWLEDGEMENTS
The paper is published with the permission of the Isle of Wight Council. The project is funded by DEFRA. The authors would like to express their thanks to all the participants in the scheme.

REFERENCES
1.   British Geological Society. Isle of Wight Memoir 1990 London HMSO
2.   Hutchinson J.N., Chandler MP. 1991. A preliminary landslide hazard zonation of the Undercliff of the Isle of Wight Slope Stability Engineering pub Thomas Telford. pp. 197-205

3.   Clark A.R., Lee E.M., & Moore R. 1994. The development of a ground behaviour model for the assessment of landslide hazard in the Isle of Wight Undercliff and its role in supporting major development and infrastructure projects. Proc. 7th Int Congress IAEG Congress. pp. 4901-4913.

Embracing environmental management in landslide stabilisation – The state of the art in South Wales, UK

J D MADDISON, Halcrow Group Limited, Cardiff, UK
I H G SHAW, Forest Enterprise, Aberwystwyth, UK
H J SIDDLE, Halcrow Group Limited, Cardiff, UK
S SOWLER, Halcrow Group Limited, Gloucester, UK

INTRODUCTION

The stabilisation of landslides frequently involves the implementation of large-scale civil engineering works, be they earthworks, drainage, structural solutions or a combination thereof, which can impact on the local environment permanently as well as in the construction phase. In recent years there has been a plethora of environmental legislation, directives and policies at international, national and local levels aimed at protecting and enhancing the environment. These have had a substantial impact on civil engineering schemes including landslide stabilisation and have brought about fundamental changes in the way schemes are developed, designed, implemented and evaluated after their completion. It is now incumbent on the designer to establish the current environmental circumstances of a site in order that appropriate mitigation and restoration measures can be implemented as part of the scheme. Stabilisation options have to be critically reviewed not only to address engineering issues but also to take account of the environment so that the design is sympathetic to the environment and provides, where possible environmental gains.

The paper presents the state of the art of embracing environmental management in landslide stabilisation in South Wales drawing on the experience from a current Forest Enterprise programme to stabilise a number of landslides on its property.

ENVIRONMENTAL LEGISLATION BACKGROUND

In June 1992, the governments of the United Kingdom (UK) and over 150 other nations across the World signed up to the Rio de Janeiro Convention on Biological Diversity. The Convention recognised the need to halt worldwide loss of animal and plant species and genetic resources and to encourage enhancement of biodiversity. Subsequent action by the UK government has included signing up to European directives aimed at protecting biodiversity, legislation to protect the national biodiversity and the preparation of the 1994 UK Biodiversity Action Plan, which set out a strategy for implementing the Rio de Janeiro Convention. The strategy has included the requirement for preparation of Regional and Local Biodiversity Action Plans; the latter being made the responsibility of local government authorities, often produced by local wildlife trusts in partnership with other conservation organisations. The key directives, legislation and action resulting from the Rio de Janeiro Convention are listed in Table 1.

Date	Key Directives/Legislation/Action
1992	Rio de Janeiro Convention on Biological Diversity
1992	EC Council Directive on the Conservation of Natural Habitats of Wild Fauna and Flora – The Habitats and Species Directive (92/43/EEC)
1994	The Conservation (Natural Habitats, &c.) Regulations 1994, UK ratification of the EU directive.
1994	UK Biodiversity Action Plan
1996	Guidance for Local Biodiversity Action Plans, vol 1, Local Issues Advisory Group 1996
1997 on	Preparation of Regional & Local Biodiversity Action Plans
1997	The Hedgerow Regulation 1997
2000	Countryside and Rights of Way Act 2000
Other protective legislation	Wildlife and Countryside Act (1981) and subsequent amendments

Table 1: Summary of Key Environmental Directives, Legislation & Action

Biodiversity is, however, only part of the overall environment issue and in addition to Biodiversity Action Plans many Local Authorities now also have planning policies relating to wider environmental issues such as landscape and heritage. Consequently, in recent times it has been necessary for civil engineering projects to incorporate new procedures and processes to fully address all environmental issues. How these affect landslide stabilisation schemes in South Wales is described in the following sections with reference to a current programme of schemes promoted by Forest Enterprise.

SOUTH WALES LANDSLIDES – GENERAL BACKGROUND
The present day landform of the South Wales valleys results mainly from the last glaciation, the Devensian, which lasted from about 60,000 to 10,000 years BP. Glacial ice spread over the area from centres in the Brecon Beacons to the north and incised the pre-glacial river valleys leaving them over-steepened and up to 300m deep. Many landslides originated within the periglacial conditions that existed in the area in late glacial times and remained as relic features within the landscape as climatic conditions ameliorated during the Holocene. The landscape of South Wales dramatically changed during the industrial revolution. The exploitation of iron and coal reserves resulted in wide spread urbanisation of the valleys, with housing and infrastructure to support the new industries and the disposal of waste mining products in extensive spoil tips. There are many instances where spoil tips have reactivated old landslides causing damage to property and infrastructure.

FOREST ENTERPRISE PROGRAMME OF LANDSLIDE STABILISATION

Forest Enterprise, an Executive Agency of the Forestry Commission, manages some 128,000 hectares of woodlands in Wales on behalf of the National Assembly for Wales. Within the South Wales coalfield area Forest Enterprise controls approximately 28,000 hectares. These areas have extensive plantations on the valley sides and inter-valley ridges. Within its property lies scores of colliery spoil heaps and ancient landslides most of which present little or no risk to people or property. However, a small number of landslides on Forest Enterprise property have had a history of extremely slow (defined after Cruden and Varnes (1996)) episodic movement and lie close to property or affect infrastructure. These sites are subject to regular inspection, monitoring and review of landslide activity.

Following assessment of identified sites within Forest Enterprise's South Wales estate considered to present significant risk and liabilities, undertaken by Halcrow in 1996, Forest Enterprise sought and obtained Government funding to investigate and remediate the sites. Experience has shown that the management of landslide events poses severe difficulties. Severing of roads can have great financial consequences and property values can suffer in areas adjacent to landslides. These adverse consequences may be long term when corrective works cannot be carried out immediately. Forest Enterprise's objective, therefore, was to carry out remediation works to actively unstable sites within a phased programme, thereby reducing the risk of having to respond to emergency situations in the future. The overall approach to the landslide stabilisation programme and the action taken to embrace environmental management issues at the sites are described in the following section. The approach has resulted in significant developments in the way landslide stabilisation schemes are designed and implemented, particular aspects of which are highlighted by way of a case history of the Ty'n-y-bedw Landslide stabilisation scheme.

LANDSLIDE STABILISATION PROGRAMME
Overall Approach

When the Forest Enterprise programme of landslide stabilisation was embarked upon in early 1999 it was recognised that there was a need to consider the environmental issues at each site and to incorporate appropriate measures in the detailed design and construction to provide the optimum engineering and environment solution. To achieve this a multi-disciplinary project team was brought together with skills including:

- landslide investigation & management, remediation civil engineering design and ecology assessment (Halcrow Group Limited)
- landscape assessment and design (WynThomasGordonLewis Limited)
- soils assessment and site restoration (Progressive Restoration Limited, together with Forest Enterprise's considerable in house expertise).

The approach to the scheme is outlined in the flow chart presented as Figure 1. A series of engineering, landscape and ecology investigations was undertaken from

522 INSTABILITY, PLANNING AND THE NATURAL ENVIRONMENT

which a design concept was developed. This represented the optimum scheme to stabilise the landslide taking into consideration the current and future form of the site within the local landscape and as far as possible maintaining and preferably increasing biodiversity at the site. Having developed a scheme concept, consultations were made with the Local Authority including Highways, Drainage and Planning Departments, the Environment Agency and Countryside Council for Wales to seek a consensus on the approach to overall scheme/landscape design and to the further phase of detailed ecology survey work and soils investigation for restoration planning. The remediation scheme detailed design was then developed taking into account the results of all these further investigations and assessments and was submitted for planning consent.

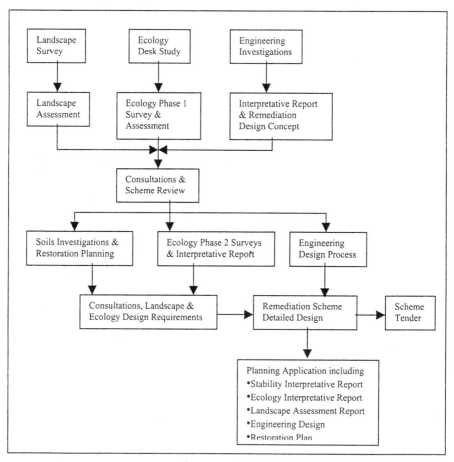

Figure 1: Integrated investigation and design process

The approach ensured compliance with current environmental legislation, directives and polices. Also, having established the existing environmental circumstances at site, it was then possible to critically review the landslide stabilisation options taking account of environmental and engineering issues to establish the most appropriate scheme overall. A design sympathetic to the environment and providing appropriate mitigation and restoration measures and where possible environmental gains, could then be developed. The methodology also ensured that the Local Planning Authority was aware of the liabilities and risks associated with the landslide and constraints on its remediation. It also ensured that the views of the local authority in respect of restoration of the site were taken into account at an early stage thereby minimising the risk of delays in the planning process or consent being refused.

Forward planning and optimum programming of environmental surveys is essential if that work is to be undertaken efficiently and completed in a concise time period. Ecology survey work, in particular, is season dependent and if a survey season is missed it can result in a scheme programme being put back by a year or more. All the surveys, many of which predate the detailed design, have to be factored into the overall financial profile of the scheme and delays can have disproportionate consequences for budgeting and scheme implementation.

'Community Engagement', ie involving the Community and Community leaders, at an early stage of the project and prior to work beginning ensures a wider understanding of reclamation proposals. It also enables the incorporation of reasonable community aspirations in both the works themselves and how they temporarily impact on a nearby community. It can also lead to a reduction in adverse community reaction that may otherwise follow if such 'engagement' does not take place.

Ty'n-y-bedw Landslide Stabilisation – Case History

Ty'n-y-bedw landslide is a deep seated complex ancient landslide that covers some 6 hectares and extends between 180 and 300mOD on the south-west facing slope of the Rhondda Fawr valley near Treorchy, ref Figure 2. The landslide became reactivated as a result of construction of a series of overlapping colliery spoil tram tips between 1880 and 1933. The tips were deposited on either side of a central incline that extended up the valley side from a plateau lying at 180mOD where Ty'n-y-bedw Colliery stood. In the mid 1970's the tips were re-profiled. However, monitoring between 1975 and present showed ongoing extremely slow displacements of up to 100mm per year, but with surges of greater movement in wet winters.

Although a colliery tip, the adjacent locally important mosaic of valleyside habitats known as Ffridd had colonised the site. The tip therefore was considered by the local authority to have the same strategic local landscape importance as the adjacent natural valley slopes. The ecological surveys showed the Ffridd zone to comprise a valuable mosaic of heath, acid grassland, gorse and bracken communities, which formed an important habitat for reptiles. Moderate populations of Common Lizard and Slow Worm were identified. Also, many notable bird and butterfly species were found to use and breed on the site, including species covered by the Local Biodiversity Action Plan. No significant habitat for aquatic flora or fauna existed at the site.

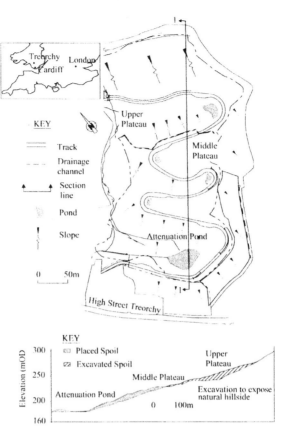

Figure 2: Ty'n-y-bedw Landslide - Site Plan and Section 1-1

There was therefore a potential conflict between the needs of a scheme to ensure public safety and the preservation of an increasingly scarce landscape resource. From the assessment procedure described earlier in the paper, the most appropriate scheme to stabilise the landslide was determined to be major earthworks reprofiling with drainage to control groundwater locally. The earthworks involved the removal of about 200,000m^3 of colliery spoil from the upper part of the landslide and its placement to form a buttress to the lower part of the landslide, constructed off the plateau formerly occupied by the colliery. In addition about 1km of surface water channels would be constructed to intercept and carry run-off from the site. Not surprisingly, the assessment also showed that there was significant potential for the loss of the environmental assets at the site, both in the short and long term. A series of measures was therefore incorporated into the scheme to mitigate the potential loss and to recreate the habitats that existed at the site, there is no proposal to establish any commercial plantation on the site. The main mitigation measures incorporated into the scheme are summarised in Table 2.

Potential Environmental Loss	Mitigation Measures
Reptile Populations	Save existing animals and encourage re-population by: • capture of reptiles and translocation to suitable receptor sites developed on the adjoining hillside • recreation of a mosaic of habitats similar to existing • monitoring population re-establishment
Mosaic of Ffridd Habitat	Re-establish a mosaic of similar habitats by: • preserving the existing seed/top soil bank by cutting down and shredding existing vegetation and its stripping together with upper 0.5m of soils keeping vegetation types separate • soils testing to establish requirements for adding nutrients to aid the re-establishment of vegetation • using stripped soils to cover specific areas of the new earthworks profile to recreate a patchwork of habitats similar to previous • cultivation of the uppermost 0.3m of spoil to redress compaction and to encourage drainage and vegetation renewal • monitoring vegetation/habitat re-establishment
Macro and micro landforms	Creation of a range of macro and micro landforms similar to existing by: • developing the basic underlying suitably stable earthworks profile incorporating slopes and areas of plateau similar to existing • creating on site as earthworks progress local variations in slope gradient, mounds and hollows in areas of plateau.
Natural appearance	Use of materials, products and finishes to provide construction sympathetic to the local surrounding including: • cellular Armorloc blocks to line watercourses on shallow gradients to promote the establishment of vegetation on the channel sides. • Local (Pennant Sandstone) blockstone to construct watercourse cascades where space allowed on steeper slopes. • Pennant Sandstone masonry facing to structures including trapezoidal channels used on steeper slopes with restricted space. • Local stone exposed aggregate finish for selected surfaces of concrete structures

Table 2: Summary of Main Environmental Mitigation Measures

In addition the scheme also looked to provide environmental gains at the site. The main environmental enhancement measures incorporated into the scheme are summarised in Table 3.

CONCLUSIONS
To accomplish environmental management in landslide stabilisation it is essential to fully integrate environmental survey and assessment work into the civil engineering

Potential Environmental Gain	Enhancement Measures
Increased Biodiversity	Provision of habitat for aquatic flora and fauna by: • Construction of shallow ponds (both permanent and seasonal) at four locations on the surface water drainage system constructed to intercept and take run-off from the landslide area.
Improved Visual Landscape	Earthworks profiling, particularly adjacent to boundaries to marry the site within the adjoining hillslope and landscape by: • Continuation of features on the adjoining hillslope into the site area, for example streamcourse valleys and hillslope benches.

Table 3: Summary of Main Environmental Enhancement Measures

assessment and design process. An appropriately skilled integrated multi-disciplinary project team is fundamental to developing an optimum overall scheme sympathetic to the environment providing where necessary appropriate mitigation and where possible environmental gains. The overall scheme programming and budgeting must make allowance for all necessary environmental survey work. This can take up to a year to complete as work such as ecology surveys is highly season dependent. This work must also be planned effectively if delays owing to missed surveys are to be avoided. Budgeting is an important consideration in any project and where possible provision for flexibility of financial expenditure over given time periods should be made.

Implementation of civil engineering works can present many conflicts with environmental and landscape issues. Full discussion with regulatory agencies from an early stage is essential. 'Community Engagement' is also seen as a key part of the consultation process, which can provide for a wider understanding of what is proposed to take place and the incorporation of reasonable community aspirations within the works. A further benefit of 'Community Engagement', as has been experienced at the Ty'n-y-bedw site, is that as a result of the process the community can taken on a proprietorial role which is absent in other areas. In other schemes within the programme of landslide stabilisation it is proposed to involve the Community in a wider range of issues, including determining final landscaping and land use.

ACKNOWLEDGEMENTS
The Authors acknowledge the work and support of their colleagues within Forest Enterprise, Halcrow Group Limited, WynThomasGordonLewis Limited and Progressive Restoration Limited to the Forest Enterprise programme of landslide stabilisation in South Wales.

REFERENCES
Cruden D M & Varnes D J, 1996. Landslides types and processes. Landslide Investigation and Mitigation Special Report 247.

Hazard identification and visitor risk assessment at the Giant's Causeway World Heritage Site, Ireland

DR B. A. McDONNELL, University of Portsmouth, Portsmouth, England

INTRODUCTION

The Giant's Causeway is the only Natural Landscape, World Heritage Site in Ireland. It is located within an Area of Outstanding Natural Beauty and contains a number of internationally recognized geological and geomorphological features. The exposed site geology is from the Tertiary period and consists of alternating layers of fractured columnar basalts and weathered interbasaltic horizons (Wilson 1972). The area is part of a dynamic coastline where a wide variety of slope failures are manifest (Prior et al 1970, Prior and Stephens 1972, Whalley et al 1982, Douglas et al 1994).

Visitors to the site can walk along a 3km cliff top path, a 1.5km lower path is also accessible to the public, however a substantial section of the lower path was closed to the general public in 1994 following a slope failure which destroyed key sections of the lower path.

In order to implement optimum management policies at the site it is essential to fully understand the impact of slope failures on visitor safety, visitor numbers are in excess of 400,000 per year, and on conservation of geological and geomorphological features. This inevitably leads to a requirement for slope failure hazard assessment and risk assessment. This paper will present and discuss a methodology for hazard assessment and demonstrate how this can be used to estimate the magnitude of the risk to which Causeway site visitors are exposed.

SITE DESCRIPTION

The cliffs at the Giant's Causeway are up to 120m in height and comprise two Tertiary basalt formations, these are the Lower Basalt formation and the Middle Basalt formation (Wilson & Manning 1978). Both formations are composed of numerous individual flows of highly jointed basalt between which periods of erosion have produced thin red bands. A red laterised basalt layer up to 15m thick separates the Lower and Middle Basalts, this is referred to as the Interbasaltic layer or Laterite layer. These red bands and in particular the Interbasaltic layer are a major structural control on the cliff morphology.

A wide variety of slope failures are present at the site including translational slides, rotational failures and mass block releases, the slope failures involve both laterite and basalt material and are the result of complicated processes.

HAZARD ASSESSMENT

At the Giant's Causeway a heuristic approach to hazard modeling using a unique condition terrain unit was adopted (Carrara 1995). The factors identified as being of particular importance in slope instability were:

- geology of the cliffs
- geotechnical properties of the material
- geomorphology, i.e. cliff geometry, slope angles
- pore pressure

The area was initially surveyed using a Leica Total Station to produce an accurate topographic map. A general survey of the site geology was then carried out at a scale of approximately 1m. The geotechnical properties of the site materials were determined using triaxial testing. The influence of these properties on slope stability was investigated by modeling generalized cliff sections using the Universal Discrete Element Code, UDEC (McDonnell in press). The information gained from modeling the generalized cliff sections was then extrapolated and applied to the rest of the site. This enabled the geological and geotechnical component of slope failure hazard for the entire site to be assessed. Each cliff section was allocated a UDEC stability number from 1 to 5, class 1 had low hazard potential whereas class 5 had high hazard potential. The information was then used to create a map in ARCVIEW of the UDEC stability classification of the cliff, Table 1 relates the UDEC stability numbers to hazard classification.

Stability Number	Hazard	Hazard Classification
1	1	Very Low
2	2	Low
3	3	Medium
4	4	High
5	5	Very High

Table 1. Slope failure hazard classification based UDEC

A digital terrain model was generated in ARCINFO and from this a map of slope angles was calculated. It was decided to base the hazard component due to slope angles on the friction angles obtained from triaxial tests which had been carried out on partially saturated samples from the site. This is because the friction angle is a measure of the maximum slope angle for which the cliff will remain stable. Table 2 shows the range of values obtained from triaxial tests on a variety of partially saturated samples.

The slope angles were classified into 5 categories with hazard assessed from very low to very high, Table 3 shows this classification. Below 5° it was assumed that all materials were stable. Between 5° and 15°, material with low consolidation e.g. scree or highly weathered laterite may destabilise, above 15° and 20° soil and laterite respectively may shear if appropriate pore pressure conditions are applied, above 30° amygdaloidal basalt may fail.

Description	Friction Angle
Laterite	20°-22°
Basalt	53°-58°
Partly Laterised Basalt	41°
Amygdaloidal Basalt	35°

Table 2. Triaxial tests results for partially saturated material

Slope Angle	Hazard	Hazard Classification
0°-5°	1	Very Low
5°-15°	2	Low
16°-20°	3	Moderate
21°-29°	4	High
>30°	5	Very High

Table 3. Slope failure hazard classification based on slope angles

It was considered that most cliff sections contain the above materials and the cliff angles are generally greater than 30° therefore the hazard component due to slope angles for the entire cliff was classified as very high. A map of the slope angle classification was created in ARCVIEW.

There were no measurements made of pore pressure at the site, it was therefore decided to infer pore pressures from a previous geomorphological survey, (Smith et al 1994), which identified the presence of percolines. The pore pressure was assessed as low or high, dependant on whether or not there was percolines in the area, this is shown in Table 4. A map of the pore pressuree classification was created in ARCVIEW.

Percolines	Hazard	Hazard Classification
None	0	Very Low
Present	5	Very High

Table 4. Slope failure hazard classification based on pore pressure

There are areas at the Giant's Causeway where site staff dump debris which has fallen from higher up the slope, this has the potential to exacerbate the slope stability problems at the site. Table 5 relates slope loading to hazard assessment, areas where loading takes place are allocated a very high hazard value, areas where loading does not take place are given a very low hazard value. A map of the slope loading classification was created in ARCVIEW.

Loading	Hazard	Hazard Classification
None	0	Very Low
Present	5	Very High

Table 5. Slope failure hazard classification based on loading

Slope failure hazard was then calculated using ARCVIEW to overlay the various hazard components of UDEC, slope angle, pore pressure and slope loading. It was considered however, that the geological and geotechnical input was of more significance to slope failure at this site than slope loading, pore pressure, or the slope angles, this is because of the high prevalence of discontinuities. The UDEC hazard component was therefore given double weighting, Figure 1.

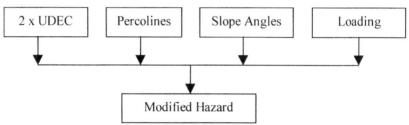

Figure 1. Overlay procedure for hazard calculation

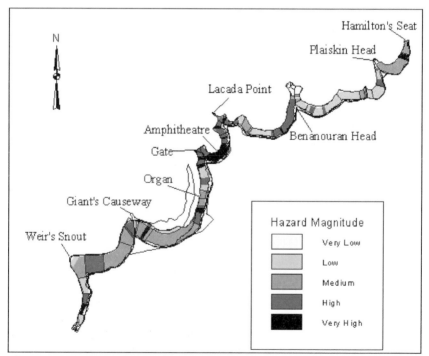

Figure 2. Magnitude of slope failure hazard at the Giant's Causeway, Co. Antrim

Table 6 relates the numerical hazard classification to the descriptive classification, the maximum and minimum numerical values possible are 25 and 3 respectively. The calculated slope failure hazard distribution shown in Figure 2 indicates that the main areas of hazard are, the extreme south, east of Weir's Snout, the Giant's Causeway and the cliff section between the gate and Lacada Point.

Value	Hazard	Hazard Classification
3-7	1	Very Low
8-10	2	Low
11-12	3	Moderate
13-17	4	High
18-23	5	Very High

Table 6. Calculated slope failure hazard classification

RISK IDENTIFICATION AND ASSESSMENT

This section deals with quantifying the risk to visitors from slope failure hazards, where risk is considered as a function of hazard and the element at risk (Alexander 1992). In order to determine risk it was first necessary to evaluate the potential for visitors to be in a given section of the path. This was evaluated from the visitor concentrations along a given section of the path and on visitor hazard perception and was based on information from a survey of visitor activity at the site.

Value	Visitor Concentration
1	Very Low
2	Low
3	Moderate
4	High
5	Very High

Table 7. Visitor concentration

The concentration of people on the lower and upper paths was given a value as indicated in Table 7 below. Using the upper path as an example the area from the extreme south of the map to east of the Giant's Causeway has high visitor concentration, from east of the Giant's Causeway to the gate there are low visitor concentrations and from the gate to Hamilton's Seat there are very low visitor concentrations.

Visitor perception of hazard was determined using the visitor's ability to identify slope failures from a series of photographs. It was assumed that if they could correctly identify an area as having failed they would consider it hazardous, large scale, disruption of vegetation was shown to be a critical factor in landslide identification for non-specialists. Table 8. lists the types of landslides shown in the photographs and the percentage of non-recognition, a hazard perception value is also quoted.

Non-Recognition	Landslide Type	Value	Classification
10%	Complex, Translational slide.	1	Very Low
10-20%	Undercutting, Slump collapse	2	Low
20-40%	Toppling	3	Moderate
40-80%	Mass block release, spalling	4	High
80-100%	Isolated block release, relict landslide	5	Very High

Table 8. Visitor hazard perception classification

Value	Risk	Risk Classification
3-4	1	Very Low
5-6	2	Low
7-8	3	Moderate
9-11	4	High
12-13	5	Very High

Table 9. Risk classification for visitors using the upper path

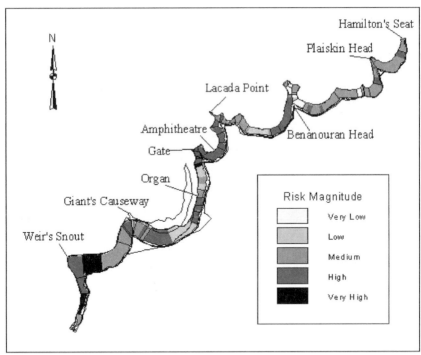

Figure 3. Magnitude of risk to visitors using the upper path from slope failures

The risk to visitors on the upper path from slope instability was then calculated by overlaying the calculated hazard, the visitor concentration and visitor hazard perception, Figure 3. Table 9 relates the numerical risk value to the descriptive risk classifications, the maximum and minimum possible values are 15 and 3 respectively. Figure 3 indicates that users of the upper path experience high to very high risk from the extreme south of the map to east of the Giant's Causeway. The area above west of the Organ is also high risk, as is the area from west of the gate to Lacada Point. A large stretch of the path west of Benanouran Head is also high risk. This applies to upper path users who are near the edge of the cliff only, those who stay away from the cliff edge will not be exposed to high hazard conditions from slope failures because failures tend to be localised.

Value	Risk	Risk Classification
6-7	1	Very Low
8	2	Low
9-10	3	Moderate
11-12	4	High
13-14	5	Very High

Table 10. Risk classification for visitors using the lower path

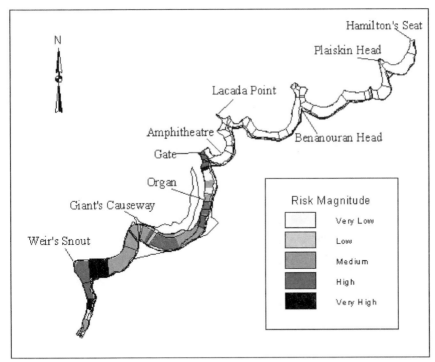

Figure 4. Magnitude of risk to visitors using the lower path from slope failures

In a similarly way the risk to visitors on the lower path from slope instability was calculated. Table 10 shows the risk classification applied to users of the lower path area, the maximum and minimum possible values are 15 and 3 respectively. Figure 4 shows the spatial distribution of risk associated with the lower path. There is a very high risk associated with users in the area from the extreme south of the site to the east of Weir's Snout, from Weir's Snout until the Organ the risk is high to medium while at the gate the risk is high to very high. The area beyond the gate is not accessible to visitors.

The results were classified on the basis of natural breaks in the data set and each map is therefore a relative zonation of hazard and risk. The maps are of value as a tool to the site managers to indicate areas of the site which have high hazard potential and high risk potential. Risk at the site can be reduced by reducing hazard, by reducing visitor concentration and/or by increasing visitor hazard perception. Limited reduction of hazard at this protected site is possible by reducing slope loading and by improving drainage. However the best method of reducing visitor risk probably lies in better information provision for visitors to enable them to make informed choices about areas where they may wish to congregate and by increasing their hazard perception, by improving their understanding and recognition of slope failures. In addition site facilities such as seating should be reviewed in order to ensure that it does not encourage visitors to loiter in high risk areas.

REFERENCES

Alexander D, 1992. On the causes of landslides: human activities, perception, and natural processes. Environmental Geology and Water Sciences. Vol. 20, No. 3, 165-179.

Douglas GR, Mc Greevy JP, and Whalley WB. 1994. Mineralogical Aspects of Crack Development and Freeface Activity in some Basalt Cliffs, County Antrim, Northern Ireland. Rock Weathering and Landform Evolution. Edited by D. A. Robinson and R. B. G. Williams, John Wiley and Sons Ltd.

Carrara A, Cardinali M, Guzzetti F, and Reichenbach P. 1995. GIS In Mapping Landslide Hazard. Kluwer Academic Publishers, Netherlands, 135-175.

Lyle P, 1997. Environment and Heritage Service Publication, Department of the Environment, Northern Ireland.

Prior DB, Stephens N, and Douglas GR 1970. Some examples of mudflow and rockfall activity in North-east Ireland. Z. Geomorph Vol. **14**, 275-278.

Prior DB, and Stephens N. 1972. Some movement patterns of temperate mudflows: examples from Northeast Ireland. Geological Society of America Bulletin, Vol. **83**, 2544-2544.

Smith BJ, Ferris TMC, Hughes DA, Greer KV and Murray MR. 1994. Footpath And Visitor Management Strategy For The Causeway Coastal Path Network. The Queen's University Belfast.

Whalley WB, Douglas GR, Mc Greevy JP. 1982. Crack propagation and associated weathering in igneous rocks. Z. Geomorph. N. F., 26, Vol. 1, 33-54.

Wilson HE. 1972. Regional Geology of Northern Ireland. Geological Survey Of Northern Ireland.

Wilson HE, and Manning PI. 1978. Geology of the Causeway Coast. Geological Survey Of Northern Ireland.

Coastal cliff retreat and instability in a weak rock, Fairlight Cove, East Sussex, UK

M.J. PALMER, Halcrow Group, Swindon, UK

ABSTRACT

A coastal cliff section in the Ashdown Beds, comprising weak claystones, siltstones and sandstones is exposed at Fairlight Cove, near Hastings. The 20 to 30m high cliffs have been receding rapidly at a rate of approximately one metre per year since 1873 and the instability threatened several private residences. This paper outlines the principal causes of cliff denudation and how this influenced the design of the protection works, which were constructed in 1990. Recent observations on the effectiveness of these works and their effect on geo-conservation are discussed.

INTRODUCTION

Instability and cliff recession has been for many years, active along the East Sussex coastline at Fairlight Cove, approximately 6kms east of Hastings. During the mid to late 1980s erosion threatened the safety of properties then close to the edge. To prevent this Rother District Council instigated protection measures constructed on the foreshore (NCE, 1989). Prior to design of this coastal defence it was necessary to investigate and determine the nature of the cliff instability to allow an adequate and safe design of the remedial works. At the same time the Fairlight Cove local resident's action group appointed a separate consultant and submitted plans for an alternative cliff-toe protection scheme.

The cliffs are approximately 20 and 30m in height. Geologically, the area under consideration lies between two major reverse faults, namely Fairlight Cove Fault southwest and Haddock's Fault northeast (Figure 1). The rocks within the cove are the Lower Cretaceous, Ashdown Beds, a sub unit of the Hastings Beds, in turn, the lowest of the Wealden Series.

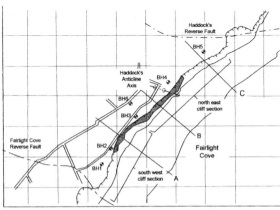

Figure 1, Fairlight Cove location plan

Instability – Planning and Management, Thomas Telford, London, 2002, 535–542

Fairlight Cove lies within the 'Hastings Cliffs to Pett Beach SSSI, first notified in 1953. The SSSI contains five separate Geological Conservation Review interests, two of which are located at Fairlight Cove, namely; 'Wealden Stratigraphy' and 'Alpine Structures of Southern England'. Furthermore, a classic type-section of the Wealden at this site would be of national and international importance for reference and research.

The cliffs comprise siltstones and fine sandstones, interbedded with weaker clays. Along a 340m section (indicated on Figure 1) a clay horizon up to 1.8m thick is present at the cliff toe. Either side of this section, the clay band disappears beneath the beach. Near vertical joints exist sub-parallel to the cliff face at many locations.

HISTORY OF CLIFF RETREAT
A history of cliff recession was determined dating from 1873. Figure 2 shows the gradual recession of the cliff, the overall rate of cliff retreat is some 0.8m per year with what appears to be an increase to 1.0m per year over the period 1979 to 1987. Cliffs within the cove are receding faster than the clays slopes west of Fairlight Cove Fault and the sandstone cliffs east of Haddock's Fault.

There was no evidence to indicate that the current day beach accumulates enough material to protect the cliff from storms. The net littoral movement is from west to east, but generally there was insufficient material currently being transported. Historically, it appeared that even when the supply was greater shingle did not accumulate.

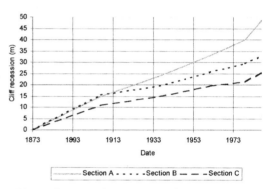

Figure 2, Recession of the cliff, 1873 to 1987

Computer modelling suggested that the formation of a stable bay resulting from continuing erosion is not expected for another 100 years. This would result in the loss of a number of residential properties and also in the long term a risk of a possible marine breach into the nearby Pett Levels.

Throughout the investigation the whole cliff face appeared in an unstable condition. During the period of site work, three cliff falls occurred; each of which brought down large masses of rock which spread out onto the foreshore distances between 2 and 9m. Moreover, individual boulders regularly fell from the cliff face.

DETAILED GEOLOGY
The Ashdown Beds comprised interbedded clays, claystones, siltstones and sandstones, which displayed vertical and lateral variability. A gentle anticlinal fold (the Haddock's

Anticline) aligned on a WNW to ESE axis affected these beds. Throughout the cliff section and between boreholes, cross stratification and facies variation were evident.

A generalised and simplified succession based on borehole information is given in Figure 3. A simplified geological cliff section around the cove is given in Figure 4. The lowest units are not visible at Fairlight. A bed of very stiff clay occasionally interbedded with silty claystones and siltstones was present at the base of the cliff over a significant length of the cove (Figure 4) and has a marked effect on the stability of the cliffs. From the north east it increases to 1.4m thick before thinning a few metres north east of the Haddock's Anticline axis. Continuing southwest the clay band obtains a typical thickness of 1.8m before disappearing below beach level again some 300 on. A level survey along the foreshore and borehole records indicated that there is a small overall rise in elevation of the clay band landward between 0.5 and 1.5°, although locally, seaward dips were observed. With time, therefore, it would appear that if erosion was allowed to continue, more of the clay bed would occur in the cliff face.

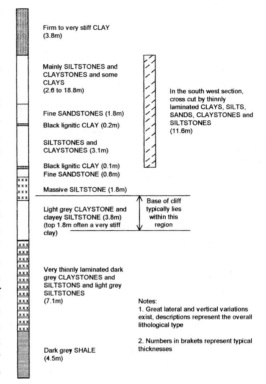

Figure 3, Generalised Stratigraphy

Above the clay band a massive siltstone (locally a sandstone) was typically overlain by a thinner sandstone bed. A variable sequence of siltstones and claystones followed often containing a thick bed of fine sandstone and thin black lignitic clay beds. Over the southwest sector of the cove, the siltstones and claystones were cut across and replaced by a large scale cross bedded unit which comprised up to 12m of thinly laminated clays, silts, sands, claystones and siltstones with much lignitic debris forming a conspicuous bed over this portion of the cliff. This unit (the Haddock's Rough Unit) is thought to be either a channel fill or erosion surface and is of importance to understanding the Wealden geology.

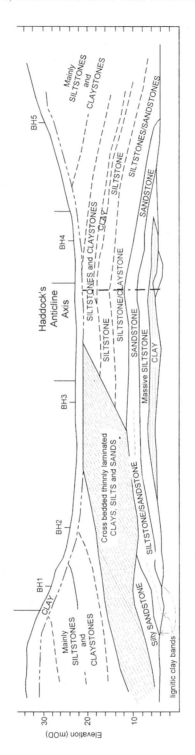

Figure 4, Simplified Geological Section

MECHANISMS OF CLIFF RECESSION

During the investigation the cliff section was studied closely and various stages of cliff erosion were observed along different portions of the cove representing an ongoing process. An account of the cliff denudation is discussed below (also Figures 5 and 6).

At Fairlight Cove, an insufficient beach permits erosion of the basal clay bed by wave action, thus undercutting the overlying siltstone bed. Sea spray and groundwater, together with the reduction in horizontal stress due to erosion cause softening of the clay to take place. Joints in the overlying beds open and blocks rotate and move forward as secondary toppling takes place. Eventually the massive siltstone blocks fall away from the cliff causing a loss of support to the overlying blocks which eventually also fall from the cliff. The talus at the base of the cliff is very quickly removed by the sea, enabling the process of events to initiate once more.

Over the southwest section of the cove where the sequence of cross bedded strata is present, the process of erosion, once the massive siltstone blocks have fallen away, continues by slaking and spalling of the sequence aided by the thinly bedded nature of the rocks and the presence of groundwater seepage (Figure 4). The stability of the overlying beds is eventually undermined by continual denudation of the underlying rock and the upper section of the cliff collapses.

Superimposed onto these erosion sequences is a second slower process, evident particularly along the northeast section of the cove. Slaking of the clays and weaker siltstones at the base and higher up in the cliff due to wetting and drying effects of sea spray, rainwater, groundwater seepage and the sun causes the gradual denudation of these horizons beneath stronger siltstone beds. Softening occurs, joints open and blocks rotate forward. Debris accumulates in the joints and the groundwater present causes this material to swell exerting pressure on the jointed blocks. Individual blocks detach and contribute to the partial or complete collapse of the cliff.

CLIFF PROTECTION

To limit the rate of cliff recession a number of schemes were considered. The cliff to protection scheme proposed by the resident's action group would have obscurred the basal 3m of the cliff section as well as beach level exposures of the Haddock's Rough Unit. Consequently the scheme was unacceptable to the Nature Conservancy Council (English Nature's predecessor). Halcrow also consulted the NCC over two proposed defence schemes, an offshore bund and a toe-protection scheme. NCC advised that in the case of the cliff-toe structure, inspection chambers would be required to help safeguard the geological interest. At an early stage it was considered that the unstable nature of the cliff meant it would not be safe to carry out works at the foot of the cliff. Consequently the most cost-effective method of reducing erosion to acceptable limits was found to be a rubble mound bund constructed on the foreshore running parallel to the cliff. The construction of a foreshore bund would firstly prevent the direct wave erosion of the basal clay layer and secondly allow the talus to accumulate at the bottom

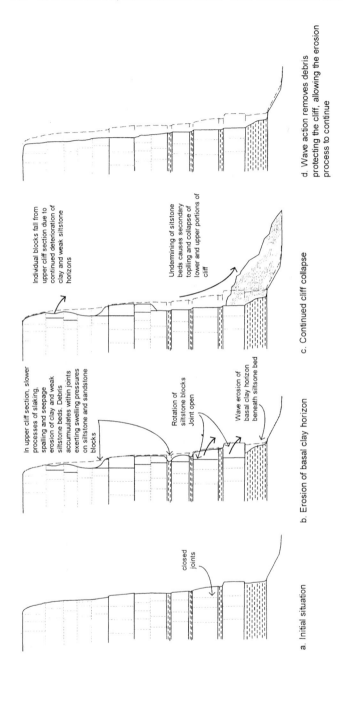

Figure 5, Cliff recession in the north east section of Fairlight Cove

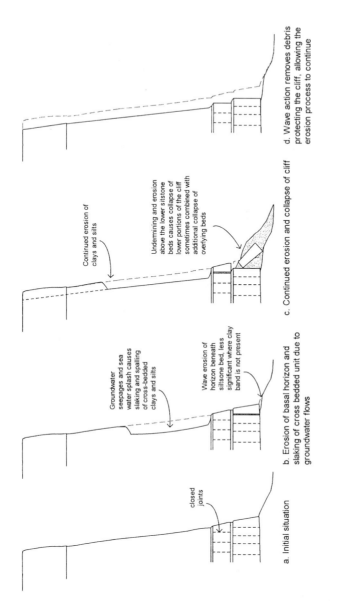

Figure 6, Cliff recession in the south west section of Fairlight Cove

of the slope and provide protection to the face against subaerial denudation. Construction of the bund was completed in 1990.

It was expected that cliff instability would continue at first until undercut and weakened blocks had collapsed. In the long term, seepage erosion, slaking and spalling were expected to continue, although zones would become protected by accumulated talus.

The overall rate of erosion was expected to be greatly reduced, although local collapses may at first occur due to the recovery of porewater pressures within the slope. In the weak sandstone cliffs protected by accumulated talus at Shanklin, Isle of Wight the rate of erosion over the period 1907 to 1981 is only 68mm/year (Clark et al 1990); the rate of recession at Fairlight Cove prior to construction was 1000mm/year. The accumulated talus at Shanklin is subject to erosion and slips. At Fairlight Cove the foreshore bund was expected to provide some protection to the talus.

RECENT OBSERVATIONS
The rubble bund has effectively stopped cliff recession rates and following the initial collapse of relatively unstable sections a talus slope has built up along the cliff toe, which is becoming increasingly vegetated. The berm together with the build up of a shingle beach (that currently appears to be accumulating) protects the talus from erosion. Consequently the geological strata are becoming increasingly obscured.

Over the last 10years since construction, there has been a number of significant changes to coastal defence policy which include the development of a strategic approach to coastal defence management (e.g. Shoreline Management Plan and MAFF's Strategy for Flood and Coastal Defence. It would seem likely that strong objections, on the grounds of geoconservation, would be met today for such a scheme (English Nature, 2000).

REFERENCES
Clark AR, Palmer JS, Firth TP & McIntyre G, 1990. The management and stabilisation of weak sandstone cliffs at Shanklin, Isle of Wight, in The Engineering Geology of Weak Rock, Geological Society Engineering Group Conference.

English Nature, 2000. Proof of Evidence of Dr Andrew King. Application and Appeal by Mrs J Fawbert & the Birling Gap Cliff Protection Association for Construction of a Rock Revetment Coastal Deense at Upper Beach, Birling Gap, East Dean, East Sussex.

Searching for solutions to conflicting interests: Lessons from the Dorset Coast

A C PRICE, Head of Planning, Dorset County Council, UK

INTRODUCTION

Avoidance of development on unstable or potentially unstable land would surely be regarded as a predictable outcome of a well-established planning regime such as that applying in the UK. It was not until 1990, however, that Government advice (PPG14: Development on unstable land) was produced, in recognition of the fact that this control of development was not common practice and that physical constraints were not being adequately considered in the planning process. It was a further five years before more detailed guidance on landslides and unstable slopes appeared (PPG14, Annex 1). Not surprisingly, given the substantial legacy of existing developments in areas at risk, and the continuing operation of natural processes, serious issues continue to arise for property owners.

High demand for coastal locations, notwithstanding extensive protective planning policies for undeveloped areas, highlights these conflicts since the effects of sea erosion present an on-going threat and challenge. A more integrated approach to coast protection has been encouraged by the development of shoreline management plans, yet this activity remains insufficiently well-related to the planning system and faces tough new challenges as European-led nature conservation measures bite harder. Finding the balance between land use, coastal engineering and environmental protection presents formidable challenges for coastal local authorities.

Dorset's 150 km coastline, located on the south central coast of England, comprises a cross-section of Jurassic, Cretaceous and Tertiary geology. East of Poole to the Hampshire border and beyond the coast is heavily developed and protected by sea defences, yet it's harbours and soft sedimentary cliffs are scheduled as Sites of Special Scientific Interest (SSSI). West of Poole, the coast is largely undeveloped, with occasional smaller holiday resorts and coastal villages; almost the entire coastline is protected as an Area of Outstanding Natural Beauty, Heritage Coast, SSSI (major parts being of international importance) and potentially as a World Heritage Site for its geology, palaeontology and geomorphology.

This paper draws on a number of case studies which illustrates how conflicting pressures are being addressed in the predominantly developed coastline of Poole and Christchurch bays, and how the balance between competing priorities is being resolved in different ways in different situations. As the search for more sustainable ways of managing our coastlines continues, the balance of priorities is slowly changing, but can perhaps never be satisfactorily resolved between all parties without greater commitment and resourcing from Government to help property owners to adjust to the consequences of a changing policy framework.

POOLE AND CHRISTCHURCH BAYS: THE CONTEXT

The first Shoreline Management Plan (SMP) for the coastal sub-cell covering these two bays, stretching from Durlston Head in the west to Hurst Spit at the entrance to the Solent, was adopted by the relevant coastal authorities in 1999. The bays have developed in the post glacial period following the breach of the chalk ridge between Purbeck and the Isle of Wight, which exposed the formerly protected river valley behind it and the surrounding soft Tertiary sands and clays to marine erosion. The harder ironstone found at Hengistbury Head, and the much later introduction of coast protection measures to hold the line in front of the international holiday resort of Bournemouth, has slowed the rate of erosion in the western part of the area. Beyond Hengistbury Head, recession of the shoreline continued much faster and carved out the secondary Christchurch Bay. Deposition of eroded materials has created spits and sand bars at the entrances to lower-lying harbours at Poole and Christchurch, and the entrance to the Solent. South of Poole the softer east-facing coastlines have been eroded between chalk and limestone outcrops to form the smaller Swanage and Durlston Bays. The harbours and headlands apart, the majority of the coastline is comprised of low sand or clay cliffs generally between 30 and 40 metres in height.

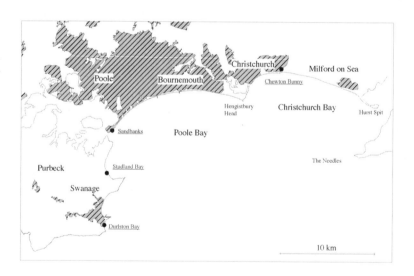

CASE STUDY 1: CHEWTON BUNNY

Where the Walkford Brook reaches the sea in Christchurch Bay, Chewton Bunny marks the boundary between Dorset and Hampshire; it also marks the meeting place between different coast protection authorities. In the small but predominantly urban borough of Christchurch an ambitious programme of coastal defence measures has been introduced since the early 70's, whilst a very different approach has been applied to the east; the much larger and mainly rural district of the New Forest has sought to protect coastal settlements at Barton and Milford on Sea but largely allowed natural processes to continue on undeveloped sections of the coast. Whilst the cliffs have been re-profiled, vegetated and drained in Christchurch, and the shoreline managed with groynes, bastions and beach replenishment, on the New Forest side none of these measures exist; as a result the clifftop has receded by more than 60 metres behind slumping clay cliffs and the shoreline itself has receded to a similar extent.

The consequences of these different management regimes extend beyond different rates of erosion. Whilst development, already set well back from the cliffs in Christchurch, now has a much more secure future, the long term prospects for the holiday camp on the Hampshire side are clearly less certain. Yet both cliffs are part of the same SSSI (Highcliffe to Milford Cliffs), designated as the stratotype of the Barton Formation and identified as a priority Geological Conservation Review (GCR) site for this reason. The value and the appearance of these two cliffs is radically different today.

By 1991 Christchurch had virtually completed its major coastal defence works, but the final bastion was in danger of being outflanked due to rapidly developing erosion on its eastern side. This posed a threat to the culverted stream behind, the loose material which covered it and therefore development above the valley sides. Emergency coastal defence works were undertaken to minimise the risk, but additional drainage was identified as necessary on the New Forest side to help reduce instability in the cliff formations. Installation of counterfort drains on the natural cliff threatened to interrupt natural processes of erosion and were objected to by the then Nature Conservancy Council (now English Nature), the Government's statutory adviser on nature (and geological) conservation.

At the ensuing public enquiry the argument centred on the need for cliff drainage on the Hampshire side, as a means of securing the major investment in Christchurch's coast protection works, against the adverse effects on the scientific value of the site affected and in particular the remaining short sections of two basal beds in the fossil rich Barton formation which were in danger of being covered or lost. In the event, the Inspector found in favour of the development, though requiring modifications to the drainage scheme. It seems curious that English Nature – a different County office – had not objected to the extensive and inevitably damaging schemes that had taken place on the Christchurch side, and that this decision was reached regardless of a different outcome just along the coast at Barton where an Inquiry Inspector had reached the conclusion in 1974 that proposed sea defence works would defeat the purpose of SSSI notification. These different decisions no doubt reflect changing values and circumstances, but nevertheless arose from the purposeful approach to coastal defence taken in Christchurch, notwithstanding the significant impacts on both the SSSI and the long term evolution of the coastline, and the perceived urgency of the situation despite the absence of geotechnical study on the Hampshire side of the boundary of the need for and the consequences of implementing the proposal.

CASE STUDY 2: DURLSTON BAY

Between Durlston Head and Peveril Point, a distance of little over a kilometre, wave attack has created a small bay out of the Purbeck Beds (clays, shales and limestones), which are also weakened by faulting. A key part of the proposed World Heritage Site, and protected by numerous landscape, ecological and geological designations, this is one of the most important fossil-bearing areas on the Dorset coast. Largely predating its protection by environmental designations, the seaside resort of Swanage had risen up the hill behind the town and extended southwards above Durlston Bay. The precedents thus established enabled infilling of plots with flats and other developments right to the cliff edge, with inevitable dangerous implications given the unstable ground.

Cliff falls in the bay are a well-known phenomenon and it should have been no surprise when in the late 1980's a significant fall removed the front garden and ground support from beneath a modern block of 12 flats. A scheme was urgently prepared and implemented by the District  Council to emplace rock armour at the base of the cliff, load the face with rockfill and install boreholes to remove ground water pressure. At a cost of £1.3 million at 1988 prices, it seems likely that the cost of relocating property was given less consideration that its protection, notwithstanding the serious effects on the value and appearance of Durlston Bay. An intrusive scheme has been established which buys more time for the residents but is not sustainable in the long term as erosion continues either side of the defence works. The SMP proposes to 'hold the line' at this point, pursuing a policy of observation and monitoring in the short term but taking no further action. North of the works a 'do nothing' policy is in place, and south of it the proposal is to observe and monitor in the short term, but to retreat the existing line in the long term.

In December 2000, following protracted heavy rainfall during the autumn and winter a further major fall took place about 100 metres to the south of the existing scheme. Taking out part of the country park's grounds and requiring diversion of the South West Coast Path around the nearby road system, this fall involved the loss of garden areas affecting further blocks of flats (44 in total). On the face of it, the SMP policy for this section needs an early rethink but the very high conservation value of the site (AONB, cSAC, SSSI, GCR, Heritage Coast and Marine Consultation Area) predicates a cautious approach. Currently actions focus on appropriate means of channelling surface water away from the slip site whilst the long term approach is investigated. The onset of winter meanwhile is viewed with anxiety by nearby residents.

CASE STUDY 3: STUDLAND AND SANDBANKS

Situated at the entrance to Poole Harbour these two peninsulas were established through accumulation of sediments probably originating in Poole and Christchurch Bays and have become strongly established, particularly in the case of Studland, in the very recent geological past. Both are now affected by erosion at their southern ends, which is thought to be influenced by a reduction in sand supply following extensive shoreline management further to the east, but very different management solutions are being followed.

Sandbanks had in recent years lost the southern end of its beach, exposing hotels, flats and houses to direct action from the sea. Since this involved some of the most expensive real estate in the UK and the town's finest beach, but had few if any nature conservation implications, the design of a beach replenishment and retention scheme was a straightforward matter.

Studland, however, is owned by the National Trust, a voluntary conservation body which acquires land and manages it sympathetically in the public interest. The majority of the peninsula is covered by ecological designations of the highest rank and the land is managed by English Nature as a National Nature Reserve (NNR). Its sandy beaches, the only undeveloped section in Dorset, attract thousands of visitors daily in the season, and the Trust seeks to manage these pressures as well as supporting the conservation of the heathland. Rapid erosion, at a rate of up to 3 metres per annum, has in recent years affected the southernmost part, adjacent to the main car park, visitor centre and beach huts which are located on the edge of the NNR. This appears to be a natural response to changing patterns of wind and wave attack, and earlier attempts to protect the area with timber seawalls, gabions and armour have had little effect. The National Trust intend to allow the natural erosion of the car park and adjoining beach to continue, in spite of the potential adverse effect on income generation from loss of the facilities and the reduction in their ability to cater for the high seasonal traffic demand.

Although neither of these examples involves landslip, both problems threaten land areas and arise as a result of interference to some extent with normal erosion processes and cliff sediment supplies elsewhere around the bay. The long term implications for the conservation management of Studland remain uncertain whilst the beach, the dunes behind and ultimately the nature reserve behind that are subjected to attack; meanwhile visitor demand to use Dorset's finest beach is unlikely to recede. The management solution proposed at Sandbanks, to protect valuable property, is evidently not considered appropriate at Studland and indeed the strategy will be to allow natural processes to continue and to promote visitor access by public transport.

DRAWING THE LESSONS

Within two centuries and mainly during the last one, the Bournemouth conurbation has developed from an area of open heathland and pine forest into one of the UK's premier coastal resorts with a population approaching half a million. A temperate south coast climate, sandy beaches and attractive views have fuelled this demand and necessitated management of its shoreline with sea walls, cliff stabilisation, groynes and beach replenishment. Some compromise has been required where protected species such as sand lizards inhabit sandy cliff faces, but this frontage has an attractive and well presented appearance.

As the case studies illustrate, the need to protect development elsewhere within the coastal sub cells has raised more difficult issues. At Chewton Bunny, the need for urgent action to protect significant investment in coast protection measures led to important geological interests being adversely affected. A similar result happened where property was in imminent danger of collapse at Durlston and emergency measures were introduced to tackle a problem that was surely foreseeable and where development ought not to have been allowed in the first place. A more recent threat close to the same site is being carefully investigated in the context of changed circumstances – the existence of an adopted though non-statutory Shoreline Management Plan, and the increased protection of the site in terms of geological and other environmental designations. The landscape and biodiversity implications are significant at the site.

The question arises as to how these situations develop. The UK has a statutory planning system which, through consultative processes, identifies where development may or may not be contemplated and determines applications to develop land. Building regulations are intended to ensure, amongst other things, that properties are constructed to avoid foreseeable insecurity and to be used safely. The onus for determining the suitability of land, and investigating ground conditions, rests squarely with the developer. On the face of it, development should be sufficiently well regulated to avoid unfortunate situations, yet it clearly is not.

In no small part this is due of course to changing circumstances, which can be particularly marked at the coast. Progressive retreat of cliffs over a period of time can threaten older property and infrastructure which was originally developed at a safe distance. Landslip events can dramatically change situations from slow, progressive loss to immediate threat to property, and climatic change may well hasten these circumstances. There is a substantial legacy of older property which can therefore become at risk although, hopefully, new developments can be better planned.

Clearly there is an important need for better information on the levels of risk which needs then to be taken into account in the process of preparing local plans. PPG14 goes as far as to say that 'coastal authorities may wish to consider the introduction of a presumption against built development in areas of coastal landslides or rapid coastal erosion.' Whilst clearly sound advice it may not rest so comfortably with pressures elsewhere from Government both to maximise use of brownfield sites and to review coast protection policies. There is a need for improved understanding of processes at work on the coast which informs decisions about which areas are to be protected and by what means, and what development can be safely accommodated in their vicinity. PPG20 advises that 'in the case of receding cliffs, development should not be allowed to take place in areas where erosion is likely to occur during the lifetime of the building. These areas should be clearly identified and mapped, and shown in development plans.' But are they and, if not, why?

I suggest that this advice is not yet properly built into the planning process, and there are several reasons for this. First is the absence of reliable information which predicts areas likely to be at risk. Second is the absence of clear policy, and evidence of pressure to change what there is, for coastal defence. Thirdly, the development plan system operates to relatively short timescales whilst property is usually intended to be there for seventy years or more. On top of this, there is an absence of public commitment to making good situations; when things go wrong it is the property owner who is left with the problem on the 'caveat emptor' principle that he must make his own investigations when purchasing and, as coast protection is a discretionary activity, there is not a public right to support. Local Authorities, unsurprisingly, may well be reluctant to 'red-line' areas that are at risk.

The Shoreline Management Plan process should help us with an improved understanding of coastal processes and a clearer indication of adopted policies.

They are, however, non-statutory documents and in their first round have probably not become well enough known, or based their findings on sufficient scientific knowledge. Establishing a comprehensive picture of ground conditions around the coast would be a costly and time consuming job, but we need to work towards it and to prioritise areas at most risk of change. Engaging communities fully in the debate about future circumstances and policy is important. With these contributions the development plan might be better equipped to anticipate and plan for appropriate development.

The environment is another dimension in the process and has profound effects on policy and decision making. In the past local authorities have understandably given higher priority to protecting human beings and their property. Ever tightening European nature conservation legislation and Government policy guidance has significantly altered this balance, and introduced tough new tests which must be met before essential development can be justified. Engineers and planners cannot ignore these requirements and nowadays need to pursue sustainable development as their principal objective. This will inevitably introduce difficult choices for decision makers, and should certainly require much closer understanding of different positions, and closer working between all the key interests, when developing local policies or deciding whether developments should proceed. For landowners, as the National Trust finds at Studland, the choices are no less difficult; their policy of managed retreat in the face of natural forces will present real challenges in being able to manage visitor pressures and, ultimately, may even threaten the nature reserve whose proper management is a key objective.

The most important element in this changing policy situation is perhaps the absence of commitment to help communities or individuals at risk. Where there is clear foreknowledge a property owner can make informed choices and these will no doubt be reflected in negotiations over purchase. This may well be more difficult for a seller who did not have this information when he acquired land, or an owner who is confronted by a sudden threat. Insurance companies continually re-assess situations in the light of new information and can protect their own positions through an annual review process. It seems perverse in those circumstances, particularly if landslip is an outcome of a changed policy stance by public authorities, for there to be no right to compensation.

The challenge if we are to achieve a more sustainable future is surely that we need organisations and professionals to work closer together, engaging with stakeholders to achieve solutions which reach an appropriate balance between necessary development, environmental protection and the needs of community and society. The coast is a dynamic environment and presents many difficult decisions if we are to achieve the right balance between man and nature. Better science, improved dialogue and communication are critical elements of the search for better solutions.

Combe Down Stone Mines, Bath - A stabilisation scheme in an area of special environmental significance

MARY SABINA STACEY - Bath & North East Somerset Council, Bath, UK

INTRODUCTION

The Combe Down Stone Mines occupy an area of approximately 18 hectares within the World Heritage Site of Bath (see Figure 1). Initially the freestone was won from quarries. Subsequently underground working extended from adits driven in from these quarries. Underground working took place between the early 1700s and the mid-19th century. Much of Georgian Bath is faced with stone from Combe Down.

The mines are now designated as derelict and are considered a risk to public safety. 1999 saw the launch of a central government funded Land Stabilisation Programme to deal specifically with non-coal mine workings. A bid was accepted and a multi-million pound project is now underway. The Combe Down Stone Mines offer a unique design problem due to a number of highly complex environmental considerations. As such, the scheme will be the subject of a planning application which will be accompanied by an Environmental Assessment and an Appropriate Assessment. The challenge for the project team is to prepare a scheme that can provide stability in a predominantly residential area without compromising the rich environmental resources of the area.

THE CONDITION OF THE MINES

There is no historic survey of the mines and no known owners. Bath City Council assumed ownership in terms of Health and Safety in 1993. The mines were surveyed in the early 1990's under the supervision of Dr Brian Hawkins, then Mine Manager. At that time there was a great deal of concern about the condition of the mines due to roof delamination and pillar failure. In some places the thickness of the roof was as little as 2m. The miners had left the discards underground and in some places the walking level is up to 3m above the floor of the mine. The pillars were classified according to condition and about 40 High Hazard areas were identified.

HISTORY OF THE PROJECT

Bath City Council and Halcrow prepared a c.£26million stabilisation proposal in the early 1990's. The scheme included pillar reinforcement and infilling with Pulverised Fuel Ash. The public opposition to this (on environmental grounds) was extreme. Subsequently, funding was withdrawn by the central government.

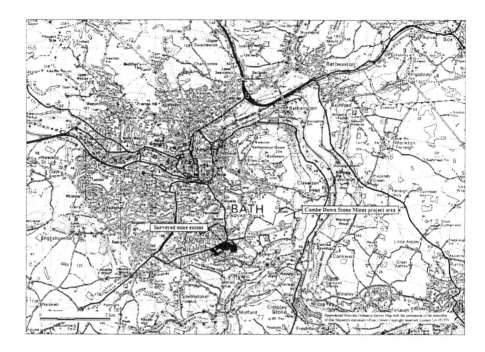

Figure 1 – Map to show the location of the surveyed mine area, Combe Down, Bath

THE LAND STABILISATION PROGRAMME (LSP)

The Land Stabilisation Programm is a 7-year government scheme which was introduced in 1999 to deal with abandoned non-coal mine workings which are likely to collapse and threaten life and property. Other projects being funded through the LSP include the salt mines in Northwich, and the chalk mines in Reading. The LSP is administered by English Partnerships. Bath & North East Somerset have a designated project team of six, in addition to appointed engineering and mines management and risk consultants. A Community Association has been formed to enable work in partnership with local people in order to prevent a repeat of the opposition to the previous scheme.

PUBLIC SAFETY AND HIGHWAY SAFETY

Public highways are considered a major hazard in the Combe Down area, as many roads are undermined. These include the A3062 which carries approximately

14,000 vehicles daily. In response to this risk, the Council as Highway Authority has implemented a series of mitigation measures including weight restrictions, chicaning and re-routing of traffic. This has proved to be an extremely complex procedure, especially in the provision of contingency routes. We are in consultation with the utilities and emergency services and have prepared an Emergency Plan in case of collapse.

ENVIRONMENTAL ISSUES.
Introduction
Combe Down is of great environmental significance being -

(a) within the World Heritage Site of the City of Bath, containing significant industrial archaeology, a number of listed buildings, and also within a conservation area

(b) of international importance for Greater and Lesser Horseshoe bats and designated as a candidate Special Area of Conservation and a Site of Special Scientific Interest.

(c) a Grade 1 aquifer

(d) adjacent to the Cotswolds Area of Outstanding Natural Beauty,

(e) the geology of Bath is of regional, national and international importance.

The final solution must reconcile these issues with the over-riding issues of safety. An environmental impact assessment addresses the additional of works related issues of: air quality, noise, vibration, transport and traffic, landscape/visual impacts, socio-economic impacts (for example on local businesses). The team has been seeking the advice of environmental agencies and specialist consultants throughout the development of the scheme.

The Bats
Significance
The mine complex is of national and international importance for Greater and Lesser Horseshoe bats, and forms part of the Bath and Bradford on Avon candidate Special Area of Conservation (cSAC). This cSAC provides the hibernation sites associated with 15% of the Greater Horseshoe Bat population in the UK. The designation is made under the Habitat Directive 1992 and the Conservation Regulations 1994 which seeks to protect and secure a network of the most important conservation sites across Europe. The legislation imposes significant restrictions and requirements upon any plan or project affecting the designated sites. The project is likely to impact upon the habitat qualities of the site. As a result, regulations require that an Appropriate Assessment must be undertaken.

The mines are also used by up to 8 other species of bat which increases the sensitivity of the site. All British bat species are specially protected under the

Wildlife and Countryside Act 1981 and under the Conservation Regulations 1994. This legislation requires that any development affecting bats must be conducted under licence from DEFRA.

The wildlife legislation affecting the mines therefore requires a very detailed and comprehensive approach to impact assessment and mitigation, including full consultation with English Nature. The intention must be to protect and sustain the integrity of the bat habitat within the mines, and to protect individual bats using the mines.

Survey
Survey work has included monitoring of temperature within the mines, winter bat census, dusk exit counts, dropping quadrants, recording winter bat activity using static surveys within the mines, swarming surveys at key mine entrances, radio-tracking studies to identify important flight paths and foraging areas used by bats.

Cultural Heritage - The World Heritage Site, the Conservation Area and the Archaeology
History of the Mines
Bath is designated as a World Heritage Site as 'an outstanding example of a type of building or architectural ensemble…'. This is often described as coming from the activities of three men in the 18th century: Beau Nash who was the master of social ceremonies, John Wood the architect who built the buildings for which Bath is so famous and Ralph Allen, a chief entrepreneur in the mining of the Combe Down stone. Ralph Allen and John Wood collaborated closely in the development of Bath. In the early 18th century, Ralph Allen developed an integrated approach whereby he extracted the stone, and transported it down the tramway he built to the bottom of the hill from where it could be shipped further afield. Jane Austen complained of 'the white glare' of Bath in reference to the number of new buildings with freshly cut stone. After Ralph Allen died mining extended further across Combe Down. The village grew, with both miners cottages and superior housing. However by the middle of the 19th century, the focus shifted to the increasingly active mine at Box Later mining included pillar robbing which contributed to the present condition of the pillars and thus the stability of the mines.

In addition to its importance for underground industrial archaeology, the area has a significant above ground cultural heritage which reflects the history of the mining and makes an important contribution to the World Heritage Site.

The Management of Archaeology in the Planning Process
Planning Policy Guidance 16 – Archaeology and Planning outlines a procedure to aid the planning authority in its decision-making. The first step is to assess how significant the site is by undertaking an assessment through desktop research and survey. The planning authority can then make a decision on the treatment of the

archaeology. Archaeological sites and structures of national importance can be scheduled. For sites of national importance there is a presumption of preservation 'in situ'. If this is not possible preservation 'by record' can be considered.

The Survey Difficulties

Conditions in the mines make it extremely difficult to conduct archaeological survey, particularly given the restrictions to stay within areas of the mines 'made safe' through Health & Safety enabling works. The presence of the bats has also restricted recording in terms of disturbance resulting from noise and flash photography. However in some ways the enabling works have helped in that for the first time there has been sufficient lighting to enable close inspection of the workings. The project has also benefited from the presence of experienced miners who have helped with interpretation.

Assessment of significance and management of the resource

The assessment indicated that the site is of national importance. This means that the planning authority (and English Heritage) will be looking to ensure that some of the site is preserved. English Heritage may also consider scheduling it. A scheduled monument cannot be altered without Scheduled Monument Consent from the Secretary of State. This clearly needs to be considered in conjunction with safety issues. It is likely that much of the mine will be infilled, however some of the mine will be preserved.

Hydrogeology

Significance

The hydrology of Combe Down is important as water infiltrating the rocks of the Combe Down plateau emerges as springs on the valley sides (see Figure 2). Some of the springs are harnessed for drinking water. The Whittaker and Tucking Mill springs have been delineated as Source Protection Zones for public water supply.

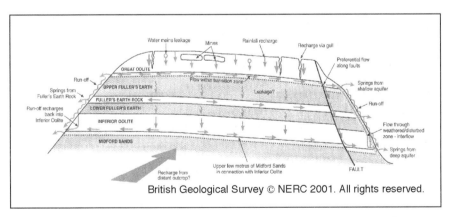

Figure 2 - Conceptual flow model for the aquifer system underlying Combe Down

Drainage from garden areas, highways, sewers and cesspits has the potential to affect the hydrology and therefore requires consideration. The majority of highway drainage enters the mines via soakaways and sewage has been encountered at a number of discrete locations.

The hydrological regime in Combe Down is intimately linked with the geology and structure of the hillside. Understanding the mechanisms which control water movements and the inter-relationships between hydraulics, geology, geomorphology and soil and rock chemistry is therefore of key importance.

Survey - Review of Existing Information
The British Geological Survey (BGS) undertook an independent review of available hydrological information. Parsons Brinckerhoff, the Council's engineering consultant, developed this work.

The review highlighted important issues in the design of a stabilisation scheme. The following gaps in information were identified:
- information regarding groundwater levels in the Great Oolite rocks which contain the mines. The groundwater catchment areas to the springs emerging on the hillsides are unknown.
- information for the lower (Inferior Oolite) aquifer which contributes to the springs at Tucking Mill.
- There are some indications that there is hydraulic interaction between the upper and lower aquifers. It is not known, however, where or how this occurs or whether it occurs in the vicinity of the mines.

Survey - The programme of additional investigation
Additional hydrological investigation was proposed to include:
- a network of boreholes of which 15 were retained for water monitoring purposes;
- *in-situ* and laboratory testing, particularly the chemical quality of any material(s) to be utilised in the stabilisation;
- hydrological monitoring;
- research with the University of Reading, to look at the infiltration of water through the roof of the accessible part of the mines.

Impact Assessment
The potential impacts of stabilisation on **water movement** and **quality** include:
- causing a reduction in the drainage potential above the mines, possibly affecting the existing drainage infrastructure or causing dampness or flooding in properties;
- causing a reduction in the drainage potential below the mines, and possibly redirecting subterranean water and changing spring yields or increasing the potential for ground instability.

- causing a reduction in the quality or quantity of water draining through the mines and entering subterranean water supplies;
- interrupting the natural equilibrium with the natural material below the mines.

Geology, Geomorphology and Soils
Significance
The geology of Bath is regarded as being of regional, national and international importance. The pioneer geologist William Smith did much of his early work in the area and the Bath area is the world type locality for part of the Jurassic System.

Knowledge of the geology of Combe Down including the spatial relationships between the various strata, their physical and chemical characteristics, and their interactions with other aspects of the natural environment, all require understanding before the efficacy and consequences of a proposed stabilisation scheme can be assessed.

Impact Assessment
The potential impacts of stabilisation include:
- destruction or impairment of important sections of geological outcrop;
- causing ground instability immediately above the workings or on the margins of the plateau;
- causing physical damage to soils;
- encountering and/or redistributing contaminated material, such as soils.

SAFETY WORKS AND EMERGENCY WORKS
Following a visit in early 2001 from Her Majesty's Inspector of Mines (HMIM), concern about the condition of the mines grew to the extent that he was minded to serve a Prohibition Notice. This would have meant that no one was allowed in the mines, which would have incurred significant project delay. In the event, the Notice was not served but the Council undertook that Health and Safety Enabling Works would be done, that no one would go in the mines except within the enabled area and also that no one except those essential to the development of the scheme would be allowed underground. The works took the form of wooden propping along approximately 900 meters. However the effect of that was that it became very difficult to undertake the work necessary to develop the scheme. There were large parts of the mines that could not be assessed.

Carrying out these Safety works involved a number of environmental issues – during the winter the bats hibernate in the mines. It was necessary to obtain a licence from +DEFRA and this involved a protocol ensuring that the works did not disturb the bats – so for example, there was to be no noise such as radios, no lights or photos inside near the bats, works had to finish before the bats came out to feed at dusk. It was also necessary to allow the archaeologists in to record the works.

Because there was a badger sett near to the mine entrance, a licence had to be obtained and constant recording and care that the compound did not encroach upon a badger run.

In September 2001, the Health and Safety Executive served an Improvement Notice concerning two small a:eas of the mines. This meant that stabilisation work had to start within 7 days. Again, this required a bat licence, archaeological recording and consultation with the Environment Agency to ensure that the solution would be acceptable.

WORKSITES
The complex environmental significance of Combe Down has meant that it was not easy to establish an above ground worksite. The main entrance into Byfield is a Site of Special Scientific Interest and a candidate Special Area of Conservation for bats, as well as having badger setts nearby. The entrance was set up according to these needs. The main entrance to Firs Mine is in the middle of Firs Field. This is the village green and is heavily used for walking and playing. It has byelaws which prevent the turning of any turf. When the compound was established for the emergency works, special feet were designed for the fencing. If Firs Field is to be used for the main scheme, the byelaws will have to be revoked. This procedure has to go to the Home Office. However there is no other possible site as the village is so densely settled. A further constraint on undertaking emergency work is that the weight limits along the road make it difficult to bring in machinery and materials.

THE DEVELOPMENT OF THE SOLUTION
At the time of writing, a stabilisation scheme that is acceptable to the regulators and the community is being developed. Consultation on the preferred option will be undertaken with the community and with the statutory agencies as the options are developed, so that by the time the preferred option is chosen, they will have had the chance to comment on it and their concerns will have been integrated into the final design. In some cases the requirements of the different environmental issues may not be easily compatible and negotiations are necessary to ensure that the final solution incorporates all concerns. There are also environmental groups interested in each of the issues and they need to be involved. The management of the scheme and development of the solution requires difficult and complex judgements on the process of reconciling the environmental issues with urgent, even emergency action.

REFERENCES
Various Combe Down Stone Mines Project documents

Session 6:

Coastal and climate change and instability

An understanding of long-term coastal change can provide vital information to improve our knowledge of coastal evolution, including instability. In view of the impacts of climate change and a potential deterioration in stability the provision of new information can assist coastal planners and managers.

Climate change and some implications for Shoreline Management in the UK

JOHN ANDREWS Posford Haskoning Limited, Haywards Heath, UK
SIMON HOWARD Posford Haskoning Limited, Haywards Heath, UK

INTRODUCTION

Much work has been undertaken on assessing scenaria for climate change. Within the "best estimates" of the various predictions there are quite wide boundaries of confidence and uncertainties.

The most significant aspect of climate change affecting shoreline management is the increase in the rate of sea level rise. Other phenomena such as variations in storm intensity and frequency, shifts in mean wind speed and deviation, and variations in far field wave generation in the Atlantic will also have an influence. Of particular importance to the stability of coastal cliffs is the prospect of increased intensity of winter rainfall and consequent higher pore water pressures. Increases in mean sea level may also have an effect on tidal ranges and extreme surge levels which are not necessarily directly related to sea level rise.

Current predictions of global sea level rise by the Inter Governmental Panel on Climate Change (IPCC) give a "best estimate" rise of 0.5 metres from the present to 2100 with variations in the estimates from 0.15m to 0.95m depending upon emission scenaria and other sensitivities (Ref. 1).

The United Kingdom Climate Impact Programme (UKCIP) Technical Report No. 1 was published in 1998 (Ref. 2). This projected climate scenaria for the 2020's, 2050's and 2080's. A further set of predictions referred to as the UKCIP02 scenaria is being prepared taking into account the Third Assessment Report (TAR) of the IPCC (IPCC 2001).

The effect of sea level rise on the shoreline has also to be balanced against uplift and subsidence caused by isostatic re-adjustment of the landmass after the retreat of the UK ice sheets. The relative rates of sea level rise for the UK coastline taking into account isostatic changes as given by the Department for Environment, Food and Rural Affairs (DEFRA) are:-

South and South East	6mm/year	
North	4mm/year	
South West and Wales	5mm/year	(Ref. 3)

This is consistent with the present IPCC and UKCIP "best estimate" predictions.

EFFECTS ON THE SHORELINE

Whilst there are many uncertainties within the various predictions of climate change, the overall picture is one of change with varying degrees of certainty on the predictions. It is against this background of uncertainty that issues on the shoreline need to be addressed.

Irrespective of these changes and uncertainties the impacts will be significantly affected by the nature of the coastline. Perhaps the most important distinction is between the effect on the unprotected and the protected frontages. This paper focuses on the presently protected frontages where there is little scope for set back and retreat and some particularly difficult issues will need to be addressed.

Unprotected Frontages

On unprotected frontages where erosion is currently taking place, i.e., most of the UK shoreline apart from areas with hard cliffs, the shoreline may be expected to retreat at a faster rate than at present. The foreshore and intertidal area is also likely to recede at much the same faster rate. Where the composition of the cliffs and foreshore is consistent, a largely unchanged beach and foreshore profile will therefore be presented to the incoming wave fronts.

Setting aside changes in offshore wave directions, which might result from other aspects of climate change, wave refraction and shoaling effects will therefore remain unchanged.

Specific effects will clearly vary with the type of frontage. Low lying areas will suffer an increased risk of flooding, drainage systems will back up, salt marshes will retreat and estuarial responses will change. Of some particular consequential significance to protected frontages is the effect on soft cliffs and shingle ridges.

Soft Cliffs

Soft cliffs will be expected to erode at a faster rate than at present. Whilst attempts have been made to correlate the rate of sea level rise directly to the rate of erosion, these need to be treated with some caution not the least because of variations in cliff type and sediment supply to the beach. Increased slope instability as a result of increased winter rainfall intensity should also be anticipated.

Increased erosion of soft cliffs will increase the sediment input to the coastal regime. However the overall effect to shoreline management will be slight. Recent estimates of the input of beach building sediment from cliff and shoreline erosion on the south coast from Beachy Head to Dawlish Warren a distance of some 320 kilometres indicate an annual input of the order of 60,000cu.m/year, (Ref. 4). This needs to be compared with approximately 830,000cu.m/year, which has been imported recently by lorry or barge over the same frontage.

Embayments between hard points or protected frontages are likely to deepen before reaching some stability with the accumulation of beach material in the mid bay positions.

Shingle Ridges
Shingle ridges and spits may be expected to roll back or breach. Particularly vulnerable areas can be at the neck of a spit. Over the past decade the neck of the Spurn Head protecting the Humber Estuary has suffered damage on a number of occasions. In many cases these are relic beaches where there is no natural re-supply of material.

Protected Frontages
The effect of sea level rise on presently protected frontages imposes a further parameter on an issue, which irrespective of climate change and sea level rise, is becoming increasingly difficult to address.

Most of the present defences stem from the mid 19th century and the growth of seaside resorts. Many of the original structures have since been refurbished and extended. These fixed frontages have been unable to respond to natural changes and now stand well forward of the natural line of the coast. Examples of this may be found in a number of locations. At Hornsea on the glacial till cliffs of the east coast the set back is some 100metres (Fig. 1).

Figure 1.
Set back of
natural coastline
compared with
defended
coastline,
Hornsea.

At Sidmouth on the south coast in the harder marls and sandstones a projection of the adjacent unprotected cliff would run through the foyers of the seafront hotels (Fig. 2).

Figure 2. Defended
frontage at Sidmouth
standing proud of the
adjacent coastline.

The effect of these changes is evident by measuring in decades rather than years and
are manifested by:-

• Beach steepening and lowering in front of the defences
• Reduced wave refraction and hence higher sediment transport rates
• Less wave attenuation with larger waves propagating inshore, exacerbating
  beach drawdown.

As adjacent unprotected frontages recede and the coastal morphology reacts to
maintain a dynamic equilibrium, the foreshore on the defended frontage is unable to
respond in the same manner. The consequence is foreshore steepening and lowering
across the defended frontage as shown in Fig. 3. The frontage becomes increasingly
difficult to protect as the adjacent coastline recedes.

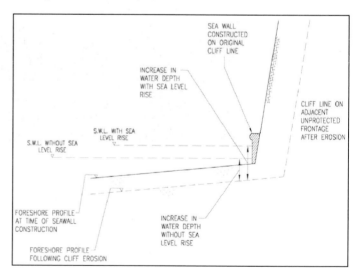

Figure 3.
Foreshore
steepening and
lowering across
a defended
frontage.

Some indication of historic foreshore steepening can be deduced by inspection of low and high water marks shown on previous editions of Ordnance Survey maps. However these interpretations should be viewed with caution bearing in mind the inherent inaccuracies in plotting the low water mark.

The addition of sea level rise clearly exacerbates this situation as demonstrated further in Fig. 3.

A steeper and lower foreshore allows larger waves to propagate further inshore and wave refraction is reduced with a consequential increase in the rate of longshore drift, regardless of changes in weather patterns and dominant wind directions which might follow as a further consequence of climate change. In general terms, beaches on defended frontages are likely to become more volatile and difficult to contain as both the gross and net longshore transport rates are likely to increase.

Much of the protected UK coastline is dependant upon retention of an adequate beach either as a natural feature or controlled by groynes or breakwaters possibly including periodic re-cycling or re-charge.

The beaches in some situations provide the only form of defence, examples can be found on the Selsey Peninsular (Fig. 4), parts of Brighton and the eastern frontage at Eastbourne. Elsewhere the beaches provide forward protection to seawalls by means of energy dissipation and wave attenuation. Otherwise such structures induce wave reflection resulting in scour of the foundations and further lowering of the upper foreshore exacerbating the propagation of larger waves inshore. The cycle repeats, leading eventually to the failure of the structure. A seawall without the benefit of a protective beach effectively becomes the agent of its own destruction.

Figure 4. Coastline protection based on beach levels, Selsey Peninsular.

In the absence of sufficient beach, means of protection are limited to structures comprising, in whole or in part, of rock or concrete armour units. The long term implications for the presently defended frontages, in order to accommodate sea level rise and the other pre-existing and concurrent morphological changes, is the increasing difficulty of retaining an adequate beach.

Beach control systems conceptually may seek to contain sediment on a chosen frontage, e.g. a groyne field is provided with a terminal structure blocking the onward drift of material and from where re-cycling along the frontage may take place. Alternatively, measures are provided to raise beach levels locally whilst allowing onward drift e.g. a groyne field tapering off with groynes of reduced height and length at the downdrift end. This latter scenario is dependent on an adequate throughput of sediment either naturally of artificially.

As a defended frontage becomes further discontinuous with the adjacent coastline the opportunity for continual throughput of material is diminished and the options, which may have to be faced are:-

i)   Is the defended frontage to be abandoned?
ii)   Is the beach to be abandoned with protection relying on a rock or concrete armour revetment?
iii)   Is the beach to be retained within large rock or concrete armour breakwaters that considerably reduce the onward movement of beach material?

Adopting Option i) would doubtless have major economic and social consequences in many cases Option ii), whilst providing adequate protection would leave the frontage without the intangible yet social significance of an amenity beach.

Proceeding with option iii) explicitly accepts that there will be a detrimental effect on adjacent coastlines leading to loss of beach material and on unprotected frontages increased erosion. Where the coastline is covered by statutory environmental designations, this option is likely to be in conflict with current United Kingdom and European Union environmental legislation including the Wildlife and Countryside Act (1981) and the Habitats Directive (92/43/EEC).

None of the above options are attractive. However, even at the mean projected rate of sea level rise of 5mm to 6mm over much of the southern half of the U.K., these issues will need to be addressed within the next 50 years at some locations, particularly where the erosion rates are already high.

There remains the issue of increased rates of overtopping. In itself this is a more straightforward issue to address although the increasing difficulties of retaining an adequate beach will clearly exacerbate the situation.

Salt Marshes
The natural response of saltmarshes in the face of sea level rise is to migrate landward providing a natural defence to higher level land. Where saltmarshes are constrained landwards by defences, the result is the net erosion of the saltmarsh with the natural progression from mud flats to low marsh to high marsh being prevented. Since the end of the last ice age (10,000 BC) saltmarshes have advanced and

retreated as relative sea levels have decreased or increased. The **presence** of sea walls is now preventing the landward migration of the marsh leading **to the** erosion of over 2% of the total remaining areas each year (Ref 5.)

WHAT MANAGEMENT STRATEGIES SHOULD BE ADOPTED?

There remain quite high levels of uncertainty within present predictions of climate change and although the predictions may well be refined the element of uncertainty will always remain.

We can never expect to be able to work to a rigorous data sheet against which coastal defence schemes can be implemented. However even with the rate of sea level rise towards the upper bound of the predictions there remains adequate time for incorporation within strategic coastal policies.

Such policies would seem to require three basic elements:-

i)   Maximising the data and understanding on coastal processes.

ii)   Flexibility of management and design that readily allows modifications to be made as climate change develops.

iii)   Addressing the impacts and uncertainties of climate change on both shoreline and coastal zone management.

A key issue with regard to understanding of coastal processes is beach and shoreline monitoring. The direct economic benefits may be difficult to quantify, however the cost in proportion to the value of existing and projected defence works is small. Monitoring will enable more robust decisions to be made on the planning and design of new or replacement structures and the effect these might have on the wider coastal regime.

There are many areas where quite straightforward modifications or improvements can be made to improve the hydraulic performance of existing defences. viz. sea wall raising, retired flood walls, supplementing existing structures with rock armour, and the addition of new groynes and breakwaters. Other more radical measures involving progressive managed realignment may become necessary. Work has been undertaken recently by the Environment Agency on the East Anglian Coast in implementing a sustainable flood defence policy including managed alignment.

Coastal zone planning may need to contemplate the concept of a buffer zone behind existing defence lines in order to provide space for secondary defences on a retired line.

SOME EXAMPLES OF RECENT SCHEMES
Sidmouth, East Devon (refer Fig. 2)

The town of Sidmouth lies on the south west coast of England. The first seawall was constructed in the 1830's; the undefended coastline to the east has since continued to erode and the natural beach feed to the frontage has declined. This led to a radical change in the coastal defence methods adopted; from a seawall with low level groynes to offshore breakwaters and substantial rock groynes in order to retain an imported beach. Rock armour was also placed at the toe of the wall. The scheme includes a regular programme of beach monitoring and management. The offshore breakwaters will provide protection for the foreseeable future. Nevertheless, the structures are broad crested and could be raised without any modification being required to the foundations. The monitoring and beach management allows information to be gathered on the performance of the beach enabling timely action to be made with regard to beach re-grading or recharge.

Eastbourne, Sussex (refer Fig. 5)

Eastbourne is a major coastal resort on the south coast of England. Since the mid/late 19th century the town has been protected by a system of seawalls and timber groynes. Following deterioration of the groyne field in the 1980's and major collapses in 1989/90 a major programme of reconstruction became necessary. The concept of the defences relied upon an adequate wave driven supply of shingle for entrapment by the groyne field. The construction of coastal and harbour works updrift had over the past decades impeded or cut off this natural supply. Following a detailed appraisal, the preferred option was similar in concept with regard to the groynes and seawalls except for a fundamental difference with the treatment of the beach.

It was accepted that the natural beach feed would be insufficient and the import of some 750,000 cubic metres of beach material formed a major part of the works. This might be compared with the natural annual supply of some 7,500 cubic metres, which is estimated to be derived from the chalk cliffs and coastal platforms updrift (Ref 4).

Figure 5. Aerial view of the completed beach reconstruction at Eastbourne.

The structure life of the groynes is probably of the order of 30 to 40 years. The profile of the groynes was designed to accommodate some 300mm of sea level rise.

Hornsea, East Riding of Yorkshire

Hornsea lies on the east coast of England where the cliffs comprise glacial till eroding at an average rate of some 1.8 metres/year. The defences comprise seawalls and groynes with an adequate natural throughput of beach material. The erosion on the unprotected frontage down drift is very noticeable (Fig. 1) and sustaining an adequate beach on the protected frontage will become increasingly difficult over time.

A particularly low length of sea wall has been upgraded in two stages, the first in 1985 provided a rear flood wall and the second in 2000 involved the raising of the front sea wall (Fig. 6). Determination of the optimum height of the defences was carried out with the aid of mathematical modelling. As cliff recession and sea level rise continues, some difficult decisions may need to be made on how the frontage is to be protected in the latter part of this century.

Figure 6.
Rear flood wall
and raised sea
wall along the
Hornsea
frontage.

Wheeler's Bay, Isle of Wight

Wheelers Bay is on the south coast of the Isle of Wight within a major landslip complex known as the Undercliff. In this case the defences are adequately robust to resist wave forces and the implications of sea level rise at least for the next 50 years. The problem lies with the stability of the coastal slope, which is particularly sensitive to pore water pressures and would therefore be vulnerable to the prospect of wetter winters. The matter came to a head in 1997 when movement was detected in the slope following a period of prolonged rainfall. The seawall was judged to be at risk from a deep seated slope failure. Options for treating the slope in its present position were limited and stabilisation was achieved by constructing a rock armour revetment forward of the line of the seawall to enable a flatter and more stable profile to be achieved (Fig. 7). The construction of the works in 1999 was timely; the rainfall in October/November 2000 was particularly prolonged and resulted in serious flooding in many parts of the country.

Figure 7.
View of the
completed
scheme at
Wheeler's Bay,
Isle of Wight.

Suffolk Estuaries
The anticipated sea level rise within the estuaries is expected to be higher than for
the open coast. A key issue in managing these estuaries is the potential flood plain,
which at present is defended, and the artificially narrow mouths. Significant change
in tidal volume as a result of sea level rise could breach the mouths. Only by taking
a strategic view of the detailed management of each defence can management of the
estuaries as a whole be properly achieved. One element built into the strategy
involved allowing defended areas to flood increasingly over time by introducing
strengthened low areas of defence and constructing sluices to retain water until low
tide. In this way the area would be abandoned over time allowing natural
environmental conversion while still managing the flow of water within the estuary.

CONCLUSION
The key issues which arise are:-
- There remains a significant level of uncertainty within the predictions for
  climate change. The most robust predictions are for sea level rise although
  within quite wide limits of confidence.
- Planning of shoreline management whilst needing to be mindful of likely
  changes should be sufficiently flexible with regard to uncertainties of the
  climate predictions.
- Retention of adequate beaches on protected frontages is likely to become a
  major issue.
- Difficult decisions may well arise where continuing the protection of an already
  defended frontage will adversely affect adjacent coastlines.

REFERENCES
1. IPCC, 2001, *IPCC WGI Third Assessment Report*, IPCC.
2. Hulme,M. and Jenkins, G.J. (1998) *Climate Change Scenarios for the UK:
   Scientific Report UKCIP Technical Report No.1*, *Climatic Research Unit,
   Norwich, 80pp.*
3. MAFF (now DEFRA), 1999, *Flood and Coastal Defence Project Appraisal
   Guidance FCDPAG3*
4. Posford Duvivier, 1999, *SCOPAC Research Project. Sediment Inputs to the
   Coastal System*, Posford Duvivier.
5. Environment Agency (1993) *East Anglian Saltmarshes*

Instability processes as a result of coastal and climate change at Grottammare (Central Italy)

MACEO-GIOVANNI ANGELI, National Research Council, Perugia, Italy
FABRIZIO PONTONI, Geoequipe Consulting, Camerino, Italy

ABSTRACT

The town of Grottammare has a long history of recorded instability extending back beyond the XIXth century. This landslide shows movements along translational slip surfaces, roto-translational slides affecting clays, and rock and debris falls from the steep escarpments of the town. There is the additional risk of seismic activity. There is also considerable evidence of damage, particularly in the medieval town.

Records of instability extend back hundreds of years and the situation at Grottammare is regarded as being a dormant landslide which may have been reactivated by the original causes. Coastal protection measures mean that effectively the town is built on abandoned slopes but this does not preclude the reactivation of the slopes due to natural and human factors. Monitoring and a detailed structural analysis of crack patterns on buildings in the landslide indicates that movements are continuing to take place within the landslide mass. The historical evidence of landsliding has contributed to the evidence supporting the ground behaviour model. The role of climate change can be considered decisive for the landslide reactivation.

INTRODUCTION

Grottammare is a small village located on the Adriatic sea; the town is densely inhabited and crossed by important communication routes. The cliff at Grottammare is a well known example of a coastal landslide system (Figures 1, 2, 3). Deep roto-translational slides affect the bedrock clays, with several landslide terraces degrading towards the sea. Rock and debris falls occur from the steepest and highest portion of the scarp, which is made up of sand and conglomerate. Superficial slides affect the detrital cover. The top of the slope attains higher altitudes (120 m a.s.l.), with the sandy-conglomerate deposits superimposed on the clayey bedrock.

Reactivations of deep movements, with regression of the landslide crown up to some tens of metres, are historically documented along the coastline north of the village of Grottammare (1843, 1928), where the coastal strip is almost absent (Figures 4, 5). More to the south the landslides seem to be rather ancient, as witnessed by the presence of sea-eroded dormant scarps ascribed to a period ranging from Roman

Fig. 1- Recent photograph of Grottammare and Colle delle Quaglie landslide sites.

times (IIIrd century B.C.) to the Middle Ages (XIIIth to XIVth century A.D.), in agreement with the evolution model of the Marche's coastline during the recent Holocene, when an active cliff was present along this coast. The present coastal strip was formed rather recently, presumably starting from the XVth century, following a massive increase of solid load carried by the rivers. This condition blocked the toe erosion of the coastal slope.

At the present time, the landslide activity seems to be affected only by the climatic fluctuations (Bruckner cycles) that can cause significant and rapid changes of hydraulic conditions within the past-failed slopes.

CANCELLI et al. (1984a; b) investigated this landslide among those found along the Italian Adriatic coast between Pesaro and Vasto: "the hypothesised landslide model is a composite one, with movement taking place along translational slip surfaces and a progressive failure mechanism within the overconsolidated clays cropping out at the foot of the cliff". ANTONINI et al. (1993) considered this mass movement a Roto-translational slide: "landslides with movement along a complex rotational or translational surface of rupture, slightly undulated at places, corresponding to lithological, structural or tectonic discontinuities", and identified the landslide crowning and accumulation zone.

The geomorphological and geotechnical investigations still in progress aim to prove the validity of the failure model proposed and the possibility of its reactivation, following the expected climatic change/fluctuations. The results of the research should lead to the identification of the highest landslide-risk areas.

GEOLOGY AND GEOMORPHOLOGY

Geological and geomorphological investigations carried out at a 1:5000 scale, also with the help of aerial photographs (both on a small and large scale; the latter being particularly useful considering the remarkable extent of the phenomenon), revealed the presence of other large coastal landslides. These movements, extending over a surface of 20 to 80 hectares, are classified as complex landslides, that is characterised by a main deep roto-translational slide affecting the bedrock clays, with rock and debris falls from the steepest and highest portion of the scarp, which is made up of sand and conglomerate; superficial slides affect the detrital cover.

Deep-seated movements have caused the formation of headlands which have stretched out from the coastline for several hundred metres. Sea currents easily eroded them within a few decades, redistributing the material along the coast. These mass movements are quite similar to the coastal landslides affecting south-eastern England (Miramar landslide, BROMHEAD 1986).

In the area of Colle delle Quaglie (Figures 1,2,4) these headlands are documented on maps going back to the XVIIIth century. The large boulders, made up of conglomerates and well-cemented sandstones which are found emerging from the sea or submerged near the coast, are the remains of landslide bodies dismantled by wave action rather than blocks which have fallen from the top scarps, as referred to by some authors. Even in the area underlying the old village (or hamlet, called from now on also with the Italian name of "Vecchio Incasato") two large rock blocks were found (mass of several thousand of cubic metres); they were dismantled in the last century following the construction of the Adriatic railway ("Sasso Piccuto" and "Sasso San Nicola" in Figure 2).

The presence of landslides is historically documented in the Colle delle Quaglie area; they resumed their activity several times and are regarded as deep-seated movements of the kind previously described (MASCARETTI 1851; PAOLI 1848; PEROZZI 1928). See also Figure 4.

Deep-seated movements are totally congruent with the models proposed by BROMHEAD (1979) which take into account the importance of the stratigraphic attitude and slope hydraulic conditions. According to the same author, in the case examined here the presence of more resistant and better drained materials at the top of the cliff (gravels and sands overlying clays) allows deep rotational landslides to take place involving the basal clays.

CURRENT SITUATION

As regards the state of activity, these processes should be considered as dormant, according to the definition reported in the UNESCO International Landslide Glossary: "dormant landslides which may be reactivated by their original causes".

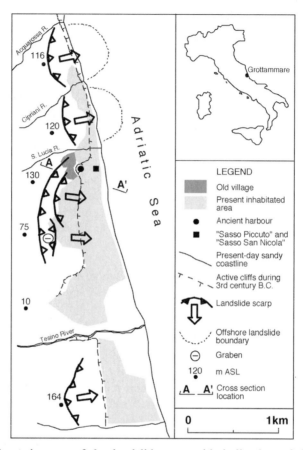

Fig. 2 – Schematic map of the landslide area with indication of the old village (hamlet or "Vecchio Incasato"), the present inhabited area and the ancient Roman harbour.

The definition "abandoned slopes" reported by ESU & GRISOLIA (1991) for these coastal slopes (owing to the regression of the coastline and the presence of protection works and transport infrastructure), does not exclude the reactivation of slope movements. This may occur as a consequence of changes of hydraulic conditions within landslide bodies, that is following slow re-equilibrium of pore-water pressures and consequent decrease of shear strength in soils, and also due to creep, softening and weathering processes (LANZO 1991; BROMHEAD & DIXON 1984; BROMHEAD 1986).

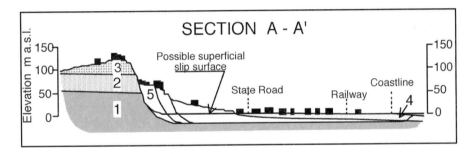

Fig. 3 - Representative cross section of Grottammare landslide complex: 1) clays (Lower Pleistocene), 2) poorly cemented sands (Lower Pleistocene), 3) Poorly cemented gravels (Middle – Lower Pleistocene), 4) alluvial and shore deposits (Olocene), 5) colluvial deposits and/or landslide bodies (Olocene – Pleistocene).

Reactivations of deep movements, with regression of the landslide crown up to some tens of metres, are historically documented along the coastline north of the town of Grottammare, between the S. Lucia and Acquarossa streams (Fig. 5).

The landslide affecting Grottammare is rather old, since it first took place before the medieval settlement of "Vecchio Incasato", predecessor of the town. A detailed structural analysis of the crack patterns on buildings clearly shows the presence of slow but extremely deep movements. More superficial landslide movements seem to have affected the ancient wall structures in a more localised way. Rock falls affecting the top conglomerates were in any case the main cause for the abandonment of the "Vecchio Incasato" settlement from the end of the XVIIIth century, with the transfer of the inhabited centre to the underlying coastal plain.

With respect to the authors previously quoted (CANCELLI et al. 1984a, b; ANTONINI et al. 1993), a greater complexity and extension has been recognised for the large Grottammare landslide, which shows a strongly asymmetrical and three-dimensional body (Fig. 1, 2, 3). Geomorphological investigations have identified the presence of a trench (or graben) which superficially assumes the shape of an elongated valley (Fig. 2), with a concave transverse profile. This phenomenon is explainable by assuming a deep-seated model of movement with a translational slip surface. The bedrock clays underlying the terraced alluvial deposits of the 2nd order seem to have shifted in block to the east.

To the north, where the top of the slope attains higher altitudes (120 m asl), with the sandy conglomerate deposits superimposed on the clayey bedrock, the failure model is ascribable to a roto-translational slide, with several landslide levels sloping towards the sea.

Besides morphological evidence, other clues have been found. At the foot of the slope, nearby the ancient village, downstream of the church of St. Augustin, basal clays outcrop with dip upstream of about 30-35°. Some deep boreholes (65 to 100 m), carried out within the framework of the consolidation works at "Vecchio Incasato", have identified a stratigraphic attitude with dismembered bedrock blocks progressively lowered and tilted backwards.

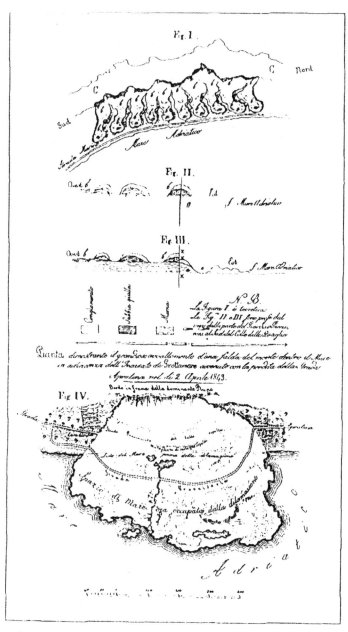

Fig. 4 - Colle delle Quaglie landslide: event of 1843.

Fig. 5.- Colle delle Quaglie landslide (9th May 1928): the soil mass flew offshore (upper photo) coming into collision with a passenger train (lower photo) and causing many casualties.

This landslide seems to be rather ancient, as witnessed by the presence of marine erosion of dormant scarps ascribable to a period ranging from Roman times (IIIrd century B.C.) to the Middle Ages (XIIIth to XIVth century A.D.), at least according to the evolution model of the Marche's coastline during the recent Holocene reconstructed by COLTORTI (1991). In that period an "active cliff" (Fig. 2) was present along this coast and, at the mouth of the River Tesino, "the sea stretched inland about 750 m from the present coastline". The cliff coastal margin was therefore formed rather recently, presumably starting from XVth century (BULI 1944), following a massive increase of solid load carried by the rivers. This was caused by intense inland deforestation and other anthropogenic activities. Environmental changes have also caused the formation of the 4th-order river terrace, which is perfectly correlated with the coastal plain (BIONDI & COLTORTI 1982). Evidence of this recent advancement of the coastline is also witnessed by an inlet downstream from "Vecchio Incasato" which, according to historians, was an active harbour from Roman times up to the XVIth century (Fig. 2). More recently, the progradation of the coastal plain is also witnessed by the available maps, which were drawn during the past two centuries.

CLIMATE CHANGE

Marks of past climate change are deeply impressed into the geological memory of the earth. Just following these signs it was possible to reconstruct the past trend of the climate.

Making reference to a geological time-scale, the Holocene must certainly be considered as a mild climate period, especially if compared with the last glacial age (Würm). Inside the general climate trend of the Holocene it is possible to distinguish some secondary climate changes, lasting for some hundreds of years. One of these is the so called "Little Ice Age", the most recent strong change which lasted some centuries. It started approximately, depending on the different latitudes, in XVth-XVIth century and terminated at the end of XIXth century. It was a very cold period, with abundant precipitation, which strongly affected the survial conditions of many people around the world.

Inside these last climate changes, though very long if compared with the length of the human life, there are also shorter cycles (or fluctuations) of intense cold or hot climate, lasting from 10 to 30 years, called "Brückner Cycles". Due to the average length of our life, the recurrence of these last cycles can be experienced by people only in extreme cases: low cycle periodicity and/or very sudden changes.

Looking at a number of rainfall stations with long recording series it was possible to recognise very important fluctuations of the climate characteristics. Carefully examining these figures it was also possible to recognise periods of strong precipitation, historically known also as periods of very frequent occurrence of large landslides.

For example, in the case of the Marche Region, the period around 1930 was very decisive for the occurrence of important landslides. Looking at the temporal trend of the yearly values of rainfall recorded in a location very close to the study area (Macerata, Fig. 6), the period around 1930 seemed to be critical also for the Grottammare landslide.

Another period lasting for all the 70's induced severe conditions of instability all through the Marche Region at the end of 70's and at the beginning of 80's. Because this period is very fresh in our memory we are able to make a series of reasonings on the planning decisions taken at the end of the 70's and also on the different (sometimes opposite) planning decisions taken at the end of the 80's.

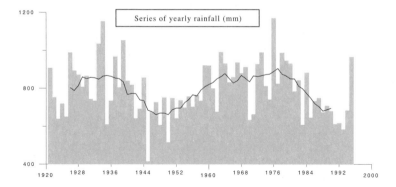

Fig. 6 - Yearly rainfall recorded in a location (Macerata) very near to Grottammare: the bold line represents the fluctuations known as Brückner Cycles.

Different solutions were adopted, forgetting or not knowing of the existence of the Brückner Cycles, in one case spending an excess of money on counter measures during the wetter cycles and nothing on prevention during the last warm and dry cycles. Therefore, the above reasoning must be developed with the aim of planning effective control measures, capable of blocking the landslides even during and after the occurrence of severe climatic conditions.

FINAL REMARKS
The geomorphological and geotechnical investigations still in progress aim to prove the validity of the failure model proposed and the possibility of its reactivation. The results of the research should lead to the identification of the areas of highest landslide risk. The aim, therefore, following the geomorphological mapping and interpretation, is to produce planning guidance.

The town of Grottammare has a long record of instability which has been recorded in books and papers for many centuries. It demonstrates the valuable contribution historical information can provide to landslide investigation. A series of early maps and engravings have illustrated how the toes of past landslides have extended out seaward of the coastline before being removed and re-distributed by coastal processes. Historical information of this kind is just one important element of landslide investigation.

The town of Grottammare has been developed over an ancient landslide complex with the old Medieval town (Vecchio Incasato) located on the steep upper slopes and the more recent development including the main town and residential areas (including high-rise buildings) on the lower tier of the landslide. Important communication routes pass north-south through the town and the landslide system including the railway and autostrada, parts of which are tunnelled. The landslide system is currently largely dormant although slight movements are being recorded with evidence visible in the more susceptible buildings and structures, particularly within the old town. The old town is a particularly attractive area situated near the back of the landslide on steeply rising ground. Although in the process of restoration, the old town could benefit from improved levels of structural and property maintenance in order to reduce the impact of landslide damage on the local community.

REFERENCES

ANTONINI G., CARDINALI M., GUZZETTI F., REICHENBACH P. & SORRENTINO A. 1993. Carta invetario dei movimenti franosi della Regione Marche ed aree limitrofe. C.N.R. - I.R.P.I. Perugia, Pubbl. G.N.D.C.I. n.580.

BIONDI E. & COLTORTI M. 1982. The Esino flood plain during the Holocene. Abstr. XI INQUA Congr., Moscow, 1982, 3, 45.

BROMHEAD E.N. 1979. Factors affecting the transition between the various types of mass movement in coastal cliffs consisting largely of overconsolidated clay, with special reference to southern England. Q.J.Eng.Geol., 12, pp.291-300.

BROMHEAD E.N. & DIXON N. 1984. Pore water pressure observations in the coastal clay cliffs of the Isle of Sheppey, England. Proc. 4th Int. Symp. on Landslides, Toronto, 1, 385-390.

BROMHEAD E.N. 1986. Slope stability. New York, Surrey University Press, 373 pp.

BULI M. 1944. Le spiagge marchigiane. C.N.R. Comit. Naz. Geogr., Roma, 95-147.

CANCELLI A., PELLEGRINI M. & TONNETTI G. 1984a. Geological features of landslide along the Adriatic coast (Central Italy). Proc. Intern. Symp. on Landslides, Toronto, Sept. 1984, vol. 2, pp.7-12.

CANCELLI A., MARABINI F., PELLEGRINI M. & TONNETTI G. 1984b. Incidenza delle frane sull'evoluzione della costa adriatica da Pesaro a Vasto. Mem. Soc. Geol. It., 27, pp.555-568.

COLTORTI M. 1991. Modificazioni morfologiche oloceniche nelle piane alluvionali marchigiane: alcuni esempi nei fiumi Misa, Cesano a Musone. Geogr. Fis. Dinam. Quat., 14, 73-86, 7 fig.

ESU F. & GRISOLIA M. 1991. La stability dei pendii costieri adriatici try Ancona a Vasto. Univ. degli Studi di Roma "La Sapienza", G.N.D.C.I. U.O. - 2.18, pubbl. n.464.

LANZO. 1991. Sability dei versanti costieri in argille sovraconsolidate. Univ. degli Studi di Roma "La Sapienza", G.N.D.C.I. U.O. - 2.18, pubbl. n.465.

MASCARETTI G.B. 1851. Memoria su l'avvallamento di parte del Colle detto Monte delle Quaglie, Ripatransone.

PAOLI D. 1848. Lettera al signor Conte Annibale Ranuzzi intorno ad alcuni slogamenti geologici. Pesaro.

PEROZZI P. 1928. La frana avvenuta a Grottammare a Cupramarittima la notte del 9 maggio 1928. Grottammare.

Managing the dynamics of the estuarine systems on the Douala Lagoon in Cameroon

DR. C. K. ASANGWE, Department of Geography, University of Buea, Buea, Cameroon

INTRODUCTION

Perhaps the most pressing need for humanity today, is in efforts geared towards sustainable development, with the environment as the centrepiece. This pertinent issue of sustainability becomes even more pervasive in the coastal areas, where about 60% of the global human population presently live. This is further heightened when it is considered that projections indicate that 75% of the world population will be living within 60 km from the coastline washed by marine waters by 2050. Coastal environments worldwide remain very crucial as areas of tremendous value to humankind as a rich ecological zone. It is also however by far the most dynamic geomorphic zone where morphogenetic processes ensure continuing evolution and changes of coastal landforms on a spatio-temporal basis. This is well exemplified in estuarine systems, which describe dynamic portions of the coastal zone where marine and freshwater come into constant flux utilising sediments for deposition to evolve fragile geomorphic features. Estuarine processes are extremely dynamic, as their impacts are clearly observable in the changing patterns of coastal forms and structures wherever they occur along a coastline. The fragile nature of landforms in the coastal configuration and the far reaching consequences on low-lying sandy formations has made management difficult of dynamic estuarine systems typical of the Wouri and Dibombari rivers which feed the Douala lagoon. This is in view of preserving and maintaining environmental quality while utilising the lagoon system. The estuaries influenced by their current changing driving forces necessitate management strategies that can be designed and implemented to function within the natural physical condition and the inherent variation effected by the influence of the Douala metropolitan area itself. Estuaries are broadly defined as "partially enclosed coastal body of water that is either permanently or periodically open to the sea in which the aquatic\ecology environment is affected by the physical and chemical characteristics of both runoff from the land and inflow from the sea". (LOICZ. 1995).

THE PROBLEM

With the mounting pressure of increased human population on the coastal areas particularly in the developing countries, tremendous developmental activity has been exerted on this geomorphologically fragile zone, witnessing increased degradation. The consequences are not only in spatial extension or growth but also significantly in their internal and area-wide functions, which have suffered deterioration in their capability and thus effecting degradation on an intense scale. This has become critical along estuarine formations, which are extremely dynamic portions of the coastal depositional environments particularly of the humid tropical areas of the world.

Changes in an estuarine environment are manifested in the mixing of wide range of temperatures and salinity, intrusions into the waters from land use effluent discharge, alongside the human-induced changes from the intense developmental activity in the area. The dynamics of the estuarine formations is thus a constant interaction of terrestrial and marine processes and the boundary zones are continuously shifting. Sediments, which remain at the centre of the physical processes, are deposited to low-lying coastal areas through fluvial activities. As they discharge into the lagoon it helps to sustain, the hydro-geomorphic character of the lagoon changes with the runoff characteristics and the rate of evaporation. This paper is focussed on the Douala barrier island-lagoon environment where climatic conditions depicts a prolonged wet season of about eight months and an equally intense dry season of about four months in a year, with great influence on the runoff characteristics. The response to coastal instability within the framework of changing runoff characteristics, nature of sediments and human induced actions from the management procedure in mitigating the degradation process. The perennial flooding and erosion menace, wetland loss, depletion of biological resources etc. in the face of demographic pressure, spatial growth of Douala necessitates the management framework asked for above for a holistic/adequate response. It becomes necessary when this flood-tide dominated zone continue to effect high influence of tidally prograding mudflats on which mangrove forest thrive, evolving marshlands, through unstable but heavily encroached particularly in the Bonaberi and Deido districts bordering the Douala lagoon complex. The mudflats compose of sand and silt far upstream into the mouths of Wouri and Dibombari rivers further makes infrastructure development intensely facing this area difficult while disrupting the hydro-geomorphic character in their drowned estuaries. The peak discharge of the fluvial systems in this zone which runs from September to May marked by average high velocities ranging from 60 to 145cm/second, do move gravel sized deposits as well as medium to fine grained sand in violent transport in the lower courses of the rivers, which drops sharply from the relatively high topographic areas of the western and south western provinces of Cameroon to the north of Douala. As the low period of runoff discharge coinciding with December –January steps in, the average of 37 to 82cm/second witness the movement of

ine grained to silt for deposition at the estuarine formations, consequently accounting
or the fast silting up of the lagoon system.

THE DOUALA COASTAL ENVIRONMENT

The Douala coastal environment describes a barrier island – Lagoon formations with the
Douala metropolitan area developed on one of such barrier island. The expansive lagoon
complex into which the estuarine systems of the Wouri and the Dibombari river mouths
have virtually been drowned due to the morphogenesis of the area dominates the Douala
metropolitan area. The Douala lagoon system evolved as a major consequence of the
Tertiary to Early Quaternary period, particularly of the Holocene (wholly recent – used
most frequently for the youngest epoch) marine transgressions, which witnessed the
inundation and submergence of coastal lowlands to form the broad embayment into the
lagoon from the Atlantic ocean. The consequence is the drowning of the estuarine
systems of the Wouri and the Dibombari rivers far inland, allowing greatest spatial
extent of the lagoon complex about 50km from the opening into the Atlantic Ocean. The
area comes under strong influence exerted by tidal movements, episodic events of
floods, storms with far reaching geomorphologic implications, which encourage further
degradation.

Broadly, lagoons are areas of shallow aquatic geomorphic attributes that have been
almost completely sealed off from the marine waters of the sea or ocean by the full
development of spits or barriers by oceanographic processes of wave action, tidal action,
alongshore drift of materials etc. The evolution of lagoons is closely associated with
sandy, swampy coastal areas of low topography, where sandy barriers, ridges or spits
develop to seal off the brackish water zone of deposition by coastal rivers. The Douala
lagoon under the influence of the Wouri and Dibombari rivers is today colonised by
mangrove swamps, wetlands, mudflats along marginal depressions, creeks, tidal inlets
etc. The massive sediment input of the estuarine systems of these rivers has evolved a
typical shoaling lagoon system, which is presently the target of continuos reclamation
embarked upon to meet the demands of the rapidly growing Douala city. Easily the most
urbanised centre in Cameroon, Douala's spatial growth continue to inflict far reaching
alteration on the hydro-geomorphic attributes of the lagoon system as well as the
continued instability of surface forms, making it imperative for planning land
management.

ISSUES OF INSTABILITY IN THE ESTUARINE SYSTEMS OF DOUALA

Estuarine processes are extremely dynamic in their changing patterns of coastal forms
and structures where they evolve along a coastline. The fragile nature of its morphology
has made it difficult to preserve and maintain environmental quality while permitting use
of its resources in the Douala area. The issues of deterioration and degradation effecting

instability in this area needs to be examined to provide knowledge for its managemen
Psuty (1992) emphasised this and stated that " this knowledge is especially important i
dealing with a system as dynamic as an estuary because there are natural changes as we
as human-induced changes which are occurring as managers attempt to apply contro
and regulations".

Estuarine systems are associated with a combination of fluvial actions performin
denudation work and oceanographic activities of waves, tidal actions in a coasta
depositional environment. Sediment dispersal and deposition by estuarine systems of th
Douala area characterised by a low-lying swampy nature has provoked hazards in th
land-water ecosystem resulting in a micro level coastal instability in this rapidl
urbanised area. The hazard scenario involves the climate related event of high magnitud
runoff form the prolonged wet season into the estuarine systems from the hinterland int
the near sea level area of Douala. In the face of saline water incursions and eventua
inundation. The Douala area therefore has the crucial problem of abundance of wate
supply in the face of scarcity of land, which poses a serious challenge of coastal zon
instability in the area.

The Douala lagoon complex located north of the city into the Bonaberi-Mabanda an
Ndobo districts cover an area of about 85sq.km though largely colonised by mangrov
forest swamps, it depicts a shoaling lagoon with its marginal depressions nov
encroached upon for human settlement. The high magnitude of runoff influenced by th
climatic conditions of constant supply of rainfall apart from ensuring vast freshwate
also carry a lot of sediments for deposition into the lagoon. The consequence is a
constantly high water table responsible for perennial flooding all year round along th
land-water interface. Flooding is a major consequence of the dynamism exhibited by th
estuarine systems of the Wouri and Dibombari on the Douala metropolitan area. Th
study revealed that though Douala is about 50km away from the Atlantic shoreline, it i
just a little over 14 metres above sea level over most parts of the core built-up area lik
Akwa, Bonanjdo, while the rapidly growing districts of the city, like Bonaberi whicl
includes Mabanda, Ndobo are at between 3-7 metres above sea level. The runoff fron
the Wouri in to the lagoon system in particular, account for the bulk of the floodin;
menace the Douala area is subjected to annually at the peak of the wet season. The low
lying nature of the city provoking floods is seen in the lack of flowing drains, resultin;
in stagnant water due to constantly high water table which has a further consequence o
increasing high rate of subsidence and tilting of residential buildings, particularly alon;
the lagoon environment. This is pronounced where poor reclamation practices have bee
employed in increasing land availability using domestic and timber industry wastes or
the loose sandy deposits of the lagoon margins. It is estimated presently that over 85% o

the residential buildings in the Mabanda and Bonendale districts of Bonaberi have experienced subsidence.

A further problem of flooding is in the sediment nature evolved from the flux from both the estuaries and the tidal flows from storm surges, which now occupy the lagoon, analysed from sampling. A unique feature of the Douala lagoon complex is that demonstrated by its shoaling nature of the high magnitude of marine source sediments far inland to the estuarine formations. This is thus a tidally induced zone evident in the reduction of sediment sizes being transported from the nearshore zone into the estuarine systems. Sediment transport and dispersal on short-term measurement show this and the finding that very fine sand and silt dominate as load for deposition in the lagoon complex. Further sediment distribution pattern show grain sizes of coarser (medium grained sand) sediments far into the estuaries of high velocity channelled flows into the lagoon, due most probably to the runoff magnitude of the prolonged wet season experienced over Douala. Generally, the tidal flats and brackish marshlands have very fine sand to silt composition over which there is a flourishing mangrove forest wherever they occur, while the intertidal creeks and channels through which the dynamic estuarine systems influence the lagoon complex have slightly coarser sediments. The implication is that flooding remains a menace in the Douala metropolis as runoff from the estuaries will continue to maintain a high water table in the area in the face of poorly consolidated sediment nature.

The Cameroon coastline of the Douala area has a broad embayment as it opens into the Atlantic Ocean, which greatly enhances tidal movements, witnessing inflow of marine saline waters. Salinity varies little from a vertical perspective in the zone, however a very dynamic surface variation ranging from between 18 to 13 ppt in the creeks has been reported in this zone. This indicates that to a large extent, the thermohaline conditions are quite homogenous. The variation in salinity that gets pronounced as one move into the mudflats colonised by mangrove vegetation, since the silt composition of deposits here tends to concentrate salinity. At high tides, the marine waters surge into the creeks at a rate of about 0.5m/s in some areas and 0.9m/s in others. The low tide of marine waters retreat at an average of about 2.7m/sec, which occur once in the face of two high tides in a 24-hour period. There is the high magnitude monthly of seasonal tides, locally called "big water" it takes about 10 days to rise, with the waters attaining between 18 metres to 20 metres in depth. These tidal movements apart from flooding the area, further progrades the mudflats and saltmarshes as salinity concentration extend further inland into the Douala lagoon with greater distortion of its hydrogeomorphic character.

Mangrove vegetation is the unique flora specie dominating the Douala lagoon environment. The dynamic nature of the estuaries in its abundance of freshwater

combined with the inflow of saline water evolve such brackish water environment o
predominantly fine sand to silt, providing the adequate ecological conditions for th
growth of the mangrove vegetation. The mangrove forest vegetation over this area doe
appear homogenous, however variations in species do occur due to responses to varyin;
degrees of salinity tolerance. A major factor in mangrove proliferation is concerned witl
the adequate supply of oxygen to roots growing in anaerobic soils, which characteris
this zone. Prominent among the species is the Rhizophora mangle and harissonii type
commonly called the red mangrove, which stands, clearly in several positions on botl
the lagoon and its marginal depressions. The Avicennia nitida, commonly called th
black mangrove are found in more restricted portions of the lagoon inlets of firme
grounds and wetlands.

This environment is also home to a vast array of fauna species, which are fas
disappearing due to encroachment, and destruction of the mangrove for development
The species of fauna found here range from Annelids, Insects, Crustaceans, Amphibians
Reptiles, Birds, Mammals, Microscopic organisms and diverse fish species. The loca
population of Douala is a predominantly fishing community and the general view is tha
of reduced catch in recent times. Most of these faunas use the mangrove fores
predominantly as a spawning and breeding ground where they share the same substrate
requirement. Further destruction of the mangrove forest will only accelerate degradatio
and increase instability, a situation that must be controlled to ensure sustainabl
development.

MANAGEMENT STRATEGIES
Cameroon today is estimated to have 4 million of its 15 million human population withir
60km of its coastline, thus accounting for more than 25% of the population over the
6.5% area of coastal land in the country. Douala metropolitan area alone has a curren
estimate of 2.5 million people and the fastest growing rate of urbanisation in Cameroon
With its environmental problem of scarcity of land in the face of abundant wate
dominated by its lagoon complex, management strategies become inevitable for it
resource utilisation and infrastructure development.

The estuarine systems into the Douala lagoon are expected to remain dynamic, as runof
magnitude under prevailing climatic conditions is not expected to change significantly.
The retention of the wetlands for their role in serving as reservoirs for the estuarine
discharge is thus desirable as a management strategy to mitigate flooding menace into
the low-lying areas due to inundation. A major way of achieving this is to keep the
creeks and tidal inlets open and linked intricately as they open both towards the estuaries
and then ocean, as this will continue to ensure wetlands into which sediments can readily
consolidate over time. The serious case of inundation in the Mabanda area is caused by

his lack of linkage where buildings now stands in hitherto wetland areas. Management implications here is engaging immediately on the ecological approach, which entails that as much as possible creeks and lagoon inlets, should be preserved for natural sediment deposition, while wetlands remain common features of the Douala geomorphic landscape.

A further management strategy should see the spatial growth of Douala into well drained old barrier islands away to the north and east from the Bekoko district and Yaounde road respectively, as this will prevent encroaching on the wetlands. A systematic process of urbanisation within this framework will evolve Douala as a city of multiple growth centres ensuring environmental quality, by reducing congestion, which is the greatest source of both domestic and solid wastes generation as well as uncontrolled industrial effluent discharge resulting in contamination and pollution. While sustaining the wetlands on one hand, the new growth centres of Douala will reduce the pressure on the core built up area and facilitate a network of roads, bridges, mass transit train service and probably subways as obtainable in modern metropolis world-wide.
Coastal tourism utilising the abundance of water of the Douala lagoon complex and its creeks and wetlands over an extensive sprawling city is expected to flourish within this management framework.

It is a common feature in developing countries to cope with lack or inadequate potable water supply and Douala metropolitan area is no exception, despite its abundance of water from the dynamic estuarine systems and its expansive lagoon. Presently, the relatively new districts which grew out of reclamation, like the Mabanda and Ndobo areas do not have adequate water, while in the older zones, multi-storey buildings do not have constant supply. A reservoir built at Soḍiko-a district of Bonaberi, as a booster by SNEC- the government owned Water Company, expected to source water from the Mungo River about 25km out of Douala has remained non-operational. Though concern is no doubt on the saline water incursion into the lagoon, it is however obvious that a process of harnessing the high magnitude freshwater discharge at the estuaries can only assist significantly in reducing the problem of domestic and industrial water supply. A strategy to reduce cost will involve mini water works designated into the multiple growth centres called for above, as the entire Douala is influenced by these two estuarine systems of the Wouri and Dibombari in their high magnitude freshwater discharge into the lagoon.

CONCLUSION
The broad aim of managing the dynamics of the estuarine systems of the Wouri and Dibombari on the Douala lagoon is to improve on the deteriorating environmental quality while subjecting it to further use. Its potentials for resource exploitation have

dwindled due to the massive hydrogeomorphic changes subjected to recently in the process of reclamation. The ecological approach to the maintenance of natural system thus remains primary in ensuring that stability in this otherwise fragile zone is achieved Definitely, the quest for economic development will continue to put pressure on the Douala area with industrial, biological resources and infrastrucural development. It therefore becomes clear that management procedure will entail accomodating the divers productive uses of a rich estuarine system like the Douala coastal environment.

In a state of coastal instability as with other environmental changes, it is often not the change but rather the rate of change that poses the problem. It is in view of this that the spatial extent and effects of estuarine changes in the Douala coastal lowlands should receive greater attention in the realm of long-term scenarios like climatic variations and global sea level rise.

REFERENCES

Asangwe, C.K.A **(1996):** Ecological Implications of Wetland Conversion and Utilisation around Metropolitan Lagos. In Eden, M and J.T Parry (Eds.). Land Degradation in the Tropics. Environmental and Policy Issues. Pinter, A Cassell imprint England. Pp 204-210.

Bird, E.C.F. (1993): Submerging Coasts. The Effects of a Rising Sea Level on Coastal Environments. John Wiley & Sons Ltd. England . 184pp

Pernetta, J.C. and J.D. Milliman (Eds.) (1995): Land-Ocean Interactions in the Coastal Zone Implementation Plan. IGBP Report no. 33, Stockholm, Sweden, 215pp.

Psuty, N.P. (1992): Estuaries: Challenges for Coastal Management. In Fabbri, P. (Ed.) Ocean Management in Global Change. Elsevier Applied Science. London. Pp 502-520.

Psuty, N.P. (1995): Dynamics and Changes in Estuarine Systems. In Duursma, E. (Ed. Proceedings of the Intergovernmental Oceanographic Commission. Workshop Report no. 105 Supplement. UNESCO.

Kuete, M. (1998): Le milleu de l'ecosysteme mangrove: le cas de Bouches du Cameroon. In Journal of Applied Social Sciences, Volume1, No.1. University of Buea Cameroon.

Monitoring the role of landslides in 'soft cliff' coastal recession

HOBBS, P.R.N., HUMPHREYS, B., REES, J.G., TRAGHEIM, D.G., JONES, L.D., GIBSON, A. & ROWLANDS, K., Coastal Geoscience and Global Change Impacts Programme, British Geological Survey, Keyworth, Nottingham. NG12 5GG, UK, and HUNTER, G.& AIREY, R., 3-D Laser Mapping Limited, Cumberland House, 35 Park Row, Nottingham. NG1 6EE, UK

INTRODUCTION

Coastal retreat and coastal flooding pose a persistent risk to the property and livelihoods of people living in coastal areas, particularly in the 'soft rock' (generally sedimentary rocks and Quaternary deposits) areas of southern and eastern England. 'Soft cliff' coastal recession is often seen as predominantly a marine erosion issue, with sea defences constructed along large parts of the coastline of south-east England to dissipate the impacts of wave attack. However, landslides, which are governed largely by terrestrial processes, make an important and often underestimated contribution to 'soft cliff' coastal recession. The role of terrestrial processes is apparent when cliffs are affected by large-scale landslide events, but gradual yet persistent erosion occurs by a variety of smaller-scale flow, fall and rainwater wash processes throughout the year. Many types and sizes of landslide are encountered due to the influence of geology and cliff height amongst other factors.

Perhaps the biggest concern to coastal researchers at the present time is the degree to which coastal recession will be exacerbated by future climate change and, in particular, by the possible impacts of extreme weather events and rising sea level (Bray & Hooke, 1997). The latest climate predictions from the UK Climate Impacts Programme (UKCIP) for the 2050s anticipate that rainfall events will become more intense in the winter months, and probably more variable (Brown, 2001). Coastal zones will bear the brunt of predicted changes to the tide and wave regimes and to any increase in storm surges, but will, uniquely, also be affected by terrestrial processes, particularly the impact of rainfall on slope stability (Polemio & Petrucci, 2000). Cliff recession should be enhanced if an increased intensity of wave attack increases instability of the cliffs and increases storm removal of talus from the foot of the cliffs, all resulting from extreme weather events in a generally wetter, stormier climate.

To date, research establishing the relationship between climate change and 'soft cliff' stability has been scant, and is considered inadequate to predict rates of recession with future climate change. Indeed, most work on monitoring cliff recession has relied either on measuring cliff top positions from published maps (e.g. Clayton, 1988) or interpretations of aerial photographs, both of which cannot provide information over short time scales. However, recent satellite and airborne LiDAR (light detection and ranging) methods have addressed this issue (e.g. Pan & Morgan, 2001).

To improve the accuracy of cliff recession modelling, and to increase our understanding of the relationship between different elements of climate change (rainfall intensity, rainfall frequency, storm events etc.) and the style and timing of coastal change, the British Geological Survey (BGS) has embarked on a programme of detailed monitoring of 'soft cliffs'. Twelve sites, located in North Yorkshire, Norfolk, Kent, Sussex, and West Dorset (Figure 1), were chosen to study the influence of geology, geotechnics, and climate on the recession processes. Surveys are regularly undertaken at 6-month intervals at all sites, with more frequent repeat surveys at selected sites where recession is particularly rapid.

CLIFF SURVEY TECHNIQUES

The monitoring techniques being used are laser scanning, terrestrial photogrammetry, time-lapse photography and GPS. Geological materials involved in the landslides are also subjected to geotechnical analysis using in-situ and laboratory techniques.

The Riegl LPM 2K long-range laser scanner (Figure 2) is being used as the main monitoring instrument. This determines the loci of thousands of points on the cliff face and foreshore to distances of up to 2 km, which are combined to form an accurate 3-D topographic model of the cliff and beach (Figure 3). Multiple laser scans, for example from the beach and cliff top, may be merged and the results referenced to the National Grid using GPS or laser sighting on control points.

Digital cliff topographies are generated from the laser scanning data which, when overlain with previous or successive surveys, provide instant identification of those areas of cliff that are stable and those experiencing change, either through a loss or addition of sediment (Figure 4). These visual images can be converted to volumetric changes using the Terrascan™ terrain modelling package. The accuracy of the volume estimates produced is a function of the density of data points. Shadow areas on the scan have to be excluded from the calculations. It is recognised that potential inaccuracies may arise

f the base station on the beach has not been located with sufficient accuracy at the start
f each survey. Presently the GPS used has an intermediate capability with a practical
ccuracy to 0.5m; improvements in accuracy are being planned in collaboration with the
Jeomatics Departments of the universities of Newcastle and Nottingham. However, in
iew of the frequency of the repeat surveys, cliff crest features have been used to help
natch successive scans. Also, targets placed on stable parts of the cliff for the
hotogrammetry work provide useful markers if they survive between surveys.

The terrestrial photogrammetry method used requires matching pairs of photographs of
he subject with 60% overlap and at least five targets common to both photographs. The
amera used is a Rollei D30 fully metric digital unit with a choice of lenses.

The combination of laser scanning, photogrammetry, and state-of-the-art image analysis
software allows images, and other data, to be 'draped' over the terrain model. The
accumulation of data over a number of years will provide tangible evidence upon which
isk assessments of landslide events and recession rates can be calibrated.

The techniques being used allow repeat surveys to be made, and allow the volumes of
material moving down the cliff faces and the residence times (temporary storage) of
slipped material at the cliff base or redistributed on the foreshore to be calculated over
varying periods of months and years. The data obtained will determine the magnitude of
the erosion processes, their relation to extreme weather events, and the seasonality of
active recession.

INITIAL RESULTS FROM NORTH-EAST NORFOLK

Much of our initial research has been undertaken in north-east Norfolk, partly because
sediment mobility on the cliffs was observed to be particularly active during 2000 in the
middle of the wettest 12 month period since records began. The cliff site chosen at
Sidestrand is 42 m high and extends 300 m along an unprotected section of coast that is
not highly populated, and is approximately mid-way between the end of the groynes
protecting the cliffs at Overstrand and Trimingham (Figure 1; Figure 3). This section of
coastline formed part of the study area for the East Anglian Coastal Research
Programme of the University of East Anglia during the 1970s and 1980s (Cambers,
1973; Clayton, 1988) in which traditional monitoring techniques and beach profile
monitoring stations were used.

The cliffs here are mainly composed of a complex Quaternary sequence of sand-rich
glacial tills and fluvio-lacustrine sediments (Lunkka, 1994). Spectacular examples of
rafts of bedrock chalk with flint aligned along bedding have been thrust up into the
glacial sequence and other glaciotectonic structures in the cliff exposures are especially

notable (Eyles et al., 1989). At the study site a sequence of alternately massive and laminated muds (the Trimingham Clays of Lunkka,1994) occupies a broad, possibly scoured, depression in underlying glacial till. This clay-rich sequence represents the fill of a small lake basin. Elongate rafts of chalky till with a marl matrix, up to 20 m long have been deposited by glaciers at the top of the lake-fill sequence.

Mechanisms of cliff recession

Terrestrial processes

Cambers (1976) reported that the mass movements on the cliffs at Overstrand comprised 73% landslides (translational and rotational slides plus sediment fall on steep slopes), 7% mudflows and 20% water erosion (sediment removed in suspension, mainly by water draining from mudflows). A large highly recessive landslide occurred at Overstrand in 1988, requiring major remediation (Anon, 2000; Frew & Guest, 1999). Our observations indicate that while complex rotational slumps provide the principal mechanism at the head of the slope, debris flows and mudflows are currently the most conspicuous manifestation of sediment transport in the central and lower slope at Sidestrand. These processes seem to occur throughout the year, almost without abatement. Erosion by rainwater is also significant, but difficult to quantify, while wind erosion is negligible.

Volume calculations derived from the laser scan data show that, between November 2000 and September 2001, there was a loss from the 'upper' part of the cliff profile of 26380 m^3. Much of this sediment would have been redistributed on the lower part of the cliff, hidden within the laser scan's 'shadow' areas (Figure 5); access restrictions prevented 'infill' scans from the cliff top. The loss to wave action from the 'toe' area of the cliff was 3,480 m^3 over the same time period. The volume of sediment moving on the studied section of cliff in a relatively short space of time is very considerable, on a par with the volumes involved in a typical large rotational slip on this coast such as at Overstrand (Frew & Guest, 1999). The loss of sediment carried in suspension by flowing water has not been estimated, but Cambers (1973, 1976) suggested this could amount to 20% of the total. Such volume estimates for specific sites provide a more accurate measure of recession than average erosion rates of 0.5 to 2.0 m/yr for the same coastal section quoted by other authors (Cambers, 1973, 1976; Clayton, 1988; Halcrow, 1995).

As documented by previous authors (Cambers, 1973; Kazi & Knill, 1969), the role of water in feeding the flows at the study site, even during the summer months, is considered critical. Rainfall indirectly feeds the seepage from the cliff sediments, and saturates and weakens the sediments. Seepage of water was particularly noticeable around the edges of the chalky till rafts high in the cliff, where it initiated small sediment flows, and from under some of the toes of the mudflows, carrying sediment out across the beach to the sea. Surface runoff also forms gullies on the non-vegetated landslides.

easonal variations in rainfall and seepage and their influence on cliff recession have lso been considered. The rainfall record for the nearest station to the Sidestrand site, 3 m away at Southrepps (Figure 6), covering the initial phases of our monitoring work hows a trend of rising rainfall from January to August 2000, which then remained high o November. During the first monitoring scan in November 2000, activity on the cliffs vas particularly marked, with active mudflows and uplifting of the sand on the beach by novement of the toes of landslides. Early 2001 saw rainfall at elevated levels compared o the same months in 2000, with some very high rainfall events in June and July 2001 Figure 6). Observations in August 2001 confirmed ongoing seepage from the cliffs and he recent formation of mudflows, reflecting recent saturation of the sediments by ainfall and groundwater seepage.

Marine processes

The seasonal activity of coastal 'soft rock' landslides is more continuous than inland equivalents due to cliff foot erosion by waves, which removes debris that would otherwise impede further slides on the cliff. At mean high water the waves wash over the oes of the most seaward slides, and wave activity is responsible for truncation of their eading edges. The existing groynes at Overstrand and Trimingham help to funnel the vave attack onto the unprotected cliffs at Sidestrand. High tide reaches the foot of the cliffs at the study site before those sections of cliff located closer to the groynes. Overtopping of some of the toes is likely during spring tides.

Wave erosion occurs throughout the year, although the wave energy is considerably greater during winter storms. Most studies have demonstrated that toe erosion by wave activity is more active during the winter months (e.g. Komar, 1998). At Sidestrand there had been wholesale removal of the leading fronts of new winter debris flows by the spring. Paradoxically, wave erosion is still significant during the drier summer months when the upper surfaces of the toes are prone to drying and cracking, because the resultant cracks provide lines of weakness to concentrate wave attack. Marine erosion had actively truncated the toes of the summer slides by early September, 2001. The tension cracks formed during movement of the debris flows, and desiccation cracks formed as the flows dry out, both provide lines of weakness which wave action can attack to prise large blocks from the toes of the slides. These blocks are subsequently rolled across the beach by strong currents, picking up flint pebbles to form armoured mud balls.

Beach levels and the foreshore profile are also highly significant; the position of the high water mark over time rather than the position of the cliff top may provide a better guide to the true level of recession. A higher beach level may temporarily protect the toes of debris flows from erosion, but may expose higher levels of the cliff to direct wave attack. Changes in the upper foreshore profile have to date been negligible at the Sidestrand site, whilst in the lower foreshore merit investigation in the future.

Cliff recession over the longer term will only actively continue if material is removed from the base of the cliff, implying that higher beach levels will ultimately slow recession. However, in the short term, landslide activity continues, for example clay slopes may remain in a state of limiting equilibrium for many years, even where total protection from the sea is provided. Also, friable sandy material at the top of the Norfolk cliffs is liable to rapid recession until it reaches its stable angle of repose.

DISCUSSION AND IMPLICATIONS

Monitoring of cliff recession processes at Sidestrand, Norfolk, has documented an enhanced frequency of landslides and cliff retreat processes since early 2000. The study has shown that cliff retreat at the study site occurs primarily due to terrestrial processes which are facilitated by marine erosion, which removes debris from the foot of the cliffs and occasionally erodes the cliff directly, maintaining the steep profiles necessary to allow further recession. The lack of coastal defence structures does concentrate wave attack in the vicinity of the study site, but marine processes have responded to, rather than caused, the enhanced recession.

The cliffs closer to Overstrand and Trimingham, lying behind coastal defences, are noticeably more vegetated. However, erosion and landsliding are still occurring, for example at Clifton Way in Overstrand between 1990 and 1994 (Frew & Guest, 1999) but not at the same rate as at the study site. Seepage and surface drainage of rainwater at the study site appears to be the key to enhanced recession. The topography of the ground behind the cliff at Sidestrand will naturally funnel drainage towards the study site and will maintain saturation of sediments in parts of the cliff. Sandier sediments at the head of the landslides and in adjacent sections of the cliff act as a reservoir for water providing seepage and hence saturation to the clay-rich study site throughout the year. The other significant factor is the lithology at the site, with a mud-filled lake basin occupying much of the study site. Cliffs on exposed coasts containing a high proportion of clay are most sensitive to changes in recession rates according to Bray & Hooke (1997).

On the evidence collected from the field to date, the relationship between local rainfall records and the frequency of landslide activity on the cliffs is striking. Previous work had noted the prevalence of landslides during the winter months. This might apply during low rainfall years, but landslides have been active throughout the year during the recent wetter weather. If higher rainfall becomes a more common feature of Britain's climate, whether in intense downpours or more prolonged wet spells, it will inevitably ncrease the frequency of landslides and the rate of cliff recession on the unprotected coastal zone of Norfolk and similar areas of south-east England. Despite the inevitability of cliff recession, the recession trends determined from detailed monitoring of cliff processes can at least provide planning authorities with a warning of the scale of future trends.

Based on our initial studies we suggest, in the case of north-east Norfolk, that increased wave height and power immediately offshore over the next decade are likely to be of esser significance to soft cliff erosion than increased rainfall. The evidence from our study is that present-day wave activity is more than sufficient to remove material derived from terrestrial processes at the front of the slides, even during the summer months, to maintain the rate of cliff recession.

Short-term monitoring of 'soft cliff' recession, such as that outlined here, will serve an important role in modelling future coastal change. Further monitoring of the site at Sidestrand, as well as other locations, will prove to be essential in assessing the coastal hazards posed by future climate change.

ACKNOWLEDGEMENTS

Paul Turner is thanked for digital cartographic assistance. This paper is published with the permission of the Director, British Geological Survey (NERC).

REFERENCES

Bray, M.J. & Hooke, J.M. 1997. Prediction of soft-cliff retreat with accelerating sea-level rise. *Journal of Coastal Research*, 13, 453-467.

Brown, I. 2001. The UKCIP02 climate change scenarios: context and application. Proceedings of the 36th [MAFF] Conference of River and Coastal Engineers, Keele University, 03.2.1-03.2.9.

Cambers, G. 1973. The retreat of unconsolidated Quaternary cliffs. Unpublished PhD, University of East Anglia.

Cambers, G. 1976. Temporal scales in coastal erosion systems. Transactions Institute of British Geographers New Series, 1, 246-256.

Clayton, K.M. 1988. Sediment input from the Norfolk cliffs, eastern England – a century of coast protection and its effect. Journal of Coastal Research, 5, 433-442.

Eyles, N., Eyles, C.H. & McCabe, A.M. 1989. Sedimentation in an ice-contact subaqueous setting: the Mid-Pleistocene "North Sea Drifts" of Norfolk, U.K. *Quaternary Science Reviews*, 8, 57-74.

Frew, P. & Guest, S. 1999. Geotechnical and other aspects of the Overstrand coast protection scheme. In: Hunstanton and North Norfolk, Engineering Group of the Geological Society Excursion Guide, 13pp.

Halcrow, Sir William & Partners Ltd., 1995. Sheringham to Lowestoft Shoreline Management Plan, Sediment Sub-Cell 3B. Phase 2B, Volume 2 – Studies and Reports. May 1995.

Kazi, A. and Knill, J.L. 1969. The sedimentation and the geotechnical properties of the Cromer Till between Happisburgh and Cromer, Norfolk. *Journal of Engineering Geology*, 2, 63-86.

Komar, P.D. 1998. Wave erosion of a massive artificial coastal landslide. *Earth Surface Processes and Landforms*, 23, 415-428.

Lunkka, J.P. 1994. Sedimentation and lithostratigraphy of the North Sea Drift and Lowestoft Till Formations in the coastal cliffs of northeast Norfolk, England. *Journal of Quaternary Science*, 9, pp209-233.

McInnes, R.G, Tomalin, D. & Jakeways, J. 2000 European Commision LIFE project LIFE - 97 ENV/UK/000510 (1997-2000), Isle of Wight Council, UK

Pan, P.S.Y. & Morgan, C.G. (2001) Monitoring of the coastal environment in South Wales using airborne remote sensing and GIS. *CoastGIS 2001. 4th Int. Symp. On computer mapping and GIS for Coastal Zone Management*, June, 2001.

Polemio, M. & Petrucci, O. 2000. Rainfall as a landslide triggering factor: an overview of recent international research. pp 1219-1225 in: *Landslides in research, theory and practice*. Thomas Telford, London.

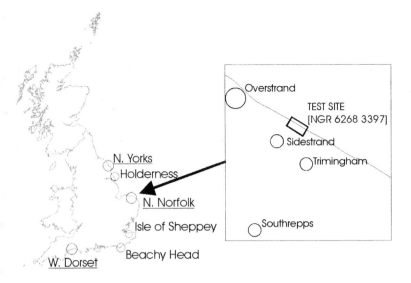

Figure 1 Project test sites (Inset: Sidestrand site, N. Norfolk)
(underlining indicates multiple sites)

Figure 2 Riegl LPM2K long-range laser scanner

Figure 3a.   Laser scan data: 22 November 2000

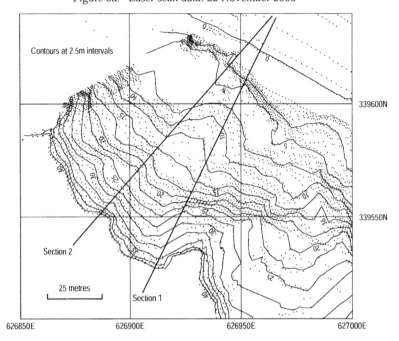

Figure 3b.   Laser scan data: 4 September 2001

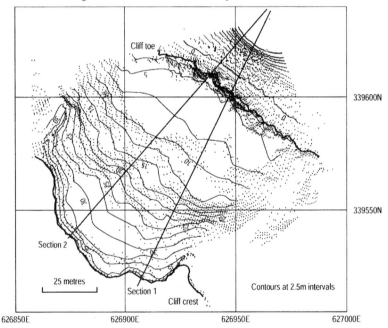

Figure 4. Height loss between November 2000 and September 2001

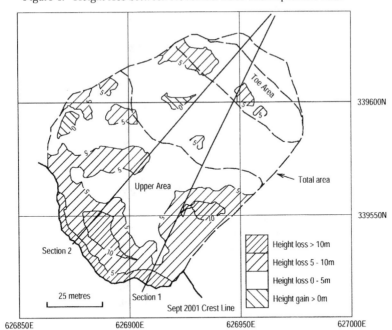

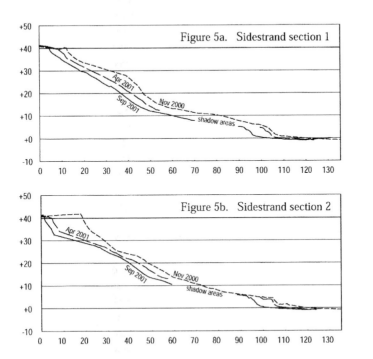

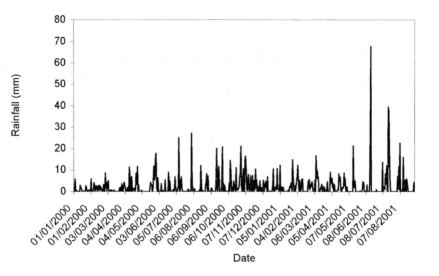

Figure 6   Rainfall, Southrepps (for Sidestrand) Jan 2000 to Aug 2001

Preparing for the impacts of climate change on the central south coast of England

A. S. D. HOSKING, Halcrow Group Limited, Swindon, UK, and
DR R. MOORE, Halcrow Group Ltd, Birmingham, UK

BACKGROUND
The potential impacts of climate change and sea level rise present a significant challenge to future coastal management. It is anticipated there will be increased levels of risk to many coastal communities and assets which will require central and local government to develop and implement policies that address the increasing risks, whilst also meeting the inevitable financial constraints.

The paper presents an overview of a study commissioned by the 'Standing Conference on Problems Associated with the Coastline' (SCOPAC). SCOPAC is the group of Operating Authorities and other interested parties responsible for shoreline management along the central south coast of England, between Lyme Regis and Shoreham (Figure 1).

APPROACH
The main objectives of the study were to:

- Derive coastal climate change scenarios for the next 80 years;
- Evaluate the potential impacts of climate change on coastal behaviour;
- Determine the vulnerability, hazards and risks along the SCOPAC coast;
- Identify the needs for coastal planning and management to address the potential climate change impacts; and
- Identify good practice for management of climate change impacts.

The approach adopted for this study essentially breaks down into two stages. The first is the identification of future climate change scenarios and their likely impacts on coastal behaviour. The second considers the consequences of these changes in terms of increased coastal risks and their implications for management.

COASTAL CLIMATE CHANGE SCENARIOS

Through the study of historical weather records and future projections with General Circulation Models (GCM), there is strong evidence that human influenced climate change is taking place, and may accelerate in the future (IPCC, 2001). Future emissions are not predictable and there are large uncertainties in forecasting future climates, so that impact studies such as this have to consider scenarios. This approach has been promoted internationally by the Intergovernmental Panel on Climate Change (IPCC) and by the UK Climate Impacts Programme (UKCIP). The UKCIP (Hulme and Jenkins, 1998) published information on four climate change scenarios known as the low, medium-low, medium-high and high, which were developed by the University of East Anglia and the Met Office Hadley Centre. More information is available on the medium-high scenario from modelling at the Met Office's Hadley Centre using GCMs and the Regional Climate Model (RCM) and this scenario that has been applied in this study.

The key climate variables considered to influence coastal behaviour are:
- Sea level;
- Nearshore waves; and
- Effective rainfall.

A 'medium-high' coastal climate change scenario was developed for the SCOPAC coast through a review of previous studies and analysis of data. The scenarios present the possible magnitude of changes to 2080, but it should be noted that this is one of many possible scenarios. There are significant uncertainties with future climate predictions and it is necessary for these to be continually reviewed as scientific knowledge and new information becomes available. In reviewing the potential impacts on coastal behaviour the actual numbers are not used, rather they provide an indication of the changing frequency and magnitude of the driving forces,

Sea level changes

The future extreme sea level scenario combines changes to mean sea level, tidal regime, local land movements and surges. The latest historical data from tide gauges available through the Permanent Service for Mean Sea Level[1] show a relative rise in sea level (including vertical land movements) for the SCOPAC region between 1.3 and 2mm/yr.

There are many published estimates of future global mean sea level rise. The latest IPCC (2001) predictions use a wide range of potential greenhouse gas emission scenarios (SRES). These give global mean sea level rise of 0.35m by 2080 for the average upper band modelled scenario, and 0.65m for the upper extreme allowing for land-ice uncertainty. The medium-high estimate of UKCIP is 0.41m by 2080. The UKCIP98 report suggests that for the south coast mean sea level rise could be expected to be 10%

---

[1] www.nbi.ac.uk/psmsl/datainfo/rlr.trends

greater than global estimates. Furthermore, it is necessary to make allowance for relative and movements, for which an additional 1mm/yr relative mean sea level rise is appropriate for the SCOPAC region (Bray et al, 1997).

Flather and Williams (2000) modelled changes in tides that would occur due to a 0.5m rise in mean sea level. The predicted change for the SCOPAC region was minor, with increases in mean high water less than 2cm. A study by Lowe and Gregory (1998), suggest that the 1 in 5yr surge height may increase by 0.1-0.2m and the 1 in 50yr surge by 0.2-0.4m, the lower values representing the western region and the higher values the eastern region of SCOPAC.

The 'medium-high' 2080 scenario rise in extreme sea level for the SCOPAC frontage is summarised in Table 1. The potential increase of 0.84m in extreme sea level is of concern to the existing coastal defences of the region, as current levels expected once in 100 years may occur on an annual basis by 2080 based on this scenario.

Global sea level rise	Local (UK) additional sea level rise	Isostatic land movement	Tidal Regime	Surge changes (1 in 50 year)	Total change for 1 in 50 year extreme sea level
0.41m	0.04m	0.08m	0.01m	0.3m	0.84m

Table 1: Factors contributing to changes in future extreme sea levels by 2080

Wind and Wave conditions
A broad scale hindcast wave modelling study of the Northeast Atlantic/North Sea area for the period 1955-1994 showed small trends for the SCOPAC region of less than 0.5cm/yr for the 90% quantile Hs (Kaas and Andersen, 2000). Differences between a control and a $2 \times CO_2$ 30-year simulation of wave climate were small, with less than 5cm increase in mean Hs by 2100. Wave direction was not considered.

A wave hindcasting analysis was undertaken for this study using wind data output from the Hadley RCM. Ten year periods of wind data for two locations were obtained, covering a control period with atmospheric forcing as at present, and a future period representing the 2080 medium-high UKCIP98 scenario. The wind data were used to derive extreme wave conditions at four locations along the SCOPAC coast (Table 2). The results demonstrate a potential increase in extreme wave heights in the west of the SCOPAC region with little change in the east region.

The offshore wave data were also transformed inshore to several example sites to derive differences between present and predicted future alongshore drift rates. The offshore waves were first transformed to nearshore points and then used to calculate potential alongshore transport rates. The results indicate the potential for changes in overall drift patterns. Such changes would clearly have significant implications for management.

Return Period	Significant wave height (m)			
	Shoreham		Lyme Bay	
	Control	2080s	Control	2080s
1:1 year	5.5	5.3	5.8	7.2
1:10 year	7.1	6.8	7.1	9.6
1:50 year	8.2	7.9	8.1	11.3

Table 2: Equivalent offshore extreme wave conditions

Effective rainfall

The importance of rainfall as a primary cause of coastal instability and landsliding across the region was demonstrated in the recent wet winter of 2000/2001. Effective rainfall (precipitation minus evapotranspiration) provides an indication of the amount of rainfall directly contributing to groundwater tables within slopes. Historical records from Ventnor (Isle of Wight) and Pinhay (Dorset) indicate that effective rainfall has been gradually increasing over the last 150 years, and this has been linked to increasing ground movement and landslide events in some locations. Table 3 presents estimated changes in effective rainfall by 2080, which is anticipated to increase over the winter months. The magnitude of change is comparable with the historic trend, which points to increasing landslide and ground movement potential.

	Sep-Nov		Dec-Feb	
	mm	% change	mm	% change
ER$_{mean}$ 2000	76.4	0	67.4	0
Med-High 2080	82.7	8	81.8	21

Table 3: Effective rainfall at Ventnor, based on UKCIP 'med-high' scenario

COASTAL BEHAVIOUR AND IMPACTS OF CLIMATE CHANGE

Climate change has the capacity to alter almost all coastal processes and landforms. There is a need to determine the extent to which climate change could affect the distribution, frequency and magnitude of flooding, deposition and erosion hazards. The approach adopted by this study involved the geomorphological characterisation and sensitivity analysis of generic coastal behaviour systems (CBS). The climate change 2080 scenarios were applied to these conceptual models to determine the possible coastal behaviour responses and the extent to which coastal hazards are likely to change.

Coastal Behaviour Systems

These were broadly defined from an assessment of coastal types and their 3D morpho-dynamic components (Table 4). By definition they comprise interlinked physical landforms that control or may influence the way in which the system responds to forcing events (process). The longshore connectivity of such systems is important and is

xplained by the sediment budget. In this way five generic coastal behaviour systems
CBS) were defined across the SCOPAC region (Table 4; Figure 1).

Landform Element	Hard Cliff	Soft Cliff	Lowlands and Barriers	Inlets, Spits and Tidal Deltas	Estuaries and Tidal Rivers
Shoreface	Steep	Gentle	Gentle	Gentle	Gentle
Shoreline	Fringing boulder beach and/or shore platform	Fringing sand, shingle or mixed beach	Fringing sand, shingle or mixed beach Free-standing shingle barrier Fronting sand or shingle beaches	Inlets Tidal Deltas Free-standing shingle beach	Tidal flat Saltmarsh Tidal River
Backshore	Hard Cliff	Soft Cliff	Lowland	Lowland	Lowland/ Soft Cliff

Table 4: Composition of Coastal Behaviour Systems, SCOPAC region

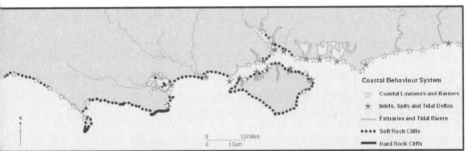

Figure 1: Distribution of Coastal Behaviour Systems, SCOPAC region

Coastal Behaviour and Sensitivity to Change

The next step was to evaluate the sensitivity to change of each system based on
historical records and expert review. The individual landform elements within each CBS
were assessed, as well as the whole CBS, allowing potential application of the
assessment to coastlines not falling within the five classifications.. The influence of
coastal defences on the natural response and development of each system was evaluated
at this stage. The conceptual CBS models were then applied with the climate change
2080 scenarios to evaluate the likely response, rates of change and potential hazards that
can be expected. The outcome of this work was presented as a series of influence
diagrams that consider the historic, contemporary and future behaviour of each CBS,
with or without coastal defences.

Summary of Potential Hazards and Climate Change Impacts

The main impact of climate change at the coast will be to increase the frequency an magnitude of hazards (Table 5). The rate of change will be progressive although, i many cases, manifest through the impact of extreme events (intense storms an prolonged wet periods).

Feature	Hazard and Impact
Beaches	Prone to breaching, decline in volume due to limited supply of sediment, and coastal squeeze due to intervention.
Saltmarshes	Accelerated loss due to increased erosion, coastal squeeze due to intervention, and potential inability to accrete at the rate of sea level rise.
Eroding Cliffs	Accelerated rates of cliff recession due to increased toe erosion.
Coastal Landslides	Increased frequency of ground movement events due to higher groundwater levels, effective rainfall and toe erosion.
Coastal Lowlands	Growing potential for combined tidal/fluvial flooding, and flash flooding in urban areas.
Existing Defences	Increased overtopping, wave loading on structures, failure of existing stabilisation measures, and increased defence exposure due to intertidal reduction.

Table 5: Key Hazards and Impacts of Climate Change

IMPLICATIONS FOR COASTAL RISK MANAGEMENT

Understanding the risk management framework and the broader social and politica context provides a basis for speculating about how climate change may impact upo SCOPAC over the next 50 years or so. It should be stressed that because of the nature o social systems, there is probably more uncertainty as to how society and politicians wi respond to these changes than their impact on coastal processes. The following are som of the key changes to the risk management framework that may be expected over th next 80 years.

As outlined above, climate change will result in an increase in the probability o damaging events along much of the south coast. Considerable investment in defenc improvements and maintenance will be required if current standards of protection are t be maintained. However, it is uncertain as to whether the funds allocated for coasta defence will match the future demand. Over the next 80 years the risk managemen framework will need to adjust to increased competition for financial resources.

It is possible that attitudes towards acceptable risks and suitable standards of protection will change. It follows that there may be a need to improve the standards of protection in high-risk urban areas to reflect these trends. This would lead to increased polarisation of risk exposure to individuals in built up and rural areas.

Climate change is likely to generate additional pressures on a variety of coastal zone uses, from tourism and amenity uses, marine aggregate extraction (e.g. for beach feeding), port and harbour operations to nature conservation and the protection of historical sites and monuments. For example, changes to the physical character of the coast and patterns of human intervention are likely to result in losses of habitats designated for protection.

It is possible that increasing reluctance of society as a whole to accept the costs (financial and environmental) of coastal defence, more restrictive environmental legislation (e.g. the Habitats Directive and its successors) and the desire to avoid tying future generations into expensive and unsustainable defence commitments will trigger a re-appraisal of the compensation issue. How such a mechanism would work and what its impact would be are unclear.

Following trends elsewhere in society it seems likely that there will be legal challenges to operating authorities decisions, especially where limited resources or environmental concerns lead to changes in the standard of protection provided to individual properties or communities. This may extend to challenging the power of operating authorities to remove an individuals right to protect their own property, at their own expense.

It should be stressed that all these possible changes could occur irrespective of climate change. Thus, despite the uncertainties associated with the changing character of natural hazards and coastal behaviour, operating authorities should be aware that their ability to manage the associated risks will not remain constant. The changes in society and the political economy that will inevitably occur over the next 80 years will be as significant as climate change in determining how coastal risks are accepted, tolerated and managed.

SUMMARY RECOMMENDATIONS
The following are some key recommendations arising from the study.
- **Driving Future Guidance**. SCOPAC can have a key role in influencing the way in which the risk management framework develops in the future. This could involve research, influencing legislation and policy, and identifying possible solutions for risk management (e.g. compensation).
- **Planning and Development**. There is a role for the planning and development process in the management of future risks. There is the capacity within the existing planning framework to adapt to climate change, although it will require better and

more consistent application of existing legislation and guidance.

- **Coastal Defence**. There are a number of key stages in the development and implementation of coastal defence policies, all of which will need to take account of the potential impacts of climate change within the existing procedures.
- **Nature Conservation**. Recommended actions are aimed at meeting statutory obligations placed on local authorities, meeting biodiversity targets and maintaining the existing nature conservation resource irrespective of climate change.
- **Monitoring and Education**. One of the central themes of climate change is the uncertainty with current predictions. In order to assess changes in coastal climate it is necessary to monitor key variables. It will also be necessary to inform the community of changing coastal risks, e.g. through future reviews of Shoreline Management Plans and advice notes.

With regard to future risk management, SCOPAC has a key opportunity to be pro-active in recognising the potential impacts of climate change at the coast and ensure that their needs are considered in future changes to the risk management framework.

ACKNOWLEDGEMENTS
The authors would like to thank the other members of the project team, much of whose work is summarised in this paper. These include Mark Lee of the University of Newcastle, Dr Malcolm Bray of the University of Portsmouth, Mark Gallani of The Met Office and Robert Harvey of Halcrow. The support of the SCOPAC Officers Research Sub-group throughout the study is gratefully acknowledged.

REFERENCES
Bray MJ, Hooke JM and Carter DJ, 1997. Planning for sea-level rise on the South Coast of England: informing and advising the decision makers. *Trans Inst Br Geogr*, NS, 22,13-30

Flather R and Williams J, 2000. Climate change effects on storm surges: methodologies and results. ECLAT-2 Report No3, Climatic Research Unit, UEA.

Graff J, 1991 An investigation of the frequency distribution of annual maxima at ports around Great Britain. *Estuarine, Coastal and Shelf Science*. 12, pp. 389-449.

Hulme M and Jenkins GI (1998) Climate change scenarios for the UK: Scientific report. UKCIP Technical Report No 1, Climatic Research Unit, Norwich.

IPCC (2001) The IPCC third assessment report. Summary for Policy Makers. Available at: http://www.ipcc.ch/index.html.

Kaas E and Andersen U (2000) Scenarios for extra-tropical storm and wave activity: methodology and results. ECLAT-2 Report No 3, Climatic Research Unit, UEA

Lowe JA and Gregory JM (1998) A preliminary report on changes in the occurrence of storm surges around the United Kingdom under a future climate scenario. Hadley Centre for Climate Prediction and Research.

Met Office (2000) An update of recent research from the Hadley Centre.

Impacts of climate change and instability – some results from an EU LIFE project

M-L. IBSEN, EU Marie Curie Research Fellow, Earth Sciences Department, University of Florence, Italy, and Lecturer, Kingston University, London, UK

ABSTRACT
The project entitled 'Coastal change, Climate and Instability' was undertaken over a three-year period between 1997-2000 and supported by the European Community Environmental Policy, LIFE (L'Instrument Fiancière de L'Environnement) (McInnes and Tomalin, 2000). It was fundamentally concerned with the potential impacts of climate change on urban landslide environments in Europe, for both coastal and mountainous areas, and how these impacts could be handled in the future. This paper summarises the section on 'impacts of climate change' which attempted to look at the following three aspects of climate and instability:
1. The historical climatic background and its relationship to instability, including the use of palaeo-environmental and archaeological landslide evidence;
2. Different future climate scenarios and how the potential changes may effect the coastal environment and instability;
3. Preparation for the future changes to the environment through possible management techniques.
The paper is set in a European context, but focuses on the southern coast of Great Britain, specifically Ventnor, Isle of Wight.

OUTLINE OF PAST CLIMATIC CHANGE & INSTABILITY FOR EUROPE
Throughout Europe, past climatic changes have undergone significant research. One aim, of which, has been to identify trends and cycles of previous climatic variations. An overall consensus is that for the Holocene, since the last glaciation there has been a general trend of climatic improvement. Within this there are identifiably periods which have been cold and humid: 3400-3300yrs BP; 2900-2300yrs BP; 1550-1250yrs BP; 850-700yrs BP and 1550-1850yrs AD (Little Ice Age) (Marabini and Veggiani, 1993). In addition, obvious warm periods have existed such as the Medieval Climatic Optimum (900-1100yrs AD).

In Europe a considerable archive exists for landslide events, but the historical information is incomplete as the recording of movement has not been continuous. Despite this the record does display some periods of landslide activity which may be associated with climate. The following summarises some of the research carried out on landsliding and climate in Europe taken from the EU projects EPOCH and TESLEC (Soldati, 1996; Schrott and Pasuto, 1999).

1. Last interglacial (125,000yrs BP) could have triggered recorded landslide events in Cantabria, Spain;
2. Late Devensian (18-11,000yrs BP) may be related to landslide movements in Ireland as a consequence of the proximity to the ice front, deglaciation and periglacial activity;
3. Younger Dryas (11-10,000yrs BP), during this cold and wet period landslide activity increased over most of Europe;
4. (11,500-9,500yrs BP) deglaciation and increased rainfall are associated with landslide occurrences in Northern Italy and Cantabria, Spain;
5. Early Atlantic (7,500-6,000yrs BP) significant landslide activity occurred in Ireland and Great Britain;
6. Sub-Boreal (6-3,000yrs BP) numerous examples of landsliding specifically on the south coast of Britain and in Northern Italy;
7. Early Sub-Atlantic (2,500-2,000yrs BP) which is a notable wet period shows some mass movement activity in Northern and Central Europe;
8. Mid Sub-Atlantic (2,000-1,500yrs BP) there have been landslide movements in both the Swiss Alps and the Pyrenees;
9. Little Ice Age (1550-1850 AD) landsliding was paramount over Europe;
10. Modern period (since 1850) throughout Europe landslide movements have been recorded almost continuously displaying substantial variations in activity.

'It should be emphasised that all these results must be regarded with caution since they are essentially describing the known documented evidence and are merely probable periods of landslide activity in Europe. Equally, it is not feasible to assess accurately the cause of a landslide event without the knowledge of independent collaborative evidence.' (Ibsen, 2000).

HISTORICAL PERSPECTIVE OF CLIMATE AND INSTABILITY AT VENTNOR, ISLE OF WIGHT

Historical data can greatly augment our knowledge of environmental systems. Detailing how things have occurred in the past improves our understanding of the processes involved in the present and the response a system may have in the future to various environmental changes. The historical data may also cover a sufficient time span to enable the identification of trends and cycles which may equally give rise to possible future scenarios. The LIFE project extensively used archaeological and palaeo-environmental techniques to gain knowledge of the past events at several locations within the European Union.

At Ventnor, Isle of Wight, as a direct consequence of the LIFE project, the landslide complex now has an event chronology which could be extended beyond 30,000 years (Tomalin, 2000). This date is revealed by the initial examination of four archaeological finds of early flint tools or 'palaeoliths' found within the Undercliff landslide complex. Previous geotechnical studies, initially considered a pre-landslide event in the Ipswichian inter-glacial period off-shore from the present site (Hutchinson, *et al.*, 1985), with the first major compound failure in Cretaceous clays and sandstones to be sometime within the Late Quaternary (Hutchinson, 1987).

Following the primary failure of the Undercliff successive events have continued almost unceasingly with numerous major and minor processes. During the Late Glacial period it is considered that a series of multi-rotational slides within the Undercliff complex were formed through both coastal and seepage erosion (Tomalin, 2000). A rapid sea level rise from c. −15m OD in c. 8,500yrs BP to c. − 6m by 6,000yrs BP occurred during the onset of the Atlantic warm phase causing major toe erosion and the initiation of significant secondary movement. Dated radiocarbon assays, of around the 5th millennium BC, at St Catherines Point and Binnel Bay coincide with this landslide movement. A further secondary phase at St. Catherines Point may be ascribed to the end of the 3rd millennium from archaeological tree and charcoal evidence, as human occupation started on the surface of the new debris apron at the beginning of the 2nd millennium BC (Tomalin, 2000). Rising post-glacial sea levels may have been the cause, but other factors may equally include widespread forest clearance and a sharp deterioration of climate.

Further archaeological evidence of a Roman bronze brooch, from the 2nd century AD found within approximately 2m of chalky hill-wash, is indicative of a typical toppling failure from the main rearward scarp at Gore cliff. Ground movements at Luccombe at the extreme eastern end of the landslide complex, between 1923 and 1931, uncovered more archaeological material relating to an early medieval settlement. Upon investigation it was noted that occupation of this settlement ceased after the 12th century which may be attributed to considerable ground movement at the time, especially since the area is known to be a particularly active part of the landslide complex (Moore et al., 1995). An early Christian cemetery, dating back to Late Saxon or Early Medieval time, was discovered at the western end of Ventnor, at Flowers Brook. The cemetery now lies very close to the cliff edge and it is considered that both the church and the community may have been lost to coastal erosion (Tomalin, 2000).

The Little Ice Age (1550-1850yrs AD) was an extremely significant phase of landslide activity which occurred during a period of extreme cold winters and very wet summers. Following this, the next important phase of landslide activity started with the present climatic warming, from about 1850, with rapid industrialisation and urban development. Throughout this modern period mass movement has shown a considerable increase over the last 100 years along with possible periodicities (Figure 1). During 1922 and 1930, the effective precipitation for Ventnor, Isle of Wight, has been directly correlated with the incidence of major landslide events (Ibsen, 1994, 2000). Further similar examples are: 1935-37; 1939-40; 1950-52; 1957-58; 1960-61; 1963-70; 1977-82 and 1986-87.

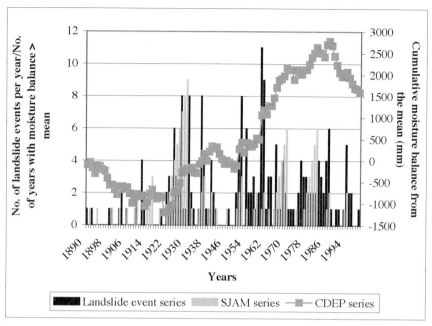

Figure 1. Moisture balance and annual landslide series for Ventnor, Isle of Wight, UK (From Ibsen, 2000)

FUTURE CLIMATE SCENARIOS AND COASTAL CHANGES

The global climate is known to have a natural variability upon which there is a discernible human influence. Current debate questions how this changing climate will alter in the future and what impacts it will have on the environment. Predictions range considerably, however, the most important factor is that of global warming. Further possibilities are an increase in extreme events and changing rainfall patterns. In Great Britain predictions for increasing temperature lie between 0.1-0.3°C per decade in the next 100 years (Hulme and Jenkins, 1998). This has been derived from Global Circulation Models (GCM's) based on different values of climate sensitivity, anthropogenic forcing and natural variations. A typical set of changes from the UK Climate Impacts Programme, 1998 (Wade *et al*, 1999) taken from a medium-high climate change scenario up to the year 2080 and calculated with respect to the 1961-90 average are the following:

- A rise in mean annual temperature of 1.2 to 3.4°C;
- An increase in mean annual rainfall of 1 to 4% with winter rainfall increasing by 6 to 22% and summer rainfall decreasing by –8 to –23%;
- Overall increase in variability of temperature and rainfall and the incidence of extreme events such as: storms, droughts and floods;
- A rise in mean sea level in the English Channel by 34cm by the 2050s.

The spatial resolution (300-400km) of these models is a significant disadvantage when relating these changes to individual landslide complexes. However, the uncertainties that are inherent in such models indicate that these results are plausible scenarios given the current knowledge of human influence on climate. Therefore, they are sufficient to apply to a local area in order to provide an idea of the possible impacts of climate change.

Adjusting the current levels of effective precipitation at Ventnor, Isle of Wight, with a rise in mean annual temperature of 3.4°C and an increase in rainfall of 4% produces an overall increase of 8.4% in the mean annual moisture balance, i.e. from 224mm to 243mm per year. Relating this to the incidence of landslide occurrence would probably increase the occurrence of landsliding by about 5% as the average annual moisture balance is lifted into almost the next class (Table 1, Ibsen, 1994).

Table 1. Relationship between moisture balance and landslide events for Ventnor, Isle of Wight, UK (1890 – 1987) (From Ibsen, 1994)

Annual moisture balance class (mm)	No. of years in each class	No. of years with slips	Years with slips as a percentage of total years
-149 - -50	1	0	0.00%
-49 – 50	7	3	42.86%
51 – 150	17	6	35.29%
151 – 250	27	20	74.07%
251 – 350	25	20	80.00%
351 – 450	15	13	86.67%
451 – 550	4	3	75.00%
551 – 650	1	1	100.00%
651 – 750	0	0	0.00%
> 750	1	1	100.00%
Totals	98	67	68.37%

In 1996 High-Point Rendel were commissioned by the former South Wight Borough Council to analyse the forecasting of landslide events on the Isle of Wight Undercliff (Sakalas *et al.*, 1996). Their method of forecasting involved the four-month antecedent effective winter rainfall totals (4AER) which correlated well with the incidence of landslide events. They identified a lower boundary of four-month total rainfall for different areas which could then be continuously monitored through a local weather station as to when it was exceeded. Applying the maximum predicted increase in winter rainfall of 22% to the model significantly increases the susceptibility of ground movement (Figure 2). For example, a 500mm four month antecedent winter rainfall (4AER) changes from being a 5 year event to a 2 year event with ground failure moving from moderate to high.

The predicted increase in extreme events and rise in mean sea level would equally have a significant impact along the coastline. Further variability in storm occurrence would increase the incidence of extreme and sudden erosion and the probability of landslide movements. The rate of sea level rise which is predicted at nearly seven times that of the last 2,000 years (Ibsen, 2000) will rapidly undercut cliffs and erode the coastline leading to numerous landslide failures.

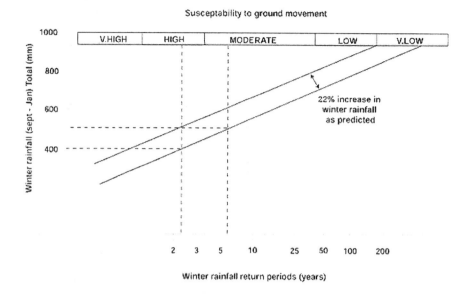

Figure 2. Predicted return periods of winter rainfall and the susceptibility of ground movements up to 2080 for Ventnor, Isle of Wight, UK (After Sakalas *et al.*, 1996)

MANAGEMENT OF FUTURE ENVIRONMENTAL CHANGES
Preparing for the challenge of uncertain climatic change is extremely difficult and it is important to analyse historical climate and instability, since it enables a probability of future events which can then be responded to by various organisations. Preparing for the future scenarios reduces the potential risks of impact and may avoid, at least, the catastrophic events. At present, there are a number of techniques for living with landslides and these are being continually updated according to the predicted changes. With coastal landsliding rising sea level and climate change are the primary issues to tackle. The former is usually controlled by various forms of sea defence whilst the later is best dealt with by using systems of land drainage. Although, any stabilisation scheme will require continual maintenance in terms of adapting to the changes and a constantly updated Quantitative Risk Analysis (QRA) would analyse the relative risks involve and provide a cost/benefit analysis for the reduction of those risks. Ultimately, however, nothing is sustainable and schemes must be applicable to the changing environment.

Institutions need to develop and implement policies to deal with the increasing risks expected by the impacts of climate change and sea level rise. Recently, the United Kingdom Department of the Environment, Transport and Regions carried out a study which recommended that a further £1.2 billion would be required to defend the coastline of Great Britain over the next hundred years. In addition to this, the United Kingdom House of Commons Agriculture Committee identified three management principles for the potential risks along the coastline (1998):

- the need to overcome a widespread public intolerance of acceptance of naturally occurring and unavoidable risks;
- the importance of developing a strategic approach at both national and regional level to coastal risks;
- the need to adjust the planning system to take account of the increasing importance of coastal defence measures.

An important problem to tackle by those examining climate change impacts is the fact that predictions are continually being updated with increasing knowledge of the environment. Therefore, it is essential that regular reviews are conducted to maintain schemes that will adapt to the current predictions. The main challenge, however, is that of uncertainty and, although, this can be reduced with strategic planning it cannot be eliminated.

An innovative approach to these problems is being carried out by the Standing Conference on Problems Associated with the Coastline (SCOPAC) along the southern coast of England between Lyme Regis in Dorset and Hove in West Sussex. The group consists of 29 coastal local authorities, nature conservationists and environmental groups who have put together a contract entitled "Preparing for the Impact of Climate Change". The aim of this study it to bring to the attention of both regional and local management the issues that must be addressed along the central southern coast of England over the next 80 years.

Despite this excellent example of authorities working together, the researchers within the LIFE project believe that the *possible changes in weather systems and patterns are likely to have a particular impact on coastal cliffs and slopes and that the question of instability remains a major and growing hazard, the implications of which are far from fully appreciated by decision-makers in many EU member states'* (Ibsen, 2000).

ACKNOWLEDGEMENTS

This work was conducted as part of a European LIFE project and the author would particularly like to thank Robin McInnes, at the Isle of Wight Centre for Coastal Environment, for inviting me to take part in this project. I would also like to thank all those involved in the project for their advice and help in carrying out this research, especially Jenny Jakeways and David Tomalin. In addition, as a Marie Curie Research Fellow, I would like to thank Prof. Canuti and the research unit in the Earth Sciences Department at Florence University for providing me with their expertise and access to their facilities. I am also extremely grateful to the European Union for funding my continuing research.

REFERENCES

Hulme, M. and Jenkins, G., 1998. Climate change scenarios for the United Kingdom: summary report. Department of the Environment, Transport and the Regions (DETR), UK Climate Impacts Programme Technical Report No. 1.

Hutchinson, J. N., Chandler, M. P. and Bromhead, E. N. (1985). A review of current research on coastal landslides forming the Undercliff of the Isle of Wight, with some practical implications', Proc. of the conference on problems associated with the coastline 17-18 April 1985. Isle of Wight Council, Newport, 1-16.

Hutchinson, J. N. (1987). Some coastal landslides of the southern Isle of Wight. In: Barber, K. E. (Ed) Wessex and the Isle of Wight, field guide. Quaternary Research Association Cambridge, 123-135.

Ibsen, M-L., 1994. Evaluation of the temporal distribution of landslide events along the south coast of Britain, between Straight Point and St. Margaret's Bay. MPhil thesis, King's College, London University (unpublished).

Ibsen, M-L. (2000). The impacts of climate change. In: McInnes, R. G. and Tomalin, D. (2000) (Eds). Coastal change, climate and stability. European Commission LIFE project, LIFE-97 ENV/UK/000510, Isle of Wight Centre for the Coastal Environment UK and partners. Vol. 1, chapter 6.

Marabini, F. and Veggiani, A., 1993. Evolutional trends of the coastal zone and the influence of climatic fluctuations. In: Grifman, P. M. and Fawcett, J. A. (Eds.) International perspectives on coastal ocean space utilization. Proceedings from the 2nd International Symposium on Coastal Ocean Space Utilization (COSU II). April 2-4, 1991, Long Beach, California.

McInnes, R. G. and Tomalin, D. (2000) (Eds). Coastal change, climate and stability. European Commission LIFE project, LIFE-97 ENV/UK/000510, Isle of Wight Centre for the Coastal Environment UK and partners.

Moore, R., Lee, E. M. and Clark, A. R., 1995. The Undercliff of the Isle of Wight – a review of ground behaviour. By Rendel Geotechnics for South Wight Borough Council (SWBC), Ventnor, Isle of Wight.

Sakalas, C., Lee, M. and Moore, R., 1996. Isle of Wight Undercliff: Landslide forecast. By Rendel Geotechnics for SWBC, Ventnor, Isle of Wight.

Schrott, L. and Pasuto, A. (Eds.) Temporal stability and activity of landslides in Europe with respect to climatic change (TESLEC). Geomorphology Special Issue, 30, Nos. 1-2.

Soldati, M. (Ed.), 1996. Landslides in the European Union. Geomorphology Special Issue, **15**, Nos. 3-4.

Tomalin, D. (2000). Ventnor Undercliff – palaeo-environmental overview, Isle of Wight UK. In: McInnes, R. G. and Tomalin, D. (2000) (Eds). Coastal change, climate and stability. European Commission LIFE project, LIFE-97 ENV/UK/000510, Isle of Wight Centre for the Coastal Environment UK and partners. Vol. 2, Palaeo-environmental study area P4.

Wade, S., Hossell, J. and Hough, M. (Eds) 1999. The Impacts of Climate Change in the South East: Technical Report. WS Atkins, Epsom, UK.

An approach to assessing risk on a protected coastal cliff: Whitby, UK

E M LEE University of Newcastle upon Tyne, UK
M SELLWOOD High Point Rendel, UK

INTRODUCTION

The construction of coastal defences has considerably enhanced the stability of many clifflines around England and Wales, and provided protection for assets that otherwise would have been lost as a result of marine erosion. However, should the defences fail it is likely that renewal of marine erosion at the cliff foot would quickly lead to landsliding and cliff recession. In order to test the economic efficiency of maintaining or improving the defences, it is necessary compare the costs and benefits associated with the defences with the consequences of a so-called "do nothing" scenario whereby no further coast protection and slope stabilisation or maintenance works are undertaken (i.e. walk away and cease all maintenance and repairs). Both the costs and benefits need to be discounted from the time they might arise in the future to their present value (PV; in England and Wales the Treasury discount rate is 6% per year).

This type of problem does not lend itself to conventional methods of simply extrapolating past cliff recession rates, because of the uncertainty over the timing of defence failure and the subsequent response of the coastal slopes. An alternative approach is presented in this paper, based on the structured use of expert judgement and subjective probability assessment. It has been developed in order to assess the potential financial risks associated with a renewal of cliff top recession along the protected glacial till cliffs at Whitby, North Yorkshire.

THE SITE

The 2.5km-long cliffline between Whitby and Sandsend is developed in a highly variable sequence of over-consolidated sandy clay tills (boulder clays) with subordinate lenses and thin discontinuous beds of sand and sandy silt. The cliffs are generally in the range of 30-40m high, reaching a maximum height of nearly 50m OD at the Metropole Cliff. The western section, between Upgang Ravine and Raithwaite Gill, Sandsend, remains unprotected. Comparison of different Ordnance Survey map editions suggests historic recession rates of retreat between 0.14m and 0.71m per year, with an average annual recession rate in the order of 0.30m/year.

The eastern section, West Cliff, (between the Whitby Spa and Upgang Ravine) has been progressively stabilised since the late 1920's. Initially a concrete seawall was constructed between the Spa and Happy Valley, which was completed in 1932.

Figure 1 Whitby cliffs: recession scenario for protected cliffs

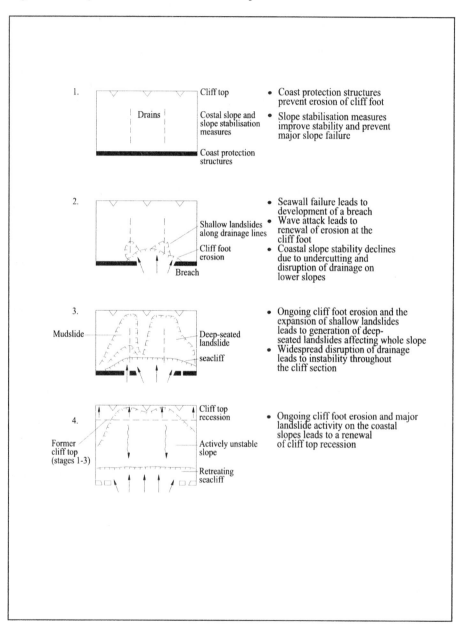

Slope stabilisation (drainage and slope reprofiling with the use of fills) between the Spa and White Point was undertaken in four stages between 1969 and 1981. Slope stabilisation and coast protection works were continued westwards along the slopes between White Point and Upgang Ravine between 1988 and 1990. Details of these more recent works can be found in Clark and Guest (1991).

POTENTIAL SEAWALL FAILURE AND CLIFF TOP RECESSION
It has been recognised that the stability of the protected coastal slopes of West Cliff are dependent on the continued structural integrity and performance of the seawalls. Failure could lead to renewed landsliding on the coastal slopes and subsequent renewal of cliff top recession. This probably will not occur as a single event i.e. there could be a *delayed response*. It is likely that localised areas will fail and become increasingly active as a result of, for example, high pore water pressures around ineffective drains, high groundwater levels or seawall failure. This activity may, in turn, promote the spread of instability onto adjacent slopes, often through the effects of loading or unloading.

It is expected that many of the changes could be sudden and dramatic, triggered during subsequent storm events after an initial breach has occurred. Marine erosion would rapidly lead to toe unloading of the slopes. It is expected that there would be a gradual deterioration over a number of years, followed by a sudden and dramatic loss of large areas of ground. Successive phases of instability after an initiating event are likely to correspond with wet years and high groundwater levels.

A plausible sequence of events (scenario) was developed for the protected cliffs, which could lead to renewed cliff top recession if the situation was allowed to deteriorate (i.e. do nothing). The scenario is based on an understanding of the causes and mechanisms of slope behaviour and involves a series of Stages (Figure 1):

- *Stage 1*; the current situation, with the slopes prone to localised small-scale failures;
- *Stage 2*; seawall failure, leading to the development of a breach. This is accompanied by the onset of wave attack at the cliff foot, with undercutting causing a decline in coastal slope stability.
- *Stage 3*; on-going cliff foot erosion causes the spread of instability across the entire cliff section, from base to crest, including the development of deep-seated landslides.
- *Stage 4*; on-going instability on the coastal slope leads to the development of first-time failures that remove part of the cliff top (i.e. cliff top recession re-commences);
- *Stage 5*; progressive cliff top recession, as on the unprotected cliffs to the west. It is assumed that once the cliff top starts to retreat it will behave in a similar fashion

to an unprotected cliffline i.e. assuming a relatively uniform retreat at a consistent average annual recession rate.

The risk assessment involved estimating both the levels of damages/losses that could result from the renewal of cliff top recession and the probability that these losses could occur in a particular year. The method comprises a combination of 2 key elements:

- *probabilistic* modelling of the likelihood of a renewal of cliff top recession (Stage 4) in any given year. This involves the use of *event trees* to estimate the *conditional probability* of Stage 4 occurring, given that Stages 2 and 3 have already taken place.

- *deterministic* modelling of the on-going cliff top recession process, at a uniform average annual recession rate along the whole frontage of the cliff section.

PROBABILITY OF THE RENEWAL OF CLIFF TOP RECESSION

Effort has been directed towards developing a reasonable model of the predicted inter-related sequences of events (Stages 1 - 4), specifically addressing the *uncertainty* over the timing of the initiating events and the subsequent responses and outcomes. Probabilistic methods are well suited to this problem and also provide the basis for the assessment of risk-based economic assessment (e.g. Hall et al 2000).

The structured use of expert judgement and subjective probability assessment using event trees can be a useful tool in addressing this type of problem. The event tree approach involves tracing the progression of the various combinations of scenario components using logic tree techniques to identify a range of possible outcomes. The individual probability of achieving a certain outcome is the product of the annual probability of the causal factor and the conditional probabilities of subsequent responses and outcomes. For example, suppose an initiating event (E) has a probability $P(E)$. Given that this event occurs, the failure mechanism M has the probability $P(M|E)$. Likewise, the outcome O has a conditional probability $P(O|M)$. The probability of this scenario, or chain of events, occurring is:

Scenario Probability = $P(E) \times P(M|E) \times P(O|M)$.

For each *cliff section* an event tree and associated estimates of scenario probabilities have been established as follows:

1. *development of event trees*; each sequence of initiating event-response-outcome was simplified to a simple event tree (Figure 2), with responses to a previous event either occurring or not occurring (i.e. yes/no options). A general indication of the range of damages/losses that might be expected to be associated with each response was separately identified.

Figure 2 Whitby protected cliffs: event tree based on recession scenario
Note: probabilities are estimated annual probabilities for Year 1

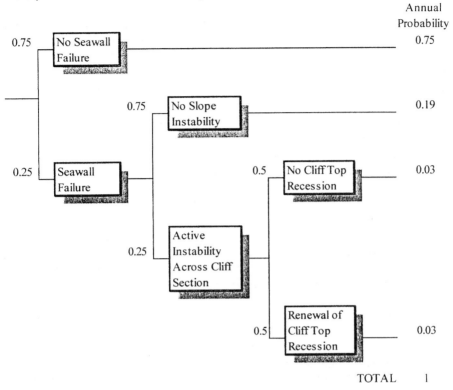

Annual
Probability

No Seawall Failure (0.75)	0.75
No Slope Instability (0.75)	0.19
No Cliff Top Recession (0.5)	0.03
Renewal of Cliff Top Recession (0.5)	0.03

TOTAL   1

2. *estimation of the annual probability of initiating events*; the probability of major structural failure of the seawall was assessed for a range of failure mechanisms, as part of a condition survey (visual inspection and review of damage records etc.). The *"do-nothing"* failure assumed that no further repair or maintenance of any kind is undertaken on the defences and that any defect or deterioration worsens until the point of failure. The estimate of when the "do nothing" failure may occur was made by the project maritime engineer, in terms of the period of time from now until there is a 95% chance that the defence will have failed. For example, at a one particular section of seawall, do nothing failure was judged by the engineer to be expected within the next 1-10 years. Assuming a normal distribution, failure in year 10 would equate to an annual probability of failure of 0.25. These initiating events were considered to be "one-off" events in the sense that they would only initiate the sequence of events defined in the event tree once.

3. *Annual probability of renewal of landslide activity across whole coastal slope (Stage 3);* on the basis of their knowledge of conditions on this coastline, the

project engineers used judgement to estimate that it would take up to 10 years for the whole slope to become actively unstable, given that a seawall **breach** had occurred. Assuming a normal distribution, this equates to an annual **probability** of failure of 0.25.

4. *Annual probability of renewal of cliff top recession (Stage 4);* on the basis of their knowledge of conditions on this coastline, the project engineers used judgement to estimate that it would take up to 2 years for the cliff top to start retreating, given that given that a seawall breach had occurred and that the whole slope had become actively unstable.

5. *Conditional probability of cliff top recession, given a seawall breach and an actively unstable slope (Stage 4)*; For Year 1 (the initiating event and response occur in year 1) the conditional probability associated with the key "branch" of the event tree was calculated as follows:

Scenario Prob. = P(Initiating Event) x P(Response 1) x P(Response 2)

Here, the Initiating Event is seawall failure, Response 1 is the whole slope becoming actively unstable and Response 2 is the onset of cliff top recession. If the initiating event and response occur in the same year, the calculation for subsequent years (Years 2 onwards) is essentially the same as the above, with the exception that the probability of a combination of scenario elements occurring in year 2 needs to take into account the possibility that the scenario actually occurred in year 1 and, hence, could not occur in year 2. However, a response might be *delayed* and occur in any year after an initiating event i.e. if the initiating event occurred in Year 1 the response could be in Year 1, Year 2 or any year up to Year 50 (the strategy lifetime). Thus, the combined probability of a response occurring in a particular year is a more complex problem. For example, the probability of the response occurring in Year 4 involves the combination of 4 possibilities: *P*(seawall failure in year 1 and the response 3 years later) + *P*(breach in year 2 and response 2 years later) + *P*(breach in year 3 and response 1 years later) + *P*(breach in year 4 and response 0 years later). For the probability of the response in Year 50 there would be 50 combinations of probabilities.

The analysis has involved the development of a sequence of related worksheets. Each worksheet comprises a 50 x 50 matrix of probabilities derived from multiplying P(Initiating event) by P(Response) for all possible combinations of timings. These provide the input data for the model on the conditional probability of Stage 4 (Annual Probability of Cliff Top Recession) occurring in a given year.

6. *Estimation of cliff top recession rate*; estimates of the anticipated average annual recession rate (i.e. Stage 5) were made, based on the historical (i.e. pre-

defences) rates and taking account of sea-level rise. An average annual recession rate of 0.625m/year was used.

MODELLING OF CLIFF TOP RECESSION

On-going cliff recession generates a *"benefit stream"* made up of a sequence of assets (e.g. property and utilities) lost at intervals determined by their distance from the cliff top and the recession rate. This is illustrated in column 1 (the left-hand column) of Figure 3. If cliff top recession commences in Year I (left-hand column), 2 houses will be lost in Year 3, 5 in Year 5 and 3 in Year 9. However, as the timing of seawall failure (Stage 2) and the expansion of instability across the whole cliff section (Stage 3) are not known, the onset of cliff top recession *could* occur in any year (or not at all). Thus, in Figure 3 if cliff top recession commences in Year II, 2 houses would be lost in Year 4, 5 in Year 6 and 3 in Year 10. Each of the columns in Figure 3 represent the *same benefit stream*, but with the renewal of cliff top recession occurring in a *different year*.

The analysis, therefore, models the probability of the renewal of cliff top recession occurring in a particular year *and* the Present Value (PV) of the losses associated with that benefit stream. With reference to Figure 3, for losses associated with the renewal of cliff top recession in Year I:

PV (Losses Year 1)   = Prob. (Stage 4) *x* (Discount Factor *x* Asset Value)
PV (Losses Year 1 -11)   = Prob. (Stage 4) *x* Σ (Discount Factor *x* Asset Value)

Figure 3 An illustration of the benefit streams associated with different timing of the renewal of cliff top recession

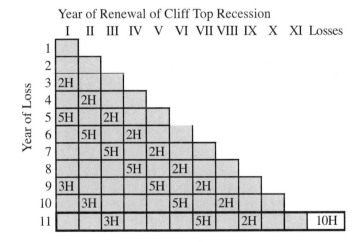

2H = 2 houses; 5H = 5 houses; 3H = 3 houses

A similar analysis can be done for each of the other *columns* (Years **II** − XI). However, the losses for a particular *row* (i.e. Year of Loss) could arise **from** 11 different combinations. For example, for Year 11 in Figure 3:

- Renewal of recession in Year I, property loss 11 years later = 0 houses;

- Renewal of recession in Year II, property loss 10 years later = 0 houses;

- Renewal of recession in Year III, property loss 9 years later = 3 houses.

And so on. Thus, all the entries for the Year 11 row in the above diagram need to be summed to produce overall PV losses for Year 11. Note that in this example the actual properties lost in Year 11 and its value would be different in each of the different combinations. In the above example, there is a possibility that either 0 houses, 2, 3 or 5 houses could be lost in Year 11 (reading across the row, depending on the timing of the renewal of recession).

The PV of losses associated with the renewal of recession over a 50-year period is the sum of the annual losses (Year 0-49).

DISCUSSION
A common problem for risk assessment on protected clifflines is that although we know that recession would follow defence failure, we do not know when i.e. there is *uncertainty*. Adopting a *probabilistic framework* for prediction can accommodate this uncertainty. Subjective probability (i.e. the approach used in this paper) is related to the *"degree of belief"* that an outcome will occur.

Expert judgement involves the use of experience, expertise and general principles to develop future cliff recession scenarios from the available historical record **and** past cliff behaviour, preferably in an explicit and consistent manner. Such judgements are usually subjective, but by proposing several possible scenarios followed by systematically testing and eliminating options it is possible to develop reliable estimates of the future cliff recession. In the example presented in this paper, the use of expert judgement offers a strategy for addressing the complex problem of defence failure and subsequent renewal of cliff recession.

REFERENCES
Clark A R and Guest S 1991. The Whitby cliff stabilisation and coast protection scheme. In Chandler R J (ed.) Slope stability engineering: 283-290. Thomas Telford.
Hall J W, Lee E M and Meadowcroft I C 2000 Risk-based assessment of coastal cliff recession. Proc. ICE: Water and Maritime Engineering, 142, 127-139.

ACKNOWLEDGEMENTS
The authors wish to thank Scarborough BC for their assistance and permission to present this paper.

Coastal Change, Climate and Instability -A European Union LIFE Environment project (1997-2000)

ROBIN M^cINNES, Isle of Wight Centre for the Coastal Environment, Isle of Wight Council, UK
JENNY JAKEWAYS, Isle of Wight Centre for the Coastal Environment, Isle of Wight Council, UK

ABSTRACT

A three year demonstration project entitled "Coastal Change, Climate and Instability" received financial support from the European Commission Directorate-General Environment in July 1997 through the LIFE Environment Programme (L'Instrument Financier pour l'Environnement). LIFE is a financial instrument which contributes towards the development and implementation of European Community environmental policy. Problems associated with deterioration of the Communities' environment, a failure to grasp sustainability issues in many sectors, the inadequacy of the legislatory strategy and the need to take responsibility for maintaining and enhancing the environment all demonstrate the need for action in this field.

The LIFE project "Coastal Change, Climate and Instability" endeavoured to encompass all these demands and considered the issues of coastal and climate change in terms of the utilisation of archaeological (palaeo-environmental) techniques to improve our understanding of the past and to help plan for the future. These studies were linked with geotechnical expertise of instability management in present and potentially unstable urban areas in coastal and mountainous locations.

Drawing upon practical experiences gained in these fields from 45 case-study areas in Great Britain, Ireland, France and Italy combined with the Swiss approach, the study aimed to provide practical assistance for those involved with both coastal zone and instability management and planning, with the objective of reducing adverse impacts on local communities and economies. This paper reviews the process undertaken through this particular study and highlights the conclusions and recommendations, with particular reference to the instability tasks.

BACKGROUND

The study has identified how instability in both coastal and mountainous regions presents significant threats to land use and development in Europe. The European

Union has a long and complex coastline, parts of which have recently been investigated in detail through thirty-five demonstration projects on the integrated management of coastal zones, as part of a concerted initiative by the European Commission's Directorate Generals Environment, Fisheries and Regional Policy. Significant lengths of the EU coastline in Great Britain, Italy, Ireland and France have been investigated as part of this LIFE study and collectively they provide a key component for the study of coastal instability within the European Union. Taking note of the fact that the population of coastal municipalities in these three Member States alone amounts to over 30 million people, there are development pressures within the coastal zone which includes significant lengths of unstable coastal cliffs and hinterland.

The LIFE study highlighted that over the last forty years major instability events have resulted in substantial loss of life and property, particularly in mountainous areas. In Italy alone there are an average of 78 deaths a year caused by hydrogeologically-related hazards. The study has identified that in a number of Member States it is difficult to quantify the costs associated with ground instability. The reason for this is that costs arising from all natural disasters (eg. floods, landslides, seismic activity and forest fires) are often combined. There is no doubt, however, that the costs associated with instability problems, both human and economic, are substantial and this is not fully appreciated in some Member States.

In coastal areas sustainable management will only be achieved if a thorough understanding exists of coastal evolution and natural processes; this is all the more important on account of the predicted impacts of climate change. This LIFE study has identified that mistakes in terms of coastal planning and management have been made in the past because of lack a of baseline information. This gap can be partially filled by the use of palaeo-environmental evidence, which allows recognition of the nature, scale and pace of coastal change over a much longer time-frame than normally considered.

Studies by the LIFE project partners in the Solent (the sea between the Isle of Wight and mainland England), Ireland and western France have revealed a rich, well-preserved heritage and palaeo-environmental archive in the coastal and intertidal zones which can be interrogated to assist shoreline management and coastal planning generally. The coastal zones and estuaries of Europe contain a rich archaeological resource which, if examined appropriately, can provide advice on coastal evolution and explain responses to past climatic change. Experiences from the past can help us to understand the possible impacts of climate change in the future. This valuable resource has long been under-estimated, which is of particular concern because as a result of climate change and sea-level rise some of these sedimentary archives will be less accessible or lost in the future. The study concluded that evidence of this kind has been underused in the past and can make a fundamental, accurate contribution to a future understanding of shoreline evolution.

Throughout this study the project partners have sought to illustrate the issues of coastal change, climate and instability by providing examples of experiences gained in the field. This has been achieved through the presentation of 45 case studies (23 geotechnical and 22 palaeo-environmental) drawing on locations from Italy, France, Ireland and Great Britain, together with the Swiss experience. Each study area is presented as both a two-page summary (highlighting key issues and lessons learnt) and as a fully detailed, illustrated description.

A particular challenge for the geotechnical project team and its specialist sub-consultants has been to show how policy can be translated into practice both in terms of landslide forecasting and risk assessments and with respect to the presentation of ideas to a wide-ranging, non-specialist audience. This has been achieved through the preparation of both a "Preferred Approach to Instability" which is aimed at those commissioning ground investigation works and a "Best Practice Guide" ('Managing ground instability in urban areas -A guide to best practice') which covers a broad range of landslide management topics.

The results of this project continue to be widely disseminated, assisted by the publication of the full report on CD-Rom, the preparation of two videos, a tri-lingual non-technical summary leaflet as well as the 'Best Practice Guide' and the 'Preferred Approach to Instability'. With respect to the geotechnical tasks this international conference on "Instability, Planning and Management" is an important aid to dissemination and a valuable forum for exchange of views.

LANDSLIDE RISK

This study has identified that the first step in understanding an instability problem in a particular region is often the gathering of existing historical data. This will assist in the establishment of a record of the number, distribution and nature of past instability events that have taken place. The inventory will ideally comprise both contemporary and relic landslide information and this will assist in investigating the relationships between the frequency and magnitude of instability events and the damage which is caused; this approach has been followed by the partners participating in this project.

For the first time, as part of this project, guidance has been provided to assist a range of interest groups involved with the management of instability in urban areas. The authors believe that the results of this study will assist those affected by instability problems in minimising costs. This will be achieved by:
- encouraging the development and refinement of meaningful landslide hazard evaluation practices for areas which lack the benefit of comprehensive geological and geomorphological baseline information;
- confirming the need for instability management to be incorporated as an element of development planning in those areas prone to instability;

- identifying the importance of increased awareness as to the range of options, both structural and otherwise, that can be adopted to ameliorate the problem;
- heightening the perception of development planners as to the nature, scale, distribution and causes of instability and significant and spatial variation of these hazards.

LEGAL AND ADMINISTRATIVE FRAMEWORKS

It should be stressed that problems have often arisen throughout Europe both inland and on the coast as a result of a lack of co-ordination between land use planning and development proposals in areas of instability. Many parts of the European Union suffer from an inheritance of unplanned communities and developments built, for example, on eroding cliff tops and other unsuitable locations. Similar spatial planning issues occur in mountainous regions.

At an early stage in this LIFE project the need for a proper understanding of the legal and administrative frameworks for addressing both instability problems and archaeological issues in each of the Member State study areas was recognised as a fundamental requirement. It was felt essential to provide an overview in order to allow the reader to understand the basis for decision-making. Differing approaches have been taken in terms of these frameworks in Great Britain, France, Italy and Switzerland which comprise a combination of legislation, advice and guidance. In addition there is the over-arching European legislation which applies a strategic perspective to these issues, particularly with respect to the concept of sustainable development, which is increasingly becoming the major influence of all aspects on planning policy.

Of particular interest is of course the success in transferring policy into practice. This was considered by each of the partners in their respective countries following the legislative and administrative review, with some of the key factors being concisely summarised in a European matrix table.

CLIMATE CHANGE

It is widely felt that instability problems across Europe are increasing. Reviews of landsliding and climate change in Europe reveal that in addition to the trend due to sea-level rise there is a variability closely related to short term climatic rhythms. The effective rainfall records for southern Britain also show an increasing availability of moisture and, therefore, soil moisture and water storage. The landslide record corresponds to this rhythm especially to the sequences of years wetter than the mean. It is, therefore, easy to speculate that the European coast will become increasingly susceptible to landsliding.

Climate change predictions have been made with reference to outputs from the Hadley Research Centre, which is a specialist unit established within the Meteorological Office in Great Britain to address climatic issues. A consistent time horizon for 2080 was selected for climate change predictions by the Centre to identify potential medium to long-term impacts: the medium-high level scenario (one of a set of four possible scenarios) must raise considerable concerns for those involved in instability management.

A particular problem facing those examining climate change impacts in the future is the fact that predictions of global warming and associated changes in the climate are likely to change as time passes. It is, therefore, very unlikely that the predictions of future climatic conditions made now, or in the next few years, will suffice for very long. In Great Britain, for example, predicted climate change scenarios include not only an increase in average summer temperature of more than 3^0C by the year 2080 but also a significant increase in winter rainfall, as well as sea-level rise and a predicted increase in storm events. Ibsen and Brunsden (1996) noted that changes in environmental controls such as climate and sea-level can lead to the acceleration or deceleration of the sequence of pre-failure - failure - reactivation.

Furthermore, Brunsden and Lee (2000) have suggested that ."the anticipated sea-level rise over the next decades, together with depleted beaches and a declining sediment supply from cliff recession, could result in significant changes in cliff behaviour which have no parallel over the historic time period".

ARCHAEOLOGY (PALAEO-ENVIRONMENTAL STUDIES) AND COASTAL CHANGE

Sustainable development in coastal areas will only be achieved if a thorough understanding exists of coastal evolution and natural processes; this is all the more important on account of the predicted impacts of climate change. This study has identified that mistakes in terms of coastal planning and management have been made in the past because of a lack of baseline information. Much coastal development has taken place since the mid-19th century -a relatively short time span. Coastal change including cliff recession has been identified largely based on a study of maps and has often not taken adequate account of coastal geomorphology and geo-archaeology. This gap can be partially filled by the use of palaeo-environmental evidence, which allows a consideration of coastal change over a much longer time-frame, up to several thousand years.

CONCLUSIONS AND RECOMMENDATIONS
LIFE Project Task One - Archaeology and Coastal Change
1.1 Archaeological and palaeo-environmental evidence has been commonly under-used and undervalued as a resource available to scientists and practitioners to assist in the understanding coastal and climate change.
1.2 The archaeological and palaeo-environmental record provides a calibrated timescale which should be used in association with geomorphological studies to ensure integrity in coastal management.

1.3 There is a need for wider European collaboration and the assembly and interpretation of geo-archaeological and palaeo-environmental evidence to assist in the illustration of global trends in coastal and climatic change.

1.4 A methodology has been devised as part of the LIFE study for assessing and prioritising the value of archaeological and palaeo-environmental sites in terms of their potential for assisting the study of coastal and climatic change. This approach should be disseminated to the European scientific community through publications and presentations.

1.5 Archaeological features on the surface of a landslide behavioural zone can assist in confirming the chronology of ground movements.

LIFE Project Tasks Two and Three - Geotechnical and Coastal Management

2.1 The impact of urban instability is wide-ranging and costly for EU Member States.

2.2 Sustainable development requires wise decision-making taking full account of past and present ground conditions.

2.3 There is insufficient accurate information on the costs of instability (both human and economic costs, including those associated with damage, monitoring, remedial works, maintenance etc.). A further study should examine how more accurate statistics can be recorded and costs calculated.

2.4 The geotechnical study areas have provided a broad spectrum of conditions and situations allowing a wide range of experiences to be highlighted and presented in the final report to the European Commission.

2.5 The research findings from this project should assist decision-makers in addressing the subject of instability at a local level. It was recognised that the subject of instability can be "politically sensitive" and the advantages of highlighting the geotechnical situation must be "sold" to decision-makers, emphasising the value derived from a positive approach to addressing instability assessment and management issues.

2.6 Instability management can be best achieved by means of a coordinated approach, thereby minimising risks to vulnerable communities by:
- guiding development away from unsuitable locations;
- ensuring that existing and future developments are not exposed to unacceptable risks;
- ensuring that development does not increase the risk for the rest of the community.

2.7 The implications of climate change and sea-level rise present a significant challenge to future instability management. Present and future policy decisions need to be placed within a long-term framework which provides a "vision" and a "direction" for policies to aim towards. An understanding of long-term coastal evolution will allow the identification of areas where management problems are likely to arise in the future.

2.8 Poor planning and siting of some forms of development in the past (both on the coast and inland) has made people, property and investment more vulnerable to natural hazards. An assessment of ground instability should be a fundamental requirement to assist the planning process.

2.9 There are concerns over the difficulties and costs of insuring property that may be at risk from sea-level rise, coastal erosion, instability or poorly planned coastal defence. The Law Barnier (France) should be reviewed by other Member States to assess its applicability.

2.10 The relevance of the issues of natural and man-made risks to the planning system is already recognised by some Member States in government guidance on planning policy. This guidance determines how much weight is attached to the issue in question. It is in the interests of insurance companies that the guidance is realistic and satisfactory in the way that it deals with risks.

2.11 The insurance industry would benefit from the establishment of procedures for reviewing draft Planning Policy Guidance issued by government departments and should respond where emerging guidance could mitigate insurance risks.

2.12 A study should be undertaken to investigate how existing policy guidance on insurance risks is influencing Structure Plans and Local Plans and to establish a framework for involvement on behalf of insurers in the development planning system.

ACKNOWLEDGEMENTS

The authors with to acknowledge the invaluable assistance of the other project partners and sub-contractors involved in the EU LIFE study 'Coastal Change, Climate and Instability': Isle of Wight Council Archaeology and Historic Environment Service; BRGM, Marseille; National Research Council, Perugia; The Discovery Programme, Dublin; Hampshire and Wight Trust for Maritime Archaeology, Southampton; University of Southampton, Université Bordeaux; Ms M-L Ibsen, Mr E M Lee, High-Point Rendel; Geoequipé; IQT Consulting.

REFERENCES

Brunsden D. 1993. Landslides and the International Decade for Natural Disaster Reduction: Do we have anything to offer? In Proceedings of the Royal Academy of Engineering Conference 'Landslide Hazard Mitigation'. London.

Brunsden D. and Lee E.M. 2000. Understanding the Behaviour of Coastal Landslide Systems: an Inter-disciplinary View. In E.N. Bromhead, N Dixon and M-L Ibsen (eds.) Landslides : In Research, Theory and Practice. Thomas Telford, London.

Brunsden D. and Ibsen M.L. 1994. The temporal occurrence and forecasting of landslides in the European Community. Report for the EU Epoch Program.

Clark A.R. Lee E.M. and Moore R. 1996. Landslide Investigation and Management in Great Britain: A Guide for Planners and Developers. Department of the Environment, UK, HMSO.

Dikau R., Brunsden D., Schrott L. and Ibsen M-L. 1996. Landslide Recognition, Report No. 1 of the European Commission Environment Programme. John Wiley & Sons, Chichester.

European Commission. 1997. LIFE Environment-Information Package 1997-1999. European Commission DGXI (XI.B.2), Brussels.

European Commission. 1992. The European Commission's Fifth Environmental Action Programme - Towards Sustainability. Brussels, Belgium. European Commission Directorate General Environment.

Glade T. and Crozier M. July 1999. Frequency and Magnitude of Landsliding: Fundamental Research Issues. Z Geomorph N F Suppl - bd/15pp 141-156 Stuttgart.

Ibsen M-L and Brunsden D. 1996. The nature, use and problems of historical archieves for the temporal occurrence of landslides, with specific reference to the south coast of Britain, Ventnor, Isle of Wight. Geomorphology, 15, pp 241-258.

Ibsen M-L. and Brunsden D. 1994. Mass movements and climate variation on the south coast of Great Britain. Report for the EU Epoch programme.

Jones D.K.C. and Lee E.M. 1994. Landsliding in Great Britain. Department of the Environment, UK.

Jones D.K.C. 1992. Landslide hazard assessment in the context of development. In Geohazards - natural and man-made. McCall et al (eds.) Pp 117-141. Chapman and Hall, London.

Jones M. 1999. Winds of Change, New Civil Engineer 28/1/99 London.

Lee E.M. 2000. Preparing for the impacts of climate change. Notes for SCOPAC study (unpublished).

Lee E.M. and Brunsden D. 2000. Coastal landslides of southern England: mechanisms and management. In E.N. Bromhead, N.Dixon and M-L Ibsen (eds), Landslides: In Research, Theory and Practice, Post Conference Tour.

Lee E.M., Moore R. and McInnes R.G. 1998. Assessment of the probability of landslide reactivation: Isle of Wight Undercliff, UK. 8th International IAEG Congress, Vancouver. Balkema, Rotterdam. Vol. 2, pp 1315-1321.

McInnes R.G. and Jakeways J. 2000. The development of Guidance and Best Practice for Urban Instability Management in Coastal and Mountainous Area of the European Union. In E.N. Bromhead, N.Dixon and M-L Ibsen (eds), Landslides: In Research, Theory and Practice, pp 1047-1052. Thomas Telford, London.

McInnes R.G. and Jakeways J. 1999. Isle of Wight LIFE Project: Coastal Change, Climate and Instability - an information leaflet. Isle of Wight Council, Newport, Isle of Wight.

Thompson A., Hine P.D., Poole J.S. and Grieg J.R. 1998. Environmental Geology in Land Use Planning. Report by Symonds Travers Morgan for the Department of the Environment, Transport and the Regions, UK.

Wade S., Hossell J. and Hough M. (eds.) 1999. The Impacts of Climate Change in the South-East: Technical Report, WS Atkins, Epsom, pps 100.

Archaeology in a drowned landscape

G MOMBER, Hampshire and Wight Trust for Maritime Archaeology, Southampton
Oceanography Centre, Southampton, SO14 3ZH, UK

INFORMATION BELOW THE WAVES

Submerged palaeo-landscapes are a diminishing and irreplaceable resource. They
represent an archive of information relating to previous glacial low sea levels, as
well as sea level-rise during the Holocene. Peat deposits and evidence of human
activity at known depths can reveal information about sea level and local
environments at that time. The formation processes of the deposits and spatial
relationships between them, can show rates of change (Long et al., 1999; Scaife,
1982). Archaeological investigation can indicate habitation responses to these
circumstances (Flemming, 1998) and their impact on the landscape.

The methods developed by the Hampshire and Wight Trust for Maritime
Archaeology (HWTMA) during the European Community L'Instrument Financière
de L'Environnement (LIFE) Project titled, *Coastal Change, Climate and Instability*
to survey, excavate, and collect samples underwater with high precision worked very
successfully. The procedures employed to interrogate the submerged archaeological
landscape could be applied to the investigation of similar underwater deposits to
help determine formation processes. This cost effective technique could be usefully
performed if high resolution detail is required from possible submerged land slide
material.

THE SUBMERGED LANDSCAPE OF BOULDNOR CLIFF

The geomorphological evolution of the coastline at Bouldnor Cliff in the western
Solent has resulted from climatic changes and sea level fluctuations. The cliff is
subject to instability that is witnessed by an active landslide complex. Yet, 500m
offshore, an area of relatively stable, drowned landscape reveals past periods of
accretion and sedimentation. Here, three layers of peat deposit have been overlain
with alluvium. This attests marine inundation which has been interrupted by
episodes of falling or static sea level, whereby allowing the shoreline to advance and
reclaim land from the sea. The archaeological material and palaeo-environmental
evidence found within these drowned deposits can provide a temporal framework for
climatic and environmental events. The site is a prime example of coastal adaptation
to climatic change, which, due to the presence of palaeo-environmental and
archaeological material can aid our interpretation of instability events within a
calculable timeframe.

The submerged forest was first identified in 1976 when local fishermen dredged up timbers and peat east of Yarmouth. Preliminary investigations by Drs D Tomalin and R Scaife recognised the significance of the find, then in 1985, during the Isle of Wight Maritime Heritage Project, the source of the material was traced to the foot of an underwater cliff by John Cross of Coastal Research, University of Southampton. Tree boles and roots were examined and active erosion of the site was noted. With the help of English Heritage, an absolute date was obtained at 6430–6120 calendar years BC (GU-5420). In 1991 this section of the Solent coast was identified as a key area to be investigated by the Hampshire and Wight Trust for Maritime Archaeology (HWTMA). In 1997 a site off Bouldnor Cliff was identified as a study area for a European LIFE Project to demonstrate the link between past environmental changes and coastal instability.

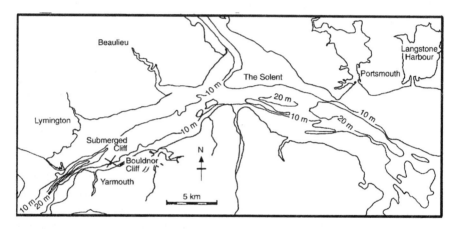

Figure 1. Local map with bathymetry showing location of submerged cliff

The date of the lowest or basal peat falls within the Mesolithic period when peoples were adapting to the changing climatic and environmental brought about by the conclusion of the Ice Age. Studies of Mesolithic settlement patterns in a culture that subsisted by fishing, hunting and gathering have shown a heavy reliance on the coast. This affinity with the maritime zone resulted in occupation sites at the edges of the advancing waters. The potential archaeological resource deposited in or above this landscape provides tangible evidence of the response by humans to sea level rise in the latter part of the Holocene. Investigation of this evidence will not only enhance our knowledge of past cultures, it may also help inform of coastal and climatic fluctuations, past human adaptations and aid interpretation of future changes.

THE RISING SEA LEVEL AND MESOLITHIC COASTAL OCCUPATION SITES

At the outset of the Mesolithic in Britain, the sea would have been many kilometres from the modern shore. About 6,000 years ago and the beginning of the Neolithic period, levels were just a few metres lower than today (Devoy, 1982; Long *et al.,* 1995; Scaife, 1980). As the water rose, lowland coastal regions in northern Europe suffered a progressive loss of land. Despite the reduction in land-mass, newly created islands and headlands would have resulted in an increased coastline. With its high levels of food productivity, this probably favoured an expansion of Early Mesolithic population (Coles, 1998; Simmons, 1996).

Studies in the Danish Storebaelt, ahead of the Fixed Link development, highlighted this phenomenon when numerous archaeological sites were discovered. These not only revealed a relatively dense population, it has also aided studies of sea level rise in the area during the Flandrian Transgression (Fischer, 1997). This was seen to peak about 5,200BC, succeeded by minor fluctuations resulting in an overall rise of under 2m in the last 7,000 years (Christensen *et al.,* 1997). Unfortunately, there are relatively few sites known around southern English shores for this period but the recent discoveries from below Bouldnor Cliff may provide material that could bridge the gaps in our knowledge.

INVESTIGATIONS OFF BOULDNOR CLIFF
Working in conjunction with Dr J Dix and Dr R Scaife, University of Southampton and lead by Dr D Tomalin, the key objectives within the LIFE project were to: investigate the submerged cliff stability and explore the implications of the Holocene transgression in the western Solent.

The HWTMA conducted dives off the Bouldnor coast to collect monolith and gouge core samples from the submerged landscape. A site was chosen for coring which was representative of the cliff section. Here, the cliff face consisted of silty clays interspersed with peat in an eroded vertical section. A seam of peat protruded from the top of the cliff and another lay a metre below it. The total height was over 10m, it beginning at 3.7 m below OD and dropping steeply before levelling off. At 8.9m below OD, the cliff is truncated by a vertical drop of over 2m. The base of the cliff lies at approximately 11.2–11.4m where the adjacent seabed is an expanse of peat and timber. The basal peat extends for about 15m to the north before it is terminated by a small cliff at 11.9-12m below ordnance datum. This then drops over another 1m into the main western Solent channel which is covered with cobble, oyster shells and sand. The lower interfaces of the three peat deposits from the shallowest to the deepest have subsequently been C14 dated to 4525 to 4330 cal BC (Beta 140102), 4920 to 4535 cal BC (Beta 140103) and 6615 to 6395 cal BC (Beta-140104) respectively.

METHODS EMPLOYED FOR CORING
The aim of coring was to gain samples from the whole cliff. Three methods of core sampling were employed. All methods were manual and necessitated a diver. The tools used were a 20mm auger or gouge core, a 40mm enclosed core with acetone liner and monolith sampler. The auger core was relatively quick and efficient so it was primarily used to identify the submerged stratigraphy while the 40mm enclosed

core and monolith were used to collect samples for radio carbon dating, diatom and pollen analysis.

Four datums were positioned running north south from the inter-tidal zone (DI), to the top of the underwater cliff (DA), the bottom of the clay cliff (DB) and to the north at the edge of the peat platform (DC). A 36m baseline was laid from DA via DB to DC. Gouge core samples were taken at intervals along the tape measure which were recovered to the surface for recording and sampling before being returned to the diver in the water. The depth at each core was recorded as was the total length of penetration. Distances between the cores were calculated to gave a continuous series of overlapping cores down the cliff.

Monolith cores were used to collect samples from peat/alluvium interfaces in the vertical sections of cliff. The cores used were open sided rectangular boxes. The box is pushed into the deposit and excavated to leave an undisturbed sample in the core. The core is then prepared for recovery underwater to be brought to the surface individually by divers.

Datum depths were calculated to a deci-metric accuracy by combining data from Aandera WLR7 depth meters and a RTK differential Global Positioning System (courtesy of New Forest District Council).

RESULTS PRODUCED BY CORING
Core samples were taken along the cliff to a maximum depth of 2m. The angle of cliff slope averaging about 30 degrees, however, made it possible to obtain representative samples from almost every depth. For the purpose of the Life Project Dr Rob Scaife, Dept. of Geography, University of Southampton, has carried out pollen and diatom analyses of the cores from this site. This work seeks to determine past vegetation environments, cliff formation processes and the progression of sea level rise (Scaife, 1999).

SURVEY OF THE PEAT PLATFORM BELOW THE SUBMERGED CLIFF
On completion of the coring, an area of the basal peat seabed was surveyed, recording the larger tree remains and any archaeological features. Underwater survey in this hostile environment presents many challenges. With each turn of the tide, water flows across the site at over 2 knots. Slack water lasts only a few minutes. Visibility averages about 1.5m but can be reduced to zero when the sediment load increases on the ebb tide. Visual difficulties are compounded by the dark peat seabed, which absorbs light. Consequently, many dives by many divers were necessary to complete the survey.

Five, 3m wide lanes in a 30m long strip running east west was searched. The results revealed tree trunks, boles and associated root systems reaching into the underlying substrate. The deposit lies between 11.1 and 11.9m below Ordnance Datum at the foot of a 2m vertical section of submerged cliff. Within the area surveyed, 13 tree stumps, extensive pieces of timber and 9 substantial boles were plotted. One such bole, orientated south to north, measured over 12 metres in length. At its southern end, where it emerged from below the cliff, its full diameter, including bark was intact. Degradation of the timber increased with distance from cliff. This demonstrated the preservation potential within the sediments before erosion, implying progressive erosion of the cliff.

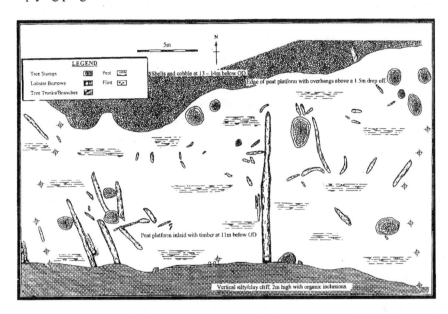

Figure 2. Plan of basal peat platform 11-12m below OD showing tree trunks and tree stumps dated to over 8,000 years old

On the northern edge of the platform, scour around tree stumps was found to be undercutting the peat creating cliffs up to 1.5m high with overhanging ledges. These were subject to failure and collapse. This eroding boundary was very well defined in the survey area. Consequently drift dives were conducted to quantify continuity. The dives traced the platform for over a kilometre east-west along an unbroken, laterally consistent deposit.

DISCOVERY OF LITHICS AND THE IDENTIFICATION OF A STRATIFIED DEPOSIT.

During the corridor search worked flints were discovered in the south west corner of the site at the bottom of the cliff. Here the flints had been excavated from the peat or underlying clay by lobsters. A lack of marine growth on the flints testified their recent exposure by these creatures.

In total, 50 flints showing signs of human industry were found in two discrete locations lying 5m apart at the entrance to lobster burrows. In addition to the surface sampling, two small 200mm deep cores were taken from the peat near the lobster burrows. Both cores contained small amounts of flint.

Figure 3. Worked flints (foreground) excavated by a lobster while burrow building beneath an 8,000 year old fallen oak tree in 11m of water

Following the discovery, a project was organised in May 2000. The principle aim was to confirm the existence of stratified Mesolithic material located at the foot of the cliff on the peat platform by excavating a small test trench..

As outlined above, the western Solent is not the easiest area to run a diving operation. To maximise productivity, the project was run from Flat Holm, Coastline Surveys' 23m vessel which was moored over the site. The ship was home to eight diving archaeologists, accommodating additional experts and volunteers on a daily basis.

The excavation was conducted using surface supply diving equipment (SSDE) with air lifts, trowels and purpose made coring boxes in an area delineated by a stainless steel grid, built specifically for the project by Analytical Engineering. All the work was recorded on a head mounted Seahawk camera system provided by Kongsberg Simrad, linking the dive supervisor and archaeological director with the excavator.
When sections of the seabed were collected in the core-boxes they were recovered to the surface for examination. The artefacts within them were then excavated and recorded on deck. The overwhelming majority of lithics were located in a dark sandy/silt context below the basal peat deposit. Analysis of their distribution revealed an increase in concentration towards the eastern end of the excavated trench, which suggests a centre for the site. To date over 300 humanly worked or burnt pieces of flint from the early to mid Mesolithic have been identified.

In addition to the archaeological and palaeoenvironmental studies, 8 samples of timber were sawn from trees within the peat for dendrochronological analysis by Nigel Nayling of the University of Wales, Lampeter. The samples were of excellent quality, providing a 280 year sequence over 8,000 years old.

Following the excavation, further monitoring was conducted towards the end of the summer when the site was revisited and the rate of erosion in the surrounding peat deposit was recorded. This was achieved by measuring calibrated rods placed across the site 10 months earlier. From the results it was clear that material was indeed being lost. This work is to be continued over the years ahead.

The methods employed during the LIFE project proved very successful and enabled the main outcomes to be realised by demonstrating that worked lithics were stratified within the seabed. The project is unique in the UK, and aims to demonstrate the archaeological potential within the prehistoric coastal zone during the Flandrian transgression.

CONCLUSION
The discovery of the submerged archaeological site off Bouldnor Cliff and the possibility that there may be many more presents a very positive outlook for future sea level studies in England. Unfortunately, in the past, the potential of Britains drowned landscape does not seem to have been fully recognised and consequently safeguards amongst regulating and sectoral bodies have been minimal (Tomalin, 1997). Many sites below the mean low water mark are undoubtedly being lost as a result of ignorance, neglect and progressing human impacts. Fortunately, this is an issue that is currently being addressed as both archaeologists and coastal managers realise that many gaps in our understanding of past cultures and coastal change could be answered with the investigation of well-preserved sites underwater.

As a result of this current LIFE programme coastal study groups in Europe are reviewing similar issues. These question; *'whether; present proposals for coastal management and protection can demonstrate wisdom drawn from a long term understanding of past natural and human events on the relevant coastline.'* They also question whether *'new field information will be required on past natural and human events, before such a question can be answered'.*

In the future the task should be twofold: firstly, to locate areas where submerged habitation sites are likely to have been located in England, identify the potential for preservation within such sites, then find them. And secondly, apply legislation to protect against needless destruction of the maritime archive and ensure adequate archaeological and palaeo-environmental recording where destructive practices are proceeding.

REFERENCES

Christensen, C. Fischer, A & Mathiassen, D. R. 1997, The great sea rise of the Storebaelt. In L Pedersen, A Fischer & B Aaby (Eds), *The Danish Storebaelt since the Ice Age - man sea and Forest*. 45 - 54. The Storebaelt Publications.

Coles, B. J., 1998, Doggerland: a Speculative Survey. *Proceedings of the Prehistoric Society*. 64: 45-81.

Devoy, R. J., 1982, *Analysis of the geological evidence for Holocene sea level change in south east England*. Proceedings of the Geological Association. 93: 65-90.

Fischer, A., 1997, People and the sea - settlement and fishing along the mesolithic coasts. L Pedersen, A Fischer & B Aaby (Eds), *The Danish Storebaelt since the Ice Age - man sea and forest*. 63 - 77. The Storebaelt Publications.

Flemming, N., 1998, Archaeological evidence for vertical movement on the continental shelf during the Paleolithic, Neolithic and Bronze Age Periods. In I. S

Long, A.J. & Tooley, M.J., 1995, Holocene sea-level and crustal movements in Hampshire and Southeast England, United Kingdom. In Jr Frinkl (Ed) Holocene Cycles: Climate, Sea Levels and Sedimentation. *Journal of Coastal Research*, Special Issue 17: 299-210.

Long, A. J., Scaife, R. G., & Edwards, R. J., 1999, Pine pollen in the intertidal sediments from Poole Harbour, UK; implications for late Holocene sediment accretion rates and sea-level rise. *Quarternary International* 55: 3-16.

Scaife, R. G., 1980, *Late-Devensian and Flandrian palaeoecological studies in the Isle of Wight*. Unpublished Ph D thesis. University of London, King's College.

Scaife, R. G., 1982, Late Devensian and early Flandrian vegetation changes in southern England. In S Limbrey and M Bell (eds), *Archaeological aspects of woodland ecology*. B.A.R. (Int. Ser.) 146: 57-74.

Scaife, R. G., 1999, Pollen analysis of the Bouldnor Submerged Peats (3): The Basal Peat Bed. Unpublished report for EC Life Project *Coastal Change, Climate and Instability* Isle of Wight Centre for the Coastal Environment.

Simmons, I., 1996, *The Environmental Impact of Later Mesolithic Cultures*. Edinburgh University Press, for the University of Durham.

Tomalin, D.J., 1997, Bargaining with nature; considering the sustainability of archaeological sites in the dynamic environment of the intertidal zone. *Preserving archaeological remains in situ: Proceedings of the conference of 1st – 3rd April 1996 at the Museum of London*. Museum of London and the University of Bradford, 144-158.

Cliff behaviour assessment and management strategy, Filey Bay, Yorkshire, UK

Dr R MOORE, P BARTER and K McCONNELL, Halcrow Group Ltd, Swindon, UK, and
J. RIBY and C. MATTHEWS, Scarborough Borough Council, Scarborough, UK

INTRODUCTION

Filey Bay is a predominantly natural undeveloped coastline (Figure 1), comprising cliffs of glacial till and bedrock of Jurassic age. Where bedrock outcrops above HWM, 2-tier or composite cliff morphology has developed due to the more resistant bedrock. Cliffs formed entirely of glacial till are prone to deep-seated landsliding.

Development is situated at Filey, Amtree Park, Hunmanby and Reighton, which is at risk from cliff instability and cliff-top recession. Coastal defences are limited to a high sea wall along the frontage at Filey, and local toe protection measures elsewhere that generally do not provide much protection to the cliffs from erosion. An important issue for the 'Coastal Defence Strategy for Filey Bay' is the need to account for potential cliff instability and recession in future planning and decision-making. Equally important is the need to assess the contribution of cliff erosion in the supply of sediment to the extensive sandy beach, which is a major feature and asset of Filey Bay. There are tourism, commercial and social pressures along this environmentally sensitive coastline. Any intervention in Filey Bay therefore needs to be informed and prioritised to avoid or minimise any potential adverse impacts.

Investigation of cliff instability requires an understanding of the coastal processes, cliff behaviour conditions and impacts of development located on marginally stable coastal slopes. The management response might involve planning and development controls, combined coastal protection and landslide stabilisation (drainage and other methods) and monitoring. The paper develops a 'cliff behaviour' approach for the investigation and management of cliff instability and cliff top recession hazards.

APPROACH

Geomorphological investigations carried out in Filey Bay have comprised a review of information, field observation and mapping, compilation of a database and reporting. These activities are described below.

Figure 1 – Filey Bay

Information Sources

Various information sources were reviewed as part of the geomorphological investigations. They provide useful background information of a geological and geotechnical nature at a 'broad' strategic scale and for specific sites. The shoreline management plan[1] and coastal planning and management[2] studies for the area consider the nature of cliff recession and coastal processes in broad terms. Various site investigation reports provide data on local ground conditions and published coastal erosion data for Yorkshire give an indication of historical rates of change[3].

Geomorphological Mapping and Cliff Behaviour Assessment

A geomorphological survey of the coastal cliffs at Filey Bay was carried out in September and November 2000. The survey extends about 10 km from Filey Brigg in the north, to the south of Reighton.

The geomorphological survey comprised observation and mapping of cliff morphology, landslides, geology, materials, current cliff activity and recession potential. The field observations and measurements were supplemented by additional information, most notably scaled measurements of distance from the base maps, interpretation of colour vertical aerial photography, oblique aerial video, and the available geological and geotechnical records.

Using the field observations and supporting information, a geomorphological interpretation of cliff instability mechanisms and processes was made (hereafter termed Cliff Behaviour Assessment). The approach provides an important spatial framework and vital clues as to the likely mechanisms, causes and consequences of cliff instability.

Database and Reporting

The outputs of the geomorphological investigation comprised a series of drawings, a database and an interpretive report. The drawings provide a summary of the main observations from the geomorphological survey and the spatial distribution of 'Cliff Behaviour Units'. Each cliff behaviour unit is coded and cross-referenced to the database and photographic record. Derivative drawings presenting the 'Cliff Recession Potential' and 'Planning Guidance' were also produced as part of the strategy study.

The database presents semi-quantitative estimates of cliff recession potential, sediment storage and supply from cliff erosion. It is recognised there are many uncertainties in estimating these parameters, as described further below. Accounting for such uncertainties, the database includes upper and lower bound estimates for these parameters, which represent credible worst-case (i.e. high erosion) and best-case (low erosion) scenarios.

CLIFF BEHAVIOUR ASSESSMENT

Cliff Behaviour Units

To understand cliff recession something must be known of the conditions and processes operating on the foreshore and cliff (and, in many cases, behind the cliff). For this reason, the concept of a 'cliff behaviour unit' (CBU) was developed for DEFRA[4], as it provides an important framework for cliff management. The study identified a range of CBU types that reflect different mechanisms and rates of sediment inputs, throughputs and outputs (Figure 2).

Geology

The solid geology of Filey Bay comprises Jurassic rocks north of Reighton and Cretaceous rocks to the south. From north to south the sequence is as follows:
- Filey Brigg: Corallian grits and limestones
- Filey Brigg to Hunmanby: Kimmeridge Clay
- Hunmanby to Speeton: Lower Greensand and Speeton Clay
- Speeton to Flamborough Head: Upper Cretaceous Chalk

Glacial till caps the exposed Jurassic and Cretaceous rocks in the northern and southern extremities of the bay and where the solid strata dip below sea level, the cliffs are formed entirely of glacial till. The glacial till deposits comprise a highly variable mixture of clays, silts, sands and gravels. They are easily eroded by wave action, and are highly susceptible to groundwater effects and mass movement.

Other superficial deposits observed during the geomorphological survey include extensive Chalk talus which separate the near-vertical Chalk cliffs and the Lower Greensand deposits at Speeton. The Chalk talus is currently being eroded and mobilised by wave action, with Chalk debris being deposited on the beaches where it is rapidly broken down into rounded Chalk pebbles.

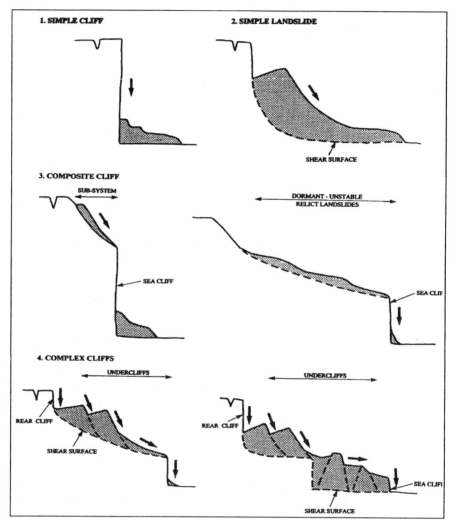

Figure 2 – Coastal behaviour unit types (after High-Point Rendel[4])

Sediment Storage on Cliffs

An estimate of the volume of sediment stored within each CBU has been calculated based on measured and estimated parameters. These are described below with the numerical formula for estimating the volume of sediment stored.

Cliff Morphology

Field measurements of cliff gradient were obtained for each CBU using a hand-held compass clinometer and by sighting from the cliff toe to cliff top, or vice versa. For composite CBU's (i.e. Filey Brigg), the cliff gradient was measured from the crest of the lower rock cliff to the cliff top. The height of the lower rock cliff was estimated from field observation. The plan length (in section) and width (longshore) of each CBU was scaled from the base maps. These dimensions and the cliff gradient were used to estimate the cliff height.

Depth of Cliff Failure

The depth of cliff failure was estimated from field observation. For simple and composite cliffs, depths of failed sediments were typically shallow ranging between 0.1m and 3.5m. For simple landslides, depths of landslide deposits ranged between 1m (i.e. for mudslides) and 6m (i.e. for rotational slips) and for complex landslides depths of 6m and 18m were estimated. Estimates were made from direct observation of exposed debris or from an appreciation of the 3D geometry of the CBU. To account for uncertainty with this parameter, two estimates of the most credible minimum and maximum depth of cliff failure were recorded.

Sediment Storage Estimation

A numerical estimation of the volume of sediment stored on the cliffs was calculated as follows:

Sediment storage = slope length * width * failure depth * 3D correction

where, 3D correction accounts for subsurface geometrical edge effects.

In the assessment of cliff storage for CBU's, account is made of potential 3D effects in subsurface geometry. For shallow planar mechanisms of cliff failure (i.e. simple and composite cliffs), subsurface edge effects are minimal and a small reduction in volume (10% upper bound; 20% lower bound) has been applied. For simple and complex landslides, which may comprise deep-seated rotational failure mechanisms, 3D subsurface edge effects can be significant and a large reduction in volume (30% upper bound; 50% lower bound) has been applied.

Cliff Recession Potential

The recession potential of each CBU was assessed from field observation and supporting information, and considers the current activity of the cliffs, the recession potential (i.e. cliff top retreat) and estimated frequency of occurrence.

Cliff Activity

Cliff activity was classified from field evidence of active landslides or erosion. A distinction has been made between simple and composite cliffs subject to surface erosion processes, and simple and complex landslides subject to deep-seated ground movements. Vegetation density (i.e. % cover) was used as an indicator of activity for cliffs subject to surface erosion processes, whereas evidence of relic or active rotational and differential shear movements and toe heave were used for the latter. In this way, the activity for each CBU was rated according to the following classification:

Activity Status	Activity %
Dormant (defended shoreline)	0
Inactive	25
Marginally Stable	50
Active	75
Very Active	100

Recession Potential
Cliff recession potential was estimated from historical records and field observation. As for cliff activity, a distinction has been made between cliffs and landslides, as the magnitude and frequency of recession events are dependent on the mechanism of cliff failure. For example, simple cliffs are generally in dynamic equilibrium, with the rate of erosion at the cliff toe in balance with the rate of retreat at the cliff top, with only minimal time-lag response. Landslides, on the other hand are rarely at equilibrium, as the presence of landslide blocks or debris storage on cliffs provides a temporary buffer (or natural protection) against the de-stabilising effects of toe erosion. Cliff top instability generally resumes once a significant portion of debris has been removed through toe erosion. For large-scale landslides this cyclical response can take many years, decades or even centuries. It should be noted, however, that coastal landslides may be subject to occasional ground movement throughout this cycle in response to groundwater and erosion at the toe of the cliff.

Based on historical records and field observation, the recession potential for each CBU has been rated according to the following classification.

Cliff-Top Recession Event	Upper Bound	Lower Bound
Low erosion	0.5m	0.1m
Moderate erosion	1m	0.5m
High erosion	2m	1m
Landslip (small <0.2ha)	20m	10m
Landslip (moderate <1ha)	50m	20m
Landslip (large >1ha)	100m	50m

Frequency of Recession Events
The frequency of cliff top recession events ranges from annual losses from ongoing erosion to episodic losses due to landslides. There are few records from which reliable estimates of landslide frequency on the cliffs at Filey Bay can be made. Field observation and anecdotal reports of recent large-scale landslides at Reighton provides some evidence that such events may not be that rare. During the period of

geomorphological survey (i.e. a period of weeks during a wet period), three mudslide surges onto the beach were observed.

Given the uncertainties with this parameter, it has been assumed that the recession of cliffs due to erosion is realised on an annual basis, whilst recession caused by episodic landslide events of various size will be realised over a 50 year period, except where coastal defences are in place.

CLIFF SEDIMENT INPUT TO BEACHES
The potential supply of sediment to the beaches from each CBU was based on the estimated cliff sediment storage, the magnitude and frequency of erosion and landslide events, cliff activity and estimated sediment grading of the various soil and rock types.

Cliff Sediment Loss Estimation
Numerical estimation of the sediment supply (annual erosion and episodic landslides) from cliffs was calculated as follows:

$$Cliff\ loss = Storage * Recession\ potential/slope\ length * Activity$$

For the composite cliffs at Filey Brigg, account is taken of the potential sediment loss from the erosion of the lower rock cliff.

Cliff Sediment Grading
Not all sediment eroded from the cliffs provides material suitable for retention on the beaches at Filey Bay. The glacial till deposits provide the main source of material, other than localised outcrops of Jurassic sandy limestones at Filey Brigg, and the argillaceous Kimmeridge and Speeton Clays, Lower Greensand and Chalk talus deposits south of Reighton. From observation of materials exposed in each CBU, an estimate of the proportion of the coarse, medium and fine sediments was recorded. It is noted that considerable variability in the composition of the glacial till exposed in Filey Bay is apparent from visual observation.

A small number of samples of sediments from the cliffs were taken and particle size analysis of these has helped reduce uncertainty with the field estimates. The results for the glacial till cliffs indicate an average coarse/medium/fine distribution of 24%, 34% and 42%, respectively.

Effective Supply of Sediment to Beaches
An estimate of the volume of sediment likely to be retained as beach material has been calculated as follows.

$$Input\ to\ Beach = Annual\ or\ Episodic\ Cliff\ loss * (Coarse\ \% + Medium\%)$$

It is assumed that all coarse and medium sediment gradings are retained in the beach and that only fine sediment is lost to the sea. The total average annual sediment supply was estimated as the sum of the annual erosion inputs, plus the episodic inputs divided by their estimated frequency (i.e. 50 years). The results are summarised below for various key locations within the bay.

All values 000 m³	Annual Cliff Erosion Inputs	50 Year Episodic Landslide Inputs	Average Annual Sediment Supply
Filey Bay north	2.74 *(0.74)*	13.42 *(1.91)*	3.01 *(0.78)*
Primrose Valley	0.12 *(0.03)*	459.72 *(72.1)*	9.32 *(1.47)*
Hunmanby	0.54 *(0.12)*	261.87 *(38.76)*	5.77 *(0.9)*
Reighton	2.2 *(0.47)*	143.95 *(28.51)*	5.08 *(1.04)*

Note: Lower bound estimates in italics

CONCLUSIONS

The cliff behaviour assessment provides the first detailed systematic evaluation of cliff instability and recession in Filey Bay. The approach combines factual data with 'best judgement' (i.e. interpretation of landslide mechanisms and depth of cliff failure) to derive semi-quantitative estimates of cliff recession potential and sediment supply to the beaches. These estimates should be regarded as provisional pending further detailed investigation and ongoing monitoring which should be used to improve and validate the findings. The initial results do, however, illustrate the important contribution of sediments from episodic landslide events. Any intervention of the episodic inputs would significantly harm the natural cliff recession process and sediment budget for Filey Bay.

ACKNOWLEDGEMENTS

The authors would like to acknowledge the kind assistance and permission of their employers, Scarborough Borough Council and Halcrow Group Ltd in the production of this paper.

REFERENCES

[1] **Mouchel Consulting Ltd (1997)** *Huntcliffe to Flamborough Head (Sub-cell 1D) Shoreline Management Plan.* Final report to DEFRA.

[2] **High-Point Rendel (1995)** *Coastal Planning and Management: Applied Earth Science Mapping – Filey to Scarborough, North Yorkshire.* Final report to the Department of the Environment.

[3] **Halcrow Group Ltd (2001)** *Filey Bay Coastal Defence Strategy.* Draft report to Scarborough Borough Council.

[4] **High-Point Rendel (In Press)** *Soft Rock Cliffs: Prediction of Recession Rates and Erosion Control Techniques.* Final Report in preparation.

Lithological and structural controls on style and rate of coastal slope failure: implications for future planning and remedial engineering measures, SW coast, Isle of Wight

DEREK RUST AND JANE GARDENER, Department of Geography and Earth Sciences, Brunel University, Uxbridge, Middx., UB8 3PH, UK

INTRODUCTION
Plans are proceeding for major engineering works to extend the life, by some 50 years, of a scenic coastal road above unstable chalk cliffs near Freshwater on the south-west coast of the Isle of Wight. These works, motivated by the importance of the road to the Island's tourism economy, draw attention to an adjacent 3-km stretch of the road running near coastal cliffs bordering Compton and Brook bays which are developed in mudstones and sandstones that underlie the chalk. In two places the road is less than 20 m from the top of these cliffs yet, because of its enormous environmental and landscape value, the local Shoreline Management Plan designates this as an area where natural processes will be allowed to continue uninterrupted. This highlights the need to understand such processes in these cliffs and to forecast how they will affect cliff stability and the coastal road, particularly over the next 50 years.

In order to approach the above understanding our study employs three interlinked strands: Firstly, estimation of the amount, rate and pattern of coastal retreat using maps and ground surveys. Second, logging of the lithologies exposed in the cliffs, noting relationships with different styles of slope failure and, finally, geochemical analysis of the rock units involved in these different styles. These strands will be considered in turn.

MAP AND GROUND SURVEYS
The area is covered by five re-surveys, spanning the period 1862 – 1976, of very large (1:1250) scale Ordnance Survey maps. These, supplemented by 19 direct tape and compass ground survey measurements from known points, provide an unusually long and accurate record of coastal change. Figure 1 depicts the results of this work and shows that, particularly in Brook Bay east of Hanover Point, the coastal cliff has retreated essentially parallel to itself, although at average rates varying from about 0.3 m a^{-1} to almost 0.6 m a^{-1}. At Hanover Point the coastline swings to a more northerly orientation in Compton Bay and here retreat is guided

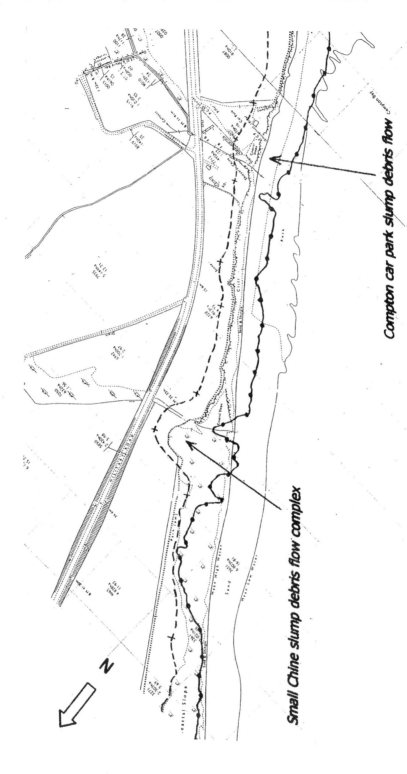

Figure 1. Adjoining reduced copies of the 1976 1:1250 scale Ordnance Survey (OS) maps showing the scenic coastal road (Military Road) where it is most threatened by coastal retreat in Brook (top) and Compton (bottom) bays. Also shown superimposed are the cliff line from the 1862 1:1250 scale OS maps and the present (2001) cliff line based on field survey data.

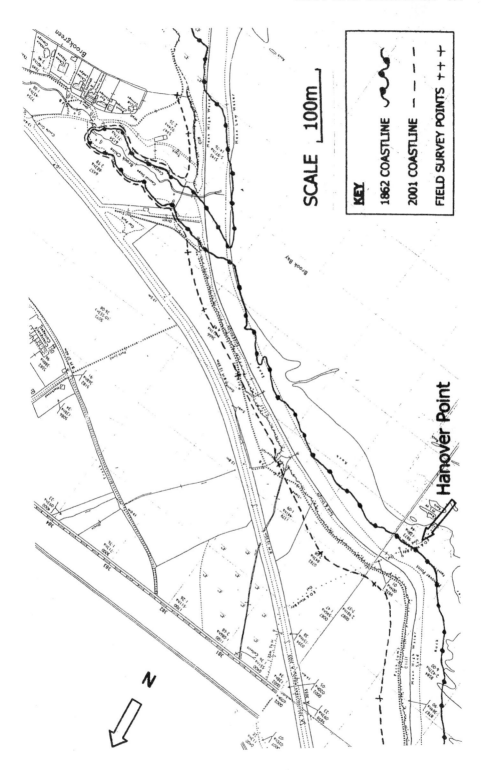

KEY

1862 COASTLINE

2001 COASTLINE

FIELD SURVEY POINTS   + +

SCALE   100m

Brook Bay

Hanover Point

N

in such a way that features move obliquely along the coastline, rather than in a simple orthogonal fashion. This is illustrated in particular by two prominent slump debris-flows, one at the car park at Compton and the other at a locality sometimes referred to as Small Chine, which have moved eastwards along the coast as retreat progresses (Fig. 1). Moreover, retreat amounts here vary from essentially nil to more than 120 m, representing rates approaching 0.9 m a^{-1}.

LITHOLOGIES AND STYLE OF CLIFF FAILURE

The mudstones and sandstones making up the cliffs lie within the Wessex and overlying Vectis formations, part of the Wealden Group, and are folded within an approximately E-W trending anticline just offshore of Hanover Point which plunges eastwards at a shallow angle. This structural arrangement means that the oldest rocks crop out in the vicinity of Hanover Point and it was here that lithological logging was commenced. It also means that lithologies remain essentially uniform eastwards to the prominent inlet of Brook Chine (Fig. 1) because the coastline is nearly parallel to the anticlinal axis and stratigraphically deeper rocks remain unexposed. Emphasis in the logging was placed on the mechanical properties of the sequence and their relationship to differing modes of slope failure; recent palaeoenvironmental and palaeoecological work in the area can be found in Radley and Barker (1998) and references therein.

Our logging (Fig. 2) shows that approximately the lower 50 m of section is dominated by a distinctive purple mottled mudstone that exhibits essentially non-plastic behaviour even where saturated in an intertidal position. Higher in the cliffs it characteristically displays a pattern of fine blocky fractures, typically on a scale of several centimetres. This unit, unexpectedly for a mudstone, is relatively resistant to marine erosion and accounts for the generally steep cliffs between Brook Chine and Compton car park (Fig. 1). North of Hanover Point, as section is gained, a number of mature fine-medium grained quartz sandstone interbeds occur within the sequence (Fig. 2). These are generally more resistant and contribute to stabilising the foot of the cliffs in this part of the coast which is more exposed to the dominant SW swell direction. Overall, this lowest part of the succession is characterised by relatively small slump failures, the broken debris from which is rapidly removed by wave action.

This pattern ends abruptly at Compton car park overlooking Shippards Chine. Here the car park is being actively destroyed as the head of a prominent slump debris-flow advances eastwards producing an asymmetric embayment in the cliff top (Fig. 1). This slope failure can be directly attributed to the occurrence of a distinctive pale grey plastic mudstone, no more than about a metre in thickness, within the sequence (Fig. 2). This unit is highly plastic throughout, with a shear vane test instrument used in the field giving a very low reading. By comparison, a similar test on a saturated part of the mottled mudstone produced a reading beyond the range of the instrument as the material eventually failed in a predominantly brittle fashion.

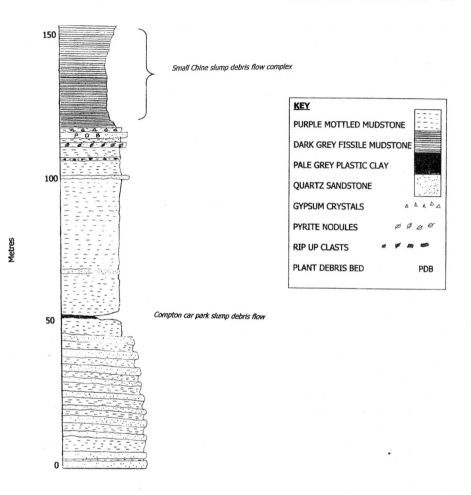

Figure 2. Lithological log of the rock sequence exposed in Brook and Compton bays. The oldest part of the sequence, at the base of the log, occurs at Hanover Point and then continues for approximately 150 m up-section northwards along the coast before being truncated by a fault forming the northern margin of the Small Chine slump debris-flow complex. See also Figure 1.

Above the pale grey plastic mudstone the succession reverts to the purple mottled mudstone, with its generally steep cliffs and small incremental slump failures. After a further approximately 50 m of section this gives way to a distinctive succession of medium-grained mature quartz sandstone beds with interbedded dark grey laminated mudstones. These mudstones, which are characteristically fissile, then dominate over the final 35 m of the logged sequence and, where they reach beach level unbuttressed by the quartz sands, give rise to a large slump debris-flow complex in the Small Chine area (Figs. 1 and 2). The northern margin of this active complex is defined by a fault striking approximately E-W. Downthrow on the south side of this fault juxtaposes the dark grey fissile mudstones with a repeated part of the mottled mudstone sequence, causing an abrupt change in style and rate of cliff failure (Figs. 1 and 3).

GEOCHEMICAL ANALYSES

Geochemical results for the three mudstones are displayed in Table 1. The purple mottled mudstone has a relatively low clay content and commensurately high quartz content, presumably as silt sized detrital grains. X-ray diffraction (XRD) peaks indicate that the clays are a mixture of illites and smectites, possibly in an approximate 50:50 ratio, although the broadness of the peaks suggests that a range of compositions may be present. In contrast the grey plastic mudstone is very clay rich and dominated by smectite. The dark fissile mudstone is intermediate in clay content, while the XRD indicates a mixture of smectite and vermiculite although their respective proportions could not be established. In other respects the analyses for each mudstone were relatively similar (Tab. 1). Loss on ignition (LOI) analyses were also carried out, indicating that the grey plastic mudstone had the highest organic content and the mottled mudstone the lowest. The fissile mudstone again exhibited an intermediate value.

DISCUSSION AND CONCLUSIONS

The results presented above demonstrate a very close relationship between bedrock lithologies, particularly within the three mudstone units, and the style and rate of cliff retreat. The purple mottled mudstone has a relatively low clay content and is generally poor in swelling smectite clays, coinciding with the field observations of non-plastic behaviour. It may be that the relatively high quartz content and distinctive patterns of fine fractures impart a degree of permeability to the unit that aids in limiting pore-pressures. These results and suggestions are consistent with the small incremental slump failures characteristic of the unit and, together with the typically steep cliffs associated with this bedrock type, indicate that the cliff edge may be allowed to closely approach the scenic road before engineering measures are required. However, at its closest approach in Brook Bay (Fig. 1) the cliff top at present is only about 18 m from the road. This particular part of the coastline has retreated historically at an average rate of about 0.3 m a^{-1}; if these rates are maintained in 50 years the road will be intolerably close to the cliff, particularly for heavy vehicles. It may be that single-lane traffic regulation could

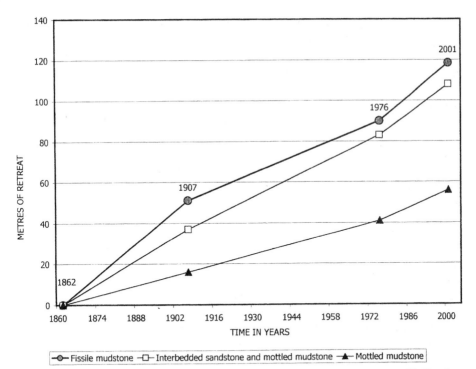

Figure 3. Plot of representative cliff retreat rates for the dominant lithologies exposed in Brook and Compton bays. The dark grey fissile mudstone is associated with the Small Chine slump debris-flow complex. The interbedded sandstone and mottled mudstone controls cliff retreat most notably north from Hanover Point to Compton car park and the mottled mudstone, with the lowest average retreat rate, forms the cliffs north of Hanover Point in Brook Bay. See also Figures 1 and 2. Data taken from 1:1250 scale OS maps surveyed in 1862, 1907 and 1976, together with field measurements made in 2001.

	PURPLE MOTTLED MUDSTONE	GREY PLASTIC MUDSTONE	DARK FISSILE MUDSTONE
BULK ROCK ANALYSIS			
Total clay	32	72	59
Quartz	60	22	35
K-Feldspar	3	2	2
Plagioclase feldspar	3	2	2
CLAY MINERALOGY			
Smectite	-	53	-
Smectite + Vermiculite	-	-	57
Illite - Smectite	53	-	-
Illite	21	16	17
Kaolinite + Chlorite	26	31	26

Table 1. Geochemical analyses of the mudstone lithologies associated with cliff retreat in Compton and Brook bays.

be installed, together with some limited realignment within the existing highway authority land immediately northeast of the road. More extensive realignment can be ruled out because the National Trust, owners of the land, have made clear their opposition to any encroachment within this designated Area of Outstanding Natural Beauty (Fig. 1). Moreover, retreat rates here in the future may increase towards the upper end of the range for this bedrock unit, perhaps in response to rising sea-levels and increased storminess, as hinted by the post-1976 retreat shown in Figure 3.

The high content of swelling smectite clays accounts for the behaviour of the grey plastic mudstone. The rate of retreat actually falls in the mottled mudstone range because this more resistant unit lies seaward of the plastic mudstone within the northward-dipping limb of the anticline. As cliff retreat continues the plastic mudstone will be progressively exposed because the coast cuts obliquely across the axis of this structure, resulting in continued along-strike (eastwards) consumption of Compton car park. It would be interesting and possibly significant from an engineering point of view to search for this highly mobile unit on the opposing, southern, limb of the anticline farther southeast along the Island's coast. If its distribution proves to be more localised it suggests that it may have originated as a secondary bentonite (Jeans et al, 1982) as Wealden drainage entrained volcanic ash and concentrated it in a floodplain or lagoonal setting. However, the similarity in the remaining mineralogy of the three mudstones (Tab. 1) may indicate a common provenance, with variations in clay minerals produced by diagenetic alteration of smectite; changes which would occur more readily in the most permeable rock units.

Vermiculite is intermediate in its swelling properties and the dark grey fissile mudstone clearly accounts for the slump debris-flow complex at Small Chine, where the head-scarp is now within 17 m of the road. The retreat rate here dictates road realignment and this must allow for retreat eastwards, as at Compton car park, rather than directly inland from the coast.

ACKNOWLEDGEMENTS
We are grateful to Mr Trevor Clayton, of the University of Southampton, for carrying out the geochemical analyses and for very helpful discussions.

REFERENCES
Jeans, C.V., Merriman, R.J., Mitchell, J.G. and Bland, D.J. 1982. Volcanic clays in the Cretaceous of southern England and northern Ireland. Clay Minerals, 17, 105-156.
Radley, J.D. and Barker, M.J. 1998. Stratigraphy, palaeontology and correlation of the Vectis Formation (Wealden Group, Lower Cretaceous) at Compton Bay, Isle of Wight. Proc. Geologists' Assoc. 109, 187-195.

Session 7:

Instability management – from Policy to Practice

It is essential to translate instability policy into good practice, to reduce risks and assist local economies. Good practice measures can include planning and building controls, advice on the management of slopes and drainage, and the encouragement of appropriate construction techniques. A co-ordinated approach involving all those with an interest in instability matters can assist this process.

Instability management in Ecuador – from policy to practice

PEDRO BASABE, DR. ès Sc., MSc., Eng. Geol., Swiss Humanitarian Aid Unit
(SHA) member. UN International Strategy for Disaster Reduction (ISDR)[1], and
CHRISTOPHE BONNARD, Civil Eng., Swiss Federal Institute of Technology of
Lausanne (EPFL)[2]

ABSTRACT

Within the framework of a large research project on natural hazards in the region of
Cuenca city, in Southern Ecuador, called PRECUPA, which was sponsored by the
Swiss Humanitarian Aid and Disaster Relief Unit (SDR) between 1994 and 1998,
several components of hazard situations were monitored, analysed and mapped,
including landslides, rainfall and earthquakes. But one main component of the
project included the strengthening of the civil defence capabilities as well as
regional and local institutional building actions, in order to improve the due
considerations of natural risks within planning and development activities.

Such efforts lead to significant practical results at four levels :
- as the national level, an agreement was signed with the National Planning
  Council (CONADE) aiming at the use of the outcomes of PRECUPA project as
  planning instruments of CONADE;
- at the regional level, the data obtained by PRECUPA were used in the
  rehabilitation works following the disaster of the Josefina landslide in 1993
  (failure of a landslide dam which caused extensive damage downstream);
- in both provinces of Azuay and Cañar in which PRECUPA developed its
  activities, the Civil Defence organisations have been strengthened through a
  professional teaching program and the improvement of the emergency plans in
  case of disaster;
- at the local level, namely the Municipality of Cuenca Canton, a new Ordinance
  for the use and occupation of the urban land has been promulgated, allowing a
  safer expansion of the city.

---

[1] SHA member. UN/ISDR Secretariat technical advisor: Palais des Nations CH-1211
     GENEVA 10 Switzerland. Phone: +41-22-9179708. Fax: +41-22-9179098. E-mail:
     basabe@un.org
[2] EPFL scientific deputy. LMS-EPFL, CH-1015 LAUSANNE Switzerland. Phone: +41-21-
     693 23 12. Fax : +41-21-693 41 53. E-mail : Christophe.Bonnard@epfl.ch

INTRODUCTION

The location of Ecuador, a small country of South America located along a convergence zone of tectonic plates, has induced its beautiful geography, but also a high susceptibility to natural disasters like earthquakes, tsunami and volcanic eruptions. Moreover, the influence of ocean currents exposes this country to intense hydrometeorological phenomena which induce landslides, debris flows, as well as floods. These natural phenomena turn into hazards as population and activities increase without due planning actions, causing a growing occupation of land in inappropriate zones, so that the vulnerability and risk factors are higher and higher.

The impact of "natural disasters" in Ecuador is very important, as it is illustrated by the last "El Niño" phenomenon which caused floods, landslides and related damage representing direct losses of 2'870 million dollars (CEPAL, 1998), i.e. 15 % of GNP, that is the highest value recorded in a South American country (Basabe, 1998).

In the South of Ecuador, the main drainage areas display loose rock formations which produce large landslides, due to intense and long-lasting rainfall as well as to inappropriate human actions; thus productive agricultural zones and inhabited areas are frequently affected. A serious event occurred in March 1993, after a period of major rainfall, when a huge landslide blocked the Paute and Jadan valleys (see Fig. 1), in he vicinity of Cuenca, the third major city of the country in the Southern Andes. Some 20 to 30 million m³ of rock which fell during the night induced the formation of a large lake causing extensive inundations upstream. After an emergency excavation of an outlet channel, some 33 days after the landslide, the huge natural dam was partially washed out by a regressive erosion mechanism, giving way to 170 million m³ of water in a few hours, which caused flooding and destruction down to 140 km downstream, in the amazonian forest (Almeida et al., 1996).

As the United Nations had declared the years 90' the "International Decade for Natural Disaster Reduction", Switzerland answered the national and international plea by sponsoring a pilot project called "PRECUPA" which represents the abreviations of "Prevention-Ecuador-Cuenca-Paute". This project gathered the national, local and university Ecuadorian institutions with counterparts from Switzerland in a joint cooperation, through the support of the Swiss Humanitarian Aid and Disaster Relief Unit (now called Swiss Humanitarian Aid Unit – SHA), which is an entity forming part of the Swiss Agency for Development and Cooperation (COSUDE in its Spanish acronym) (Basabe et al., 1998).

PRECUPA PROJECT AIMS AND COMPONENTS

The objective of the project included first the assessment of the different natural hazards which could affect the high and central drainage area of Paute river, over an area of 3'700 km² (see Fig. 2) in which some 700'000 inhabitants are living.

Figure 1. Major landslide of "la Josefina" which took a heavy toll of some 80 persons and then caused intense destruction by flooding the valley downstream

Moreover, this region is important as, in the lower part of Paute drainage area, the major hydroelectric scheme of Ecuador has been built, producing some 70 % of the national electricity needs (Basabe et al., 1996).

The PRECUPA project developed in six components or fields of action (see Fig. 2, lower part), following a systematic methodology :

- Establishing complementary mapping and creating a database
- Investigating the geodynamic hazards
- Analysing the hydrometeorological hazards
- Establishing the relevant data for the seismic hazards
- Gathering information on the contamination of the residual lake which stayed upstream of the slide after the natural dam failure.

These five first components required the planing, installation, operation and maintenance of several specific monitoring nets in order to obtain the data necessary for the identification and assessment of the corresponding hazards.

But the major aspect of the project included a large interinstitutional cooperation with national, regional and local partners, in a multidisciplinary approach. Above all a large weight was given to the professional training, in order to ensure the continuation of the monitoring and analysis activities at the end of the project; this training included also the civil defence activities, the land planning work of the regional and local authorities, in order to apply the obtained results, as well as the promotion of regulations to be introduced in development projects.

As the geodynamic phenomena represented the major hazard, the detection of landslides was extensive, including geological and geotechnical mapping, landslide and hazard maps as well as pilot studies on vulnerability aspects. This lead to the production of seven complete maps at 1:25'000 which cover the most populated areas (see Fig. 2).

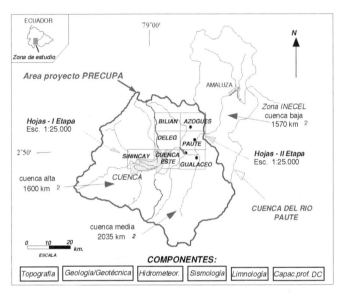

Figure 2. Study zone of the project and location of areas where landslide and hazard maps were produced. The six components of the project are indicated below

The aspect of the project related to landslides, as well as the component of management of hazards in land planning will be developed in the following chapters.

ASSESSMENT AND MAPPING OF LANDSLIDE PHENOMENA AND HAZARDS

The stability of a slope is directly conditioned by the geological nature of the materials encountered, by their geomechanical behaviour as well as by the impact of external factors like rainfall that causes the saturation of the ground, earthquakes and anthropic factors (Bonnard, 1994). Thus a slope will become unstable when the intrinsic preparatory causes will combine with triggering causes, inducing a landslide phenomenon; even a small triggering cause may be sufficient to induce a marked acceleration of the instability phenomenon (Bonnard & al., 1995).

In the PRECUPA project, the most important preparatory causes were determined by the geological, morphological and geotechnical features of the slopes (presence of finely fractured marine siltstones, deep cuts in slopes due to rivers –canyons–, intense weathering and unfavourable dip of geological structures, leading to low geomechanical parameters).

The main triggering causes are the intense and long-lasting rainfall, as well as the anthropic actions like deforestation, inappropriate cuts and fills for roads, quarry explotation and undue use of soil.

Thus 8 specific types of instability phenomena were identified, which can be grouped into three main categories (UNESCO, 1993):
- Slides, characterised by relatively slow movements, to which creep and superficial erosion can be associated.
- Rockfall, characterised by discontinuous fast movements developing along planar discontinuities (dip, cracks), implying fresh or weathered rock.
- Flows, formed of mud or debris, which can develop at a high velocity along natural drainage channels.

Methodology applied
In order to follow a well established methodology for the detection of landslide-prone areas, as well as to detect their level of hazard, the Swiss recommendations published by the federal offices of water economy (OFEE), of land planning (OFAT) and of environment, forest and landscape (OFEFP) were used (Lateltin, 1997; Loat & Petraschek, 1997; Kienholz & Krummenacher, 1995). These recommendations were adapted to the local geological context for the PRECUPA project. Thus some 28 maps at a scale 1:25'000 were produced, covering 900 km^2; 20 specific landslide zones were studied and monitored, leading to maps at a scale 1.5'000.

The methodology implied four phases :

1 - Detection of landslide-prone areas and characterisation of rock outcrop and soil cover
2 - Improvement of available maps (especially at a scale 1:5'000) and analysis of air and satellite photographs; comparison of photographs of different epochs in highly endangered zones; planning and installation of geodetic monitoring networks in most active zones (64 landmarks), including crack opening measurements (see Fig. 3).
3 - Preparation of maps of phenomena, including location of scarps, limits and accumulation zones, and determination of level of activity (active, dormant relict) and of probable depth (Turner & Schuster, 1996). The main mechanisms are specified by a letter and the major phenomena studied in detail are identified by a code number.

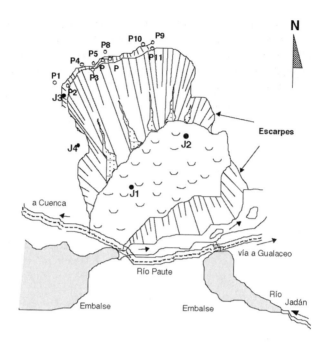

Figure 3. Location of geodetic landmarks [J] and crack opening devices [P] at La Josefina Landslide

4 - Assessment of danger level depending on the preparatory and triggering factors, the observed level of activity, the intensity of the phenomenon, allowing to determine the level of hazard in relation with the assessed probability and intensity (see Fig. 4).

The reference values of probability and intensity obtained are introdruced in standard diagrams which include both variables (Petraschek, 1995), and which were adapted for their use in the PRECUPA project.

Final product
The hazard map legend includes three colours as shown in Figure 4, as well as three letters for each zone. The first letter identifies the type of phenomenon; the second, as an index, indicates the probability of occurrence and the third, also as an index, states the intensity of the phenomenon. Although this codification is qualitative, the values of probability and intensity are backed by characteristic values determined within the project, that are based on quantitative conditioning or triggering parameters.

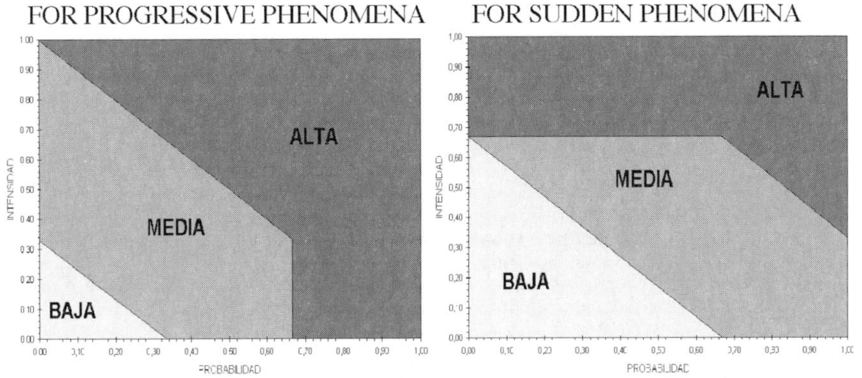
FOR PROGRESSIVE PHENOMENA
FOR SUDDEN PHENOMENA
ALTA
MEDIA
BAJA
ALTA
MEDIA
BAJA
INTENSIDAD
PROBABILIDAD

Figure 4. Assessment of level of hazard for landslides according to (Petraschek, 1995), adapted for the project, considering relative values of probability and intensity

"Baja" means "low"
"Media" means "medium"
"Alta" means "high"

The index letters have the following meaning:

	Qualitative	Quantitative
A	high	$0.66<x<1.0$
M	medium	$0.33<x<0.66$
B	low	$0 \; <x<0.33$

Vulnerability studies for geodynamic hazards

Looking at more significant parameters to introduce in natural disaster prevention studies, the PRECUPA project developed a methodology applying a vulnerability factor (Mora & Wahrson, 1994). It was used in particular in a case study in which the unstable zones were moving up to 84 m/year, affecting a rural zone of some 13 km^2 with a high hazard level, especially as it was in a process of urbanisation; this zone called Paccha is located 10 km East of the town of Cuenca.

For the computation of the vulnerability factor, a first assessment of the various hazard levels was carried out in the Paccha area (see Fig. 6) dividing the landslide into 6 zones of potentially high vulnerability, in which a population census was carried out, as well as a census of houses, economic activities and social conditions, with verifications in the field. Such an assessment allowed a better knowledge of the exposed zones that was expressed through 4 specific vulnerability factors, namely human, socio-economic, physical and preparation factors the values of which vary from 0 (not vulnerable) to 1 (totally vulnerable). In the hazard map, these 4 factors, established by a standard characterisation, are shown as a dial (see Fig. 5).

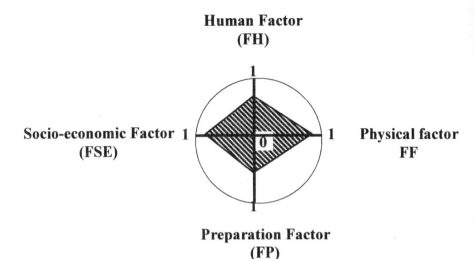

Human Factor
(FH)

Socio-economic Factor 1   1 **Physical factor**
(FSE)   **FF**

Preparation Factor
(FP)

Figure 5. Graphical representation of the factors used in the vulnerability study

The human factor considers the level of knowledge of the population when facing an eventual hazard, considering that the landslide may be slow (as in the Paccha case) or violent (as in the Josefina case),inducing thus a different level of impact. The socio-economic factor analyses the potential damage in the field, in terms of daily work carried out by the population, as well as by the impact on housing, health, education, employment, economic activities and standard of living. The physical factor considers potential damage to buildings and infrastructures, namely roads, electric lines, sewage and drinkable water systems, as well as schools, churches, health centres and houses. The preparation factor considers the level of knowledge of the population with respect to the Civil Defence programs and the preparatory activities that may save their lives, like evacuation routes (see Fig. 6).

Finally it may be sometimes necessary to adopt a global vulnerability value with a weighted average, considering specific coefficients established according to the experience gained in the studied area. This global value was assessed as low when included between 0 and 0.4, as medium if it varied between 0.5 and 0.7 and as high when it reached a value between 0.8 and 1.0.

EFFECTS OF THE PROJECT ON LANDSLIDE MANAGEMENT
From the beginning of the project, the application of its results was duly considered. First a major effort of education was made to train more than 80 young professionals working in national institutions and many of them were directly active in the project. Thus the gained know-how could lead to a practical application of the results at different levels.

PACCHA

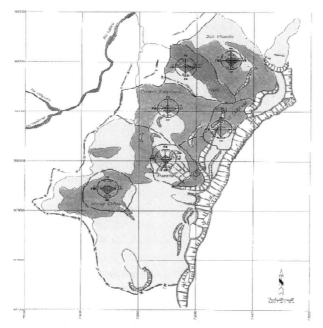

Figure 6. Landslide-prone area hazard and vulnerability map in the zone of Paccha

National level
The studies carried out, as well as the maps, the GIS and the monitoring systems were handed over to the national institutions participating in the project, namely the National Direction of Civil Defence, the National Institute of Electrification and the National Meteorological and Hydrological Institute. These institutions have continued to use the obtained results and to manage the monitoring networks.

Furthermore, the gained information was delivered to higher education centres and to the National Planning Council, called now National Planning Secretary of the Presidency of the Republic (ODEPLAN).An agreement was signed with this institution so that the obtained information and maps could be used and the risk factor included in the development agenda and planning. Presently ODEPLAN leads a ministerial and institutional working group in order to carry out a national plan for the disaster prevention and risk reduction.

Regional level
The studies carried out by PRECUPA project allowed the development of other projects in the region of Cuenca, in particular the one, financed by the European Union, which was set forth by the institution created for the rehabilitation of the zone affected by La Josefina Landslide and which used the results obtained by the

PRECUPA Project (see Fig. 1). Thus the destroyed road was rebuilt, considering the landslide maps and recommendations produced, so as to connect the cities of Cuenca and Azogues, upstream of La Josefina site, with the towns of Gualaceo and Paute downstream of this point. In the same project levees and dams were built to stabilise the riverbed. The obtained results were also useful to the Development and Redeployment Centre for Southern Ecuador (CREA) which used the installed monitoring systems and data.

Provincial level

The provincial council of Azuay, namely the province government, received a constant support by the PRECUPA projects in order to assess the situation of landslide-prone areas in rural surroundings, even outside of the zone of study. Such advice was directly useful to the population for opening new roads, building protection works or providing support to local communities. The results and maps of the project were also handed to the Provincial Council of Cañar, North of Azuay Province.

University level

Several activities of the project were carried out with the participation of the regional universities, which also contributed to the training of students and assistants. For instance, the seismic monitoring network and investigation on its results are managed by the Cuenca University.

Up to now three specialisation courses, in which the head of PRECUPA project intervened, were organised on disaster assessment, risk management, effect on environment and resisting structures, so that a deeper conscience was developed in the sense of risk reduction.

Local level

The main local counterparts of the PRECUPA Project were the Municipality of Cuenca Canton and the Telecommunication, Water and Sewage Company of this canton.. First they participated in the studies, then they profited of the given courses, obtained useful results, satellite photographs and equipment of the project in several fields. Both mentioned institutions reinforced the Direction of Environment Management and with the help of the project, the Interinstitutional Commission of Environmental Management was created which also deals with the risk aspects. This commission implies public and private institutions working in the canton.

Finally the Cuenca Municipality used the results obtained by the PRECUPA Project in order to define ordinances for the use of the land at the level of the plot, in rural as well as in urban areas. These ordinances were considered in the process to obtain the decision of classification of Cuenca City as Humanity Patrimony by UNESCO in 1999, as well as cultural capital of America in 2001.

CONCLUSIONS
The participation of national and local institutions in the project, the training of its staff and the diffusion of the obtained results has guaranteed the continuity of activities of a development research project funded by Switzerland, especially as many data could be incorporated in municipal ordinances, thanks to the good relationship built with the authorities. The conscience of the importance of natural risks in planning activities has clearly increased, allowing a better use of the land. This is expressed by a widespread popular slogan "The knowledge and respect of Nature is fundamental to protect ourselves".

REFERENCES
Almeida, E., Basabe, P. Jaramillo, H., Ramón, P., Serrano, C. (1996). "Desestabilización de laderas", por la crecida debido a la ruptura de ""La Josefina". " Libro: "Sin plazo para la esperanza", reporte sobre el desastre. EPN, 1993. Quito-Ecuador, p. 213-223.

Basabe, P. (1998). "Amenazas por inundaciones y plan de contingencia ante el fenómeno "El Niño". Comité de Emergencia Nacional del Paraguay y PNUD, Asunción–Paraguay, UNDHA, 69 p.+anexos.

Basabe, P. et al. (1996). "Prevención de desastres naturales en la cuenca del río Paute". Libro: "Sin plazo para la esperanza", reporte sobre el desastre de ""La Josefina"-Ecuador. EPN. Quito-Ecuador, p. 271-287.

Basabe P., Neumann, A., Almeida, E., Herrera, B., García, E.. Ontaneda, P., (1998). "Informe Final del proyecto PRECUPA" de cooperación entre el CSS, DNDC, INECEL, INAMHI, Munic. de Cuenca/ETAPA, U. Cuenca y Consejo de Programación por ""La Josefina"". para la prevención de desastres naturales en la cuenca del río Paute. Temas: Topografía/geodesia, Geología/geotecnia, Hidrometeorología, Sismología, Limnología, Capacitación/difusión y apoyo a Defensa Civil., CSS, 1998, 369 p. más anexos, Cuenca–Ecuador.

Bonnard Ch. (1994): Los deslizamientos de tierra: Fenómeno natural o fenómeno inducido por el hombre? Memorias 1er simposio panamericano de deslizamientos de tierra. Soc. ecuat. mec. suelos y rocas. Guayaquil, Ecuador.

Bonnard Ch., Noverraz F., Lateltin O., Raetzo H. (1995): Large Landslides and Possibilities of Sudden Reactivation. Proc. 44th Geomechanics Colloquy, Salzburg. Felsbau No 6/95, pp. 401-407.

Kienholz, H. & Krummenacher, B. Dangers naturels, Recommandations: Légende modulable pour la cartographie des phénomènes. Office fédéral de l'économie des eaux (OFEE), Office fédéral l'environnement, des forêts et du paysage (OFEFP), Berne, 1995. 19 p, 18 annexes.

Lateltin, O. Dangers naturels, Recommandations: "Prise en compte des dangers dus aux mouvements de terrain dans le cadre des activités de l'aménagement du territoire Offices fédéraux suisses: de l'économie des eaux (OFEE), de l'aménagement du territoire (OFAT) et de l'environnement, des forêts et du paysage (OFEFP), Bienne, 1997. 42 p.

Loat, R., Petrascheck, A. Dangers naturels, Recommandations: "Prise en compte des dangers dus aux crues dans le cadre des activités de l'aménagement du territoire", Offices fédéraux suisses: de l'économie des eaux (OFEE), de l'aménagement du territoire (OFAT) et de l'environnment, des forêts et du paysage (OFEFP), Bienne, 1997. 32 p.

Mora, S. & Vahrson, W. (1994): Macrozonation methodology for landslide hazard determination. Memorias 1er simposio panamericano de deslizamientos de tierra, Sociedad Ecuatoriana de Mecánica de Suelos y Rocas, Guayaquil, p. 406-431.

Petrascheck, A (1995): Mapas de peligros: Encuesta respecto de los niveles de peligrosidad. Offices fédéraux suisses: de l'économie des eaux (OFEE), de l'aménagement du territoire (OFAT) et de l'environnment, des forêts et du paysage (OFEFP), Berna. Inédito. 16 p., y anexos.

Turner A.K., Schuster R.L. (1996). Landslides-Investigation and Mitigation. Special Report 247. Transportation Research Board, Academy of Sciences, Washington D.C., 673p.

UNESCO (1993): Multilingual landslide glossary. The International Geotechnical Societes. UNESCO Working party for world landslide inventory. BiTech Publishers. ISBN 0-920 505-10-4. Canada.

The IMIRILAND project – Impact of Large Landslides in the Mountain Environment: Identification and Mitigation of Risk

M. CASTELLI, Politecnico di Torino, Torino, Italy
C. BONNARD, École Polytechnique Fédérale de Lausanne, Lausanne, Switzerland
J.L. DURVILLE, Laboratoire Central des Ponts et Chaussées, Paris, France
F. FORLATI, Regione Piemonte, Torino, Italy
R. POISEL, Technische Universität Wien, Wien, Austria
R. POLINO, Consiglio Nazionale delle Ricerche, Torino, Italy
P. PRAT, Universitat Politècnica de Catalunya, Barcelona, Spain
C. SCAVIA, Politecnico di Torino, Torino, Italy

SUMMARY

The paper relates the activities carried out within the IMIRILAND project, funded by the European Commission within the fifth framework programme (Research and Technological Development, Activities of a Generic Nature: The Fight against Major Natural and Technological Hazards). The project has a duration of 30 months, starting from March 1st, 2001.

The project focuses on aspects of management of risk in the case of large landslides, in its scientific, technical and land-planning aspects. Eight large landslide sites have been selected in the countries involved in the project, where geological, geomechanical and monitoring data already exist, with different levels of knowledge, in order to show how risk can be assessed in different situations. These data, after a suitable organisation, are the basis for the following phases of the project. New methods or new applications of the existing methods are worked out in order to obtain the scenarios of the landslides evolution. Reference is made to geological, numerical, statistical and neural network based methods. Then the developed methodologies are applied to the management of the endangered landslide zones. Actions for risk mitigation are proposed on the basis of risk assessment and the legal and economical framework. Finally, dissemination of risk management methodologies among the users is a major task of the project.

INTRODUCTION

Large landslides affect many mountain valleys in Europe. They are characterised by a low probability of evolution as a catastrophic event but may have very large direct and indirect impacts on man, infrastructure and environment. This impact is becoming more and more dangerous due to increasing tourism development and the construction of new roads and railways in mountainous areas. Methodologies for the identification and mitigation of risk are therefore a major issue.

As a matter of fact, many experiences during critical development of landslides have shown a lack of methodologies and above all a non-systematic approach of interpreted risks. Risk management is in practise accomplished by local and regional authorities only during the critical event in a necessarily improvised way. This "reactive" approach induces negative consequences on the identification procedure. For instance, very expensive monitoring systems have been installed on several large landslides without any well-established methodology linking the interpretation of the measures and the understanding of deformation mechanisms to the practical questions concerning the management of the risk.

Furthermore, risk can extend well beyond local damage (for instance, risk of river damming which may induce major hydrological hazards: floods, inundation of sewage plants, loss of drinking water resources), so that it must be considered in a wide perspective.

What do we know about landslide risks? A first theoretical definition was given by Varnes in 1984: risk is the probability of an event of a given magnitude multiplied by consequences. Since then a lot of technical and scientific papers have been written on this topic (i.e. Einstein, 1988, Cherubini *et al.*, 1993, Canuti & Casagli, 1994, CALAR, 1999). Recent works (IUGS Working Group on Landslides, 1997) propose the following multidisciplinary procedure for quantitative risk analysis of slopes:

- **Hazard analysis**: analysis of the probability and characteristics of the potential landslides;
- **Identification of the elements at risk**, i.e. their number and characteristics;
- **Analysis of vulnerability** of the elements at risk;
- **Calculation of the risk** from the hazard, elements at risk and vulnerability of the elements at risk.

However the evaluation of hazard is still made through different approaches leading to results which differ from the qualitative and quantitative point of view. Besides, only a few works have been done (i.e. Leone et al., 1996) on vulnerability. Finally, a well defined procedure does not exist to include the results of the risk analyses in land planning, taking the legal framework into account.

The present project intends thus to deepen the hazard analysis of different types of landslides, by the application of new and updated geological, geomechanical and statistical methods. Particular attention is given to numerical models where finite elements, boundary elements and finite difference methods will be applied in three dimensional conditions on the same slope stability problems. This multidisciplinary approach will thus allow a comparative analysis of different hazard prediction techniques to be carried out. Then, it is foreseen to develop the vulnerability and risk analysis for several landslides considering short and long term perspectives, direct and indirect consequences, as well as technical and social impacts, using tree event techniques. The combination of the results obtained in relation with hazard analysis and through the risk approach will allow the development of a new practical and quantified risk assessment programs which will be applied to several sites.

The partners involved in the project are:

- Regione Piemonte - Direzione Regionale Servizi Tecnici di Prevenzione (Italy)
- Politecnico di Torino - Dipartimento di Ingegneria Strutturale e Geotecnica (Italy)
- CNR - Centro Studi sulla Geodinamica delle Catene Collisionali (Italy)
- Laboratoire Central des Ponts et Chaussées - Division for Soil Mechanics, Rock Mechanics, Engineering Geology (France)
- Universitat Politècnica de Catalunya - Departamento de Ingeniería del Terreno (Spain)
- Technische Universität Wien - Institute for Engineering Geology (Austria)
- École Polytechnique Fédérale de Lausanne - Département de Génie Civil-Laboratoire de Mécanique des Sols (Switzerland).

PROJECT STRUCTURE

The multidisciplinary programme of activities is built on a four-stage structure:

- **Phase 1**: Data collection.
- **Phase 2**: Development of risk assessment *methodologies.*
- **Phase 3**: Application to management.
- **Phase 4**: Dissemination of risk management methodologies.

It is subdivided into 9 workpackages (WPs, see figure 1).

Data-collection and organisation

The eight selected cases imply a complex risk situation, for which no reasonable stabilisation works can be achieved. The partners involved in the project have already some knowledge about the sites (some of them are very well known, while some others are in an initial stage of study), the data of which are already available but very dispersed (data banks, reports, publications...). In this way it will be relatively easy to collect a large amount of information without dispersion of energies. A general scheme for data description has been created and compiled for each site (WP1). Each partner is then able to access data, which can be managed in a homogeneous manner. The objective is also to check the reliability of the available data, to indentify which ones are to be considered.

Selected sites are summarised in table 1 and their location is shown in figure 2.

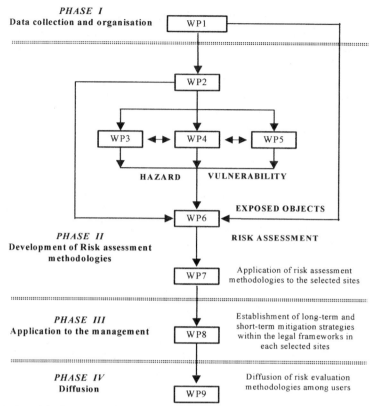

Figure 1. Graphical presentation of the project's components

Development of risk assessment methodologies

The purpose of Phase 2 is to provide the overall framework for risk assessment methodologies. **Hazard analysis** is the first step, and consists of the following activities:

- models to calculate the probability of preparatory or triggering events,
- models of progressive development of failure,
- interpretation of monitoring data,
- effect of tectonic or weathering action on the rock mass,
- models of propagation of the moving mass, in order to determine a certain affected zone for a certain probability of occurrence.

As a result hazard analysis methods give the scenarios needed for risk assessment. Several methodologies have been proposed in the past for the establishment of hazard. Scenarios for the evolution of large landslides obtained through such methodologies provide different results, so it is difficult to express hazard in a unique way. It is therefore fundamental to compare and organise the different approaches so as to give to the EU all the elements necessary to practise an effective co-ordination role.

Within the project the following methodologies will be considered:

- **field analysis**: the ability of geology, geomorphology and tectonics will be shown in order to obtain qualitative scenarios of the evolution of large landslides (WP2);
- **mechanical modelling for the landslide triggering**: application of numerical methods. Partners have specific experience in Boundary Element, Finite Element and Finite Difference methods. Existing computer codes will be improved and applied to the study of the selected large landslides in 3D. The results obtained through the different methods will be at the end compared on some cases (WP3);
- **mechanical run out modelling**: application of numerical methods. When the possibility of instability has been established through the methods of WP2 and WP3, the landslide run-out is studied through two different specific numerical methods, taking into account the dynamic character of the phenomena. Scenarios of evolution of the slide will be the result of this WP, and in some cases the scenarios obtained through the different methods will be compared (WP4);
- **black box models**: application of neural network and statistical methods. This application will complement the numerical analyses of landslides, for which it may be difficult to get the input parameters with a sufficient degree of accuracy. The statistical and heuristical methods are applied by using a large amount of previously gathered environmental data, which may influence the behaviour of landslides, in an undetermined way (WP5).

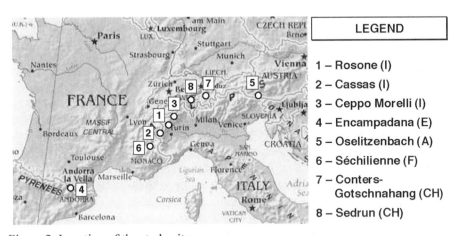

LEGEND

1 – Rosone (I)
2 – Cassas (I)
3 – Ceppo Morelli (I)
4 – Encampadana (E)
5 – Oselitzenbach (A)
6 – Séchilienne (F)
7 – Conters-Gotschnahang (CH)
8 – Sedrun (CH)

Figure 2. Location of the study sites

Next step is the **risk analysis**, which includes:

- the consideration of endangered population and objects, their value and their vulnerability, i.e. the degree of loss from a particular danger;

- the risk analysis itself, through the evaluation of costs due to direct and indirect impact as well as the assessment of variable warning time which is a general characteristic of a well monitored pre-failure phase

Table 1. Investigated landslide sites

Site	Country	Volume or affected area	Elements at risk
ROSONE	Italy	$(30 - 50) \cdot 10^6 \, m^3$	• *Rosone* village • Main road SS 460 • Hydroelectric power plant • Downstream valley
CASSAS	Italy	$(20 - 30) \cdot 10^6 \, m^3$	• Highway A32 (*Frejus*) • International railway *Torino- Lyon*
CEPPO MORELLI	Italy	$(4 - 6) \cdot 10^6 \, m^3$	• National Road N. 549, only connection to the *Macugnaga* tourist resort • *Prequartera* and *Campioli* hamlets
ENCAMPADANA	Spain	Several tens of millions m^3	• Important ski resort • *El Tarter* small village • Main road *Andorra - France* • Downstream villages
OSELITZENBACH	Austria	$150 \cdot 10^6 \, m^3$	• *Tröpolach* and *Watschig* hamlets • Main road *Naßfeld – Bundesstraße*, important connection to the valley *Gailtal*
SÉCHILIENNE	France	$25 \cdot 10^6 \, m^3$	• National road RN 91 • Downstream valley
CONTERS-GOTSCHNAHANG	Switzerland	$25 \, km^2$	• New A28 national road • *Lanquart* town • Downstream valley
SEDRUN	Switzerland	$1 \, km^2$	• Regional railway line • National road • *Sedrun* touristic resort

Workpackage WP6 is dedicated to the state of the art in the evaluation of vulnerability and risk analysis. In order to quantify risk levels, three components will be analysed, related to the exposed objects, which may be influenced by landslides:

- the nature, structure and spatial distribution of exposed objects on the landslide mass or within the exposed zones, with due consideration of their long-term behaviour;
- the sensitivity of these objects to movements, in order to assess how they are directly or indirectly affected by movements;
- the value of these objects, in a wide sense, as well as the indirect economical impact of the changes they may suffer.

Finally, the results of phase 2 will be exemplified on the selected sites in workpackage WP7, which will summarise the scenarios and give the risk evaluation for each site. Limits and applicability of the proposed risk assessment methodologies will be illustrated.

Application to management

Phase 3 is related to the application of the developed methodologies to the management of endangered landslide zones (WP8). Mitigation strategies will be formulated in order to:

- allow the local or regional authorities responsible for the affected areas to control the evolution of the studied dangerous phenomena;
- predict the potential development of a critical event that might induce serious damage to infrastructures or cause victims;
- take the necessary safety or preparation measures so as to limit the direct and indirect consequences of the foreseen disaster.

Corresponding types of actions are proposed in relation to the degree of urgency of the problem. For example, if a real crisis is expected within some weeks or even days, an evacuation plan must be prepared including large buffer zones and alarm systems must be placed to stop the traffic in case of a sudden failure. On the contrary, if the expected event is uncertain and might occur within months or years, alert systems must be installed and threshold values have to be determined for different types of possible landslide behaviours; it is also wise to determine alternative routes and to check if they present limitations or hindrances.

All the suggested actions related to safety criteria depend of course on the legal framework and on the powers that local or regional authorities have to solve critical situations or to manage minor or residual risks (i.e. limited risks that are induced by the later or final degradation of a failed slope, which can affect the exposed objects after years). This legal framework depends on the national or regional legislation, so that management policies have to be duly adapted. A catalogue of typical mitigation actions to face several hazard situations is also established.

Possible actions will be also proposed on the basis of the relevant economical conditions. Direct or indirect consequences are considered. For example a road leading to a major industrial site or tourist resort endangered by rockfalls will justify major protection investment to ensure a permanent transit even though the exposed infrastructures themselves are hardly affected by the stones falling on the road. The cost of works and labour force will also be a significant criterion for the selection of mitigation solutions.

The assessment of acceptable risks will also be based on the type and extent of hazard that might occur. A large sudden landslide causing the death of one hundred persons has not the same psychological impact as one hundred small rockfalls, each inducing one victim; this effect, called the aversion factor, has to be included in the determination of mitigation strategies.

Diffusion

The diffusion of risk management methodologies is the natural conclusion of the project (WP9). As regards this topic:

- information dissemination will be promoted, which improves public awareness of landslide hazard, through hazard and risk maps, information on typical damage and risk to life and goods, using not obscure and univocal terminology (relevance of metadata);
- understanding of public response establishing better knowledge will be improved, as an instrument to take the proper decisions about landslide mitigation actions.

CONCLUSION

The project IMIRILAND has been presented briefly in this paper. The project, carried out by a multidisciplinary European consortium, is focused on the assessment, management and mitigation of large landslide risks. Different methodologies are used and their results will be compared on some selected landslide cases. A special diffusion action of risk management methodologies and mitigation strategies among decision-makers (e.g. politicians and administrators) and emergency-managers (e.g. civil protection) will form the conclusion of the project, in summer 2003.

REFERENCES

Canuti P. & Casagli N. (1994), Some considerations about the evaluation of landslides risk (in Italian), Proc. of Convegno Fenomeni Franosi e Centri Abitati, pp. 1-58, Bologna

Cherubini C., Giasi C.I. & Guadagno F.M. (1993), Probabilistic approaches of slope stability in a typical geomorphological setting of Southern Italy, Risk and reliability in ground engineering, pp. 144-150, Thames Telford, London

Concerted action for forecasting, prevention and reduction of landslides and avalanche risks (CALAR) (1999). X-Calar'99 Proceedings, Innsbruck

Einstein H.H. (1988), Special lecture: Landslide risk assessment procedure, Proc. 5th Int. Symp. on Landslides, Lausanne, Switzerland, Vol. 2, pp 1075-1090

IUGS Working Group on Landslides, committee on risk assessment (1997), Quantitative risk assessment for slopes and landslides – The state of the art, Proc. of Int. Workshop on Landslide Risk Assessment, Honolulu

Leone F., Asté J.P. & Leroi E. (1996), Vulnerability assessment of elements exposed to mass-movement: working toward a better risk reception, Proc. of 7th Int. Symp. on Landslides, Trondheim, June 1996, p. 263-270

Varnes D.J. (1984), Landslide hazard zonation; a review of principles and practice, Natural Hazards 3, UNESCO, 63 pp.

Managing coastal instability - a holistic approach

K. COLE, West Dorset District Council, UK
G. M. DAVIS, West Dorset District Council, UK
A. R. CLARK, High-Point Rendel, London, UK
D. S. FORT, High-Point Rendel, London, UK

INTRODUCTION

The town of Lyme Regis in West Dorset (Figure 1 & Plate 1) faces serious instability problems. The town is located on an actively eroding coastline and it is believed that a large part of the original mediaeval settlement has been lost to coastal erosion and landslipping. Over the past 100 years in particular, many properties have been destroyed or severely damaged due to landslide movements and there has been considerable disruption to infrastructure, including the loss of the main coastal road in the 1920's. Historically, approaches to deal with coastal instability in West Dorset, in common with much of the UK coastline, have tended to comprise isolated studies and stabilisation schemes, carried out in response to discrete landslide events. For example, a breach in a sea wall would lead to the repair and strengthening of the sea wall, or if a house fell off the edge of a retreating cliff some boreholes might be sunk in an attempt to discover what mechanisms had caused the disaster. Nowadays in West Dorset, in common with some other areas, a more proactive and larger scale philosophy is taken, with the objective of preventing or managing landslide problems, rather than reacting to a landslide event after it has caused the damage.

The area is one of environmental importance, both in terms of natural history and historic structures, and the area is a major tourism centre. This has generated much interest both from the public and statutory authorities.

COASTAL INSTABILITY STUDIES AT LYME REGIS

In the development of coastal instability management options at Lyme Regis one of the principal initial objectives has been to gain an understanding of the nature of the instability problem itself. To this end, a holistic approach has been employed in which a major series of multidisciplinary studies have been carried out (Brunsden 2002, Clark et al. 2000). This has involved a considerable investment in acquiring and interpreting data on ground stability, coastal erosion and other issues that may have an influence on the implementation of stabilisation schemes, such as economic and environmental matters. The principal features of the studies, in contrast to the traditional approach, have been:

Instability – Planning and Management, Thomas Telford, London, 2002, 679–686

1. Consideration of the entire coastal system from the top of the coastal slope through the cliffs and the intertidal zone to the sea bed about 1km offshore. (Plate 1).

2. Consideration of the whole of the urban coastal area of Lyme Regis including several kilometres of undeveloped coastline to both the east and west

3. Multidisciplinary studies both fairly detailed and wide in scope carried out in order to address the relatively complex nature of the instability problems in the area.

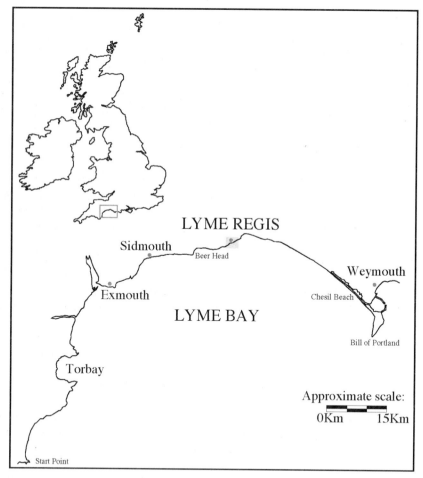

Figure 1: Location of Lyme Regis, Dorset

Plate 1: Aerial Photograph of Lyme Regis taken from the East.

The nature and relationships between the individual principal studies are summarised in Figure 2. The main groups of investigation carried out were historical studies, condition surveys of coast protection structures, analysis of coastal processes, large scale geotechnical investigations, economic analyses and environmental studies. The detailed results of the investigations, and the geotechnical investigations in particular, have enabled the development of ground models for the various forms of instability existing within the town (Sellwood et al. 2000). These models form the basis of the future control of landslipping, enabling a strategic approach to be taken rather than emergency responses to failures. As the studies developed, there was found to be a strong degree of concordance and interrelationship between the individual study elements, with each one complementing the others.

The main benefits of such a holistic approach are considered to be:

- A deeper understanding of instability and erosion mechanisms and their relationship with coastal defence structures than could have been gained from the study of isolated problem areas alone.

- An appreciation of the context of the instability problems affecting the West Dorset coastline generally. For example, much useful information on the landslide mechanisms operating beneath the town was gained by studying the landsliding taking place in the well exposed cliffs on either side of the town.

- Interpretation of the instability affecting the town in terms of the evolution of the coast, both in the long-term and short-term. Patterns of landslide evolution and coastal retreat could be established, which then served as models for prediction of future landslide behaviour.

- Economies of scale - the cost of studying coastline as a whole (as a single project) being less that that of individual studies of isolated sites at different times, by different parties.

- A pro-active philosophy, which is much more cost effective in the long term than a short-term reactive approach - economic analyses have established that prevention is better than cure.

- Additional insights into the problem from the multidisciplinary approach. Studies carried out in a particular sub-discipline often yielded benefits in another area. For example, geophysical surveys carried out to investigate the nature and distribution of sea bed sediments for the analysis of sediment budgets also gave information on the geological structure beneath the sea bed. This information could then be extrapolated onshore to assist in the interpretation of landslide mechanisms, with a consequent saving in ground investigation costs.

- Spin off benefits, for example the establishment of a large landslide monitoring network (Fort et al 2001, Davis et al 2002), principally to gain information on landslide mechanisms, allowed the establishment of a type of early warning system to alert residents to accelerations in ground movement which could be the precursors of destructive landsliding (Cole & Davis 2002).

- An in-depth understanding of the coastal processes operating in the area which enables the selection of engineering solutions for the instability which will not have a detrimental long term impact on other sections of the coast down drift.

- Provision of high-quality information to better inform interested parties such as statutory consultees, the planning process and the public.

- It helps to ensure that management options and stabilisation schemes are concordant with sensitive natural and built environments and the marine and landslide processes operating on adjacent sections of coastline.

- It determines the very best coast protection management options and the optimum programme for their implementation, which could not have been done had the studies been more superficial. Hence a relatively low factor of safety can be adopted in the design of remedial works because of the high confidence in the data on which the designs are based. Only the minimum engineering works need be implemented, with consequent cost savings.

CLIENT-CONSULTANT RELATIONSHIP

The client for the project is West Dorset District Council in its role as the local coast protection authority. The Council employs an in-house group of specialists, based locally, who manage the project on a day-to-day basis, and carry out some elements of the work. Where the nature of the work is beyond the expertise or mantime resources of the in-house staff, consultants are engaged for particular assignments. The consultant's and the client's staff work in close collaboration as part of an integrated project team. The benefits of the approach are as follows:

- The client maintains full control of the project through in-house management.

- The use of in-house specialist staff allows the client to fully appreciate the technical aspects of consultant's reports and proposals, and to explain the information and its implications to elected members of the Council and local people on a day-to-day basis (Davis & Cole 2002).

- The client gains a greater understanding and 'feel' of the problems, on which to base management decisions.

- The on-site presence allows matters such as the monitoring of slope instrumentation, organisation of contractors and local consultations to be carried out rapidly and efficiently by personnel who are 'in tune' with local issues and opinion, leaving the consultants free to concentrate on the more technical aspects.

ADDED BENEFITS OF THE HOLISTIC APPROACH

The historic approach to coast protection is consistent with the recommendations of the Government's Department for Environment, Food and Rural Affairs (DEFRA) who are responsible for coast protection in the UK. This approach embraces a strategic review of all the issues affecting a managed stretch of coastline, which ensures that manpower and financial resources are focused on high priority sites. The holistic approach enables all aspects of alternative remediation schemes to be rigorously tested to ensure that they are technically, environmentally and economically sound. This approach also informs the planning process. The

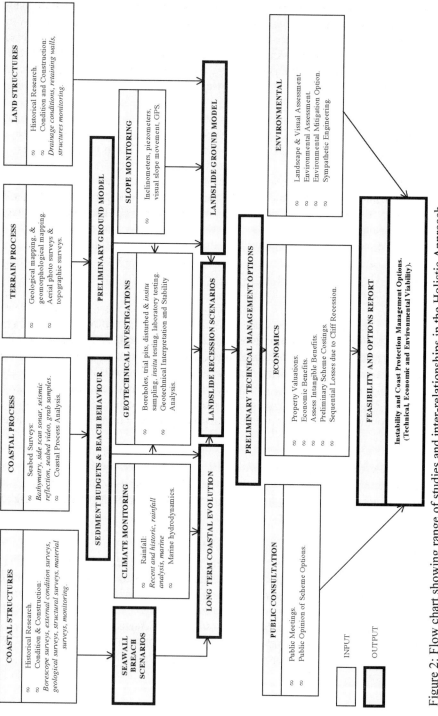

Figure 2: Flow chart showing range of studies and inter-relationships in the Holistic Approach.

consideration of preferred remediation schemes by the Planning Authority during granting of planning permission prior to construction needs to review the complex interrelationships between the proposed scheme and the potential social, environmental, economic and visual impacts. The holistic approach ensures that these issues (see Figure 2) have been explored, so that the planning process is fully informed.

The holistic approach facilitates the extension of coast protection and stabilisation management into related spheres. For example, the large body of data obtained from the instability studies may be used in the preparation of planning guidance maps for developments on potentially unstable land, which may then be extended to regions further inland. The information gained on environmental and social issues, as well as the instability, may be employed in coastal management more generally such as in long term coastal environmental improvement and economic regeneration plans.

CONCLUSIONS

The coastal slope instability and coast protection problems at Lyme Regis are complex and the consequences of continued instability and deterioration of the defences are considerable. The area is one of environmental importance, both in terms of natural history and historic structures, and the area is a major tourism centre. This has generated much interest both from the public and statutory authorities. The investigation of the technical issues and identification of scheme options to address the problems has been extensive and has included a wide range of interrelated and complementary studies. This holistic approach to the investigation of the coastline at Lyme Bay has proved to be particularly successful in identifying the key issues, in the selection of the most appropriate remediation approach and ratification of the chosen scheme solutions by close consultation with statutory authorities and key interest groups.

ACKNOWLEDGEMENTS

The paper is published by permission of West Dorset District Council. The principal consultant is High-Point Rendel. The project is funded by DEFRA. Naomi Healey-Cathcart helped to prepare the figures.

REFERENCES

Brunsden D. 2002. The Fifth Glossop Lecture - Geomorphological roulette for engineers and planners: some insights into an old game. *Quarterly Journal of Engineering Geology and Hydrogeology*, 34.

Clark A.R., Fort D.S. and Davis G.M. 2000. The strategy, management and investigation of coastal landslides at Lyme Regis, Dorset, UK. *Landslides in research, theory and practice* 1, 279-286. Thomas Telford, London.

Cole K. and Davis G.M. 2002. Landslide warning and emergency planning systems in West Dorset, England. Proceedings of the International Conference on Instability - Planning and Management. Thomas Telford, London

Davis G.M. and Cole K. 2002. Working with the community - public liaison in instability managemen. at Lyme Regis, Dorset, England. Proceedings of the International Conference on Instability - Planning and Management. Thomas Telford, London

Davis G.M., Fort D.S. and Tapply R.J. 2002. Handling the data from a large landslide monitoring network at Lyme Regis, Dorset, UK. Proceedings of the International Conference on Instability - Planning and Management. Thomas Telford, London

Fort D.S., Clark A.R., Savage D. T. and Davis G.M. 2000. Instrumentation and monitoring of the coastal landslides at Lyme Regis, Dorset, UK. *Landslides in research, theory and practice* 1, 573-578. Thomas Telford, London.

Sellwood M., Davis G.M., Brunsden D. and Moore, R. 2000. Ground models for the coastal landslides at Lyme Regis, Dorset, UK. *Landslides in research, theory and practice* 3, pp1361-1666. Thomas Telford, London.

Hazard assessment and management experience in the Principality of Andorra

J. COROMINAS, Universitat Politècnica de Catalunya. Barcelona, Spain
R. COPONS, Euroconsult S.A., Andorra la Vella, Andorra
J. M. VILAPLANA, Universitat de Barcelona, Barcelona, Spain
J. ALTIMIR, Euroconsult S.A., Andorra la Vella, Andorra
J. AMIGÓ, Eurogeotècnica, S.A., Cerdanyola, Spain

INTRODUCTION

The Principality of Andorra is a mountainous country located in the Central Pyrenees, which main valleys are densely populated. The urban growth has spread through areas threatened by natural hazards (especially, landslides, snow avalanches and floods). Last few years several rock falls and landslide reactivation events affected residential areas and infrastructures. Since 1989, the Government of Andorra has promoted landslide hazard studies and maps. More recently, in December 2000, a new landslide hazard map of the whole Principality has been completed, that will be included as a basic documentation in building codes that are under preparation. The hazard map has been prepared at 1:5,000 scale. This scale is considered as appropriate for regional landslide hazard management.

In this paper we present the procedure used to assess landslide hazard and for hazard zoning. It has been made taking into account that different landslide mechanisms are involved. Finally several technical guidelines have been proposed for inclusion in the building codes.

LANDSLIDE HAZARD ASSESSMENT

Landslide hazard is a space-time concept that is defined as the probability of occurrence, within a specific period of time and within a given area, of a potentially damaging phenomenon (Varnes, 1984). This definition means that hazard assessment has to include not only the location of the potential landslide source but the affected runout area and the appraisal of the imminence of the event. Furthermore, this has to be made for all the landslide mechanisms existing in the region. In practice, obtaining all the parameters involved in the concept of hazard is not an evident task. Particularly, in what respects the determination of the temporal activity.

We are going to describe a simple procedure for assessing landslide hazard and consequent mapping. The procedure follows several steps directed to the appraisal of the fundamental parameters involved in the concept of hazard (Corominas et al.

2002). The first one is the susceptibility analysis, by which the locations that are more prone to produce slope failures and their volume are established. Susceptibility analysis also involves the determination of the area that may be affected by landslide runout. Both, landslide source and runout area are the only ones that will be considered in the consequent hazard assessment and zoning. The second step is the estimation of the probability of landslide occurrence from its frequency. Finally, hazard is evaluated by considering the magnitude and frequency at the landslide susceptible sites. The GIS has become an indispensable tool for manipulating the large number of data and layers generated. The detailed procedure chart is shown in the figure 1.

The complete procedure requires the use of large amount of basic geological and geomorphological information. The main lithological units, past landslide events, and landslide activity indicators were mapped by means of air-photo interpretation and intensive field reconnaissance, and stored in the GIS. Topographic and geometric attributes of the slopes were derived from available Digital Terrain Model (DTM).

Susceptibility analysis
The first step of the hazard assessment is the identification of areas potentially affected by slope failures (susceptibility analysis). These areas include both the landslide source and the runout zone. Even though both items are related, as the runout depends on the location and volume of the landslide source, the procedures followed for determining them have been independent.

Because different landslide types occur under different geological and topographical conditions and each one may produce different runout distances, susceptibility analysis has been performed separately for each type of landslide. If several mechanisms may take place at the same slope, all of them have been considered.

Determining Landslide source
The criterion used to assess the proneness of the terrain to fail considers the slope lithology, the nature of the potential landslide and the critical slope angle. The inventory of slope failures triggered during the extreme rainfall event of November 1982 in the Pyrenees, allowed the establishment of threshold slope angles for different small-size landslide types. On the other hand, all the existing active and non-active large landslides are considered as being susceptible to reactivate. A map with the critical slope angle intervals has been created with the GIS from the DTM and overlapped to the lithological map. The areas having slope angles below the landslide threshold angles were considered as non-susceptible and have not been included in the subsequent hazard analysis. These areas will be considered in the category of low landslide hazard except for those locations that can be reached by landslides coming from neighbouring slopes.

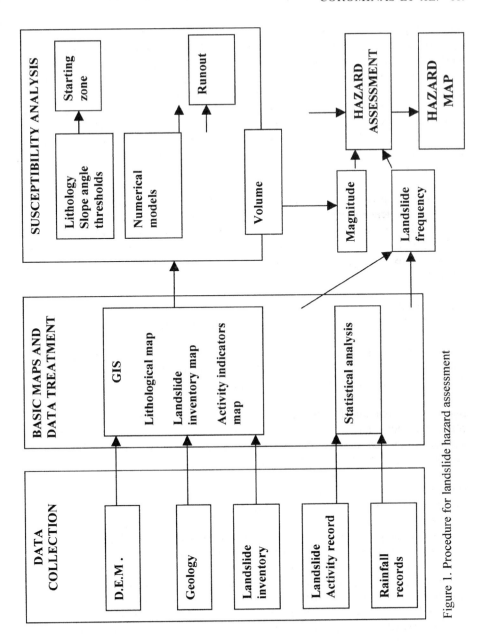

Figure 1. Procedure for landslide hazard assessment

Runout estimation
Harzardous areas are defined by the presence of landslide sources and their expected runout. Two different procedures have been used to determine the travel distance of potential landslides: an empirical expression (the reach angle) and a deterministic model (the Eurobloc code).

The angle of reach, introduced by Heim and used by Hsü (1975) as mobility index, may be also used to delineate the area potentially reached by a landslide (Corominas et al. 1990). Given the location of a potential landslide source with a known volume, the travel distance is determined as the intersection of the line dipping α degrees (reach angle) from the source with the ground surface. For the same landslide source different reach angles may be obtained according to the type of landslide, potential volume and obstacles expected in the path (Corominas, 1996). The results of the reach angle were checked with a numerical code, Eurobloc (Lopez *et al.*, 1997; Copons *et al.*, 2000a) and with the observation of the extent of former landslide deposits. Because the use of numerical models is more demanding in terms of data requirements and time, the analysis with Eurobloc has been performed only in slopes threatening urban areas and main infrastructures. The results have shown an excellent agreement with those obtained with the reach angle. Travel distances from all the identified potential landslide sources have been delineated with the GIS.

HAZARD ASSESSMENT
The procedure of landslide hazard assessment integrates the identification of the landslide susceptible zones with the estimation of both landslide magnitude and frequency.

Landslide magnitude
Magnitude of the landslide is the capability to produce damage. It may be expressed as the energy associated to the detached mass. Ideally, the size of the mobilised mass along with the propagation velocity will provide an estimation of the landslide magnitude. In fact, the expected kinetic energy along the path is a fundamental parameter in the design of protective measures against rock falls.

Three magnitude (energy) ranges, expressed in KJ have been considered: low (0 to 2,000 KJ), medium (2,000 to 10,000 KJ) and high (more than 10,000 KJ). These ranges have been defined according the presently available protective structures (commercial rock fall fences and earth embankments). Energies over 10,000 KJ are considered as non-manageable.

Rock fall magnitudes have been established in two steps. First step consists in the assessment of a starting volume and then, in a second step, we have obtained the potential path, velocities and energies at several selected points by means of Eurobloc numerical model. Volume of the blocks observable at both the rock cliff and on the surface of talus deposits, provided the starting volumes (Copons et al., 2000b). Rock fall simulation was performed in representative slopes for different lithologies and potential volumes. The results were extrapolated to all slopes of similar lithology by means of the GIS.

The magnitude of rock falls, shallow landslides and debris flows was inferred from the potentially movable volume at the identified susceptible locations (Corominas et al. 2002). Instead, all large landslides have been considered as always having high magnitude. Even though they often display small displacement rates, remedial works are expected to be both inefficient and economically unaffordable.

Frequency assessment
The probability of occurrence of a landslide has been expressed in terms of recurrence period. High frequency corresponds to landslides with recurrence intervals of less than 40 years, moderate frequency between 40 to 500 years, and low frequency when landslide events may occur with recurrence periods higher than 500 years. Different approaches have been followed to determine frequency for each landslide type. Thus, rock fall frequency is derived from silent witnesses and historical records. Frequency of shallow landslides and debris flow has been obtained from temporal distribution of triggering events (heavy rainfall). The activity of large landslides has been inferred from the degree of morphology preservation; presence of activity indicators (open cracks, deformed structures); dendromorphological analysis, and from instrumented landslides.

HAZARD ZONING
The assessment of magnitude and frequency for all landslide types has been carried out in a GIS at each susceptible cell. From each potential landslide source cell, the area affected downslope has been estimated as well. As a result all cells have been classified according to the hazard matrix (table 1) and then the consequent hazard zoning map may be prepared.

Table 1. Landslide hazard categories

	Frequency (return periods)	< 40 yr	40-500 yr	> 500 yr
Magnitude	High	High hazard	High hazard	Moderate
	Medium	Moderate	Moderate	Low hazard
	Low	Low hazard	Low hazard	Very low hazard
Non-susceptible areas		Very low hazard	Very low hazard	Very low hazard

The map has four hazard categories: (a) very low, in which no potential hazard has been observed; (b) low, in areas that may be affected by small-size slope failures with moderate-high frequency and that can be mitigated at low cost; (c) moderate, assigned to areas where either frequent landslides of small magnitude or large landslides with low frequency may take place. Landslide countermeasures are feasible; and (d) high hazard is assigned to areas where large landslides may reactivate or are active. Landslide countermeasures are not feasible.

Hazard category of a particular area can be reconsidered in the future if detailed studies (at 1:1,000 scale or larger) are undertaken. In practice, these studies are required in most of the hazardous areas to check the stability conditions, the landslide paths and in order better locate the boundaries between hazard categories. These new studies may be promoted by either the local and state administration or the private initiative. In fact, a list of dangerous sites, where hazard studies are too costly for private owners, have been proposed to be analysed by the administration in a plan for the next 10 years.

ADMINISTRATIVE PROCEDURE
An administrative procedure has been envisaged to help decision-makers. Several documents will be asked in case of new developments or infrastructures according to the degree of hazardness, which are synthesized in figure 2. For very low landslide hazard areas no specific document will be required. For low hazard areas, the owner or developer must fill a form of acknowledgement of the type of threat that may affect the property. This form is signed by the responsible engineer or architect, mentioning that the possible hazard has been taken into account in the project design. For moderate hazard areas, besides the acknowledgement form it will be required a technical report. This document must include specifically the countermeasures that will be undertaken in order to avoid or mitigate the potential hazard along with an estimation of the residual risk (particularly for those events of large return periods). Finally, for high hazardous areas no new constructions or facilities will be allowed. A few exceptions, however, are envisaged. Warehouses with no permanent activity, linear infrastructures (i.e. water pipes) that will not threat population or the environment in case of failure, or roads without alternative corridors might be allowed and, in this case, both acknowledgement form and technical document will be required to justify that technically the project. The procedure lets open the possibility of authorization to build in high hazardous areas if the promoter demonstrate with the adequate technical studies that countermeasures to avoid or mitigate instability are feasible.

The integration of the hazard map in a GIS allows the knowledge of the type of hazard that is threatening a particular site. Therefore, developers of this site may know in advance what kind of technical report they will be asked for.

In areas where detailed studies have been carried out by the administration or private promoters, the technical report might not be required if countermeasures suggested in the detailed study are incorporated in the project or the degree of hazard has proven to be smaller than that indicated in the hazard map.

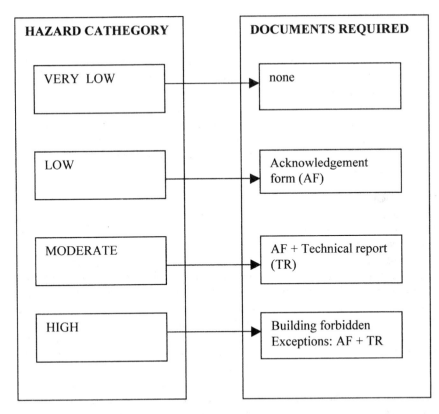

Figure 2. Administrative procedure for building permission

FINAL REMARKS

The experience of Andorra, has shown that preparation of hazard maps requires large amount of information., particularly, good quality geological data (thickness of superficial formations, quality of the rock mass, etc) and of previous landslide activity (necessary for frequency analysis), which is difficult to obtain. This information will affect the accuracy and applicability of the map, which in any case, must be permanently updated.

The administrative procedure for building authorisation has been conceived not as an additional constraint (difficulty) to the developers but as a guidance. By means of the GIS and its data base, they may know the type and nature of the hazard, if any, they are coping with. Furthermore, they know in advance the estimated magnitude and frequency of the event , thus allowing a preliminary cost-benefit analysis of the intended development. On the other hand, with this map and the administrative procedure, local authorities are expected to have better tools for land use planning and hazard management.

REFERENCES

Copons, R.; Altimir, J; Amigó J; Díaz, A and Vilaplana, J.M. 2000a. "EUROBLOC: Un modelo de simulación de caída de bloques y su máxima adaptación a la realidad". Geotemas, Vol 1: 219-222

Copons, R.; Vilaplana, J.M.; Altimir, J. y Amigó, J. 2000b: Estimación de la eficacia de las protecciones contra la caída de bloques. Revista de Obras Públicas, 3394: 37-48.

Corominas, J. 1996. "The angle of reach as a mobility index for small and large landslides". *Canadian Geotechnical Journal*, Vol 33: 260-271.

Corominas, J.; Esgleas, J. and Baeza, C. 1990. "Risk mapping in the Pyrenees area: a case study". Hydrology in Mountainous Regions II. IAHS Publication 194: 425-428

Corominas, J.; Copons, R.; Vilaplana, J.M.; Altimir, J. And Amigó, J. 2002. "Integrated landslide susceptibility analysis and hazard assessment in the Principality of Andorra". Natural Hazards (in press)

Hsü, K.J. 1975. "Catastrophic debris streams (sturzstroms) generated by rock falls". Geological Society of America Bulletin, 86: 129-140

Lopez, C., Ruíz, J., Amigó, J. and Altimir, J. 1997. "Aspectos metodológicos del diseño de sistemas de protección frente a las caídas de bloques mediante modelos de simulación cinemáticos" *IV Simposio nacional sobre taludes y laderas inestables*, Granada. Vol. 2: 811-823

Varnes, D.J. 1984. "Landslide hazard zonation: a review of principles and practice". Natural Hazards, 3. UNESCO. Paris. 63 pp.

Working with the community - public liaison in instability management at Lyme Regis, Dorset, England

G. M. DAVIS, West Dorset District Council, UK
K. COLE, West Dorset District Council, UK

INTRODUCTION
The popular seaside town of Lyme Regis is situated on an actively eroding stretch of the Dorset Coast and faces considerable challenges arising from coastal erosion and landsliding. The active landslipping has created spectacular coastal scenery which attracts hundreds of thousands of visitors to the town annually, making tourism the principal industry in the area. Major coast protection works and slope stabilisation schemes are required to ensure the integrity of the coastal areas of the town in the long term. The impact of the works on residents and the tourist industry, both in their completed form and in terms of the disruption they may cause during construction, are critical issues to be dealt at the design stage. This paper describes some of ways in which the District Council has invested in public liaison and consultation during the development of the coast protection and land stabilisation project.

Lyme Regis Environmental Improvements Project
The purpose of the project is to safeguard the coastal areas of the town in the long term by reducing the risk of destructive landslipping and coastal erosion, principally through major slope stabilisation and coast protection works (Brunsden 2002, Cole et al. 2002). The scheme is being implemented in phases in order of urgency (Fig. 1). Phase 1, which comprises a new seawall and promenade protecting the central part of the town, was completed in 1995. At the time of writing, Phase II is in detailed design stage and the remaining phases are in preliminary studies stage.

The schemes for the remaining phases will comprise major civil engineering works including beach replenishment, the provision of new foreshore structures, and extensive piling, drainage and earthworks on unstable slopes above. This will involve clearing large areas of land in order to install the works, construction on and adjacent to private land and considerable movement of construction traffic within the town. Due to the unstable nature of the coastal area, and the difficulty and risk of operating in the winter when groundwater levels are high, a large proportion of the works will need to be carried out in the summer season when the town is at its busiest.

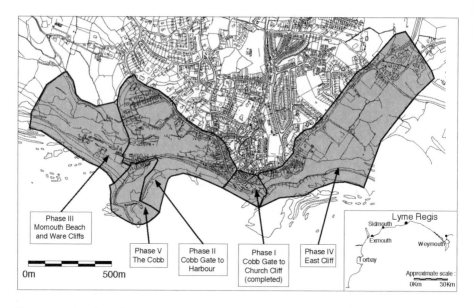

Figure 1. Location plan

PUBLIC LIAISON

West Dorset District Council invests heavily in public liaison to help the smooth running and long-term development of the project. The principal aims of the liaison are:

1. to keep the public, landowners, businesses and other organisations with an interest in the project up to date with developments

2. to discuss, answer questions and receive feedback on scheme proposals

3. to encourage debate on critical issues

4. to gather information relevant to the project and the design of the works

5. to reduce potential sources of conflict, particularly during the later stages of the project.

6. to encourage a spirit of openness and cooperation.

The principal methods by which the Council aims to achieve these objective are summarised in Table 1.

Method	Advantages	Disadvantages
Local office	Local presence greatly encourages communication and spirit of cooperation.	A considerable investment which does not attract grant-aid from Central Government.
News releases	Inexpensive, reach wide audience locally.	May be modified by media editor in preparation of final news story.
Leaflet drops	Council have complete control over content. May be delivered to all households in the town, if necessary.	Relatively expensive to prepare and deliver.
Local forum	A great medium for exchange of information and discussion. Allows communication with many different organisations at the same time.	Provides opportunity for antagonists to pillory Council. Firm control of meetings required to avoid sidetracking by individuals or groups pursuing objectives not related to coast protection.
Questionnaire surveys	Provides a definitive measure of public opinion.	Large surveys, such as those seeking opinions form the whole town, are relatively expensive to prepare, deliver and analyse.
Web site	Reduces officer time spent providing information and answering general queries.	Needs continual updating to be most effective.

Table 1. Principal methods of liaison and communicating with the public with advantages and disadvantages.

Local Office

The people of Lyme Regis have a strong tradition of independent thought and historically have resisted actions which were deemed to have been imposed upon them by 'outside' agencies. West Dorset District Council has, therefore, made a considerable investment in establishing within the town a team of specialist staff specifically to develop the Lyme Regis Environmental Improvements project. In time, this has given the feeling that the District Council team are 'part of the town' rather than faceless bureaucrats in a remote head office. The office provides a point of contact for communication and members of the public are encouraged to 'come in for a chat'. In this way, interested parties may obtain information quickly by face-to-face contact with the most appropriate staff. Copies of reports on various aspects of the coast protection project are available for examination and members of staff are on hand to answer specific queries A benefit to the project team is that locals often call in with information potentially relevant to the project, such historical data or the location of fresh ground movements. The office also serves as a point of contact for more general queries relating to slope instability such as those concerning the purchase of property or proposed new developments on potentially unstable land, both on the coast and inland. The local presence proved to be particularly valuable during winter 2000-2001 when accelerating ground movements threatened destructive landslide activity within built up areas of the town (Cole and Davis 2002).

A disadvantage of running a local office is that Council must spend additional money from its own purse to cover the office and staff overheads, which do not attract coast protection funding from UK Central Government. However it is considered that this is outweighed by the benefits gained in terms of improved public liaision.

Local Forum

A local coastal forum has been established, in one form or another, for many years with meetings usually held within the town every 6 to 8 weeks, or more frequently if critical issues are under discussion. A typical meeting (Fig. 2) would include an update on the overall progress of the project, a detailed presentation on a particular topic by Council staff or a specialist consultant, followed by a question and answer session and discussions. Exhibition material is often put on display. The forum attracts representatives of numerous organisations within the town as well as businesses people and interested residents, reducing the need for meetings to be held with each of those organisations individually. Audiences of between 60 and 100 are common. The comments of members of the forum on specific options for stabilisation and coast protection works are particularly welcomed as this helps to identify, at an early stage, which proposals would be acceptable, or unacceptable, to the town.

On the negative side, the forum offers an ideal platform for opponents of the Council to criticise it public, often on spurious grounds. There is also a tendency for the discussions to be dominated by a few vociferous individuals not necessarily representing the overall view of the town, and sometimes pursuing issues unrelated to coast protection or slope stabilisation. Firm control of the meetings by the chairman of the forum is required to keep these distractions to acceptable levels.

Leaflet Drops and News Releases

The issue of news releases to the various local media and delivering leaflets to households and businesses within the town, provide ways of getting information on the project to large numbers of people quickly. Media releases are inexpensive to prepare and will reach a large proportion of the local population. However, the context in which the information is ultimately presented to the public is controlled by the media editors who will often wish to impose a certain 'slant' on the story. For example, the announcement of the plan to implement a £20M coast protection project received just a few column inches of coverage in one particular local newspaper. Information leaflets (Fig. 3), distributed around the town, although more expensive to produce and to deliver, have the advantage of providing the public with unedited information. Leaflet drops carried out across the whole town or may be targeted at specific areas, for example those that would be affected by a specific set of stabilisation measures.

Figure 2 A coastal forum meeting.

Figure 3. Example of an information leaflet distributed to households in Lyme Regis

Questionnaire Surveys
Questionnaire surveys are used from time to time to gain a quantitative measure of public opinion on critical matters which may be unobtainable by any other means. Questionnaires may be distributed at major coastal forum meetings or by door-to-door delivery. Occasionally, surveys are carried out which aim to obtain the opinion of visitors to the town, as well as residents, and this may require canvassing holiday makers in public areas such as the seafront. Care must be taken in the analysis and interpretation of the result of the surveys so that judgements may be made on their representativeness and meaning.

Web Site
West Dorset District Council maintains a home page on its web site specifically for the Lyme Regis Environmental Improvements project. This includes notes on the background to the project, landslide information, recent news releases and links to other relevant web sites. As well as providing a useful source of information for the public, the web site reduces the staff time required answering basic queries on the project or instability in the area generally.

CONCLUSIONS
West Dorset District Council invests heavily in public consultation for the Lyme Regis coast protection and stabilisation project, and instability issues in the area generally. Although there are some disadvantages to this approach, such as relatively large amounts of mantime required and providing opportunities for negative criticism, these are outweighed by substantial benefits. High levels of liaison help to progress the stabilisation projects in a positive and constructive manner in which the local individuals, traders and organisations may be involved in the development process throughout, thereby minimising potential conflict in the later stages of the project.

ACKNOWLEDGEMENTS
The paper is published by permission of West Dorset District Council. The project receives grant-aid funding from DEFRA.

REFERENCES
Brunsden D. 2002. The Fifth Glossop Lecture - Geomorphological roulette for engineers and planners: some insights into an old game. Quarterly Journal of Engineering Geology and Hydrogeology, 34.
Cole K. and Davis G.M. 2002. Landslide warning and emergency planning systems in West Dorset, England. Proceedings of the International Conference on Instability - Planning and Management. Thomas Telford, London
Cole K., Davis G..M., Fort D.S. and Clark A.R. 2002. Managing coastal instability - a holistic approach. Proceedings of the International Conference on Instability - Planning and Management. Thomas Telford, London.

Landslide hazard in Garhwal Himalaya (India): Strategies for roadside landslide management

DR VARUN JOSHI, G.B. Pant Institute of Himalayan Environment and Development Garhwal Unit, Post Box 92, Srinagar (Garhwal) U.A. 246 174, India, and
DR A.K. NAITHANI, Dept. of Geology, HNB Garhwal University, Post Box 68, Srinagar (Garhwal) U.A. 246 174, India

ABSTRACT
Geologically the Himalayan belt is very young and complex. Its rugged topography, in combination with commonly occurring earthquake/other natural hazards and occasional high intensity rainfall, areas of weak earth material and complex geological structures, contribute to significant mountain hazards. In the wake of these hazards the Himalayan region is leading towards degradation. The repercussions of such changes decipher from depleting resources and environment. The occurrences of various types of landslides/mass movements, depleting spring sources etc. are few examples of these changes. The frequency of landslides in Garhwal Himalaya varies from area to area depending on the underlying structure, physiographic setting and anthropogenic changes taking place. It is well established that the developmental activities like construction of roads, dams, tunnels, infrastructures, etc. are encouraging the incidences of landslide frequency in Garhwal Himalaya many fold. Generally the people feel the damage or trouble whenever there are losses of life or property. Every year, the highway network in the Himalayan region sustains damage at hundreds of locations during the rainy season, due to incidence of landslides. Two to five landslides have been reported in every kilometer of a Himalayan road. Sometimes because of the lack of remedial measures or management a small landslide is turned into a huge one in the due course of time and causes huge losses in terms of money or delay in vehicular traffic. If the landslides triggered by any developmental activities are taken into account for management practices well in time, then the severity of landslide may be reduced. An extensive survey in different parts of Garhwal Himalaya has been made regarding to road landslides. The remedial measures so far applied on few landslides have also been studied in detail. It is observed that if the engineering and bioengineering measures are put together for the landslide management, then the severity of the landslide can be controlled upto a considerable level even with involvement of less amounts of money.

Such types of measures are long lasting in nature. In this paper, the detailed engineering and bioengineering remedial measures have been suggested for various roadside landslides in the study area.

INTRODUCTION

Landslide hazards in the Himalayas are due to some complex processes and are most commonly triggered by precipitation. The combination of geological, geomorphic, meteorologic, seismic factors and human activity control these. It is well known that the Himalayan region is the youngest mountain and is also tectonically very active and hence inherently vulnerable to hazards. Roads and other developmental activities are important contributory factors responsible for various types of landslides. It has been observed that the frequency and instability of landslides and mass wasting have increased substantially in the last five decades. This increase is mainly attributes to two main causes i.e. deforestation and anthropogenic activities (Valdiya, 1987). Landslides are also considered to be an unfortunate consequence of increased mountain development activity (Eisbacher, 1982).

The road network in a hilly region, especially in the Indian Himalayan region, is considered to be a 'lifeline' since it provides the only mode of transport and communication. A road network is also said to be a marker of economic development, as it provides the basic infrastructure of any kind of investment and for harnessing the economic potential of the area. Apart from this, during natural catastrophies i.e. earthquakes, floods, cloud bursts, forest fires, etc, hill roads play an important role for rehabilitation and relief operations. The compilation of available information indicates that more than 70,000 km long of road network is under operation and many more are under construction in the Indian Himalayan region. Every year, the highway networks in this region sustains damages at hundreds of locations. Each km of road cutting generates 40 to 80 hundreds cubic meter of debris in Himalayan region (Valdiya, 1987). Road construction ranks with and perhaps ahead of surface mining and deforestation as the major human factor in the initiation of new landslides in Himalaya (Bruijnzeel and Bremmer, 1989). The damage caused by landslides was estimated to cost more than US $ one billion, causing more than 200 deaths every year which is about 30 percent of the total of such losses world wide (Li, 1990). The maintenance expenditure of landslides and for keeping roads open to vehicular traffic costs the Government exchequer approximately 100 crore rupees per year for Himalayan region (Thakur, 1996).

Various roads in Garhwal Himalaya are either of strategic importance or pilgrimage routes. Therefore, all the major roads have their own significant importance. Due to frequent landslide incidences most of the roads suffer vehicular problems especially during rainy season. It is observed during the survey that if the measures are in conjugation with engineering and bioengineering put together, they are most successful. Therefore, an attempt is also made to give some suggestions to stabilize landslides using engineering and bioengineering measures.

GEOLOGY AND GENERAL GEOMORPHOLOGY

The area tectonically comprises of two separable major lithostrigraphical units i.e. Garhwal Group (Lesser Himalaya) and Central Crystalline Group. These groups are separated from each other by a major tectonic contact know as the Main Central Thrust (MCT). The Lesser Himalayan zone is composed of low to medium grade metamorphics. While the Central Crystalline is composed of medium to high-grade metamorphic units. The main rocks in the area are dolomitic limestone, calcareous quartzite, quartzite, metavolcanics, garnetiferous schist, micaschist, biotite schist, amphibolite, gneiss, granites etc. The area has suffered multiple deformations, accompanied by thrusts and faults (Valdiya, 1980).

Highly rugged conical peaks and steep sided valleys are the prominent physiographic features in the higher Himalaya, whereas, the lower Himalaya contains comparatively less steeper slopes with truncated spurs and almost flat hilltops. The main ridges or major water divides generally follow the Himalayan trend i.e. NW-SE and conform to the major tectonic trends. Thus, the area is geologically controlled by the major structures and lithology as well. The valley generally contains well developed fluvial and fluvioglacial terraces and the maximum population lives in the valleys.

METHODOLOGY

In order to identify the hill slope instabilities, an extensive survey in different phases was conducted along some road sections of Garhwal Himalaya. Details of location specific studies on causes, years of trigerrance and measures under taken/implemented on different slides, were collected during the field visit. The staff of the concerned department has also been interviewed to know about remedial measures being implemented on the slides. The other slides, which are not undertaken for such measures, were studied in detail, in order to know the triggering year, probable causes, meteorology and geology etc. Depending upon the causative factors and other conditions of the slides, suggestions were made for landslide stabilization, which mainly include engineering and bioengineering techniques. A comparison between engineering and bioengineering or the conjugation of engineering and bioengineering was also made during the study.

STUDY AREA

The study area was taken up mainly in the Alaknanda and Bhagirathi river catchments in Garhwal Himalayan region, because of extensive expansion of roads and settlements in these areas during the last few decades. The rainfall generally induces landslide occurrence during the monsoon (mid-June to mid–September). The mean annual rainfall is about 1734 mm, the maximum rainfall recorded at Ukhimath is 2257 mm and minimum 700 mm at Tehri (Joshi and Negi 1995). In general, the humidity in the area is found throughout the year causing moistening and chemical, mechanical weathering of rocks, under day and night temperature rhythms.

DISTURBANCES CAUSED BY ROAD CONSTRUCTION

The construction of roads has created many types of ecological disturbances and environmental degradation, though it is considered a key factor for development of any area. The construction of roads creates ecological disturbances in the hills, mainly in the following ways:

Loss of vegetation

The prerequisite of any road construction activity is the felling of standing trees along the alignment of the roads and the adjoining areas. Beside this, a large number of trees/shrubs/herbs get uprooted either by fall of debris on hillside slopes below the road alignment or by the landslides occurrence after the road construction.

Geological disturbances

The road construction activity in the hills, specially in unstable areas, preoccupied by thrust/fault/joint etc. invariably, involves operations of blasting. The blasting results in the exposures of the faults, fissures/joints, which were earlier covered, after this water seepage increases and creates further instability on the hill slope. Due to this disturbing effect, activation of a large number of landslides has been observed specially during rainy seasons (Haigh , Rawat and Bartarya, 1989).

Change in natural drainage systems

The construction of roads in one or other ways interrupts the natural drainage system. The runoff from the hill slopes is uniformly distributed over the entire hill slope, but it gets concentrated at the point where cross drainage works are provided on a road. The cross drainage works are often located without considering the adverse soil erosion likely to be caused by the flow beneath the road alignment. The debris from the hill slide cutting and landslides sometimes blocks the channels and streams leading to different types of problems around the roadside as well as on the hill slope.

Drying of springs

The cutting of roads exposes the hillsides, which leads to greater evaporation of water from the area. The debris spread over the hill slope also reduces the infiltration of rainwater resulting in greater runoff. Subsequently in due course of time the natural springs dry up due to non availability of ground water (Valdiya and Bartarya, 1981).

Silting of river basin

The road construction activity in a catchment area generates a tremendous amount of debris. The debris generated by cutting goes down due to runoff on the hill slopes through nearby streams and eventually into the rivers. When the river reaches the foothills, due to its low velocity, it cannot carry the silt load, which gets deposited. If a dam is being constructed down stream, this silting results in a loss of capacity of the reservoir.

LANDSLIDE HAZARD ON HILL ROADS

Various landslide hazards occur on the hill roads of Garhwal Himalaya every year. These roads are blocked at many locations due to rock falls, landslides, avalanches, bank erosion, debris flows, collapse/washing away of bridges, etc. The annual rate of debris generation from the seasonal mass movements is 724 m^3/km/yr in Jammu-Srinagar (Kashmir) Highway, 411 m^3/km/yr in the Tanakpur-Tawaghat road (Kumaun), 591 m^3/km/yr in Kameng Highway (Arunachal Pradesh). On an average every year, the mass movements on the roads in the entire Himalayan region produce about 2.4 million cubic meters of debris. In 1983, a 12 km stretch of Dehra Dun-Mussoorie road suffered from 10 landslides, in Dehra Dun-Tehri road about 10 slides each generated about 550 m^3 debris, and one chronic landslide for each 10 km stretch of road has occurred. 102 landslides were counted within a stretch of 11 km along Ghansali-Tilwara road in the Bhagirathi river catchment. More than 40 major landslides were counted in NH-58 from Kaudiyala-Joshimath (Alaknanda river valley, Sikdar, 2000). Since not a single road is free from landslides or related phenomenon, it could be advised that a detail survey for construction and probable measures to check the landslide in the initial stages may be suggested or implemented. During the construction of Banepa-Sidhulli road in Nepal all the possible problems were identified and measures were advised and executed in the primary stage of slope stabilization (Figure 1).

STRATEGIES FOR ROADSIDE LANDSLIDE MANAGEMENT

It is evident from the above study that the roadside landslide plays an important role in creating several types of environmental problems in hill areas. Management guidelines should be prepared, to ensure that environmental considerations are integrated with the road surveys, design, landslide stabilization and monitoring etc. The guidelines are to be used for new road construction and maintenance of old roads. If these guidelines are properly implemented, the environmental impact resulting from road construction, operation and maintenance could be minimized. Ultimately, it is hoped that the implementation of the guideline measures will improve the road performance and reliability, increase benefits to local residents and will maximize cost-effectiveness. Therefore, it is important to develop a strategy to control these landslides, which could be economically viable and long lasting. It is observed from literature by other parts of the world, the measures used in conjugation with bioengineering and engineering are most successful, economic and long lasting (Li, 1996).

The basic requirement for bioengineering measures is to evolve a site-specific detail. Each site in which measures are to be taken is illustrated by a drawing giving a brief description of its characteristics including the causes of failure and the bio-engineering/engineering techniques to be implemented. This illustrates the need for a flexible approach to bioengineering and engineering measures ensuring that design matches the distinct geological, meteorological and geomorphological conditions of each site. The following site details are required for the bioengineering/engineering measures:

Slope and Altitude
A profile of the slope showing the slope angles, the position of structures, length of slope and land use around the slope which effects the type and rate of flow over the surface and the subsurface, as well as the likelihood of erosion caused by surrounding land use practices.

The altitude affects the temperature and therefore, rate and type of weathering. It has an important influence on the occurrence of natural vegetation types and selection of plant species. On the basis of climatic zones, the vegetation associated with a particular zone can be selected easily.

Year of failure
The year of failure means that in which the slope failure was first noticed. The year of failure indicates the length of time the slope has had to repair itself. Each year has its specific reasons i.e. drought, heavy rainfall, earthquake, road construction etc. Some of the slope failures might have a combination of factors related to road construction, monsoon, weathering, etc.

Aspect
Aspect is important because it directly influences (a) the amount of rain falling on slope, and (b) the dryness of a slope, particularly in winter and spring. In general, north facing slopes receive less rain because they lie on the leeward side of the ridges, but retain moisture for much longer time because they receive less direct sunlight. The higher temperature on south-facing slopes will increase evapotranspiration. East-facing slopes receive the morning sun while the west-facing slopes receive afternoon sun (when

temperatures are high). The magnitude of the effect depends on slope angles and the position of the sun behind the horizon.

Rainfall
The rainfall data of the nearest recording station is required. The data are important because (a) the probability of slope failure is directly related to the total rainfall, intensity, duration and timing of precipitation, (b) based on the trend of rainfall, species of plant/grass can be selected.

Geology and soil type
Both geology and soils are of vital importance in determining the type and likelihood of landslide and erosion on a site. The amount of fractures, joints and the size of gaps in the rocks should be noted. Soil types and underlying parent materials will have a direct effect on the ability of vegetation to establish on a site.

Causes of failure
It is important to know the causative factor of slope failure, time of occurrence and where the problems are large and likely to recur. The identification of site specific causative factors are most important for further suggestion of suitable measures to be undertaken.

Engineering works
The suggested engineering measures to be taken are retaining structures, catch drains, etc above and below the site. These works are carried out before planting. The plan of engineering work should be shown on main drawing before application on site.

Bioengineering works
The bioengineering works involve selected site specific species of trees, shrubs and grasses on the site in various combinations and patterns depending on site characteristics and the engineering role the vegetation is designed to fulfill. All the vegetative material should be readily grown in the nearby nursery.

Other measures
Other measures depending upon the site-specific problems may include bioengineering, engineering in conjugation or separately, jute net covering, protection planted sites, etc. It is imperative to protect the plantation site from grazing, in which protection measures are mainly undertaken by watchmen who guard the sites, also reseed, replant, weed, water under certain circumstances, provide mulch and apply compost collected during weeding. All these activities greatly increase the growth rate of vegetation thus reducing the time between planting out, establishment and canopy closure of the vegetation.

CONCLUSIONS

Himalayan landslides are attributed to the combined effect of intense human activity, rainfall, deforestation and seismicity of the region. Landslide problems of the Himalayas are complex and defy simple solutions. The chronic landslides cannot be checked but the further expansion may be reduced upto some extent. The bioengineering and engineering measures are successful for small/shallow landslides. It is observed that the small landslides in due course of time convert into the chronic ones. Therefore, it is imperative that the measures taken well in time may reduce their further expansion. The detailed field survey indicates that the measures in conjugation with bioengineering and engineering are more successful than those which were taken individually (Figure 2a, 2b and Table 1). For the implementation of vegetative work, a calendar is followed in Nepal (Anonymous, 1993). Similar types of calendar may be prepared and followed for the Garhwal Himalayan region also.

In general, multidisciplinary approaches for the implementation of landslide control measures are not in practice all along the Indian Himalayan region. There are few examples where bioengineering, engineering and geological measures are put together to control the slide (Joshi and Krishna, 2000; Joshi et al., 1998; Joshi et al., 2001). If we can put all the efforts together to get the multidisciplinary approach we can get success.

The ultimate solution is believed to lie in making the best use of local resources, skills and of the technology that cuts down both the time and the cost of construction, and will be long lasting. It is suggested that if the measures taken into consideration at construction time of the road, the recurring expenditure may be reduced to zero. During construction with measures (engineering and bioengineering measures) the increment of cost is only 10 % of the total cost. In due course of time the recurring expenditure on these roads will be zero.

ACKNOWLEDGEMENTS

The author (VJ) gratefully acknowledges the encouragement and facilities provided to carry out this work by the Director and Core Head, LWRM of our Institute. The author (AKN) expresses thanks to Head, Department of Geology, HNB Garhwal University for discussion.

REFERENCES

Anonymous (1993). Bioengineering for road protection in Nepal. Published by Eastern Region Road Maintenance Project, Nepal.

Bruijnzeel, L.A. and Bremmer, C.N. (1989)- Highland-lowland interactions in the Ganges Brahmputra river basin: a review of published literature. International Centre for Integrated Mountain Development, Kathmandu, Nepal, Occasional

Paper 11: 136pp.

Eisbacher, G.H. (1982)- Slope stability and landuse in mountain valleys. Geoscience Canada, Vol. 9(1): 14-27.

Haigh, M.J., Rawat, J.S. & Bartarya, S.K. (1989)- Environmental indicators of landslide activity along the Kilbury road, Naini Tal, Kumaon Lesser Himalayas. Mountain Research and Development, Vol 9 (1): 25-33.

Joshi V., Naithani, A.K. and Negi, G.C.S. (2001). Study of Landslides in Mandakini River Valley, Garhwal-Himalaya, India. In GAIA. (May issue).

Joshi, V. & Negi, G.C.S. (1995)- Analysis of long term weather data from Garhwal Himalaya. *ENVIS Bulletin*. Kosi, Almora, India: Gobind Ballabh Pant Institute of Himalayan Environment and Development, Vol. 3 (1 & 2): 63-64.

Joshi, V. and Krishna, A.P. (2000). Control measures for soil erosion, landslides and debris flow in Hindu-Kush Himalayan belt of People's Republic of China. Indian Journal of Soil Conservation. 28(1), 1-6.

Joshi, V., Naithani, A. and Negi, G.C.S. (1998). Study of landslides caused by natural and anthropogenic reasons in Garhwal Himalaya. Proceedings of National Conference on Disaster and Technology (Eds.) Mehrotra, N. & Panicker, B.G., Manipal Institute of Technology, Manipal-Karnataka. pp.48-54.

Sikdar, P.K. (2000). Planning and management of road network in Himalaya. National Seminar on Geodynamics and environmental management of Himalaya. HNB Garhwal University, Srinagar (Garhwal), India.

Tainchi, Li (1990)- Landslide management in the mountain area of China. International Centre for Integrated Mountain Development, Kathmandu, Nepal, Occasional paper No 15: pp.35.

Tainchi, Li (1996). Landslide hazard mapping and management in China. . International Centre for Integrated Mountain Development, Kathmandu, Nepal. 33pp.

Valdiya, K.S. (1980): Geology of Kumaun Lesser Himalaya. Wadia Institute of Himalayan Geology, Deharadun, India.

Valdiya, K.S. (1987)- Environmental Geology Indian context. Tata McGraw Hill Publication.New Delhi, 583 pp.

Valdiya, K.S. and S.K. Bartarya (1991). Hydrological studies of springs in the catchment of the Gaula river, Kumaun Lesser Himalaya, India. Mountain Research & Development 11(3): 239-258.

Table 1. Details of some landslides studied along road section in Garhwal Himalaya

Name of Slide	Location on Road side	Causative factors	Remedial measures	Rate of success
Barasu	213 km on Haridwar-Kedarnath	Cloudburst, toe cutting, road construction, earthquake tremor	Construction of toe wall, Diversion drain	Partly Success
Davidhar	215.5 km on Haridwar-Kedarnath	Reactivation of strike slip fault, heavy rain, seepages, toe cutting	Construction of Drum diaphragm & Retaining wall	Partly Success
Kaliasaur	147 km on Haridwar-Badrinath	Presence of the plane of shearing in quartzite & road construction	Retaining wall, Drum diaphragm wall, Geotextiles	Partly Success
Tangri	252 km on Haridwar-Badrinath	Presence of fault zone, crushed and fragmented zone, toe erosion	Retaining wall	Unsucc essful
Patal Ganga	255 km on Haridwar-Badrinath	Reactivated of old slide, three sets of joints & plane of shearing	Retaining wall	Unsucc essful
Nakurchi	51km on Haridwar-Badrinath	Heavy rain, weathered & fractured rocks, road cutting, debris accumulated on slope	Retaining wall	Unsucc essful
Devprayag	95.9 km on Haridwar-Badrinath	Heavy rain, weathered sheared & jointed rocks, road cutting	Retaining wall	Unsucc essful
Chinialisaur	117 km on Rishikesh-Uttarkashi	Weathered formation, water seepage along drains, heavy rainfall, toe erosion by the river	Retaining wall	Partly Success
Matli	139.6 km on Rishikesh-Uttarkashi	Weathered formation, presence of fault, road construction, toe erosion by river.	Retaining wall, Drainage control	Partly Success
Srinagar	2.8 km on Srinagar – Pauri	Weathered, fractured & shear planes & thick alluvial matter, road construction	Retaining wall & bioengineering	Success ful

Figure 1 Banepa–Sindulli road in Nepal with engineering and bioengineering measures

Figure 2a Successful stabilization of a landslide by engineering and bioengineering measures in Garhwal Himalaya

Figure 2b Engineering measure for landslide stabilization in Garhwal Himalaya

A dynamic framework for the management of coastal erosion and flooding risks in England

E M LEE, University of Newcastle upon Tyne, UK

INTRODUCTION

A basic and long-standing principle of British law is that individuals have the right to protect their own property against flooding and coastal erosion, under common law. Hence, the primary responsibility rests with the landowner, not with the state. However common law rights have been altered and reduced over time by statute law to allow state intervention in the interests of the common good. Individuals do not have to exercise their rights, although case law has indicated that landowners or occupiers have a general duty to their neighbours to take reasonable steps to remove or reduce hazards if they know of the hazard and of the consequences of not reducing or removing it.

Individuals or private businesses have either avoided high-risk areas, accepted the losses as the price to pay for living and working in such areas, or have sought to "improve" the conditions through private engineering works. Maintenance, repair and clean-up are often a central element of most strategies for dealing with natural hazards (Lee 1995). Insurance is available for mitigating the losses associated with flooding or landslip (but excluding landslide losses caused by marine or river erosion). Occasionally compensation has been sought through litigation.

Although the ultimate responsibility for managing risks rests with individual property owners, the state has gradually acquired a key role in addressing a number of specific problems (e.g. Lee 1993). These include:

- the provision of publicly funded flood defence works and coast protection works to prevent erosion or encroachment by the sea, under permissive (not mandatory) powers;
- the provision of flood warning systems;
- funding and co-ordinating the response to major events;
- controlling development in areas at risk and minimising the impact of new development on risks elsewhere, through the land use planning system.

Instability – Planning and Management, Thomas Telford, London, 2002, 713–720

THE BROADER CONTEXT FOR RISK MANAGEMENT

The development and functioning of the risk management framework needs to be viewed in the broader social, economic and political context within which it operates (e.g. Palm 1990). In addition to understanding these macro-scale relationships, it is also necessary to consider the individual response to risks (micro-scale responses). Macro-scale risk management involves the planning of public expenditure to increase social welfare by reducing flood and erosion losses. As only a minority of the community is affected, the use of public funds can be viewed as a subsidy (e.g. extending the property life or safeguarding investments). It has been argued that this intervention is typically directed towards those least capable of withstanding the financial impact of losses and the least able to use political pressure to reduce their vulnerability (e.g. Penning-Rowsell et al 1986). Coastal defence is, therefore, a means of safeguarding the most vulnerable in society and helping towards the redistribution of wealth.

The pervading attitude amongst the general public has been that floods and erosion losses are "Acts of God" and not the fault of those at risk. Thus, it can be argued that the State intervenes on behalf of the unaffected community to assist the blameless, unfortunate victims of circumstance. At a broader level, floods can have a significant impact on public health (although loss of life is fortunately rare) and cause widespread financial damage (it has been estimated that a failure of the Thames Barrier could result in property losses in London alone, in the order of £10B; Clement 1995). Flood defence, therefore, can be seen as contributing to the delivery of social welfare objectives of prosperity, health, security and opportunity. For example, flood or erosion prone areas provide society with the opportunity for exploitation, such as urban and industrial growth; these opportunities can only be realised with coastal defence investment. There are, of course, other mechanisms for delivering improved social welfare (e.g. education, health and efficient infrastructure). Allocation of public resources for coastal defence is fundamentally a political decision, influenced by the need to find an acceptable balance between investments in a wide range of competing public services. As coastal flooding and erosion only affect a minority, but are funded by the wider public, the relative perception of the investment priorities will vary across society.

The differing perspectives on the relative merits of public investment in coastal defence as opposed to other public services is compounded by the fact that defences may result in environmental losses. To the individuals affected by coastal hazards, the benefits of coastal defence may far outweigh these losses. To others, the losses represent an unacceptable price to pay for subsidising the lifestyle of a few. Others, of course, may remain indifferent to both perspectives. It follows, however, that reducing coastal risks does not necessarily improve social welfare; for many it may lead to deterioration in their quality of life.

An appreciation of the impact of climate change and sea-level rise on how risks are managed needs to take a broader perspective than simply the quantification of how certain processes may change. Indeed, as Hewitt (1983) has suggested, awareness of and response to natural hazards are less dependent on the physical environment than on the "*ongoing organisation and values of society and its institutions*".

Figure 1 The macro-scale risk management framework

A MACRO-SCALE FRAMEWORK FOR RISK MANAGEMENT

Figure 1 presents a framework for risk management that emphasises that macro-scale decision-making (i.e. risk evaluation) will reflect four *arenas*: *the physical environment*, the *human environment*, the *cultural environment* and the *legal and administrative framework*. Risks arise when the natural hazards associate with the physical environment (i.e. flooding and erosion/instability) overlap with the human environment (i.e. land use and people). However, all four arenas factors interact and set the overall context for risk management in any country or area. These arenas are considered below in terms of their influence on the decision-making process.

It has been suggested that the need to make management decisions arises from *"stress"* (Kasperson 1969). These can include flood or erosion events. One such event was the 1953 east coast floods, during which many thousands of homes were flooded and some 300 lives were lost, with estimated losses of around £5B (at current prices; Willis Faber and Dumas Ltd, 1996). A Departmental Committee, the Waverley Committee, was set up to investigate the causes of the disaster, consider standards for future defences and the need for a flood warning system. The committee recommended, in 1954, that the water level reached in 1953 should be the general maximum standard of protection that could reasonably be afforded. As a consequence, coastal authorities embarked upon an extensive programme of reconstruction and improvement of defences. On a much less dramatic scale the Holbeck Hall landslide, Scarborough has had a significant impact on coastal management practice, because of both its scale and timing (Lee 1999).

In other circumstances, the *"stress"* might be a change in legislation (e.g. the implementation of the Habitats Directive through the Conservation (Natural Habitats &c Regulations 1994), policy (e.g. the introduction of High-Level Targets, planning guidance or Biodiversity Targets) or the availability of resources (e.g. a change in the eligibility for grant-aid or EC funds).

Whether or not an operating authority *responds* to these stresses depends on their legal position. A key aspect of the coastal defence framework is that powers to act are permissive, not mandatory. However, there may be other legal responsibilities and duties that will dictate whether action is required (e.g. common law duties, covenants, Occupiers Liability, the Highways Acts etc.; see Lee et al 2001). How the authorities respond is constrained by the legislative framework, as it provides the context for what can and cannot be achieved (see Lee 1995). For example, at present, communities are either protected or they are not. This tends to intensify the pressure during the lobbying for defences. Often the argument is polarised between safeguarding local homes and businesses or contributing to national nature conservation targets. To many it is clear that alternative risk management options are required. For example, the House of Commons Agriculture Committee wrote: *"we are firmly convinced of the need to put in place a robust financial mechanism for the reimbursement of property holders and landowners whose assets are sacrificed for the wider interests of the community"* (Agriculture Committee 1998).

Pressures exerted from interest groups and the broader socio-economic and political context will also influence the response. For example, there can be powerful lobbying for new coastal defence works (Lee et al 2001). As O'Riordan (1981) and Penning-Rowsell e al (1986) have both suggested, *"the invisible political power of influence through social and political connection is far more significant than the publicly documented expressions of pressures on decision-making"*.

The *perception* of coastal risks and the management response varies across society and is a central issue in trying to understand how conflicts between coastal defence and conservation have arisen. For many, coastal erosion is unacceptable, raising visions of a loss of national resources to a hostile invading power. As John Gummer (2000), former Minister at MAFF recently wrote:

"Of course it has happened before. It's just that the last time our shores were successfully invaded was 1066. Now the east coast is being crossed again, and more effectively than by Norman or Dane. East Anglia is threatened as far inland as Bedfordshire. Already, more than 70 per cent of its beaches are in retreat. If erosion goes on at its present rate, my own constituency of Suffolk Coastal will simply continue falling into the sea...Britain is proud of being an island, and her people of being an island race. The time has now come for us to pay the price of defending this blessed plot from the very sea which has been our defence so often in the past".

This *"fortress Britain"* view is in direct contrast to the environment-based view of erosion as an essential process in the functioning of the coastline and necessary for maintaining and sustaining many nationally and internationally valued landforms, landscapes and habitats. From this perspective (the *"living coast"*), erosion is an acceptable price to pay for conserving the wilderness and natural beauty of the coastline; public intervention should be limited to providing only essential and sustainable defences. The "fortress Britain" view has great public, media and political (at least at a popular level) support, whereas the "living coast" view is supported by European and national legislation and targets. Trying to reconcile these two contrasting perspectives in managing coastal erosion risks is a major challenge for coastal managers.

THE NATURE OF THE RISK MANAGEMENT FRAMEWORK
The key features of the framework include:

1. it is extremely *complex*, reflecting the need to reconcile the demands for publicly funded defences with broader concerns over environmental consequences and competing demands on limited resources. Statute law has, therefore, introduced (Lee 1995):
   - consenting arrangements which ensure that management measures do not affect other interests or increase the level of risk elsewhere;
   - provisions to ensure the conservation and enhancement of landscape and nature conservation features, involving the protection of designated sites and areas of national and international importance;
   - consultation arrangements between key interest groups whose interests may be affected by risk management measures.

The complexity of existing arrangements is highlighted by the wide range of bodies and authorities with responsibilities that address or can be influenced by erosion and flooding problems. The variety of policies, decisions and actions made by these numerous bodies with an interest in aspects of management can influence the way in which management objectives are realised or achieved. In this context, neither the *coastal defence system* nor the *planning system* - the cornerstones of risk management in England - can be considered to operate in isolation and, hence, their interrelationships with other "systems" are of particular significance.

2. *the current English framework is the product of slow evolution,* over the last 1000 years or so. The present administrative framework for the management of natural hazards must, therefore, be seen to be the product of the way they have presented problems to society in the past.

3. *different countries have different risk management frameworks,* reflecting different attitudes to how the State should become involved in social welfare. For example, France has developed a mechanism for compensating affected parties. In France, the Law Barnier (2nd February 1995) authorises the ex-appropriation and compensation by the Government of all property threatened by natural risks when the remedial works are too expensive to undertake. Compensation is funded from a State Surcharge of 9% that is added to all property insurance premiums. A Risk Prevention Plan (PPR) determines the areas where a natural risk is foreseeable. The PPR is intended to allow action to be taken in advance by the proprietor and the local authority.

4. *The framework is dynamic and is continuously adjusting to changing circumstances.* Changes in any of the arenas (the physical environment, the human environment, the cultural environment and the legal and administrative framework) can lead to adjustments in the framework and, hence, the ways in which risks can be managed. For example, The Coast Protection Act 1949 was in response to limited local authority resources after WWII.

5. *The framework will continue to change in the future.* However, there is considerable uncertainty over both the stimuli and the direction of change. Possible causes of change over the next 50 years include:

    *   Changing perception of tolerable/acceptable risks (e.g. the impact of the recent Paddington & Hatfield rail crashes), could lead to pressure for improved standards of defence;
    *   Litigation; greater public challenging of local government/EA decision-making on risk (e.g. some areas sacrificed for the common good);
    *   Enhanced nature and geological conservation legislation.

Understanding the risk management framework and the broader social and political context provides a basis for *speculating* about how climate change and sea-level rise will impact upon SCOPAC over the next 50 years or so. It should be stressed that because of the nature of social systems, there is probably more uncertainty as to how society and politicians will respond to these changes than their impact on coastal processes.

SPECULATIONS ON THE IMPACT OF CLIMATE CHANGE AND SEA-LEVEL RISE ON RISK MANAGEMENT

Over the next 50 years, climate change and sea-level rise will result in an increase in the probability of damaging events along much of the coast of England. However, it is uncertain as to how the operating authorities will be able to manage the increased risks. To maintain the current standards of protection along the whole coastline will require considerable investment in defence improvements and maintenance. As publicly funded risk reduction measures are essentially a form of social welfare, in "competition" with other public services for limited resources, it remains to be seen as to whether the funds allocated for coastal defence will match the future demand. Although the overall funds may grow, the rate of growth could be significantly lower than the actual demand.

One possible consequence is that defences that are currently protecting marginally economic and clearly uneconomic sites will either be abandoned or maintained at a lower standard of protection. It should be appreciated, however, that society has become less risk tolerant, especially from public/private services as illustrated by the reaction to the recent Hatfield and Paddington rail disasters. It follows that there may be a need to improve the standards of protection in high-risk urban areas to reflect these trends. This would lead to increased polarisation in the exposure to risk experienced by individuals in built up and rural areas.

Climate change and sea-level rise is likely to generate additional pressures on a variety of coastal zone uses, from tourism and amenity uses, marine aggregate extraction (e.g. for beach feeding programmes), port and harbour operations to nature conservation and the protection of historical sites and monuments. For example, the long-term implications of shoreline management policies (in association with a rising sea-level; Full Retreat Scenario) to coastal habitats (all European sites: possible, potential, candidate or designated SAC/SPAs, and Ramsar sites) along the England and Wales coastline are predicted by Lee (1998, 2001) to include:

- a net loss of freshwater and brackish habitat of around 4000 ha, primarily wet grassland (c3200 ha) but also including significant areas of coastal lagoon (c500 ha) and reed bed (c200 ha). These losses would all be associated with managed retreat strategies.
- a net gain of intertidal (saltmarsh and mudflat/sandflat) habitats of around 2220 ha. The overall net gains associated with managed retreat (c12500 ha) are important in compensating for the expected losses due to coastal squeeze and erosion on the unprotected coast.

These, and other pressures will be manifest in heightened competition between different interest groups over how best to manage coastal resources. As society's values and political attitudes towards social welfare change (as illustrated by the pressure for private investment and involvement in the National Health Service and education), so the rationale behind the public subsidy of private property may be challenged.

All these possible changes could occur irrespective of climate change and sea-level rise. Thus, despite the uncertainties associated with the changing character of natural hazards, operating authorities should be aware that their ability to manage the associated risks will not remain constant. Plans and actions that are acceptable today may not be acceptable or deliverable in the future as values, politics and the availability of resources change. The changes in society and the political economy that will inevitably occur over the next 50 years will be as significant as climate change and sea-level rise in determining how coastal risks are accepted, tolerated and managed.

REFERENCES
Clement D 1995. Property insurance and flood risk. *Proceedings of the 30th MAFF Conference of River and Coastal Engineers.*
Gummer J 2000. *Country Living*, 57-60, March 2000.
Hewitt K (ed.) 1983. *Interpretations of Calamity*, Allen and Unwin, Winchester, Mass.
Kasperson R E 1969. Environmental stress and the municipal political system. In R E
Lee E M 1993 *Coastal Planning and Management*: A Review. HMSO.
Lee E M 1995. *The Investigation and Management of Erosion, Deposition and Flooding in Great Britain*. HMSO.
Lee E M. 1998 The implications of future shoreline management on protected habitats in England and Wales. Environment Agency R&D Technical Report W150.
Lee E M 1999. Coastal Planning And Management: The Impact Of The 1993 Holbeck Hall Landslide, Scarborough. *East Midlands Geographer*, 21, 78-91.
Lee E M 2001 Coastal defence and the Habitats Directive: predictions of habitat change in England and Wales. Geographical Journal, 167, 39-56.
Lee E M, Brunsden D, Roberts H, Jewell S and McInnes R 2001 *Restoring biodiversity to soft cliffs.* English Nature Research Report 398, Peterborough.
O'Riordan T 1981. *Environmentalism*. London: Pion.
Palm R I 1990. *Natural hazards: an integrative framework for research and planning.* John Hopkins University Press, Baltimore.
Penning-Rowsell E.C., Parker D.J. and Harding D.M. 1986. *Floods and drainage.* George Allen and Unwin.
UK Biodiversity Group 1999. Maritime cliff and slopes Habitat Action Plan. In *Action Plans, Volume V Maritime Habitats and Species*, 99-104.
Willis Faber And Dumas Ltd 1996 *Research Report UK east coast flood risk*. Willis Faber and Dumas Ltd.

Managing ground instability in urban areas, Cognito: a tool for a preferred approach

DR ERIC LEROI, GEOTER International, Parc d'activités de Napollon, La ferme de Napollon, 280, avenue des Templiers, 13400 Aubagne, France, e-mail: eric.leroi@geoter.fr

ABSTRACT

The conduction of an investigation on an unstable site, the installation of a monitoring system, and the initiation of risk reduction operations, are all choices specific to the site investigated. Given the evolution of society and the need for accountability, the search for the "best" solution can only be made on the basis of a comprehensive approach integrating scientific as well as technical considerations, and social and political ones too. The search for an optimal solution, whether in the dramatic aftermath of a crisis situation, or over the long term (for example for the definition of construction zones), must necessarily be based on an analysis, a structuring and a clear and precise definition of the need or the problem, as well as on a structured and comparative construction of the response in close cooperation between the technician and the decision maker. This approach has led to the development of the decision-aid software, 'Cognito', an application particularly designed for the analysis of unstable sites, which allows the step by step building of the solution taking into account financial and technical constraints.

INTRODUCTION

The consideration of environmental constraints in town and country planning, as well as the growing public demand for protection, should lead decision-makers and technicians alike into putting forward optimal and integrated management approaches for natural hazards, in order to offer sustainable, harmonious and safer development. Natural events which cause damage, whether they are catastrophic or not, are no longer accepted by western societies as inevitable, and the occurrence of such a phenomenon will lead almost systematically towards research into responsibility and culpability.

Over the past three decades we have helped to bring about a formidable change in the law relating to danger and disasters. The obligations of local councils and senior civil servants have changed in nature to become an obligation of safety punishable under criminal law. From now on it should be stated that the criminal judge has become an obligatory partner of those responsible.

In order to face up to this development in society, political leaders have put in place a well-structured regulatory framework. However, regardless of the capability of the public sector to legislate, the legal texts allow for considerable interpretation. Confronted with a serious situation, notably in the case of loss of life, decision-makers as well as technicians will have to justify their actions, be it before a Court, the media or management. At this stage, it will be necessary to prove that the best solution had been adopted.

Because the system of reference for the analysis of the problem will have changed, the exercise would be difficult by its nature. When the technician and the politician have to make choices (technical, financial, environmental, social etc.) they have to be based on predictive or prospective models as well as on uncertainties. Their decision will therefore contain a probability and a period regarded as "trust". When the judge has to analyse a recognised situation the explanatory models and the chain of events or the causal factors will necessarily be more complete and the field of uncertainty will be lifted. It will then be necessary to convince him that the adopted solution did correspond to the optimal choice.

To do this, it will be necessary to be in a position to explain the procedure followed, to justify the criteria chosen and to move towards a more quantitative approach to risk.

Ground movements are complex and diffuse processes. The manifestations observable at the surface (deformation, pluck niche) are merely partial witnesses of mechanisms which occur in three dimensions, whose parameters are neither visible nor often accessible. The variety of the rupture mechanisms and site conditions makes diagnoses as well as prognoses difficult to formulate. At best, they contain a large uncertainty.

Yet the consequences of these processes can be dramatic (property destruction, fatalities) and must be minimised while remaining reasonable in terms of the investment to be made. The cautionary principle underlying any natural risk prevention policy can lead to situations that are difficult to manage, both for the decision makers and the technician.

Conventional engineering approaches are based on the evaluation of a safety factor F. Many standards have been developed over the years to define procedures and to interpret the results. But these approaches have their limits:
- adapted to "routine" situations, they may soon prove deficient on more complex cases;
- needing information on the geotechnical characteristics of the materials set in motion, they usually require large scale investigations unfeasible on large areas; conventional geotechnical approaches are hence mostly reserved for occasional events, declared and relatively well pinpointed in space;

- the experience of the technician in charge of the analysis is crucial; the technician's quality and expertise will condition the accuracy of the diagnosis and prognosis;
- based on cause analysis, they fail to assess the risk as a whole and particularly its consequences; the choices concerning applicable solutions very seldom integrate political or social analyses on risk reduction, on the notion of residual and/or acceptable risk.

Recent approaches based on quantitative risk assessment have emerged in the field of geotechnical engineering, in the wake of the approaches developed in the field of technological risk. The analysis of the consequences and the cost/benefit analysis of the investments allocated to reduce the risk accordingly become of major importance.

The response to a problem of slope instability, whether potential or declared, is more than ever based on a necessary and enlightened dialogue between the technician and the decision maker. Given the evolution of society and the need for accountability, the search for the "best" solution can only be made on the basis of a comprehensive approach integrating scientific as well as technical considerations, and social and political ones too.

The search for an optimal solution, whether in the dramatic aftermath of a crisis situation, or over the long term (for example for the definition of construction zones), must necessarily be based on:
- an analysis, a structuring and a clear and precise definition of the need or the problem,
- a structured and comparative construction of the response in close cooperation between the technician and the decision maker,
- experience gained in the past on similar cases.

This approach has led to the development of a decision-aid software, Cognito, an application designed particularly for the analysis of unstable sites, for the choice of an investigative method, a monitoring system, or a risk reduction solution (in terms of groundwater control).

THE COMPREHENSIVE APPROACH
The conduct of an investigation on an unstable site, the installation of a monitoring system, and the initiation of risk reduction operations, are all choices specific to the site investigated. These choices depend not only on the site as such (area, morphology, access, type of event), but also on the immediate environment (urban, rural, broad or limited implications) and the socio-political context (population and decision makers familiar with the process, municipality with large or small financial resources, availability of local technical skills).

This precludes the provision of a catalogue of "ready made" measures or "turnkey" methods to the decision makers or the technicians. Each solution must be constructed step by step to optimally "match" the realities to the constraints of the site.

It is proposed, with Cognito, to provide aid for constructing an optimal solution. The aid is provided at two levels:

- basics: presentation of a structured approach, a methodology for approaching the problem. This means an understanding of the complexity of managing risks associated with ground movements, and familiarising the technicians and decision makers with a more rigorous, more conceptual preliminary approach, but one that is in many cases financially and politically "profitable".
- at the operational level: provision of a tool box, a kit to enable the technician to build his own solution, working alongside the decision makers faced with a problem of instability.

The comprehensive approach to be followed to plan an investigation campaign, install a monitoring system, and pick a risk reduction solution, is the following:

1. Definition of the problem, context and constraints
2. Analysis of the problem
3. Inventory of feasible solutions
4. Selection of acceptable solutions
5. Choice of the optimal solution

Definition of the problem, context and constraints
Every unstable site is unique. Risk reduction solutions must be constructed step by step, rigorously and structured. To do so, the problem to be addressed must be clearly defined and the facts accurately analysed (disassociating them from assumptions and interpretations), the potential consequences analysed in terms of the challenges identified, and an inventory of site constraints be ompiled and ranked. Whenever possible, attempts must be made to identify similar situations that may have occurred in the past.

Information will therefore be provided on (fig 1):
- the type of process observed (landslides, rockslides).
- the possible and probable evolution of the movement in the short and medium term. This means estimating whether the process may evolve so as to endanger people and property.
- the type, number and size of the exposed property, to rank the different risk reduction operations.

The constraint analysis will identify:
- financial resources available and mobilisable in accordance with the endangered property.
- deadlines for installing stopgap solutions.
- environmental, social and political constraints that must be integrated and ranked.

Analysis of the problem
In all cases, regardless of the situation at hand, it is necessary to conduct a number of preliminary investigations rapidly (fig 1):
- site inspection; this phase is unfortunately too often ignored although vital;
- consult the existing topographic, geological, geomorphological and hydrogeological maps.

The synthesis of the information obtained should help to:
- confirm the preliminary impression of the site (type of process, challenges identified, motion kinematics);
- justify the choice of the phase (expertise, surveillance, operations).

Inventory of feasible solutions
The response to a given problem can be provided, schematically, either directly or iteratively :
- the direct response implies that complete and immediate solutions exist, and that they fully solve the problem. This is the case, for example, of the expert opinion generally demanded during crisis situations;
- the iterative response is based on the fact that the problem to be analysed is complex and that it must be broken down into distinct sub-problems for which a direct solution seems conceivable. If a sub-problem retains a complex character, it is broken down iteratively in turn, until all the sub-problems identified have been solved directly.

This approach is only feasible if:
- at each step of the breakdown of the problem into elementary sub-problems, all the feasible direct solutions can be presented and qualified;
- the breakdown into elementary sub-problems is made in a structured and explicit manner;
- each feasible solution is described and characterised so that a fully informed and justified decision can be made.

This work of structuring knowledge, ranking sub-problems and describing and characterising solutions, is the core of the approach followed with the Cognito software. The outcome of this work is organised into matrices. Using the information stored in the matrices, the technicians and decision makers can then:
- selection the type of solution to be applied to each problem;

- have a complete inventory of existing solutions;
- back their choices on the basis of synthetic and comparative information on each solution.

This approach to knowledge structuring and iterative breakdown of the problem into elementary sub-problems associated with a complete inventory of conceivable solutions, helps to clarify the "realm of the possible". Clearly, if each sub-problem can be solved by numerous solutions, the "realm of possible solutions" for solving the overall problem rapidly becomes very large and grows exponentially. Besides, all the elementary solutions end to end do not necessarily represent comprehensive, realistic, economically viable or pertinent solutions. In this "realm of the possible", it is essential to identify the "realm of the acceptable".

Inventory of acceptable solutions
An inventory of feasible solutions can be compiled at each step of the analysis of the problem (fig 1). However, not all solutions are necessarily acceptable, for two reasons:
- a solution provided to a sub-problem may contain specifics incompatible with the constraints of the overall problem (e.g. a particular technique can only be used if power line is available on the spot, which is inconceivable for one particular site);
- even if an elementary solution meets the requirements of the sub-problem, the sum of all elementary solutions may lead to an overall solution whose characteristics do not fit into the outline of the overall constraints (e.g. the budget available is 100,000 €, and each of the five elementary solutions costs 25,000 €, adding up to more than the total budget).

Thus the constraints specific to each site can be defined by the user. There are three overall constraints : total amount of available resources, maximum operating time and difficulty of operational implementation. At each selection of an elementary solution, the overall project level is reviewed. Insofar as one of the constraints has not been fulfilled, a message alerts the operator who can continue or interrupt the construction of the solution.

The operator also has a second level of analysis and selection to adopt or reject a proposed solution. Each solution is characterised by a set of technical descriptors such as accuracy, spatial representativity, and capacity to collect information in real time. In light of each descriptor, the operator can decide whether or not to select a given solution (e.g. the need to have 3D data, the desire for off-site information transmission). Further to the simple description of the technical characteristics of the solution, it is possible to get more detailed description from a Web file as shown in figure 2.

Furthermore, the redundancy principle must be applied in many cases. This means that to make a judgement, the technician will try to rely not on the information supplied by a single technology, but on a set of convergent elements deriving from many techniques. Thus the answer to an elementary sub-problem may be provided by several solutions simultaneously. In any case, the operator must ensure that the overall constraints are satisfied.

Choice of the optimal solution

Beyond the selection of acceptable solutions, the optimal solution must be sought. Given the many factors to be considered, automatic processing has not been proposed in this version of Cognito. This is a deliberate decision based on the following two reasons:

-   at the technical level, such a module is difficult to develop and the algorithms require multivariant analysis with variables of which the ranges of variation are not comparable. This demands a ranking of the factors with respect to one another, and the weighting coefficients are difficult to calibrate;
-   a prototype monovariant analysis was nonetheless developed to determine the optimal solution in light of a single factor (e.g. choice of the cheapest solution, choice of the fastest solution), but excessively rapid use of this module can lead to untimely decisions. The monovariant analysis is often easier to conduct, but is inherently extremely reducing. The analytical mode is generally used for an intellectual analysis, which is perfectly legitimate because it is difficult to mentally rank functions with several variables, of which the values are not quantitative. Hence this module was not integrated in Cognito.

Thus the optimal solution must be constructed step by step, by making choices at each step. This selected orientation may appear cumbersome, and it certainly is! Yet it is safer and more realistic in view of the many uncertainties on the ranges of values supplied, but above all, it is better adapted to the complexity of the processes to be analysed and the decisions to be taken. It makes it possible to devise a constructed and supported answer in all cases.

Presentation of the result

The results of the modelling exercises can be presented in three way, as shown in figure 3: tree, summary and detailed document.

The summary document comprises the name of the project, the characteristics of the site, the constraints and the assessment (cost, timescale, difficulty) and the list of the chosen laws and parameters. The detailed document is structured in the same way as the summary document, except the characteristics of each law and parameter have been added.

THE PROSPECTS

The decision aid supplied by Cognito is based on structured knowledge within matrices. This knowledge is inherently ranked and organised, but the basic data, particularly relative to deadlines, cost and difficulties, can evolve rapidly. They also depend largely on the context and the country, as well as the local skills available. Thus the data supplied are merely indicative and must be used with caution.

This is the reason why an "empty" matrix is provided within Cognito so that the user could build his own matrix. If necessary, the user can simply modify the basic matrix filled in, and save it to use it later. So far, however, it is not possible to modify the structure of the matrix within Cognito. Only the content can be modified.

In any case, a learning process is essential for the Cognito software, and any operational use on site must be preceded by a calibration phase. This calibration must be conducted by a competent technician.

Cognito has development prospects at three levels:
- supplementation of knowledge bases on the softwares, techniques and methods available in order to offer a complete range of solutions at each step;
- develop a module for modifying the structure of the matrices within Cognito in order to integrate new data on the softwares, techniques and methods;
- develop a module to automatically search for the solution based on multivariant analysis, with realistic restriction of the field of acceptable solutions. This task initiated with monovariant analysis (not released) can only be carried out with feedback on the use of Cognito.

Cogito is a knowledge structuring software. The module developed concerns the analysis of unstable sites in urban areas. Beyond this context, the entire body of knowledge can thus be structured. This work was also initiated for risks associated with underground quarries as well as risk connected with shrinkage-swelling processes in sensitive clays. The internal structure of Cognito allows these developments with tremendous flexibility, regardless of the level of analysis and the complexity of the problem.

ACKNOWLEDGEMENTS

Cognito was partly developed with the financial support of the European Union as part of the LIFE Environment program (EC DG Environment), and with the aid of Bureau de Recherches Géologiques et Minières from its own research funds. The three-year Demonstration Project entitled "Coastal Change, Climate and Instability" was coordinated by Robin McInnes from the Isle of Wight Centre for the Coastal Environment, UK).

Figure 1 Main screens of Cognito

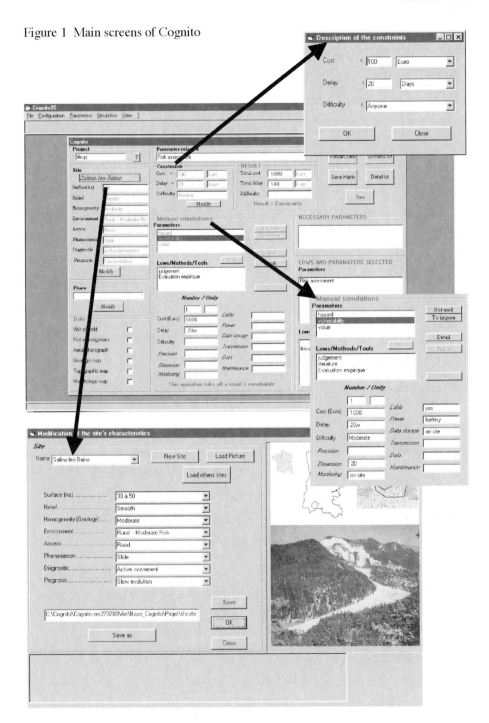

Figure 2 Example of a Web file

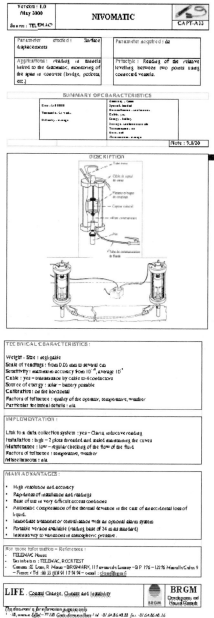

Figure 3 Presentation of results

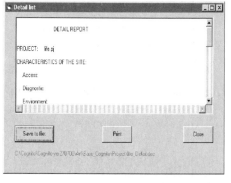

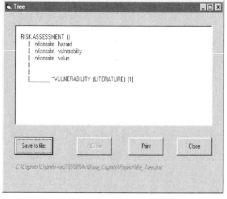

Land instability and Building Control

J A LUTAS MRICS, Principal Building Control Surveyor, Isle of Wight Council, UK

INTRODUCTION
Building Control in the United Kingdom is concerned principally with securing the health, safety, welfare and convenience of people in or about buildings, and this is achieved through the administration and enforcement of the Building Act 1984 and the Building Regulations 2000. These regulations relate to matters such as structural stability, fire safety, sound insulation, drainage, thermal insulation and access and facilities for disabled persons. In areas of known land instability, the role and function of the Local Authority Building Control Surveyor may be extended to offer advice to the planning department as to whether proposed development is considered suitable.

Around the coastline of the Isle of Wight, there are certain areas of known ground instability, and this paper will review the roles of the Building Control Section of the Isle of Wight Council in relation to the further development of these areas.

ADVICE TO THE PLANNING DEPARTMENT
Building Regulations do not cover all factors relevant to slope stability and indeed, certain works are not covered by this legislation. Ground stability issues therefore need to be taken into account at the planning stage, and government guidelines have been issued to advise local authorities, landowners and developers on the role the planning system should play in ensuring suitable and safe development. Planning Policy Guidance note 14, 1990, "Development on unstable land" (PPG 14), explains briefly the effects of instability on land use and the need for this to be taken into account at the planning stage. PPG 14 (Annexe 1), 1996, "Development on unstable land : Landslides and Planning" takes the above guidance further, with particular reference to problems caused by landslides and unstable slopes. The aim of the guidance is to ensure that unstable land is identified as early as possible in the planning process, in order that the risk of undesirable consequences such as property damage, injury, personal distress and damage of the physical environment is reduced to a minimum. PPG 14 makes it absolutely clear that the responsibility for safe development of a site rests with the developer and/or landowner. It is therefore necessary for the developer to appoint a competent person to carry out a thorough investigation and assessment of the ground to ensure that it is suitable for the proposed development. Whilst the local planning authority does not owe a duty of care to individual landowners when granting planning permission and is not liable for loss caused to an adjoining landowner by permitting development, it should ensure that the proposed development will not be

affected by instability. In addition, it should ensure that the proposed development will not adversely affect the stability of adjoining land. To this end, the local authority is entitled to rely on the expert advice provided by the developer. A check should be made though, to ensure that the content and scope of the stability report is sufficient, and that all relevant factors that may affect slope stability have been considered.

In any decision-making, consistency is a key factor. In terms of proposed development in areas of unstable land, this has been aided considerably on the Isle of Wight by ground stability studies that resulted in the production of a suite of maps indicating landslip potential. Geomorphological maps have been produced, showing the positions of the geomorphological units that occur and identifying the nature and extent of individual landslide units. From these maps and considerable research, a ground behaviour map summarising the nature and extent of the different landslide processes was developed. A planning guidance map was then produced which identifies in broad terms those areas most suitable for future development, and those where development should be avoided. These planning guidance maps are useful to give an initial indication as to whether a site can be safely developed, and also to determine what level of site investigation will be required to accompany a planning application. This information can be given to a proposed developer by planning officers.

On receipt of the planning application and the accompanying stability report, it is necessary to carry out a site inspection and review the ground behaviour and geomorphological maps in order to assess the factors that should be included in the stability report. This is a more technical aspect, and as Building Control are closely related in the development process to planning, it is only natural that they should be consulted. In addition, the Planning department also consult with the Coastal Management section, and this dual approach helps to ensure that unsuitable development is avoided.

In assessing whether the stability report has addressed all relevant factors, consideration is given to the following aspects, depending upon the scale of the proposed development and potential impact upon slope stability.

- Has the report been prepared by a suitably competent person?

- Has a sufficient desktop study been carried out?

- Is the walkover survey adequate, and does it identify the condition of adjacent properties and structures or any other indications of ground instability, and what will be the effect of the development on existing retaining walls?

- Is the sub-surface investigation adequate to assess the stability of the site and ability to support the development loads, and has laboratory testing of samples been carried out? Will it be necessary to monitor the stability of the site?

- What will be the impact of the development on the landslide system and adjacent properties? Consideration should be given to the effect of cut and fill operations and re-profiling slopes, the siting of the proposed structure, its scale and mass. Will deep excavations across the slope for foundations or services be necessary, and is a method statement required? Has consideration been given to the groundwater levels, and will the flow of ground water be obstructed? Also, how will foul and surface water drainage be dealt with?

- Will the unstable land affect the proposed development? Can the building be designed to accommodate anticipated ground movement, and is there a risk of rock falls?

- Have calculations been prepared to prove that the site has an adequate margin of stability, and is the design philosophy considered acceptable?

- What conclusions are made in the report, and are recommendations for slope stabilisation works enforceable?

Provided these factors have been considered in the stability report and have been investigated sufficiently, it will generally be concluded that PPG 14 has been satisfied. If, however, it is considered that further information should be requested to accompany the stability report, then Building Control will advise planning accordingly. More often than not, most of the problems highlighted can be overcome by further investigation and re-design. However, there are instances where such concerns exist as to the safe development of the site in respect of land instability, that a recommendation is made to the planning authority that the application should be refused. Occasionally, Building Control will advise planning that an external geotechnical consultant should be appointed to appraise the scheme, particularly if the site is considered to be in an extremely sensitive location.

Whilst PPG 14 advises that geotechnical specialists should prepare stability reports, comments are occasionally received from agents suggesting that the local authority surveyor should also be qualified to the same standard. Whilst all Building Control Surveyors commenting on development proposals are qualified professionals with considerable experience, the claim that they are not classed as competent persons for the purposes of carrying out a stability report is not disputed. However, as stated previously, the role of the surveyor is to ensure that all material factors affecting site stability have been taken into account, and if that is the case then the engineer's conclusions will not be disputed. We are also fortunate in our section to have an experienced structural engineer with whom any findings or concerns can be discussed and who can advise on any parameters used in the stability calculations, such as soil characteristics and consequential factors of safety.

One method of ensuring that stability reports submitted will be sufficiently comprehensive is to recommend that a standard format is adopted. This would certainly assist in the assessment of such reports. Another recommendation is that such stability reports are accompanied by a declaration form signed by the engineer and confirming that he has taken into account all relevant factors. Whilst this will not negate the need for the reports to be checked, it should at least indicate to the engineer the level of site investigation considered necessary and avoid having to request further information, with the accompanying delays in issuing decisions this inevitably brings.

BUILDING REGULATIONS

In addition to obtaining planning permission, it is usually necessary for developers to submit an application for the proposed building works under the Building Regulations. In order to meet the requirements of these regulations, the building control body must be satisfied *inter alia* that the building will be constructed so that ground movement caused by landslip or subsidence will not impair the stability of any part of the building. It is therefore necessary to ensure that the foundations, building structure and services are designed to either resist or accommodate foreseeable ground movement, and designs submitted under the Building Regulations must be supported by structural calculations. These calculations are checked by our structural engineer for design philosophy and accuracy. The most suitable design for buildings in areas of land instability depends largely on the cause and extent of the instability. Where shallow/ medium slips have developed, a piled solution may be appropriate. However, where the scale of the landsliding is such that engineering solutions to prevent further ground movement are infeasible, such as on the southern coast of the Isle of Wight, buildings must be designed to accommodate anticipated future ground movement. In such cases, factors that must be considered in the design of buildings include;

Structural form

Buildings need to be designed at the outset with structural integrity in mind, and because of the possibility of tilt occurring, the height of structures needs to be restricted. In addition, simple plan shapes can be more easily designed to accommodate ground movement than complex plan shapes. More complicated layouts may therefore need to be sub-divided into smaller bays. Internal buttressing walls will also add to the stiffness of the structure.

Foundation design

Whilst many different types of foundation design have been utilised, stiff reinforced raft foundations have generally proven to be the most practical, economic and successful, as they are robust, can absorb minor ground movements and span over voids in the ground. On some sites particularly vulnerable to ground movement, jacking points below the foundation slab may be considered necessary in order that re-leveling of the structure can be carried out should tilt occur.

Structure

Timber and steel framed structures have the ability to absorb slight ground movement, whereas masonry will show signs of distress at an earlier stage. In addition, in order to maintain an equilibrium within the landslide system, it is often necessary to keep the load of the new structure to a minimum. Lightweight framed structures are therefore the most suitable form of construction. If an independent masonry outer-leaf is preferred on aesthetic grounds to weatherproof cladding fixed directly to the frame, as is often the case in towns with a victorian heritage, then frequent movement joints should be provided.

Services

It is essential that all service pipes are flexible, and that their entry into the building allows for sufficient movement. Gradients of drainage systems should be increased to allow for ground movement, and gutters should also be laid at greater than the recommended falls. Artificial re-charging of the ground water has a great influence on the rate of ground movement and must be avoided. All storm-water should be connected to the public sewer, or else discharged to an established watercourse. Foul drainage should be connected to a sealed cesspool if it is not possible to connect to the public sewer. Septic tank drainage systems utilising soakaways are not permitted.

Retaining walls

As retaining walls are often required at the abutment of different landslide units, they are most likely to be affected by ground movement. They should generally not form part of the building and if constructed of concrete or masonry should be designed as short panels, separated from adjoining panels by movement joints. Rock anchoring has also been employed successfully. Preferably, however, an alternative method of supporting the ground should be adopted, such as crib-lock walls or gabions. Retaining walls that do not influence the structure of the building are, however, outside of the scope of the Building Regulations and cannot therefore be controlled under this legislation.

These requirements can have a significant influence on the building design, and it is therefore recommended that the designer consults with the building control section before the detailed design has been finalised. This is advantageous to everyone as it helps to avoid unnecessary work and consequential delays.

DANGEROUS BUILDINGS AND STRUCTURES

Urban developments in areas prone to ground movement are susceptible to structural damage. This may result from the cumulative effects of minor settlements, or as a result of a major landslip or cliff fall. The legal provisions for dealing with dangerous structures are contained in the Building Act 1984, and the Isle of Wight Council Building Control Section is on call 24 hours a day to respond to any dangerous structure.

On arrival at the site, the first consideration of the building control surveyor is the safety of all persons. This includes the general public and occupiers of properties, and also the emergency services, contractors and himself.

In areas of unstable land, localised failures of retaining walls are common. Whilst repair is the responsibility of the landowner, it is often necessary to initially fence off the danger and possibly close the highway.

Major ground movements affecting property are the most dramatic events, and very distressing to the property owners. Sometimes failure is sudden and without warning, and properties have to be vacated immediately. At other times, whilst large scale ground movement may have occurred in the grounds of the property, tension cracks in the ground affecting the structural stability of the building develop over a period of time. In such instances the building may not be considered to be in imminent danger of collapse, and therefore enforcement action is inappropriate. The role of the building control surveyor is then to monitor the structural stability of the property and offer advice to the occupiers.

Cliff falls also present a major danger to occupiers of properties, and if a threat of a fall is perceived, occupiers are advised to vacate property.

Making the decision as to whether a building or structure is considered to be dangerous is not straightforward. Whilst decisions must always err on the side of caution, over reaction to a situation is of no use to anyone. Building control surveyors gain considerable knowledge and experience of building structures and the way different elements perform and fail. The decision as to what action to adopt is based on the circumstances in each particular case and through previous experience gained. The keeping of adequate records, including photographs, is essential if legal action is considered necessary, and in any case contributes to expanding the knowledge base.

CONTROL OF WORK

The dual controls of planning legislation and Building Regulations make a major contribution in ensuring only suitable development is permitted to be built. However, this is of little value if there is not adequate control of work on site. It is essential that the contractor adheres strictly to the approved specification, and it is recommended that the structural designer or geotechnical engineer is also engaged to supervise critical elements of the work. This is particularly important in respect of excavation for foundations. Whilst building inspections by the building control section are made at various stages of construction to ensure that the work is adequate, it is important to note that little control can be exerted by the local authority regarding the manner in which the works are carried out, and many instability problems have been caused by unsuitable site practices. An example of this is the excavation by a contractor of a large section of soil from the toe of a critically stable slope during a period of prolonged wet weather, in blatant disregard to the approvals and engineer's method statement.

A potential breakdown in the coordinated approach between planning and building control also exists when the building control function is to be carried out by an approved inspector. The approved inspector is unlikely to be aware of any discussions carried out at the planning stage, and may not have had sight of any reports that have been submitted. The approved inspector may not even be aware that the land is potentially unstable, and suitable design and detailing in the construction may not be adopted. The experience of such inspectors in dealing with slope stability issues is also likely to be significantly lower than the local authority building control surveyor, who has gained experience at first hand of the effects of ground movement on buildings and understands the major factors that influence slope stability.

Another problem exists in relation to works that do not require planning permission and are also exempt from the requirements of the Building Regulations. For example, the construction of a swimming pool in the grounds of a domestic property could have a major affect on the stability of the area, and the local authority may be powerless to prevent it, unless permitted development rights under planning legislation have been withdrawn.

CONCLUSION

The construction of new buildings in areas of known ground instability presents challenges to all those involved in the development process. Whilst the developer is ultimately responsible for ensuring that the development will not cause land instability problems, the local authority has a duty to ensure that sufficient investigation into this topic has been carried out, and that new property will not present a hazard to the health and safety of people in or about the building and the surrounding area. This is carried out through the determination of Planning and Building Regulation applications.

A coordinated approach between all parties is essential, and from the local authority's point of view, the technical knowledge and experience gained by local authority building control surveyors through their training, building inspections and dealing with dangerous structures makes them most suited to providing a lead role in this process.

By ensuring that only suitable development is permitted under planning controls and that the building has been designed such that anticipated ground movement can be accommodated to comply with the Building Regulations, the likelihood of unsuitable development can be significantly reduced. However, even with this approach there are still problems regarding the control of certain works for which the local authority have limited powers under current statutory legislation.

Managing ground instability in the Ventnor Undercliff, Isle of Wight, UK

ROBIN M^cINNES, Isle of Wight Centre for the Coastal Environment, Isle of Wight Council,UK
JENNY JAKEWAYS, Isle of Wight Centre for the Coastal Environment, Isle of Wight Council, UK

ABSTRACT
Drawing upon experiences gained in the field of instability management from study sites in Great Britain, France, Italy, Switzerland and other locations as part of a European Union LIFE Environment Project entitled "Coastal Change, Climate and Instability" (1997-2000), this paper highlights the practical assistance that is available for those involved with instability planning and management, the objective being to reduce the impacts of ground movement on local communities and economies. The paper describes good practice and highlights the value of a co-ordinated approach to reducing the impacts of landsliding through the involvement of local government staff, architects, surveyors, builders, estate agents, insurance companies as well as local residents.

The paper outlines the growing awareness of the need to implement preventative planning and landslide management policies to counter a problem which is likely to escalate as a result of both future development pressures and climate change impacts. It describes the importance of developing improved methods of managing natural hazards and landslide mitigation to prepare for these changes. A key aim of landslide management is to minimise costs through encouraging the development of landslide hazard evaluation practices, particularly where detailed baseline information is not available, as well as assisting with the incorporation of instability considerations into the planning process. The paper also describes the response of local residents to management initiatives being promoted by the Isle of Wight Council.

INTRODUCTION
In "A review of landsliding in Great Britain", Jones and Lee (1994) concluded that landslide problems are not 'Acts of God', unpredictable, entirely natural events that can at best only be resolved by avoidance or large-scale engineering works. The role of human activity in initiating or reactivating many slope problems should not be underestimated. In areas where urban development has taken place on

degraded mudrock landslides, the problems tend to be related to very slow ground movement and progressive damage to property, services and infrastructure. In such circumstances many problems can be reduced if there was a programme of active landslide management where the local community is able to come to terms with the situation and learn to "live with landslides" (Lee and McInnes 2000).

VENTNOR UNDERCLIFF LANDSLIDE MANAGEMENT STRATEGY

An example of this approach is the case of the Ventnor Undercliff on the south coast of the Isle of Wight, UK, where a Landslide Management Strategy has been in place since 1993. The coastal town of Ventnor and the nearby villages of Bonchurch, St Lawrence, Niton and Blackgang are built on a large landslide complex, the Undercliff, but fortunately the geological setting and the style of landsliding is such that movements are often concentrated in a few locations, and the intervening areas show negligible or no movement (Lee and Moore 1991; Moore et al. 1995). Of the areas affected by movements, a number of sites are already public open space or have been adapted for this purpose. The Undercliff Landslide Management Strategy aims to reduce the likelihood of future movement by seeking to control the factors that cause ground movement and limit the impact of future movement through the adoption of appropriate planning and building controls (Lee et al. 1991).

A variety of approaches have been adopted to address the ground movement problems, including:
- Improving ground conditions through the control of water in the ground and coast protection measures;
- preventing unsuitable development through planning control and building control;
- monitoring ground movement and weather conditions at automatic and manual recording stations;
- raising professional and public awareness through displays and meetings.

A Landslide Management Committee meets up to twice a year to enhance professional and public awareness of how the strategy is being implemented and to monitor its effectiveness. This Committee comprises technical staff from the Isle of Wight Council including the Coastal Manager, planners, building control officers and highway engineers, the water company and other service industries, the Builders Employers Confederation and local estate agents; the Association of British Insurers has observer status.

An important element of the Landslide Management Strategy for the Undercliff is to ensure that future development is compatible with ground conditions and is discouraged where the likelihood of movement is high. New property within the Undercliff must be sustainable, capable of withstanding movement and not lead to a worsening of slope stability at the site or on adjoining land. These requirements are overseen by the Council through the planning system and application of the Building Regulations.

PUBLIC CONSULTATION AND FEEDBACK ON THE VENTNOR UNDERCLIFF LANDSLIDE MANAGEMENT STRATEGY 1991-2000

Residents of the town of Ventnor were aware that they lived within a landslide complex. The complex, which is probably over 8,000 years old, was initiated as a result of sea-level rise during the Flandrian Transgression. Although considerable areas of the Undercliff are quite stable some locations are close to a threshold of instability, as through various processes and mechanisms the landslide slowly degrades. Historically ground movement had been slow and episodic with more severe movements concentrated in a quite limited number of locations and occurring particularly after exceptionally wet winters.

An analysis of historical information suggested that ground movement problems had increased over the last century or so. A key factor was no doubt the rapid urban development that took place in Victorian times as Ventnor established itself as a popular seaside resort and spa; this development continuing on through the Edwardian period and up until the middle of the last century. The increased urbanisation of the Undercliff, centred particularly on Ventnor, resulted in an impact on the landslide through human activity. Excavations for building sites, inappropriate cut and fill operations affecting slope geometry and additional drainage into the ground from sewage and leaking pipes were all contributing to this increase in instability.

In 1987 the former Department of the Environment (DOE -now Department for Transport, Local Government and the Regions, DTLR) was investigating suitable sites for research that could lead to the publication of advice and guidance to assist the planning process relating to development on unstable land. The DOE decided that the town of Ventnor would form an excellent case study and the results and publications arising from this three year study proved to be of enormous benefit for the town. In particular it assisted in developing cost-effective approaches to mitigating against landslide problems through the development of planning controls and other procedures which take ground instability into full account.

The results of the DOE study were presented as:
- a technical report which included a series of 1 to 2,500 scale maps (land use, geomorphology, ground behaviour and planning guidance).
- a summary report which was aimed at non-specialist professionals (including for example local authority engineers and planners, insurers, designers and estate agents) and the "educated layman".

Following the launch of the report, which had comprised a presentation to the Association of British Insurers, the Council and other interest groups, the former

South Wight Borough Council commissioned its consultants to establish and staff a 'Geological Information Centre' to be run from an empty shop in the town of Ventnor. This shop was opened on 18 March 1991 for a period of approximately three months and was staffed by two members of the geomorphological team directly involved in the project (Lee et al. 1991).

The Geological Information Centre comprised a static display (a series of wall-mounted display boards) which provided a wide range of information contained in the DOE report on The Undercliff. Each board was arranged to provide a clearly understandable and well-illustrated account of the landslide problems.

In addition to the display boards a four page explanatory leaflet entitled "Land Stability in Ventnor and You" was prepared jointly by the DOE and the Council for distribution to both residents and businesses.

Over the 2½ month period approximately 2,000 visitors, representing one-third of Ventnor residents, came to the Information Centre. The majority of the visitors appeared to be either homeowners, especially elderly people with retirement homes, as opposed to younger professional people. A second group who attended the Centre were professionals involved in the housing market including developers, builders, insurance companies, local estate agents and loss adjustors; the dissemination of results to these interest groups was regarded as particularly important.

Following the completion of the public consultation on the Ventnor report and the closure of the Geological Information Centre, the Council commissioned its consultants to develop the findings of the DOE study and formulate a Landslide Management Strategy. In due course a strategy was developed which included the following main elements:
• Civil engineering measures (coastal protection works and drainage).
• Public education and information dissemination.
• A strategy for control of water in the ground (leaking pipes, septic tanks, etc).
• Avoidance of areas affected by more serious instability.
• Further research and monitoring.

Furthermore the Council, supported by the Association of British Insurers, extended the original Department of the Environment study of central Ventnor town by including the area of Bonchurch village (to the east) and by working westwards along the Undercliff to include the villages of Steephill and St Lawrence and towards Niton. On completion of this further study in 1995, an additional six maps were produced together with a further information leaflet. This Extension Study was launched in a similar way but in addition workshops were held with groups such as estate agents, insurers, architects and builders.

The following year, in 1996, the final phase of the Undercliff geomorphological mapping was undertaken between Niton and Blackgang and a further more comprehensive leaflet was launched in 1996 called 'Advice to Homeowners in the Undercliff'. This four page leaflet aimed to provide specific information to assist the public in problems that they might experience with maintenance of slopes and vegetation cover, retaining walls, drainage systems and also advice with respect to insurance cover. The leaflet also explained what the Council was doing to address major infrastructure issues such as drainage and coastal protection.

This leaflet was distributed to each of the 2,600 property owners in the Undercliff, accompanied by a programme of publicity and information. Over the intervening period an ongoing effort was made by the Isle of Wight Council to develop the Landslide Management Strategy by protecting the toe of the landslide system with coast defences and by encouraging the water company to replace public drainage systems to prevent leakage; the provision of water meters in Ventnor provided a further incentive to homeowners to ensure that their service pipes were not leaking.

Questionnaire for Undercliff residents
In February 2000, as part of the EU LIFE project 'Coastal Change, Climate and Instability', a further questionnaire was prepared by the Council and was distributed to all local residents within the Ventnor Undercliff.

It's purpose was to obtain feedback from residents of a range of instability, planning and management initiatives undertaken by the former South Wight Borough Council and the Isle of Wight Council.

The results of the survey confirmed a high degree of satisfaction with the approach being adopted by the Council. A high percentage of residents (over 60%) had lived in the Undercliff over ten years and the majority were aware of ground instability issues at the time of moving to the area (82%). Approximately half of those intending to move to the Undercliff sought professional advice on ground stability, the majority of which was obtained from surveyors, consulting engineers or local estate agents. It is encouraging to note that of those who sought advice from the Council over 90% found the advice very helpful or helpful and two thirds of those who responded had read the report "The Undercliff of the Isle of Wight - a review of ground behaviour" (High-Point Rendel, 1995). Again, it was pleasing to note that 100% of those who read the report found it to be either very informative (55%) or informative (45%).

In relation to the four page leaflet that had been distributed by the Council in 1996, 20% of those who responded to the latest questionnaire had actually moved to the Undercliff area since the leaflet had been published. Of those who had received a copy over 70% found the leaflet to be very helpful or helpful. The questionnaire also sought advice on whether further instability guidance would be of value (such as the dissemination of results from the EU LIFE project). 81% of residents felt that further advice would be very useful and a further 15% felt it would be useful.

A particular area of concern had in the past been the difficulty in obtaining insurance cover because of lack of knowledge of the true extent of ground instability conditions and blight caused by insufficient information being available. In 1988 when the Department of the Environment carried out its study of central Ventnor insurance was difficult to obtain in many cases. The Council was, therefore, encouraged to note that in this latest survey that 85% of those questioned had been able to obtain full insurance including subsidence cover. It is believed that a significant contribution to this change has been the availability of better information and guidance for local residents and for insurers and the perception that the Council was actively working with other key organisations to try and reduce the impact of ground instability for the benefit of local residents. Finally, residents were asked whether they considered ground stability in the Undercliff to be an issue of concern and 96% of those questioned regarded this as of great importance.

The Isle of Wight Council is far from complacent over these results, recognising much is still to be achieved in a potentially worsening scenario as a result of predicted climate change impacts. A sustained effort will be required to even maintain a 'status-quo' in terms of ground stability in the Undercliff.

Isle of Wight Coastal Visitors' Centre
A key initiative has been the establishment in Ventnor of the 'Coastal Visitors' Centre' which opened in August 1998. The Centre provides a comprehensive display of ground instability issues as well as a valuable resource for education and scientific research. The Council regards the Centre as a vital component of its Landslide Management Strategy.

MANAGING GROUND INSTABILITY IN URBAN AREAS –A GUIDE TO BEST PRACTICE
To assist the implementation of sustainable landslide management strategies the Isle of Wight Centre for the Coastal Environment (within the Isle of Wight Council), which led the EU LIFE Project 'Coastal Change, Climate and Instability' (1997-2000) prepared a guide to management of ground instability in urban areas as one of the project outputs, called 'Managing ground instability in urban areas – A guide to best practice'.

Drawing upon experiences gained in the field of instability management from sites in Great Britain, France, Italy, Switzerland and other locations, the 80-page guide is intended to provide practical assistance for those involved with instability planning and management with the objective of reducing the impact of ground movement on local communities and economies.

The guide is written in a format that, it is hoped, will be of interest to both technical and non-technical users, well illustrated and making extensive use of practice experiences. In addition to scientists working in the fields of engineering geology and geotechnics, the publication is also aimed to assist local government staff, architects, surveyors, builders, estate agents and insurance companies. The Council has received very positive feedback from a range of users and interest groups.

CONCLUSIONS
The EU LIFE project, 'Coastal Change, Climate and Instability' drew on experiences from across Europe and elsewhere, illustrating that there are wide-ranging areas of common interest and concern in terms of management of ground instability. It has also been shown that although the total cost arising from instability is difficult to quantify, the economic and social effects have a significant impact across the European Union.

Sustainable development requires wise decision-making taking full account of past and present ground conditions. This can be achieved most effectively by means of a co-ordinated approach to instability management, thereby minimising risks to vulnerable communities by :
- identifying and understanding the nature and extent of instability
- guiding development towards suitable locations
- ensuring that existing and future developments are not exposed to unacceptable risks
- ensuring that development does not increase risk for the rest of the community
This forms the basis for a Landslide Management Strategy.

There is a need for closer integration between the actions of engineers, planners and the construction industry. Local authorities can play a pivotal role by co-ordinating landslide management strategies "on the ground" and they are probably best placed to maintain the momentum following the development of a strategy.

It is hoped that the advice and information provided in 'Managing ground instability in urban areas –A guide to best practice' will be of practical assistance in reducing the impact of instability on local communities and economies in both coastal and mountainous areas within the European Union and elsewhere.

REFERENCES
Doornkamp, J et al 1991. 'Coastal Landslip Potential Assessment : Isle of Wight Undercliff'. Technical report by Geomorphological Services Ltd for the Department of the Environment, UK. Research Contract PECD/7/1/272. Department of the Environment, UK, London.
Jones, D K C, 1992. 'Landslide hazard assessment in the context of development'. 117-141, in Geohazards - natural and man-made. McCall et al (Eds).

Chapman and Hall, London.

Jones, D K C and Lee, E M, 1994. 'Landsliding in Great Britain'. Department of the Environment, UK.

Lee, E M, 1997. 'Landslide Risk Management : Key Issues from a British perspective'. 227-237, in Landslide Risk Assessment. Cruden and Fell (eds). Balkema, Rotterdam.

Lee, E M, 'Landslide Hazard in Great Britain'. Geoscientist Vol. 5. No. 4.

Lee, E M, Brunsden D, Moore R and Siddle H J, 1991. The assessment of ground behaviour at Ventnor, Isle of Wight. In R J Chandler (ed.) Slope Stability Engineering, developments and applications, 207-212.

Lee, E M, and McInnes, R G, 2000. 'Landslide hazard mapping for planning and management: the Isle of Wight Undercliff, UK'. In Living with Natural Disasters, Proceedings of the CALAR Conference, Vienna, January 2000 (in press).

Lee, E M, Moore, R and McInnes, R G, 1998. 'Assessment of the probability of landslide reactivation : Isle of Wight Undercliff, UK'. 8th International IAEG Congress, Vancouver. Balkema, Rotterdam. Vol 2, 1315-1321.

Lee, E M and Moore R, 1991. Coastal Landslip Potential : Ventnor, Isle of Wight. GSL Publications.

McInnes, R G, 1996. 'A Review of coastal landslide management on the Isle of Wight, UK'. In Landslides - Glissements de Terrain, Proceedings of the 7th International Symposium on Landslides, Trondheim. Balkema, Rotterdam. Vol. 1, 301-307.

McInnes, R G, 1993. 'A Landslide Management Strategy for the Isle of Wight Undercliff'. Report to South Wight Borough Council, Newport, Isle of Wight.

McInnes, R G and Jakeways, J, 1999. 'LIFE Project : Coastal Change, Climate and Instability' - an information leaflet, Isle of Wight Council, Newport, Isle of Wight.

McInnes, R G and Jakeways, J, 2000. 'The development of guidance and best practice for urban instability management in coastal and mountainous areas of the European Union'. In 'Landslides : Research, theory and practice', Thomas Telford.

McInnes, R G, Jewell, S and Roberts, H, 1998. 'Coastal Management on the Isle of Wight, UK'. The Geographical Journal, 164, 291-306.

McInnes, R G, and Lee, E M, 2000. 'Living with Landslides: the management of the Isle of Wight Undercliff, UK'. In Living with Natural Disasters, Proceedings of the CALAR Conference, Vienna, January 2000 (in press).

Moore, R, Lee, E M and Clark A R, 1995. 'The Undercliff of the Isle of Wight - a review of ground behaviour'. Report by Rendel Geotechnics for South Wight Borough Council, Ventnor, Isle of Wight.

Thompson, A, Hine, P D, Poole, J S and Grieg, J R, 1998. 'Environmental Geology in Land Use Planning'. Report by Symonds Travers Morgan for the Department of the Environment, Transport and the Regions, UK

Stability study relating to the Communal Territory of Belmont S/Lausanne

F. NOVERRAZ, Karakas & Français SA, Lausanne, Switzerland

INTRODUCTION

The commune of Belmont, situated on the eastern outskirts of the city of Lausanne, has witnessed an important development in construction, essentially of residential property, during the past forty years. At present its population is about 2'500 inhabitants.

As is often the case, the original settlement was situated on a crest of stable land. Subsequent development, however, has been carried out on a much wider stretch of land, a large area of which is subject to landslides or other precarious geological conditions.

The reason for this is that the area of the communal territory of Belmont coincides precisely with a geological formation which is extremely unfavourable from the point of view of stability, known as the "Lower Freshwater Molasse (Chattien, Upper Oligocene)", a formation consisting predominantly of clayey marl layers with thin to metric layers of sandstone.

In the light of the numerous problems faced by the construction industry as a result of this situation, above, the communal authorities of Belmont appointed for the first time in 1959 a firm of geologists to prepare a report, with maps, illustrating the dangers of the communal territory.

In 1990 an important accident occurred, in which a landslide destroyed three houses (Noverraz et al.,1994). Following this expensive catastrophe, the communal authorities again decided to appoint a geological/geotechnical firm (the authors of the present article) to prepare a revised and more detailed report on the stability conditions of the communal territory.

This report, and the circumstances of its origin and its utilisation, are described here below.

THE GEOLOGICAL CONDITIONS OF THE COMMUNAL TERRITORY

The communal territory of Belmont is situated between two practically parallel rivers, each of which corresponds to the position of a large tectonic overlap, forming the borders of a rocky marl-sandstone formation dating from the Oligocene period : the Lower Freshwater Molasse. This rocky formation is richer in marl and clay than other types of Molasse, and is consequently more sensitive to slope instability characteristics.

Moreover, the tectonic overlap of this formation on the Aquitanian Molasse of the Swiss Plateau during the period of formation of the alpine chain had the effect of tilting the rock layers by some twenty to forty degrees, so that these now dip strongly towards the south-east, i.e. with a main direction towards the average slope and the Lake of Geneva. This unfavourable dipping of the layers further increases the precarious conditions of stability of this formation, and consequently the stability of the communal territory of Belmont practically throughout the area on which construction has taken place.

Furthermore, the loose topsoil which includes highly clayey morainic relics, tends to slide on the often marly top of the rock formation.

Where rocky layers are oriented parallel to the slope (dip slope areas), which is the case in a part of the communal territory of Belmont, a double risk of instability does exist : by sliding of the topsoil on the rock, which is rather smooth when corresponding to a surface layer, or dipslope sliding inside the rock massif.

The result of this very particular situation is that more than half of the communal territory of Belmont consists of old landslides, either dormant or slightly active and even active close to watercourses ; they are always extremely sensitive to the changes in stability conditions generated by any construction work (see fig.1).

BACKGROUND TO THE STUDY

The study was carried out from 1994 to 1995. It responded to presentation of e resolution to the communal council calling for the undertaking by the authorities of an exhaustive and official report on the conditions of stability of the communal territory.

In fact, several studies had already been published related more or less directly to the problems of stability within the region: maps from the 1960 report on the stability of the territory of Belmont, studies relating to the construction of the N9 motorway, a 1976-1979 inventory of unstable zones in the canton of Vaud (devoted solely to those zones affected by communal extension plans), and finally the 1:5000 instability maps drawn up in 1981-1984 as part of a study undertaken by the Swiss Federal Institute of Technology of Lausanne (project DUTI, Noverraz,1984). These maps were integrated in the 1986-1993 global inventory of unstable zones in the canton of Vaud commissioned by the Cantonal Department for Land Planning (Noverraz et al.1993).

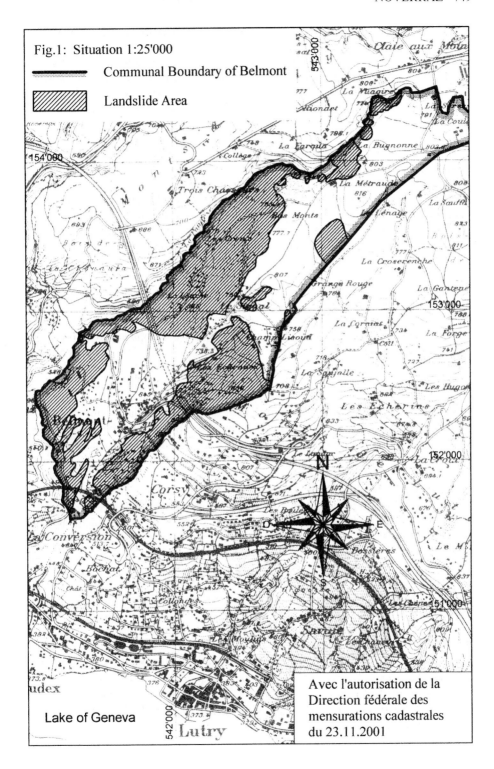

Fig.1: Situation 1:25'000

Communal Boundary of Belmont

Landslide Area

Avec l'autorisation de la
Direction fédérale des
mensurations cadastrales
du 23.11.2001

Lake of Geneva

The comparison of these different reports, however, revealed a lot of sometimes important discrepancies. The situation was rendered even more unsatisfactory by the attitude of the Cantonal Insurance against Fire and Natural Disasters - the sole and obligatory building - insurer which continued to use the old 1960 study as a basis for establishing its conditions and regulations relating to construction, or for excluding the landslide risk from the global insurance contracts.

Therefore, the idea of appointing a specialist firm to undertake an updating of these reports, with the objective of producing a single and official study, had already been discussed since 1992 at the Communal Council (the legislative authority). The principle having been accepted, the Municipality (the executive authority) prepared the specifications of a study to be drawn up, which were finalised in 1993. The study was awarded to the authors of the present article, and began in 1994, being completed in 1995, and published at the beginning of 1996.

CONCEPT
The study is based on an examination of all of the files of the geotechnical and geological reports in the possession of the Commune. More than 120 files relating to the construction of buildings, motorway work and different public works were examined. These reports included more than 300 boreholes spread over an area of approx. 25 km^2. The cartography used as a base remained that prepared by the DUTI study of 1981-1984, published in 1986.

The report consists of three elements:
.   A map of the landslides with all the relevant geological details at 1:2,000 on a precise topographic map (fig.2), mentioning all logged boreholes.
.   A map of the unstable zones at 1:1,000, in seven parts, on a cadastral map.

Each of these two maps specifies the areas of differentiated instability, based on the average annual known or estimated velocity of movement, according to the DUTI legend. For each borehole, the depth of the slide or of the slidden mass, and of the bored bedrock, are indicated whenever relevant. They also detail the positions of all of the ancient underground coal-mining tunnels, together with the extent to which they were exploited, in the beginning of the XXth century, and even those of the XIXth century, as all these galleries were reputed, largely erroneously, of perhaps being responsible for subsidence at surface level. These 1:1,000 maps additionally show those areas considered as potentially unstable - though not yet subject to sliding - on account of the geological and structural conditions that would apply in the event of earthworks being undertaken.

A text describing, for each of the approx. 600 communal plots, **the geological context** deriving from the cartographic documents and from the nearby related studies, together with an assessment of the **practical consequences for construction**. For the large plots, the description is given, when appropriate, sector by sector.

This text first summarises the bulk information contained in the maps, in particular the type and characteristics of the existing ground, then also draws attention to the geological contexts that are potentially dangerous for structural reasons : for example, in the event of a change in the conditions of natural stability resulting from works in those sectors for which the maps do not mention any existing slide (potentially unstable zones in the 1:1,000 maps). The catastrophic landslide of 1990 resulted from such a context.

The mapping documents are freely available for examination in the technical department of the commune. The report describing the geological conditions and practical consequences for each plot is submitted to the owner or to his representatives only, on account to the principle of data protection.

A concluding chapter lists a number of proposals regarding the additional investigations that would be required in the event of new planned construction, taking into account the depth of the unstable earth and rock on site, the proximity of existing buildings, and also where the map indicates no pre-existing conditions of instability.

It is to be noted, however, that the information contained within the scope of the present study is not intended to replace any specific geotechnical reports relating to a given project : its unique purpose is to describe the area and delimit its stability problems. It never reduces the responsibilities of any future builders.

It is accepted that it will periodically be necessary to bring this study up to date, without defining a specific time limit: one could agree that this period should not exceed ten years.

The creation of a "user-friendly" computerised version of the cartographic documents, adapted to the communal equipment, has been analysed but not yet undertaken.

IMPLEMENTATION BY THE AUTHORITIES OF THE RESULTS OF THE STUDY

The communal authorities intervene to request a geological or geotechnical study whenever a public enquiry is held in respect of a new construction project on land indicated as being unstable according to the 1996 study. The building permit is granted subject to such study being undertaken. However, the communal authorities do not question the results or recommendations of the geotechnical study, nor do they control by themselves its implementation. However, they require that a geotechnician supervises the building.

In practice, a geotechnical report is required, as it was already the case before the study took place, for **all** requests for building permits. On the other hand the communal authorities tip Article 106 of the CAMAC (Centrale des Autorisations en Matière d'Autorisation de Construire", a cantonal control office for building permits) file, which states : "The construction is situated in a sliding or inundation zone."

The relevant extract from the 1996 study is delivered to the person or firm requesting the permit, who in any event has free access to the cartographic documentation. The petitioner and his agents are also advised at the outset of the nature and importance of the instability situation as far as it relates to his particular project. The communal authorities also insist upon the holding of an "opening of works" meeting, and the appointment of a geotechnical/geological firm as advisors.

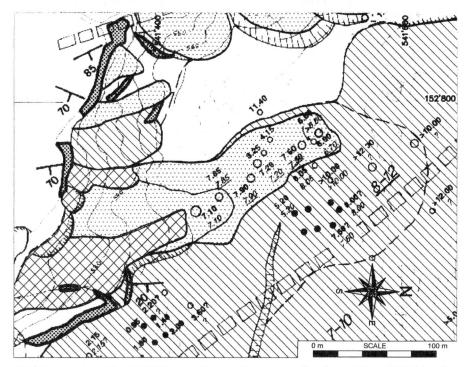

Fig.2: extract of the 1:2'000 map illustrating areas of active sliding (▨▨), of very slow sliding (▦) and of substabilized landslide (▨▨), existing boreholes with reference to the depth of the slide and of the bedrock (for details, see text), and different geological signs.

The same applies to the Cantonal Assurance Organisation as a requirement for issuing cover against the "risk of sliding on construction sites in hazardous areas". The Organisation utilises an evaluation procedure for unstable land updated in 1996 with the assistance of various geotechnical firms. This procedure, in the form of a questionnaire requiring marks, is based on four criteria: the quality of available information, an appreciation of the degree of danger relative to the plot of land, the potential danger relative to the planned buildings, and finally an appreciation of the planned construction controls. The questionnaire is required to be completed by an authorised geologist or geotechnician.

Up to now, however, the Cantonal Assurance Authority continues to utilise the DUTI maps of 1981-1984 as a base of reference rather than the survey of the instability of the communal territory of Belmont of 1996.

RESULTS ARISING FROM USE OF THE SURVEY

Thanks to well-known present economic conditions, there has been little progress in the construction industry during the years following the completion of the survey. Building activity recovered to some degree in the year 2000, thereby giving a brief opportunity for the exposed survey to be tested. Twenty new bulding licences have been delivered in 2000, and seven in 2001.

The survey has not basically changed the way in which the communal authorities handle requests for construction permits. It does, however, enable them to give precise information to builders as to the extent to which their properties, and/or projects, are likely to be subject to instability problems.

By making this information available, the authorities are complying with the requirements of article 89 of the Cantonal Law on Land Development of 1989, which stipulates:

Any construction on land which is either lacking in stability or exposed to specific risks such as avalanche, collapse, flood or landslides, is forbidden until works as may be necessary has been undertaken by properly specialised firms either to rectify or divert the risks in question: the authorisation to build does not create any liability on the part either of the commune or the State.

On the other hand, the availability of the unbiased facts contained in the survey, to the extent that they apply to a particular project, can assist in resolving any disagreements with future neighbours in the event of objections that may be put forward during the period of the public enquiry.

One of the merits of the 1996 survey that has been recognised by the communal authorities is to have shown that none of the plots situated in the approved constructible zone were by definition unfit for construction from a geological/geotechnical point of view, contrary to the belief of certain citizens.

CONCLUSIONS

The Commune of Belmont is to be congratulated on having tackled the problem of the risks arising from land slides, driven no doubt by the various problems to which the difficult geological conditions of their territory have given rise. Their approach may be compared with that of many other communes, which have continued to adopt a "laissez-faire" attitude, and a tendency to shift responsibility on to the shoulders of private individuals, contrary to the Federal Law of 1979, which requires both cantons and communes to take natural risks into account when formulating their development plans.

The concept of unstable land continues to be regarded above all as having a depreciative effect on the value of land devoted for construction. Often, it is either ill-perceived or quite simply ignored.

It does not appear, however, that the action of the Commune of Belmont in commissioning this survey, and the requirements relating to construction that have been formulated as a result, have in any way reduced the pressure to undertake new construction projects within their territory.

REFERENCES
- Noverraz F., Weidmann M. : Le glissement de terrain de Converney-Taillepied (Belmont et Lutry, Vaud, Suisse). Bulletin de Géologie de Lausanne No 269 (1983)
- Bonnard Ch., Noverraz F. : Instability risk maps : from the detection to the administration of landslide prone areas. IVth Int. Symp. on Landslides, Toronto, 9/1984, Vol. 2, pp. 511-516
- Noverraz F., Bonnard Ch. : Mapping methodology of landslides and rock-falls in Switzerland. VIth Int. Conf. and Field Workshop on Landslides, Milan, 9/1990, pp. 43-
- Noverraz F. : Essai de recensement cartographique des glissements de terrain et écroulements rocheux sur le territoire suisse. Proc. Int. Conf. on Water Resources in Mountainous Regions II Water and Slopes, IASH Publ. No 194, Lausanne, 1990, Vol. 2, pp. 429-436
- Noverraz F. : Répartition géographique, origine et contexte géologique des glissements de terrain latents en Suisse. Proc. Int. Conf. on Water Resources II, Water and Slopes, IAHS Publ. No 194, Lausanne, 1990, Vol. 2, pp. 437-446
- Noverraz F., Bonnard Ch., Giraud A. : Environmental impact of a large landslide near Lausanne, Switzerland. Proc. Int. Conf. on Slope Stability Engineering-Development and Applications, Isle of Wight, 1991, pp. 101-106
- Noverraz F., Bardet L., Bonnard Ch. : Cartes d'instabilité du Canton de Vaud dans le cadre du Plan Directeur Cantonal. C.R. Symposium FEANI / IDNDR, Lausanne, 9/1993, pp. 111-121
- Noverraz F., Schopfer & Karakas SA : Glissement rocheux sur un quartier de villas; processus, mesures d'urgence et de confortation; 7eCongrès AIGI, Lisbonne, 9/1994, Vol. 3, pp. 1517-1526
- Noverraz F., Karakas & Français SA : Commune de Belmont, étude de stabilité du territoire communal, vers. mod. 5/1999, unpublished

Hazard of debris flows on slopes and its control in China

PROF. WANG SHIGE, Institute of Mountain Hazards and Environment, Chinese Academy of Sciences, China

INTRODUCTION

Debris flows on slopes are a common hazard, of which the distribution is widespread, especially in hills in the northeast and the south, in mountains in the southwest and Qing-Zang Plateau (Tibet) of China. They often occur suddenly in groups and zonally, damaging and destroying villages, railways and highways. For example, typically more than 300 debris flows occur in Hong Kong every year, and a debris flow destroyed many houses and killed 71 persons on June16, 1972 (Z. L. Su 1988 and D.O.K Lo 2000). There are about a hundred debris flows gullies along Chengdu—Kunming railway in the southeast of China, the hazards resulting from them have blocked the railway ten times; particularly, the railway bed, small bridges and culverts are often buried and damaged by slurry and stone (X. Q. Xie etc.1994).

CHARACTERISTICS OF THE DEBRIS FLOWS ON SLOPES

Debris flows on slopes (Fig.1) are defined as those that take place on slopes according to debris flow taxonomy in China; an opposite kind is the valley debris flow. The debris flow often occurs on a relatively long slope covered by thin unconsolidated rock and soil debris (2~3m, in general), especially where the vegetative cover has been removed by logging or fires. No gully is evident on most of the slopes before the debris flow, but there is usually a small flute or a water outlet on the upside of the slope in some situations. Some special characteristics of the debris flow are discussed as follows:

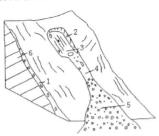

Figure 1. Debris flows on slopes: 1.Colluvium 2.Scarp 3.Formation Zone 4.Passage Zone 5.Deposition Zone 6.Bedrock

Small scale

A debris flow on a slope -small scale one without deep channel and branches. The basin area of most gullies is less than $0.4km^2$, and the basin consists of the formation zone and the deposition zone without a passage zone as usual. In general, sediment transport is $10^2 \sim 10^4 m^3$ every time in a gully, sediment from the upstream deposits directly nearly the mouth of gully and forms a small cone or a steep fan with $10° \sim 15°$ of gradient.

Steep gradient and tremendous force

The debris flows usually take place on steep slope of $20° \sim 40°$, the channel gradient is almost the same as the slope's. When fluid with heavy density flows along steep slope, its velocity increases quickly, the fluid obtains tremendous force, enough to destroy or damage anything on the way. After flowing, a long channel may result from erosion by the fluid. Ecosystems will be damaged severely in the areas where debris flows occur in groups and zonaly. Serious debris flow hazards usually ruin vegetation, soil may be removed thoroughly and bare bedrock may outlet in large area, so it is very difficult to resume primary environment in natural condition.

Sudden burst

Most debris flows on slopes are trigged by rarely heavy rain. First, the earth in the flute in the upside of slope is saturated by rainfall; then the pore water pressure in the earth increases and friction between the earth and below bedrock decreases as rainfall continuously infiltrates the earth; finally, while the pore water pressure is rising the highest, the earth suddenly slides down suddenly, and transforms into a debris flow immediately. It only takes a few minutes to finish this process. Because it is difficult to predict weather exactly in a small district in the mountains in time and there is little auspice before the debris flow, the debris flow often bursts unexpectedly and results in serious hazard.

COUNTERMEASURES AGAINST THE DEBRIS FLOWS

A number of debris flows on slopes have been controlled, which threaten villages, railways and highways in China. Most projects are successful. In general, there are three common measures to control the hazards, viz. prevention, forecast-warning and engineering.

Prevention measures

Selecting new highway and railway routes should keep away from the areas where debris flow distribution is dense and fast or will develop in the future. If the route must pass through the area, some engineering works preventing the debris flow should be designed, for example, controlling the gully and slope near a tunnel and construction, converting old culverts into bridges.

New villages to be built should keep away from the sites where there is potential for a disaster, such as banks between the gully, the mouth of the gully, the site below the flute and water outlets on slopes, the site below earth deposit, etc.

Environment, especially forests and natural slopes, should be protected near residential areas in mountains, protecting forests from fire and cutting, forbidding farming on steep slopes and the excavation of slopes.

Some preventative engineering would be made if there are cut slopes and fill slopes, or artificial outlets near the site. The retaining walls should be made under cut slopes and filled slopes, and the surface of the slope should be covered by trees and grass. The artificial outlet should be combined with a canal, draining water to a safe site.

Forecast-warning measures

We know from experience that forecast-warning measures are sufficiently lifesaving when debris flows will occur immediately or have just occurred up stream of a gully.

The basic knowledge about debris flows is propagandized constantly to increase the population's awareness of preventing disasters. The dwellers are requested to become familiar with the environment around their houses in order to understand if their houses are in the risk area. The dwellers living in a risk area should understand what they can do at ordinary times, such as understanding debris flow alarms, selecting a safe refuge and a safe path to the refuge.

Forecast-warning systems consist of ombrometers, weather radar, weather satellite, computer, communication system and the thresholds that describe the minimum rainfall rates needed to trigger abundant debris flows on slopes in risk areas in general.   Making use of the system, forecast-warning information can be sent to the local population in the rainy season by radio, television and telephone etc., to mitigate death and loss of property. Forecast-warning systems have been researched since the 1980s, a complete forecast-warning system was finished in Peijing City in 2001. The complete system not only includes the above contents, but also a series of methods to organize, retreat and save local dwellers. Making use of those ordinary methods, hundreds people were saved in the 1991 rainy season. Several simple systems with ombrometers, computers and communication systems have been installed along railways in the southwest of China since 1980s. In order to ensure security of the railway, the railway department uses both the system and the person who guards the railway line nearby to emphasise debris flow gullies in the rainy season, in this way, some accidents of derailment were avoided successfully.

Engineering measures

In general, engineering measures include civil engineering and biology engineering. The civil engineering concerning the debris flow hazards control includes drainage ditches, check dams and retaining walls etc. in common use.

A drainage system often consists of catchwater and canal. Catchwater established on stable slopes above origin of the gully or flute can head off runoff from the top of hill, reducing the water into gully, preventing the debris flow from occurring. Canal is used to drain the debris flow and flood to safe site, in order to protect the railway, highway and village against the hazard.

Check dam applied to controlling the debris flow widely is an important engineering construction technique in China. Most check dams made of masonry are 2~3m high, and lay in channel in series. They can stabilize the channel and protect the channel and banks from eroding, reducing sediment provided for debris flows. After making the check dam, trees and grass can be planted on the channel bed and banks to fix the soil. A Biology Check dam with trees or bamboo is used to protect a very small gully.

A retaining wall made of masonry or concrete is used to reinforce an excavated slope and natural slope above a developed area, an important construction to protect them against damage.

In most sites, a project with both civil engineering and biology engineering is always more effective than a single one.

AN EXAMPLE OF A CONTROLLED DEBRIS FLOW ON A SLOPE
A sanatorium of the Chinese Academy of Sciences' is located in a hill of the Lushan Mountain in Jiangxi Province. An intense rainstorm with total rainfall of 246.3mm in 4 hours fell on Oct. 15 1995 in the mountains; two debris flows on slope (25˚) above the sanatorium were trigged. The basin areas, called Dong gully and Xi gully, are only 0.025 km^2 and 0.021km^2 respectively, but the debris flows hazards were severe. To begin with, a small landslide formed in small groove with about 30~40m^2 near the top of the hill, then it slipped down quickly and transformed into a debris flow at once. It made a dash for a road and the sanatorium, the road passed the submontane area was blocked by deposition immediately, the fence and retaining wall behind the sanatorium were damaged, then the debris flow went into the court , a number of boulders and a great amount of sand was deposited there. There are two deep flutes on slope that were slick, because the debris flows eroded extremely. The Lushan Mountain is famous scenery in China, the hill is located near the center of the mountain, and so the scenery was damaged severely. In order to control the hazard and resume the scenery, a project has been carried out since 1996.

Aim and principle for controlling the hazard
The aim of the project is efficient, that is, reducing formation conditions of debris flows, controlling the hazard, protecting the sanatorium and resuming scenery as quickly as possible. The design standard of the debris flow is 2% (the return period is 50 years) in the project. The principle is a comprehensive one with various works, that is, making some check dams for stabilizing the channel bed and banks of the

gullies, decreasing sediment affluent, establishing drainage channels for draining floods through the sanatorium safely and covering slopes with trees and grass for fixing surface soil and resuming beautiful scenery.

Planning for controlling the hazard

The planning consists of two parts, viz civil engineering and biology engineering. Civil engineering includes check dams, retaining walls, draining systems and fender piers (Fig. 2).

Check dams distribute on middle stream of the gullies, 11 and 13 on Dong gully and Xi gully respectively. Dams are 2~3m high and the foundation is 1~2m depth. According to our experience, the check dam is usually made of sand, stone and cement, but dry masonry dams aren't used for controlling debris flows in China. They are laid along channels, forming a series dams like a ladder. It is a key that a series dams keep the bed and banks stable to preventing them from eroding and slipping.

The masonry retaining walls 355m long are laid alongside the road, preventing the foot of the hill from slipping and blocking the road or the ditch beside the road. The gravity wall is 1.5~2.0m high, the top is 0.6m width; the back slope and front slope are 1:0.0 and 1:0.3 respectively. Inverted filter with 0.5m thickness is made behind the wall and a ditch in front of wall.

The drainage system is a small and complete one with advantage function. It consists of drop chutes, plunge wells, ditch beside road, culverts, flood canals and a sediment pool. The flood from check dams runs through a drop chute into a plunge well and direct runoff from the road runs through the ditch into the plunge well yet, then fluid converging on the well goes through a culvert into cannels in the court of the sanatorium. The flood through the sediment pool flows down into a lake at the end. The system with functions both draining and trapping sand would protect the sanatorium and the lake efficiently.

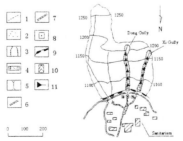

Figure 2. Sketch map of engineering for controlling debris flow hazards:
1.Borderline of Basin 2.Deposition Area of Debris Flow 3. Avalanche 4.Sediment Pool 5.Culvert 6.Old Flood Canal 7.New Flood Canal 8.Fender Pier 9.Retaining Wall 10.Drop Chute 11.Check Dam

There are some fender piers between the road and the sanatorium. The piers would be used for trapping big boulders and floating timber in large debris flows and protect the sanatorium from serious hazard if the debris flow return period overran the design standard (2%).

Biology engineering is used in common ways, viz planting tree and grass, it is useful for conserving the surface soil and resuming scenery rapidly.

Benefit of the project
The project has been carried out successfully, most works were completed from 1996 to 1998. No hazard has occurred since 1998, neither debris flows nor landslides. In the rainy season in 1998, a very heavy rain (return period of over one hundred years) in down stream Yangtze River resulted in a severe flood, very serious hazards have occurred commonly in the area, as you know. Great debris flows and landslides occurred in Lushan Mountain, which destroyed many roads and buildings, but there wasn't any hazard in the sanatorium, there was even no sand from the hill, since the project protects the environment against damage. Now you will not see ugly flutes in the hill if you go there.

REFERENCES
1. D.O.K. Lo (2000): Review of natural terrain landslide debris-resisting barriers design. Hong Kong: Special project report, SPR 1/2000, p.6.
2. Z. L. Su (1988): The reason, damage and control of instable slope in Hong Kong. China: Soil and water conservation in China. 1988(8), p. 46~51.
3. X. Q. Xie (1994): Causes of Slope Debris Flow and its Prevention Alone Cheng-Kun Railway Line, in Proceedings of the 4th Notional Symposium on Debris Flow, Lanzhou. China: Culture Press in Gansu, Lanzhou, P. 241~250.
4. Institute of Mountain Hazard and Environment, Chinese Academy of Sciences (1989): Research and prevention of debris flow. China: Science Press, Beijing, p.121~133.
5. D. J. Li (1997): Theory and practice for hazard reduction of debris flow. China: Science Press, Beijing, p. 52~53.
6. Richard Dikau, Denys Brunsden, Lothor Schrott and Maia-Laura Ibsen (1996): Landslides Recognition. England: John Wiley & Sons Ltd, p. 97~102, 161~178.
7. Y. N. Xu, S. F. Kuang, W. W. Li and L. Wang (1998): A preliminary study of the effect of morphology on avalanche, in Researches on Mountain Disasters and Environmental Protection across Taiwan Strait. China: Science and Technology Press in Sichuan, Chengdu, p. 256~262.
8. U.S. Department of the Interior (1995): Debris –Flow in San Francisco Bay Region. U.S.: Geology Survey, p. 1~4.

Author index